NanoScience and Technology

For further volumes:
www.springer.com/series/3705

The series NanoScience and Technology is focused on the fascinating nano-world, mesoscopic physics, analysis with atomic resolution, nano and quantum-effect devices, nanomechanics and atomic-scale processes. All the basic aspects and technology-oriented developments in this emerging discipline are covered by comprehensive and timely books. The series constitutes a survey of the relevant special topics, which are presented by leading experts in the field. These books will appeal to researchers, engineers, and advanced students.

Juan Bartolomé • Fernando Luis •
Julio F. Fernández
Editors

Molecular Magnets

Physics and Applications

Editors
Juan Bartolomé
Institute of Material Science of Aragón and
Department of Condensed Matter Physics
CSIC–University of Zaragoza
Zaragoza, Spain

Julio F. Fernández
Institute of Material Science of Aragón and
Department of Condensed Matter Physics
CSIC–University of Zaragoza
Zaragoza, Spain

Fernando Luis
Institute of Material Science of Aragón and
Department of Condensed Matter Physics
CSIC–University of Zaragoza
Zaragoza, Spain

ISSN 1434-4904 ISSN 2197-7127 (electronic)
NanoScience and Technology
ISBN 978-3-642-40608-9 ISBN 978-3-642-40609-6 (eBook)
DOI 10.1007/978-3-642-40609-6
Springer Heidelberg New York Dordrecht London

Printed on acid-free paper

Springer is part of Springer Science+Business Media (www.springer.com)

Preface

This book aims to provide a coherent and pedagogical collection of articles on the physics and applications of molecular magnets. All contributors have played a major role in either (1) discovering or elucidating the physics that underlies molecular magnets, or in (2) the present exploration of avenues toward their applications. Issues that are by now well understood as well as open questions are covered. Inevitably, overlaps among some chapters do occur, but we are sure that the reader will find them complementary rather than repetitious.

Molecular magnets are made up of chemically identical molecules with high–spin cores. The cornerstone for the rise of present day interest in molecular magnetism was the discovery of magnetic quantum tunneling in Mn_{12}-acetate molecules. There was before 1996 some indirect evidence for quantum tunneling of large spins. In 1993, magnetic hysteresis (and thus magnetic memory) at liquid Helium temperatures was shown to come from single molecular clusters of Mn_{12}-acetate. However, a clear imprint of magnetic quantum tunneling was only observed in 1996. Then, experiments revealed that magnetic hysteresis in Mn_{12}-acetate is rather unconventional, in that the magnetization jumps at equally spaced values of the applied magnetic field. The gist of this effect is that spins can tunnel between different magnetic states as they are brought on and off resonance by an external magnetic field. Mn_{12}-acetate molecules thus behave as "single molecule magnets" (SMMs). "Resonant spin tunneling" in molecular magnets illustrates beautifully quantum physics at the mesoscopic scale, that is, in the crossover region between the macroscopic and microscopic worlds, where quantum and classical physics meet. Finally, SMMs are a variant of magnetic nanoparticles, which are at the basis of magnetic recording. Much interest in SMMs arises from this fact.

The field has expanded considerably in the last two decades, owing to the creativity of molecular chemists (who have crafted high and low spin clusters and single chain magnets), to the observation and elucidation of interesting phenomena (e.g., hole burning, spin avalanches and deflagration, as well as dipolar long-range ordering), and to the development of experimental techniques (e.g., single molecule manipulation on substrates). Finally, there is the vibrant ongoing work on applications. Most of it has to do with the fact that single molecule magnets are potential

2-level qubits for quantum computation. There are other applications for molecular magnets, such as to magnetic refrigeration (making use of the magneto–caloric effect) of electronic devices at cryogenic temperatures.

A brief historical account of the discovery of stepwise hysteresis loops, which are the hallmark of molecular magnets, as well as the physics of the underlying magnetic-quantum-tunneling phenomena, can be read in the first section, *Tunneling of Single Molecule Magnets*. How stepwise hysteresis loops were discovered, and first reported early in 1996, is the subject of Chap. 1. How the existence of stepwise hysteresis loops was subsequently corroborated in Mn_{12}-acetate single crystals, and more, can be read in Chap. 2. The theory of magnetic quantum tunneling that takes orbital angular momentum into account is given in Chap. 3. There is however more in the first section. Interesting effects that cannot be accounted for assuming each SMM acts as a single spin S are reported and explained in Chap. 4.

The second section, *Beyond Single Molecules*, covers various collective phenomena. Deflagration is one of them. It has been found to proceed in molecular magnets by rapidly moving magnetic-quantum-tunneling fronts, much as ordinary deflagration takes place by chemical combustion processes. Experimental and theoretical accounts are given in Chaps. 5 and 6, respectively. A rather different sort of collective phenomenon, equilibrium magnetic phase transitions, have been observed in some of the best known molecular magnets. Magnetic ordering is brought about by magnetic-dipolar interactions. Because system-wide ordering processes cannot bypass slow quantum tunneling processes, the realization of magnetic ordering was not a foregone conclusion. Order can either be destroyed by heating, through a classical phase transition, or by applying a transverse magnetic field, through a quantum phase transition. This is the subject of Chap. 7. *Single Chain Magnets*, the subject of Chap. 8, resemble SMMs in that they can relax extremely slowly. Their underlying physics is however rather different. In single chain magnets relaxation proceeds through thermal excitation of domain walls. Models are also discussed in this chapter. Metal-phthalocyanine (MPc) are uniquely suited for the exploration of the intrinsic mechanisms which give rise to molecular magnetism. The structural and magnetic properties of bulk crystals, thin films and single MPcs molecules adsorbed on different substrates are covered in Chap. 9. The Kondo interaction, tunneling processes, switchability and spin control are reviewed.

Most of the section on *Applications* is devoted to issues that arise from the role molecular magnets can play in information technology. How to control and exploit the quantum properties of SMMs, achievements of recent years and foresight for their near future are all weaved into *Molecular Nanomagnets for Information Technologies*, which is Chap. 10. In Chap. 11, *Molecular Magnets for Quantum Information Processing*, a brief introduction into quantum computing is given. DiVincenzo's criteria for its successful physical implementation are introduced and used as a guideline throughout. Utilization and control (mainly, through the spin-electric effect) of the spin degrees of freedom in SMMs as qubit states is considered. The various decoherence mechanisms which affect SMMs and their advantages on this point over more traditional qubits are examined. Finally, a proposal to implement Grover's algorithm using molecular magnets is discussed. In Chap. 12,

Single Molecule Spintronics, recently developed techniques that can measurements of electronic transport through a SMM are discussed. Spe information, obtained from measurements on spin-transistor-like three-term ups, confirms the high-spin state and magnetic anisotropy of the robust Fe_4 SM The experimental observation that electric gate fields drastically modify the magnetic properties of an oxidized or reduced molecule is discussed. The main aim of molecular quantum spintronics, that is, to bring together concepts from spintronics, molecular electronics and quantum computing for the purpose of fabrication, characterization, and study of molecular devices—such as, molecular spin-transistors and molecular spin-valves—are reviewed in Chap. 13 (*Molecular Quantum Spintronics*). Finally, Chap. 14 is devoted to a totally different topic, the application (by means of the magnetocaloric effect) of molecular magnets to very low temperature refrigerants in microdevices.

In closing, the Editors wish to express their pleasure at having worked with the authors, and we would like to thank each and everyone of them for their warm response and full co-operation.

Zaragoza, Spain

J. Bartolomé
F. Luis
J.F. Fernández

Preface

This book aims to provide a coherent and pedagogical collection of articles on the physics and applications of molecular magnets. All contributors have played a major role in either (1) discovering or elucidating the physics that underlies molecular magnets, or in (2) the present exploration of avenues toward their applications. Issues that are by now well understood as well as open questions are covered. Inevitably, overlaps among some chapters do occur, but we are sure that the reader will find them complementary rather than repetitious.

Molecular magnets are made up of chemically identical molecules with high–spin cores. The cornerstone for the rise of present day interest in molecular magnetism was the discovery of magnetic quantum tunneling in Mn_{12}-acetate molecules. There was before 1996 some indirect evidence for quantum tunneling of large spins. In 1993, magnetic hysteresis (and thus magnetic memory) at liquid Helium temperatures was shown to come from single molecular clusters of Mn_{12}-acetate. However, a clear imprint of magnetic quantum tunneling was only observed in 1996. Then, experiments revealed that magnetic hysteresis in Mn_{12}-acetate is rather unconventional, in that the magnetization jumps at equally spaced values of the applied magnetic field. The gist of this effect is that spins can tunnel between different magnetic states as they are brought on and off resonance by an external magnetic field. Mn_{12}-acetate molecules thus behave as "single molecule magnets" (SMMs). "Resonant spin tunneling" in molecular magnets illustrates beautifully quantum physics at the mesoscopic scale, that is, in the crossover region between the macroscopic and microscopic worlds, where quantum and classical physics meet. Finally, SMMs are a variant of magnetic nanoparticles, which are at the basis of magnetic recording. Much interest in SMMs arises from this fact.

The field has expanded considerably in the last two decades, owing to the creativity of molecular chemists (who have crafted high and low spin clusters and single chain magnets), to the observation and elucidation of interesting phenomena (e.g., hole burning, spin avalanches and deflagration, as well as dipolar long-range ordering), and to the development of experimental techniques (e.g., single molecule manipulation on substrates). Finally, there is the vibrant ongoing work on applications. Most of it has to do with the fact that single molecule magnets are potential

2-level qubits for quantum computation. There are other applications for molecular magnets, such as to magnetic refrigeration (making use of the magneto–caloric effect) of electronic devices at cryogenic temperatures.

A brief historical account of the discovery of stepwise hysteresis loops, which are the hallmark of molecular magnets, as well as the physics of the underlying magnetic-quantum-tunneling phenomena, can be read in the first section, *Tunneling of Single Molecule Magnets*. How stepwise hysteresis loops were discovered, and first reported early in 1996, is the subject of Chap. 1. How the existence of stepwise hysteresis loops was subsequently corroborated in Mn_{12}-acetate single crystals, and more, can be read in Chap. 2. The theory of magnetic quantum tunneling that takes orbital angular momentum into account is given in Chap. 3. There is however more in the first section. Interesting effects that cannot be accounted for assuming each SMM acts as a single spin S are reported and explained in Chap. 4.

The second section, *Beyond Single Molecules*, covers various collective phenomena. Deflagration is one of them. It has been found to proceed in molecular magnets by rapidly moving magnetic-quantum-tunneling fronts, much as ordinary deflagration takes place by chemical combustion processes. Experimental and theoretical accounts are given in Chaps. 5 and 6, respectively. A rather different sort of collective phenomenon, equilibrium magnetic phase transitions, have been observed in some of the best known molecular magnets. Magnetic ordering is brought about by magnetic-dipolar interactions. Because system-wide ordering processes cannot bypass slow quantum tunneling processes, the realization of magnetic ordering was not a foregone conclusion. Order can either be destroyed by heating, through a classical phase transition, or by applying a transverse magnetic field, through a quantum phase transition. This is the subject of Chap. 7. *Single Chain Magnets*, the subject of Chap. 8, resemble SMMs in that they can relax extremely slowly. Their underlying physics is however rather different. In single chain magnets relaxation proceeds through thermal excitation of domain walls. Models are also discussed in this chapter. Metal-phthalocyanine (MPc) are uniquely suited for the exploration of the intrinsic mechanisms which give rise to molecular magnetism. The structural and magnetic properties of bulk crystals, thin films and single MPcs molecules adsorbed on different substrates are covered in Chap. 9. The Kondo interaction, tunneling processes, switchability and spin control are reviewed.

Most of the section on *Applications* is devoted to issues that arise from the role molecular magnets can play in information technology. How to control and exploit the quantum properties of SMMs, achievements of recent years and foresight for their near future are all weaved into *Molecular Nanomagnets for Information Technologies*, which is Chap. 10. In Chap. 11, *Molecular Magnets for Quantum Information Processing*, a brief introduction into quantum computing is given. DiVincenzo's criteria for its successful physical implementation are introduced and used as a guideline throughout. Utilization and control (mainly, through the spin-electric effect) of the spin degrees of freedom in SMMs as qubit states is considered. The various decoherence mechanisms which affect SMMs and their advantages on this point over more traditional qubits are examined. Finally, a proposal to implement Grover's algorithm using molecular magnets is discussed. In Chap. 12,

Single Molecule Spintronics, recently developed techniques that can be applied to measurements of electronic transport through a SMM are discussed. Spectroscopic information, obtained from measurements on spin-transistor-like three-terminal set ups, confirms the high-spin state and magnetic anisotropy of the robust Fe_4 SMM. The experimental observation that electric gate fields drastically modify the magnetic properties of an oxidized or reduced molecule is discussed. The main aim of molecular quantum spintronics, that is, to bring together concepts from spintronics, molecular electronics and quantum computing for the purpose of fabrication, characterization, and study of molecular devices—such as, molecular spin-transistors and molecular spin-valves—are reviewed in Chap. 13 (*Molecular Quantum Spintronics*). Finally, Chap. 14 is devoted to a totally different topic, the application (by means of the magnetocaloric effect) of molecular magnets to very low temperature refrigerants in microdevices.

In closing, the Editors wish to express their pleasure at having worked with the authors, and we would like to thank each and everyone of them for their warm response and full co-operation.

Zaragoza, Spain

J. Bartolomé
F. Luis
J.F. Fernández

Contents

Contributors

Marco Affronte Universitá di Modena e Reggio Emilia, Modena, Italy

Bernard Barbara Institut Néel, CNRS & Université Joseph Fourier, Grenoble Cedex 9, France

Juan Bartolomé Instituto de Ciencia de Materiales de Aragón and Departamento de Física de la Materia Condensada, CSIC–Universidad de Zaragoza, Zaragoza, Spain

Enrique Burzurí Kavli Institute of Nanoscience, Delft University of Technology, Delft, The Netherlands

Eugene M. Chudnovsky Department of Physics and Astronomy, Herbert H. Lehman College, The City University of New York, Bronx, NY, USA

Enrique del Barco Department of Physics, University of Central Florida, Orlando, FL, USA

Marco Evangelisti Instituto de Ciencia de Materiales de Aragón and Departamento de Física de la Materia Condensada, CSIC–Universidad de Zaragoza, Zaragoza, Spain

Marc Ganzhorn Institut Néel, CNRS & Université J. Fourier, Grenoble Cedex 9, France

D.A. Garanin Department of Physics and Astronomy, Lehman College, City University of New York, New York, USA

Dante Gatteschi Department of Chemistry, University of Florence, Sesto Fiorentino, Italy; INSTM, Florence, Italy

Stephen Hill Department of Physics and National High Magnetic Field Laboratory, Florida State University, Tallahassee, FL, USA

Junjie Liu Department of Physics, University of Florida, Gainesville, FL, USA

Daniel Loss Department of Physics, University of Basel, Basel, Switzerland

Fernando Luis Instituto de Ciencia de Materiales de Aragón and Departamento de Física de la Materia Condensada, CSIC–Universidad de Zaragoza, Zaragoza, Spain

Carlos Monton Center for Advanced Nanoscience, Department of Physics, University of California San Diego, La Jolla, CA, USA

Myriam P. Sarachik City College of New York, CUNY, New York, NY, USA

Ivan K. Schuller Center for Advanced Nanoscience, Department of Physics, University of California San Diego, La Jolla, CA, USA

Dimitrije Stepanenko Department of Physics, University of Basel, Basel, Switzerland

Javier Tejada Dpto. de Física Fonamental, Facultat de Física, Universitat de Barcelona, Barcelona, Spain

Filippo Troiani Institute NanoSciences, CNR, Modena, Italy

Kevin van Hoogdalem Department of Physics, University of Basel, Basel, Switzerland

Herre S.J. van der Zant Kavli Institute of Nanoscience, Delft University of Technology, Delft, The Netherlands

Alessandro Vindigni Laboratory for Solid State Physics, Swiss Federal Institute of Technology, ETH Zurich, Zurich, Switzerland

Wolfgang Wernsdorfer Institut Néel, CNRS & Université J. Fourier, Grenoble Cedex 9, France

Part I
Tunneling of Single Molecule Magnets

Chapter 1
From Quantum Relaxation to Resonant Spin Tunneling

Javier Tejada

Abstract A brief historic review of the research on quantum tunneling of magnetization is given and the story behind the discovery of resonant spin tunneling in Mn-12 acetate is told with the emphasis on the underlying physics. Important areas of studies in this field are mentioned and thoughts are presented on the future of research on molecular magnetism.

1.1 Historic Notes

My interest in quantum tunneling of magnetization was instigated by the 1988 Physical Review Letter of Chudnovsky and Gunther [1], in which they proposed that superparamagnetic behavior of small magnetic particles may not freeze down to absolute zero due to quantum tunneling of the magnetic moment. In 1990 Chudnovsky came to Barcelona on my invitation and we discussed at length possible experiments on quantum tunneling of the magnetization. I chose to study magnetic relaxation (which is sometimes called magnetic after-effect) in arrays of magnetic particles and other magnetic systems and to see if it persists down to very low temperature and eventually becomes independent of temperature when quantum effects take over thermal fluctuations.

From 1990 to 1995 we mounted a large effort at the University of Barcelona to observe these effects in systems of small particles, random magnets, and superconductors. This effort was reviewed in the book of Chudnovsky and myself "Macroscopic Quantum Tunneling of Magnetization" (Cambridge University Press, 1998) [2]. Published experimental results demonstrated that non-thermal magnetic relaxation was, indeed, present in all of the above systems. Moreover, theoretical estimates of the temperature of the crossover from thermal to quantum relaxation agreed well with our findings. Similar results on small particles were obtained by Berkowitz at the University of California—San Diego [3]. Barbara at CNRS-Grenoble [4], O'Shea at the University of Kansas [5, 6], and Arnaudas at the University of Zaragoza [7] observed non-thermal magnetic relaxation in bulk materials

J. Tejada (✉)
Dpto. de Física Fonamental, Facultat de Física, Universitat de Barcelona, 08028 Barcelona, Spain
e-mail: jtejada@ubxlab.com

J. Bartolomé et al. (eds.), *Molecular Magnets*, NanoScience and Technology,
DOI 10.1007/978-3-642-40609-6_1,

as well. These findings created a significant boost for research on quantum tunneling of magnetization that culminated in our discovery of resonant spin tunneling in Mn-12 acetate [8].

The discovery of magnetic bistability of Mn-12 by Sessoli et al. [9] in early 1990s made us think about definitive experiments on this system that would prove the existence of spin 10 tunneling beyond a reasonable doubt. The beauty of Mn-12, as compared, to systems of small particles, was in that the molecules, unlike magnetic particles, were identical, so that the tunneling rates had to be nearly the same for a macroscopic number of molecules in a solid. This implied that the decay of the magnetization of the Mn-12 sample would be similar to nuclear decay, that is, exponential in time. (This turned out later to be only approximately true because of the distribution of dipolar and nuclear hyperfine fields, as well as due to solvent disorder inside individual molecules and due to crystal defects.)

In 1995 I came to New York for three months on the invitation of Chudnovsky to do magnetic measurements with the group of Myriam Sarachik at City College. They had just bought a Quantum Design Magnetometer that they were not sure how to use. Our goal was to observe magnetic tunneling at low temperature. Previously, Chudnovsky and I had chosen to work on Mn-12. To have samples of Mn-12 I invited Ron Ziolo, head chemist at Xerox Corporation, with whom I had extensive previous collaboration on small particles, to join our group. I was also pleased to work with Jonathan Friedman, a talented Ph.D. student of Prof. Sarachik. Jonathan came up with a brilliant idea to align microcrystals of Mn-12 (that we had at the time) in a stycast using a high magnetic field. As far as magnetic measurements were concerned, this made the array of microcrystals equivalent to a large single crystal and allowed us to observe, for the first time, the equidistant steps in the magnetization curve of Mn-12.

Our findings and their correct explanation in terms of resonant spin tunneling, were first reported and publicly discussed at the MMM conference in Pittsburgh in 1995. Same year we submitted a paper to Physical Review Letters that was published [8] in May 1996. During spring of that year my group at the University of Barcelona, together with the group of Juan Bartolome at the University of Zaragoza, confirmed equidistant tunneling resonances in measurements of the ac susceptiblity of Mn-12 [10]. Barbara's group reported in a follow-up article [11] experiments performed on single crystals which corroborated these results.

From 1996 research on spin tunneling in molecular magnets picked up. Experimental landmarks worth mentioning are topological interference in spin tunneling demonstrated by Wernsdorfer and Sessoli [12], determination of the tunneling terms in the spin Hamiltonian of Mn-12 by the groups of Kent and Hill [13], and the works of New York and our groups on magnetic deflagration [14, 15]. There also has been a remarkable effort by chemists to synthesize new molecular magnets (Christou, Hendrickson, and others) with hundreds of them now available. With low temperature physics suffering from the rising cost of helium the future of this field depends, in my opinion, on whether research on molecular magnets finds any applications.

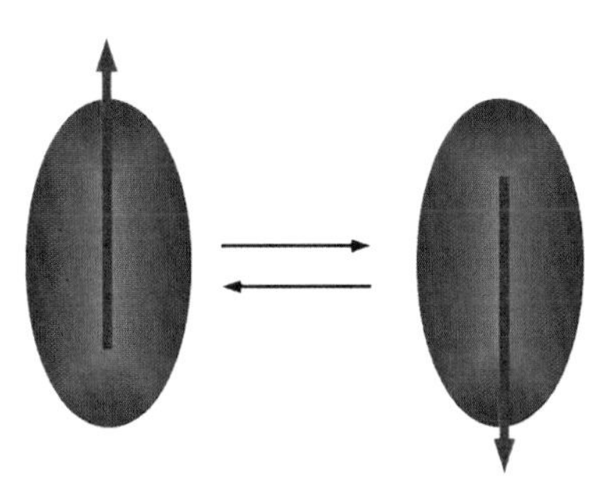

Fig. 1.1 Quantum tunneling of the magnetic moment in a single-domain particle

1.2 Early Experiments on Magnetic Tunneling at the University of Barcelona

In the absence of the magnetic field, single-domain magnetic particles have two opposite directions of the magnetic moment that correspond to the minimum of the magnetic energy. This is true independently of the shape or other anisotropic factors and is a consequence of the time-reversal symmetry. Magnetic field breaks time-reversal symmetry, making the two minima different in energy. Still, under a certain field strength, there is an energy barrier between the two states. It is due to the combined effect of shape and magneto-crystalline anisotropy of the particle. In large particles the barrier is high and the magnetic moment of the particle is frozen in a certain direction. In smaller particles, thermal fluctuations may kick the magnetic moment over the barrier, leading to the phenomenon of superparamagnetism. As temperature goes down, thermal fluctuations die out and magnetic moments of even the smallest particles would be frozen in the absence of quantum transitions. If, however, there were quantum underbarrier transitions (tunneling) between energy minima, the superparamagnetism would persist down to absolute zero, see Fig. 1.1.

Similar effects must exist in bulk magnetic materials where the change of the magnetization occurs through motion of domain walls. The barriers for such a motion are provided by pinning of the domain walls by defects in a solid. In the absence of the external magnetic field, the lowest energy state of a macroscopic magnet has zero total magnetic moment owing to the breakage into magnetic domains of opposite magnetization. Motion of domain walls separating magnetic domains is required to achieve this lowest energy state. At high temperature, thermal fluctuations kick the domain walls out of potential wells created by the pinning rather effectively, thus providing good mobility of the walls. Consequently, a permanent magnet would lose its magnetic moment rather fast. At low temperature, however, thermal processes slow down and the only reason for domain walls to escape potential wells created by the pinning would be quantum tunneling. This also applies to type-II superconductors, where change in the magnetization requires motion of Abrikosov flux lines pinned by defects.

The earliest mentioning of the possibility of quantum tunneling of the magnetic moment, probably, belongs to Bean and Livingston [16] who noticed that relaxation of the magnetization in systems of small particles did not disappear completely down to very low temperature. Later, various authors observed similar effect in bulk magnetic materials and superconductors, and speculated about tunneling of domain walls and Abrikosov flux lines [2]. Sustained progress in this direction was impeded by the absence of the general theory of such phenomena. Chudnovsky in 1979 noticed [17] that quantum tunneling of magnetization is given by the imaginary time solutions (instantons) of the Landau-Lifshitz equation that had been traditionally used to describe classical micromagnetic phenomena. This suggestion received further development in several papers published by van Hemmen and Sütö [18], Enz and Schilling [19], and Chudnovsky and Gunther [1, 20] between 1986 and 1988, the latter two papers being a major breakthrough in the understanding of the tunneling of single domain particles at low temperatures. In the first of these papers they computed the WKB exponent and estimated the temperature of the crossover from thermal to quantum superparamagnetic behavior for single-domain magnetic particles. The second paper dealt with quantum nucleation of domains in a magnetic film. These papers triggered modern interest, including my own, to quantum tunneling of magnetization.

The most important theoretical prediction for experiment was that the temperature of the crossover from classical to quantum regime scaled with the components of the magnetic anisotropy for both, uniform under-barrier rotation of the magnetic moment in small particles [1] and quantum diffusion of domain walls in bulk materials [20]. The challenge for experiment was to test this and other predictions in the presence of the broad distribution of energy barriers in systems of small particles and bulk magnetic materials. Exponential dependence of the tunneling rate on the barrier height stretches the lifetimes of metastable magnetic states from nanoseconds to the lifetime of the Universe. This, in fact, is a general situation in nature whether one studies magnetic relaxation or relaxation of the elastic stress in solids. The main point is that only those barrier heights, U, which have associated a flipping time of the order of the experimental time scale t, contribute to the relaxation process towards the global thermodynamic equilibrium. That is, measuring the slow evolution of magnetization with time allows us to sweep the barrier height distribution: the bigger the time is, the higher the barriers are.

Observation that allows one to investigate such situation quantitatively is that only certain barrier heights, U, contribute to the relaxation at a time t that elapsed since the preparation of the material.

The rate of thermal decay of the metastable states with a barrier U at a temperature T is given by

$$\Gamma_U = \nu \exp\left(-\frac{U}{k_B T}\right) \tag{1.1}$$

where ν is some pre-exponential factor (the attempt frequency) and k_B is the Boltzmann constant. The characteristic time of the decay depends exponentially on the barrier height: $\tau_U = 1/\Gamma_U = \tau_0 \exp(U/k_B T)$ where τ_0 is the so-called microscopic attempt time. The barriers that contribute to the relaxation at a time t are determined by the equation $t = \tau_U$, which gives $U = k_B T \ln(t/\tau_0)$. By the time t metastable states that have lower barriers have already decayed, while metastable states with

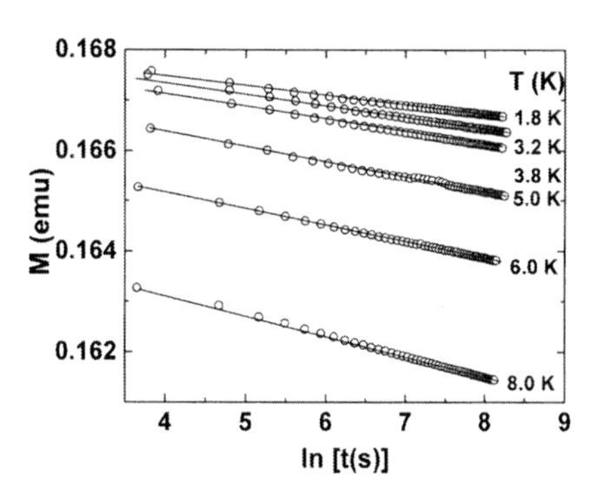

Fig. 1.2 Logarithmic relaxation of the magnetic moment of the array of $CoFe_2O_4$ nanoparticles measured at various temperatures [21]

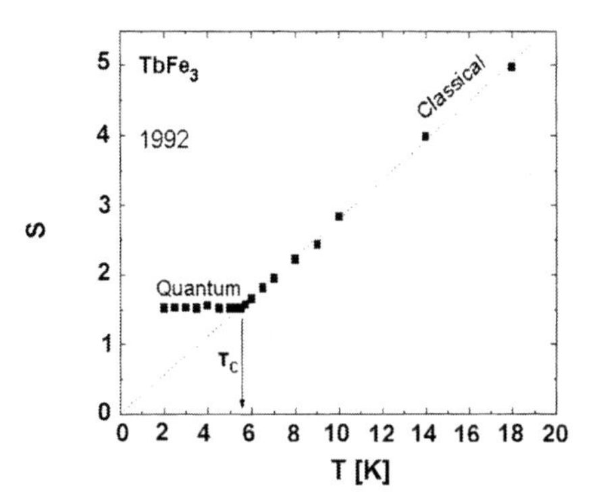

Fig. 1.3 Temperature dependence of the magnetic viscosity of $TbFe_3$ nanoclusters (1993) [22]

higher barriers have not decayed yet. Analysis of this situation [2] shows that the change in the total magnetization, M, depends linearly on $T\ln(t)$. Such logarithmic relaxation is known in many systems, magnetic or non-magnetic, and is related (through Fourier transform) to the notorious $1/f$ noise. An example of the logarithmic time relaxation in a system of $CoFe_2O_4$ nanoparticles [21] is shown in Fig. 1.2.

Thermal logarithmic relaxation implies that the derivative $dM/d\ln t$ (called magnetic viscosity) is proportional to temperature and must go to zero as temperature is lowered. Failure of the magnetic viscosity to go to zero in the limit $T \to 0$ would be an indication of non-thermal magnetic relaxation. This was the basis of our early experiments on quantum tunneling of magnetization. Temperature dependence of the magnetic viscosity in a system of $TbFe_3$ nanoclusters [22] is illustrated in Fig. 1.3. It clearly shows the plateau in the magnetic viscosity below 6 K, in agreement with theoretical expectation for quantum tunneling in these clusters having very high magnetic anisotropy. Another theoretical prediction is that the crossover temperature goes down with the decreased energy barrier. The latter can be achieved by applying the magnetic field since the field of a certain strength destroys anisotropy barriers altogether. The observed decrease of the crossover temperature on the magnetic field in $TbFe_3$ nanoclusters is shown in Fig. 1.4.

In the early 1990s similar results were obtained by us and other groups in various systems, including small particles, mesoscopic wires, thin films, bulk magnetic materials, as well as type-II superconductors [2]. Tunneling in antiferromagnetic clus-

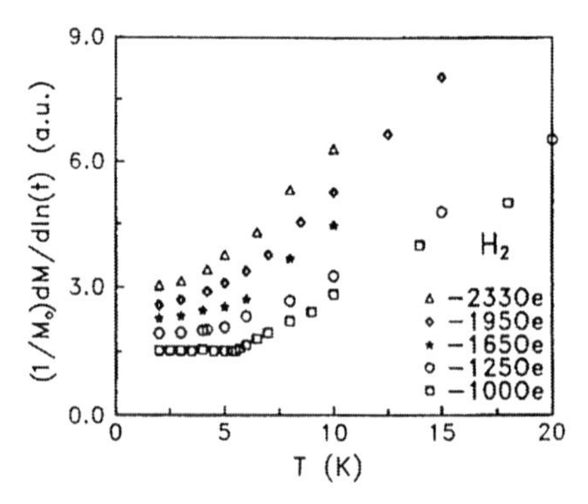

Fig. 1.4 Field dependence of the low-temperature magnetic viscosity of $TbFe_3$ nanoclusters (1993) [22]

ters has been studied (Chudnovsky and Barbara, Awschalom, Tejada). All data for magnetic materials were combined by us in a table showing the dependence of the crossover temperature on the strength of the magnetic anisotropy. It proved correlation between the two quantities, with a clear tendency of the crossover temperature being proportional to the anisotropy [2].

As years went by, measurements of individual nanoparticles had become possible. Wolfgang Wernsdorfer pioneered such measurements in Grenoble. In 1997 he reported [23] evidence of non-thermal magnetic relaxation in a single magnetic nanoparticle of barium ferrite below 1 K. The reduction of the energy barrier needed to provide a significant tunneling rate was achieved in Wernsdorfer's experiment by application of the magnetic field that was close to the field destroying the barrier.

1.3 Experiments on Mn-12

In the early 1990s Roberta Sessoli from Gatteschi's group in Florence discovered [24] that Mn-12 acetate molecules had a 65 K barrier between two lowest energy states with opposite directions of spin $S = 10$. Consequently, a crystal of Mn-12 molecules was equivalent to a system of identical superparamagnetic particles. This removed the challenge of broad barrier distribution that clouded interpretation of experiments with magnetic particles and placed Mn-12 at the top of candidates for observation of quantum tunneling of magnetization.

The spin Hamiltonian of the Mn-12 molecule subjected to the magnetic field, **B**, in the direction of the easy magnetization axis z is

$$H = -DS_z^2 - g\mu_B S_z B_z + H_\perp \tag{1.2}$$

where D is the anisotropy constant, g is the gyromagnetic factor, μ_B is the Bohr magneton, and $H_\perp$ is a perturbation that does not commute with S_z. If **S** were a classical vector, the dependence of the classical energy of the molecule on the orientation of this vector, up to the small perturbation due to $H_\perp$, would be given by

$$E = -DS^2 \cos^2\theta - g\mu_B SB \cos\theta \tag{1.3}$$

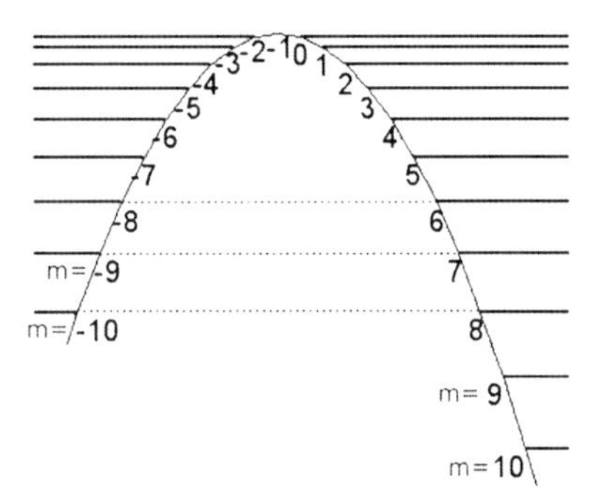

Fig. 1.5 Quantized spin energy levels of Mn-12 molecule at the second resonant field

where θ is the angle that **S** makes with the z-axis. The dependence of E on $\cos\theta$ is shown by the parabolic line in Fig. 1.5. It represents the magnetic anisotropy barrier discussed in the previous section. In quantum mechanics, however, the projection of **S** onto the z-axis is quantized, $S_z|m\rangle = m|m\rangle$, where m is an integer between -10 and 10 for $S = 10$. Consequently, the energy is quantized too,

$$E = -Dm^2 - g\mu_B Bm \tag{1.4}$$

The corresponding 21 quantized energy levels for $S = 10$ are shown by horizontal lines in Fig. 1.5.

Imagine now that one magnetizes the crystal of Mn-12 molecules in the negative z-direction, making all molecules to occupy the level $m = -10$. If the field is now applied in the positive z-direction the $m = -10$ level becomes metastable, while $m = 10$ becomes the ground state, see Fig. 1.5. If no other interactions are involved, a transition, due to $H_\perp$ in (1.2), from a level on the left of the parabolic barrier to a level on the right of the barrier is possible only if the energies of the two levels coincide. Equating E of (1.4) for two levels, m and m', one obtains that energies of pairs of m-levels coincide at discrete values of the magnetic field,

$$B_z = -\left(m + m'\right)B_0, \quad B_0 = \frac{D}{g\mu_B} \tag{1.5}$$

Since m and m' are integers, it is easy to see that the corresponding resonances are equidistant on the magnetic field, separated by the field B_0. For Mn-12 this field is about 0.45 T.

The above consideration makes it clear that at $T = 0$ the change of the projection of **S** on the easy magnetization axis may occur only via quantum tunneling between resonant m-levels at the discrete values of the magnetic field. Consequently the hysteresis curve of the Mn-12 acetate crystal will not be as smooth as the conventional hysteresis curve of magnetic materials because the bulk of the change of the magnetization should occur at the field that is multiple of B_0. Finite temperature makes this effect less dramatic because of thermal overbarrier transitions. However, the steps in the magnetization at the values of the field that are multiple of B_0 should be apparent at a finite temperature as well because quantum tunneling adds to thermal

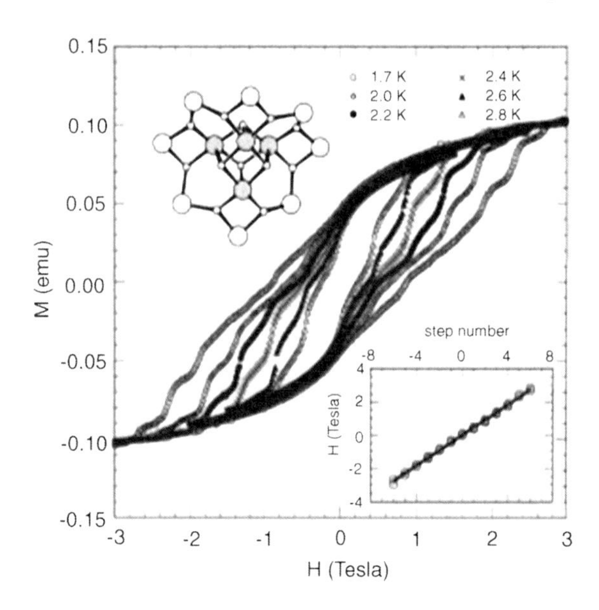

Fig. 1.6 This figure reports stepwise magnetization curves in Mn-12 acetate

activation. (In Mn-12 thermal effects completely take over quantum transitions at temperature above 3 K.)

Early experiments on quantum tunneling of magnetization in Mn-12 performed in Grenoble tried to detect non-thermal magnetic relaxation at very low temperature. They were inconclusive because of the very low tunneling rate in Mn-12 at $T = 0$. Our first measurements of Mn-12 in New York and Barcelona were less ambitious. They were aimed at the accurate magnetic characterization of Mn-12 acetate through measurements of its magnetization curve. As it turned out this was all one needed to demonstrate unambiguously the existence of quantum spin tunneling in Mn-12. Quantum hysteresis curve of Mn-12, first observed by us [8] in dc magnetization measurements performed in New York in 1995, is shown in Fig. 1.6. Soon thereafter we performed in Spain the ac susceptibility measurements of Mn-12 [10] that confirmed the existence of resonant spin tunneling and permitted to extract the tunneling prefactor $\tau_0 \sim 10^{-7}$s in the formula for the relaxation rate, $\tau = \tau_0 \exp[U/(k_B T)]$, see Fig. 1.7.

The importance of our first works on Mn-12 was not simply in the observation of equidistant magnetization steps but also in their correct interpretation as resonant spin tunneling along the lines of the theoretical argument presented above. That interpretation of the data was not straightforward at the time, it took us some time to sort things out. The problem was in the strong dependence of the height of the magnetization steps on temperature and on the rate at which the magnetic field was changed in experiment. Different steps kept appearing and disappearing on us depending on experimental conditions. Theoretical works appearing after our experiments suggested that this striking behaviour was based upon two main mechanisms: Dobrovitski and Zvezdin [25] noticed that the height of the magnetization step was determined by the Landau-Zener theory of adiabatic quantum transitions between

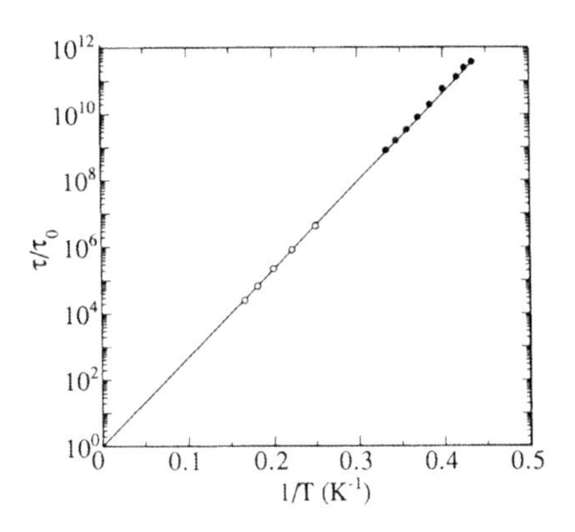

Fig. 1.7 Original figure [10] (1996) that permitted extraction of the tunneling attempt time from ac susceptibility measurements (*open circles*) and dc magnetic relaxation measurements (*closed circles*)

crossing energy levels, which explained the dependence on the field-sweep rate. On the other hand, the temperature dependence of the resonant tunneling effect was explained by means of the concept of thermally assisted spin tunneling, in which spins were first excited by phonons to higher energy levels and then tunnelled from these levels across the energy barrier [26–29].

For ten years after our discovery of resonant spin tunneling in Mn-12, one mystery about Mn-12 remained unsolved. In early 1990s Paulsen and Park, working in Grenoble, reported that sufficiently large crystals of Mn-12 exhibited abrupt reversal of the magnetization that did not seem to follow any clear pattern on temperature and magnetic field. This phenomenon received the name of "magnetic avalanche". For quite a while it was considered an impediment to the measurements of quantum magnetization steps. In 2005 the group of Sarachik in New York placed micro-Hall sensors along the length of the Mn-12 crystal and observed that the avalanche propagated through the crystal as a narrow front moving at a constant speed [14]. Chudnovsky came up with an explanation in terms of the magnetic deflagration.

Deflagration is a technical term for slow combustion. A mixture of hydrogen and oxygen in a pipe would burn via a slow (compared to the speed of sound) propagation of a burning front inside which the chemical energy is released. In a Mn-12 crystal the role of the chemical energy is played by the Zeeman energy of the magnetic moment in the magnetic field. The rest is similar to the burning of a flammable substance. In Sarachik's experiments deflagration was always triggered by the magnetic field sweep and occurred at random values of the field. This prevented Sarachik's group from observing magnetic avalanches at resonant fields. In Barcelona we triggered avalanches by surface acoustic waves at a fixed value of the field. This allowed us to observe an interesting phenomenon that was not known for chemical combustion. We found that the "flammability" of the Mn-12 crystal increased at the resonant fields due to the quantum enhancement of the rate of the magnetization reversal [15], see Fig. 1.8. We called this phenomenon "quantum deflagration".

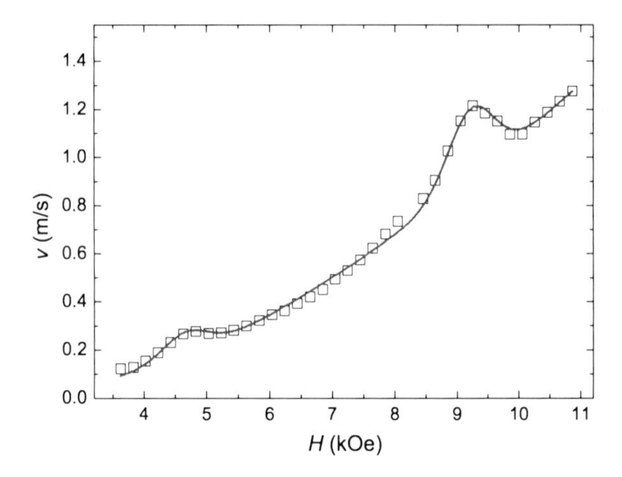

Fig. 1.8 Quantum peaks in the field dependence of the velocity of the magnetic flame in a Mn-12 crystal [15]

1.4 Conclusion

As years went by, Mn-12 has been researched almost exhaustively, or has it been?

Chemists found that Mn-12 acetate molecules were not entirely identical as previously thought, but slightly different in $H_\perp$ due to solvent disorder. The exact form of $H_\perp$ was established in a series of precise measurements performed by the groups of Hill and Kent [30]. Another spin 10 molecular magnet, Fe-8, was discovered and soon thereafter chemists began producing new molecular magnets at an astonishing rate. They have been investigated theoretically by diagonalization of spin Hamiltonians [31, 32] and by density functional theory [33]. Topological interference in spin tunneling phenomena has been predicted [34] and observed [35]. Tunneling in antiferromagnetic and ferrimagnetic molecular clusters has been studied [36–38]. Effects of nuclear spins and dipolar interactions in molecular magnets were addressed [39–43] and magnetocaloric effects have been investigated [44, 45]. Crossover between quantum and thermal regimes has been investigated in some detail [46–50]. Resonant interactions of molecular magnets with electromagnetic radiation has been measured [51–59]. Also long-range dipolar ordering has been studied [60–65]. Behavior of Mn-12 in ultrafast pulses of the magnetic field has been investigated [66, 67]. Magnetic molecules on surfaces have been studied [68] and conduction through single molecules placed between conducting leads has been measured. This account is, to the best of my knowledge, accurate. As the field of quantum magnetic relaxation and spin resonant tunneling goes back to the end of the 1980s, there will doubtless be some recollections that I have failed to fully acknowledge. For these mistakes, I sincerely apologize in advance.

Are any important discoveries left in this field for the younger generation of researchers? One area that remains cloudy is the possibility of superradiant emission of electromagnetic radiation by a crystal of molecular magnets. While we and others have published papers in this field [69, 70] the definitive proof of this possibility (or impossibility) is still absent. The distances between the spin levels of many molecular magnets are in the sub Terahertz range which has multiple applications. It is not clear, however, whether random dipolar and nuclear fields would allow strong coherent effects. Experiments inside a resonant cavity may help to figure this out. Up to date the reversal of the magnetic moment in molecular magnets has been explained claiming it occurs at the level of individual molecules [71]. However, it may be the case that this reversal process could occur at the level of clusters of molecules induced by spin-phonon transitions.

Since the energy barrier between opposite orientations of the magnetic moment is formed by weak relativistic interactions, a crucial question would be whether stable molecular magnets can ever break liquid nitrogen temperature of 77 K. Molecules with big magnetic moments, such as those containing rare earth atoms, may have their magnetization frozen at 77 K because of the strong magnetic anisotropy. Making identical molecules comparable to mesoscopic magnetic particles will be a challenging task for chemists. Another challenging question would be whether magnetic molecules can ever become ultimate memory units of conventional computers or even elements of quantum computers. I hope to see answers to these questions in the near future.

References

1. E.M. Chudnovsky, L. Gunther, Phys. Rev. Lett. **60**, 661 (1988)
2. E.M. Chudnovsky, J. Tejada, *Macroscopic Quantum Tunneling of the Magnetic Moment* (Cambridge University Press, Cambridge, 1998)
3. R.H. Kodama, C.L. Seamon, A.E. Berkowitz, M.B. Maple, J. Appl. Phys. **75**, 5639 (1994)
4. C. Paulsen, L.C. Sampaio, B. Barbara, R. Tucoulou-Tachoueres, D. Fruchart, A. Marchand, J.L. Tholence, M. Uehara, Europhys. Lett. **19**, 643 (1992)
5. M.J. O'Shea, P. Perera, J. Appl. Phys. **76**, 6174 (1994)
6. P. Perera, M.J. O'Shea, Phys. Rev. B **53**, 3381 (1996)
7. J.I. Arnaudas, A. Del Moral, C. De la fuente, M. Ciria, P.A.J. de Groot, Phys. Rev. B **50**, 547 (1994)
8. J.R. Friedman, M.P. Sarachik, J. Tejada, R. Ziolo, Phys. Rev. Lett. **76**, 3830 (1996)
9. R. Sessoli, D. Gatteschi, A. Caneschi, M.A. Novak, Nature **365**, 141–143 (1993)
10. J.M. Hernandez, X.X. Zhang, F. Luis, J. Bartolome, J. Tejada, R. Ziolo, Europhys. Lett. **35**, 301 (1996)
11. L. Thomas, Fl. Lionti, R. Ballou, D. Gatteschi, R. Sessoli, B. Barbara, Nature (London) **383**, 145 (1996)
12. W. Wernsdorfer, R. Sessoli, Science **284**, 133 (1999)
13. E. del Barco, A.D. Kent, S. Hill, J.M. North, N.S. Dalal, E.M. Rumberger, D.N. Hendrickson, N. Chakov, G. Christou, J. Low Temp. Phys. **140**, 119 (2005)
14. Y. Suzuki, M.P. Sarachik, E.M. Chudnovsky, S. McHugh, R. Gonzalez-Rubio, N. Avraham, Y. Myasoedov, E. Zeldov, H. Shtrikman, N.E. Chakov, G. Christou, Phys. Rev. Lett. **95**, 147201 (2005)
15. A. Hernandez-Minguez, J.M. Hernandez, F. Macia, A. Garcia-Santiago, J. Tejada, P.V. Santos, Phys. Rev. Lett. **95**, 217205 (2005)
16. C.P. Bean, J.D. Livingston, J. Appl. Phys. **30**, 1205 (1959)
17. E.M. Chudnovsky, Sov. Phys. JETP **50**, 1035 (1979)
18. J.L. van Hemmen, A. Sütö, Europhys. Lett. **1**, 481 (1986)
19. M. Enz, R. Schilling, J. Phys. Condens. Matter **19**, 1785 (1986). Ibid., **19**, L711 (1986)
20. E.M. Chudnovsky, L. Gunther, Phys. Rev. B **37**, 9455 (1988)

21. J. Tejada, R.F. Ziolo, X.X. Zhang, Chem. Mater. **8**, 1784 (1996)
22. J. Tejada, X.X. Zhang, E.M. Chudnovsky, Phys. Rev. B **47**, 14977 (1993)
23. W. Wernsdorfer, E. Bonet Orozco, K. Hasselbach, A. Benoit, D. Mailly, O. Kubo, H. Nakano, B. Barbara, Phys. Rev. Lett. **79**, 4014 (1997)
24. R. Sessoli, D. Gatteschi, A. Caneschi, M.A. Novak, Nature (London) **365**, 141 (1993)
25. V.V. Dobrovitski, A.K. Zvezdin, Europhys. Lett. **38**, 377 (1997)
26. E.M. Chudnovsky, D.A. Garanin, Phys. Rev. Lett. **79**, 4469 (1997)
27. D.A. Garanin, E.M. Chudnovsky, Phys. Rev. B **56**, 11102 (1997)
28. F. Luis, J. Bartolomé, J.F. Fernández, Phys. Rev. B **57**, 505 (1998)
29. A. Fort, A. Rettori, J. Villain, D. Gatteschi, R. Sessoli, Phys. Rev. Lett. **80**, 612 (1998)
30. E. del Barco, A.D. Kent, S. Hill, J.M. North, N.S. Dalal, E.M. Rumberger, D.N. Hendrickson, N. Chakov, G. Christou, J. Low Temp. Phys. 140(1/2) (2005). doi:10.1007/s10909-005-6016-3
31. M.I. Katsnelson, V.V. Dobrovitski, B.N. Harmon, Phys. Rev. B **59**, 6919–6926 (1999)
32. J.J. Borrás-Almenar, J.M. Clemente-Juan, E. Coronado, B.S. Tsukerblat, Inorg. Chem. **38**(26), 6081–6088 (1999)
33. K. Park, M.R. Pederson, C.S. Hellberg, Phys. Rev. B **69**, 014416 (2004)
34. A. Garg, Europhys. Lett. **22**, 205 (1993)
35. W. Wernsdorfer, R. Sessoli, Science **2**, 133–135 (1999)
36. O. Waldmann, C. Dobe, H. Mutka, A. Furrer, H.U. Güdel, Phys. Rev. Lett. **95**, 057202 (2005)
37. W. Wernsdorfer, N. Aliaga-Alcalde, D.N. Hendrickson, G. Christou, Nature **416**, 406–409 (2002)
38. C.M. Ramsey, E. del Barco, S. Hill, S.J. Shah, C.C. Beedle, D.N. Hendrickson, Nat. Phys. **4**, 277–281 (2008)
39. A. Garg, Phys. Rev. Lett. **70**, 1541–1544 (1993)
40. N.V. Prokof'ev, P.C.E. Stamp, J. Low Temp. Phys. **104**, 143–210 (1996)
41. A. Morello, F.L. Mettes, O.N. Bakharev, H.B. Brom, L.J. de Jongh, F. Luis, J.F. Fernández, G. Aromí, Phys. Rev. B **73**, 134406 (2006)
42. W. Wernsdorfer, R. Sessoli, D. Gatteschi, Europhys. Lett. **47**, 254 (1999)
43. E. Burzurí, F. Luis, B. Barbara, R. Ballou, E. Ressouche, O. Montero, J. Campo, S. Maegawa, Phys. Rev. Lett. **107**, 097203 (2011)
44. F. Torres, J.M. Hernandez, X. Bohigas, J. Tejada, Appl. Phys. Lett. **77**, 3248 (2000)
45. M. Evangelisti, A. Candini, A. Ghirri, M. Affronte, E.K. Brechin, E.J.L. McInnes, Appl. Phys. Lett. **87**, 072504 (2005)
46. E.M. Chudnosvky, D. Garanin, Phys. Rev. Lett. **79**, 4469–4472 (1997)
47. D. Garanin, X. Martínez Hidalgo, E.M. Chudnovsky, Phys. Rev. B **57**, 13639–13654 (1998)
48. L. Bokacheva, A.D. Kent, M.A. Walters, Phys. Rev. Lett. **85**, 4803–4806 (2000)
49. K.M. Mertes, Y. Suzuki, M.P. Sarachik, Y. Paltiel, H. Shtrikman, E. Zeldov, E.M. Rumberger, D.N. Hendrickson, G. Christou, Phys. Rev. B **65**, 212401 (2002)
50. W. Wernsdorfer, M. Murugesu, G. Christou, Phys. Rev. Lett. **96**, 057208 (2006)
51. R. Amigó, J.M. Hernandez, A. García-Santiago, J. Tejada, Phys. Rev. B **67**, 220402(R) (2003)
52. S. Takahashi, R.S. Edwards, J.M. North, S. Hill, N.S. Dalal, Phys. Rev. B **70**, 094429 (2004)
53. S. Hill, S. Maccagnano, K. Park, R.M. Achey, J.M. North, N.S. Dalal, Phys. Rev. B **65**, 224410 (2002)
54. L. Sorace, W. Wernsdorfer, C. Thirion, A.-L. Barra, M. Pacchioni, D. Mailly, B. Barbara, Phys. Rev. B **68**, 220407(R) (2003)
55. M. Dressel, B. Gorshunov, K. Rajagopal, S. Vongtragool, A.A. Mukhin, Phys. Rev. B **67**, 060405(R) (2003)
56. J. van Slageren, S. Vongtragool, A. Mukhin, B. Gorshunov, M. Dressel, Phys. Rev. B **72**, 020401(R) (2005)
57. E. del Barco, N. Vernier, J.M. Hernandez, J. Tejada, E.M. Chudnovsky, E. Molins, G. Bellessa, Europhys. Lett. **47**, 722 (1999)
58. E. del Barco, A.D. Kent, E.C. Yang, D.N. Hendrickson, Phys. Rev. Lett. **93**, 157202 (2004)

59. M. Bal, J.R. Friedman, Y. Suzuki, K.M. Mertes, E.M. Rumberger, D.N. Hendrickson, Y. Myasoedov, H. Shtrikman, N. Avraham, E. Zeldov, Phys. Rev. B **70**, 100408(R) (2004)
60. A. Morello, F.L. Mettes, F. Luis, J.F. Fernández, J. Krzystek, G. Aromí, G. Christou, L.J. de Jongh, Phys. Rev. Lett. **90**, 017206 (2003)
61. J.F. Fernández, Phys. Rev. B **66**, 064423 (2002)
62. X. Martinez-Hidalgo, E.M. Chudnovsky, A. Aharony, Europhys. Lett. **55**, 273 (2001)
63. M. Evangelisti, F. Luis, F.L. Mettes, N. Aliaga, G. Aromí, J.J. Alonso, G. Christou, L.J. de Jongh, Phys. Rev. Lett. **93**, 117202 (2004)
64. D. Garanin, E.M. Chudnovsky, Phys. Rev. B **78**, 174425 (2008)
65. P. Subedi, D. Kent, B. Wen, M.P. Sarachik, Y. Yeshurun, A.J. Millis, S. Mukherjee, G. Christou, Phys. Rev. B **85**, 134441 (2012)
66. J. Vanacken, S. Stroobants, M. Malfait, V.V. Moshchalkov, M. Jordi, J. Tejada, R. Amigó, E.M. Chudnovsky, D. Garanin, Phys. Rev. B **70**, 220401(R) (2004)
67. W. Decelle, J. Vanacken, V.V. Moshchalkov, J. Tejada, J.M. Hernández, F. Macià, Phys. Rev. Lett. **102**, 027203 (2009)
68. M. Mannini, F. Pineider, P. Sainctavit, C. Danieli, E. Otero, C. Sciancalepore, A.M. Talarico, M.-A. Arrio, A. Cornia, D. Gatteschi, R. Sessoli, Nat. Mater. **8**, 194–197 (2009)
69. E.M. Chudnovsky, D.A. Garanin, Phys. Rev. Lett. **93**, 257205 (2004)
70. J. Tejada, E.M. Chudnovsky, J.M. Hernandez, R. Amigó, Appl. Phys. Lett. **84**, 2373 (2004)
71. J.F. Fernández, J.J. Alonso, Phys. Rev. B **69**, 024411 (2004)

Chapter 2
Quantum Tunneling of the Collective Spins of Single-Molecule Magnets: From Early Studies to Quantum Coherence

Bernard Barbara

Abstract This chapter presents a review of the discovery and study of the phenomenon of mesoscopic quantum tunneling in Single Molecule Magnets (SMM, such as Mn_{12}, Fe_8) and Single Ion Magnets (rare-earth ions) starting from the first roots in the 70's and ending with the most recent studies of coherent Rabi oscillations and of decoherence mechanisms.

2.1 Introduction

The possibility of observing quantum phenomena "at the macroscopic scale" has been discussed from the earliest times of quantum mechanics (see, e.g. the Schrödinger's cat paradox [1]). Experiments, clearly devoted to the possible observation of "Macroscopic Quantum Tunnelling" (MQT) started in the 70's or 80's, in particular under the impulse of A.J. Leggett [2] who developed the concept of quantum tunnelling of a "collective order parameter" associated with the ground-state of systems of intermediate sizes—we now say, "Mesoscopic". In some sense this is, in a more general approach, the quantum counterpart of Néel's classical order parameter of super-paramagnetic nanoparticles, which are nanomagnets with a continuous order parameter equal to their magnetic moment They were intensively studied between the 40's and the 60's, under the name of "thin particles" [3]. As far as I know, the first clear observation of a classical to quantum crossover in MQT was made on single micrometer-size Josephson junctions, at IBM Yorktown Heights in 1981 [4]. The "switching current", thermally-activated at high temperature, was independent of temperature below a certain crossover temperature, in agreement with the expectations of quantum tunnelling in weak dissipation regime [5]. At nearly the same time, other quantum phenomena, such as those associated with a single atom placed in the quantum field of a cavity, showed evidence of quantum coherence of a single small object (atom, photon) [6], opening the door for quantum computing with, in our context, spin systems [7, 8]. The search for MQT in Magnetism (MQTM) apparently started in the early seventies (for a short historical review, see Ref. [9]) but

B. Barbara (✉)
Institut Néel, CNRS & Université Joseph Fourier, BP 166, 38042 Grenoble Cedex 9, France
e-mail: bernard.barbara@grenoble.cnrs.fr

J. Bartolomé et al. (eds.), *Molecular Magnets*, NanoScience and Technology,
DOI 10.1007/978-3-642-40609-6_2, © Springer-Verlag Berlin Heidelberg 2014

the experimental situation evolved slowly due to the technical impossibility to work with single-nanoparticles or single-spins [10–20] (which is no longer the case now, see e.g. [21–24]). Those works paved the way to the unambiguous demonstration of MQTM, which was obtained over the period of 1994–1996 on the Single Molecular Magnet Mn_{12}-ac, which is a single-crystal made of an ensemble of identical molecular magnets [9, 25–28]. An intensive multi-disciplinary research on the quantum behaviour of magnetic molecules followed and is still very active all over the world. The observation of MQTM in nanometer size single molecule magnets (SMMs) was later extended to the case of rare-earth ions diluted in insulating non-magnetic matrices, known as "Single Ion Magnets" [29–32], showing that the mesoscopic scale in magnetism has nothing to do with the spatial extensions of wave functions (size of the object) but depends on the value of the spin only. It ranges up to a few hundreds of spin units. Above quantum effects, even if they are undoubtedly present, are more difficult to evidence due to the loss of measurable quantization (continuum of states, see Sect. 2.2.1).

2.2 Prehistory and History

To my knowledge, the first publications clearly devoted to the search of MQT in magnetism were about magnetic relaxation experiments performed in highly anisotropic rare-earth inter-metallic single-crystals (Dy_3Al_2, $SmCo_{3.5}Cu_{1.5}$ [11–20]). They provided magnetic relaxation experiments, thermally-activated above a certain crossover temperature (a few kelvins) and independent of temperature below it. Due to the high anisotropy to exchange ratio of rare-earths, the domain walls are "narrow" and their thickness is a few inter-atomic distances only leading to intrinsic pinning [11, 12] by a "magnetic Peierls potential" analogous to the Peierls potential of dislocations. As this had been observed with dislocations [33–36] and obtained in a theoretical attempt to interpret our first experiments [37] we attributed this non-thermal relaxation to the de-pinning of small portions of domain-walls by tunnelling through their magnetic Peierls potential. More precisely, the tunnelling effect was considered to be a quantum nucleation on the wall surface (irreversible local wall deformations) followed by a 2-D soliton-like propagation on the wall surface. Interestingly, tunnelling of dislocations was recently brought back to light to give a possible interpretation of the controversial phenomenon of the super-solidity of ^{4}He [38]. Those first results on rare-earth inter-metallic systems [10–13] motivated the first theory of quantum depinning of domain walls by T. Egami [37], leading, in particular, to the first evaluation of the crossover temperature T_{co} between quantum and thermally-activated relaxation regimes (see also [10]). Our experimental studies, showing that spin reversal takes place within independent spin-blocks of $\sim$ 1–2 nm, were followed by more focussed studies on magnetic thin films, ensembles of ferromagnetic nanoparticles with narrower size distributions (15nm-TbCeFe, 2nm-FeC, ...) and tri-layer systems with a single domain-wall pinned in the centre layer, which we called "Domain-Wall Junctions" (Co/CuCo/Co, $GdFe_2/SmFe_2/GdFe_2$,

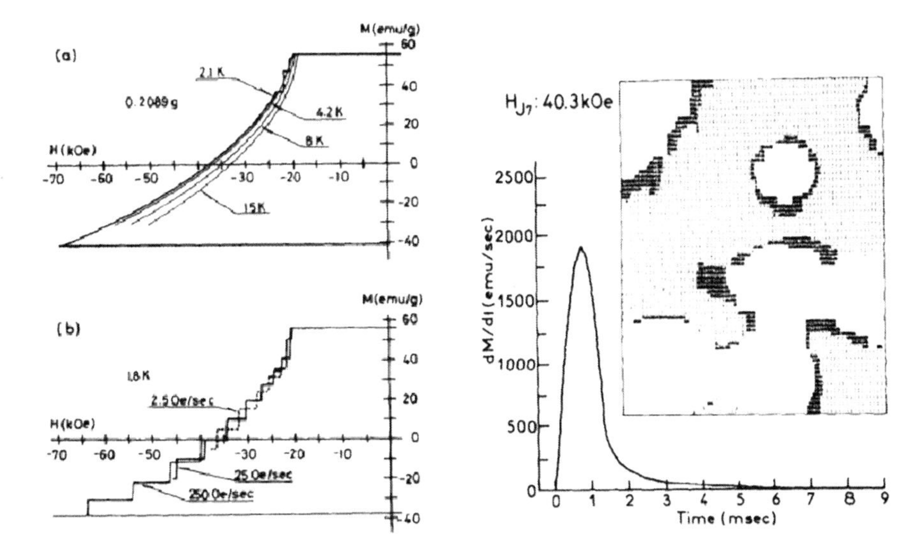

Fig. 2.1 (**a**) Staircase hysteresis loop of a $Sm_{3.5}Cu_{1.5}$ single-crystal. The magnetization jumps below 2.1 K were measured at constant sweeping field rate. (**b**) Faraday observation of macroscopic domain walls jump and the related variations of raising and damping time [18]

FeTb/Cu/FeTb...) ([19, 20] and references therein). Spin-reversal volumes were found to remain constant below a certain $T_{\rm co}$ (non-thermal reversal mechanism attributed to MQTM) and to linearly increase with temperature above it (thermally-activated reversal) [19, 20, 39]. Those $T_{\rm co}$ were in good agreement with the ones calculated theoretically according to the emerging theories of MQTM [10, 37, 40–44]. Note that the prediction of MQTM in an antiferromagnetic nanoparticle [43] showed that this effect should be easier to observe than with ferromagnetic nanoparticles. This is an obvious consequence of the fact that the antiferromagnetic order parameter (Néel vector) does not commute with the anisotropy Hamiltonian. As this is now well known, this aspect of quantum mechanics of a single order-parameter is not valid for a macroscopic antiferromagnet, which has been at the origin of what was probably the most important controversy in the history of magnetism, ending with the discovery of antiferromagnetism (B. Barbara, "L'œuvre de Louis Néel").

Those magnetic relaxation studies in which spin-reversals of classical or quantum origin take place at the nanometer-scale were sometimes followed by large magnetization jumps which we called "magnetic avalanches" or "macroscopic Barkhausen jumps" [17–20, 45, 46]. First observations of this phenomenon were made on Sm-based single crystals above 1 K [17, 19] (Fig. 2.1(a)). Interestingly those jumps were large but of finite size, leading to staircase hysteresis loops (Fig. 2.1).

Each avalanche results from macroscopic domain-wall jumps, initiated by 2-D nucleations (of classical or quantum origin) randomly distributed over the existing domain walls and followed by classical domain-wall motions spreading through the crystal over about 1 μm with linear raising time $\tau_1 \sim 1$ ms (constant velocity ~ 1 μm/ms) and exponential slowing-down times of $\tau_2 \sim 5$ ms (associated with

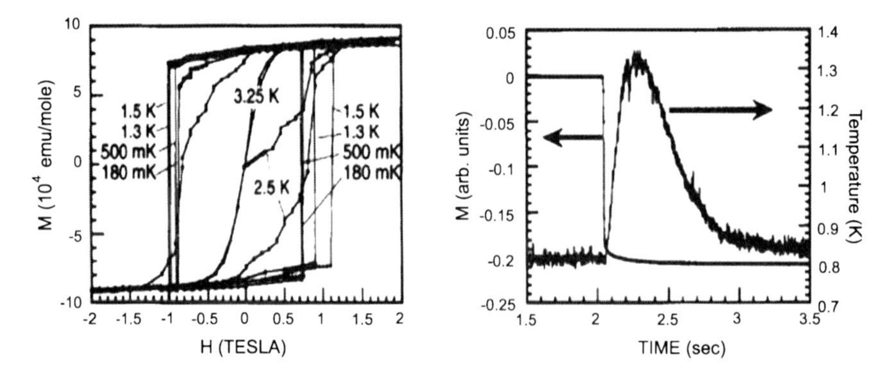

Fig. 2.2 *Left*: avalanche observed in Mn_{12}-ac and *right*: associated temperature variation (the sample was at 850 mK and tilted by 45° away from the field) [47, 48]. Similar temperature variations, with a crossover to a regime where the temperature oscillates have been observed in Sm-based systems in the limit of low sample-bath coupling [19]

heat release, Fig. 2.1(b)) [17, 19, 20]. The same phenomenon was also observed later in Mn_{12}-ac [47, 48] (Fig. 2.2), especially when the sweeping field rates exceed a certain value or/and the crystals are not very small (for reasons which are made obvious below).

Further studies of this phenomenon on Mn_{12}-ac compared this phenomenon to the propagation of a flame-front through a flammable substance "magnetic deflagration" (see reference [49] and the chapters by D. Garanin and M. Sarachik in the present book). Our explanation for those avalanches was, and still is, related to the transfer of a part of the magnetic energy stored in the hysteresis loop $H\mathrm{d}M/\mathrm{d}t$ to phonons, plasmons, spin-waves, etc, after the fast motion of a domain wall. Avalanches were observed only if that heat could reach the cryostat at a rate slower than the rate at which it was produced.

All that showed that unambiguous proof of MQTM was lacking, especially after we realized that unavoidable energy barrier distributions could be a real problem [28, 39]. On the basis of very general arguments, we actually showed that in the presence of size or/and switching field distributions, the energy distribution function of non-interacting switching blocks is a power-law $f(E) \propto E^{1/\alpha-1}$ where α is a parameter depending on details of the initial distributions. The measured magnetic viscosity $S = \mathrm{d}M/\mathrm{d}\ln t = Tf[E = k_{\mathrm{B}}T\ln(t/\tau_0)]$, should therefore be independent of temperature if $\alpha \gg 1$, an effect which could be mistaken for the expected temperature-independent plateau of MQTM. The application of this model to the systems that we studied showed that (i) with large α, ensembles of Ba-ferrite nanoparticles exhibit such a distribution-plateau below 10 K, thus hindering the observation of MQTM and (ii) with $\alpha = 1$, large assemblies of $TbCeFe_2$ or FeC nanoparticles, and amorphous-FeTb multi-layers do not show such a "distribution-plateau", suggesting that in those cases the plateaus observed were likely of MQTM origin [19, 20] (Fig. 2.3).

Regarding interacting switching blocks, we developed a numerical model which also ended up with a power-law distribution for $f(E)$ [28, 45]. This was a 3-D

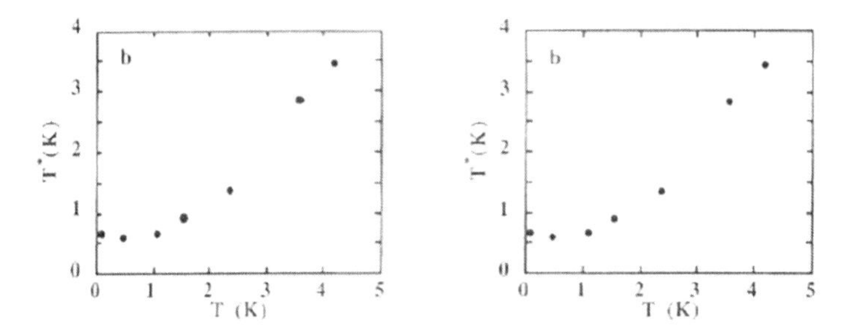

Fig. 2.3 Variations of the effective temperature T^* (defined as $\tau = \tau_0 \exp[-E(H)/k_B T^*] \sim$ measuring time scale) vs. the measured temperature T, in particles of $TbCeFe_2$ (**a**) and FeC (**b**) of a few nm [20, 39]

Ising model with ferromagnetic interactions J_{ex} and a random anisotropy distribution of width $K_0\Delta$. The simulations showed the occurrence of spin avalanches induced either by self-heating (as above) or by exchange-induced cooperative spin-flips. In order to simulate heat transfer (first case) we assumed that each switched spin transfers a fraction f of its energy dissipated during the switching to its first neighbors. In the second case, exchange-induced cooperative spin-flips simply increase with J_{ex}. When f or/and J_{ex} increase, more and more cooperative spin-flips take place, leading to larger and larger spin-avalanches. In particular, at $f = 0$ spin-flips are non-correlated, giving a continuous hysteresis loop when $J_{ex}/K_0\Delta < 1$ and a single infinite avalanche when $J_{ex}/K_0\Delta \gg 1$. In between, we obtained a hierarchy of avalanches with a power-law size distribution, suggesting the occurrence of Kohlrausch dynamics. As an example, the low temperature magnetic viscosity $S(T)$, calculated at constant J_{ex}, shows a plateau for $f \sim 0.2$, and a divergence for $f \sim 0.4$ because the self-heating becomes too important. This model allowed us to make a connection between our low temperature "distribution plateaus" (which should not be mistaken for MQTM) and the occurrence of "critical self-organized avalanches", showing quite interestingly that the former is at the origin of the latter. These distribution models helped us considerably to discriminate between the plateaus of "quantum origin" and those of "distribution origin". Despite the fact that it was not much developed, this study [45] was among the first models of this type which were later popularized under the name of cracking noise models [46].

In order to minimize the effect of distributions on our MQTM studies, we started, in the early nineties, two simultaneous projects with the search for (i) measurements of single nanoparticles and (ii) measurements of ensembles of nanoparticles with the narrowest possible size distributions. This led to (i) the development of the micro-SQUID magnetometer for micro- and nano-magnetic detection [24], still one of the most important existing tools for nano and molecular magnetism and (ii) the study of Single Molecule Magnets where each "nanoparticle" is a magnetic molecule with a collective spin $S = 10$ [28–31].

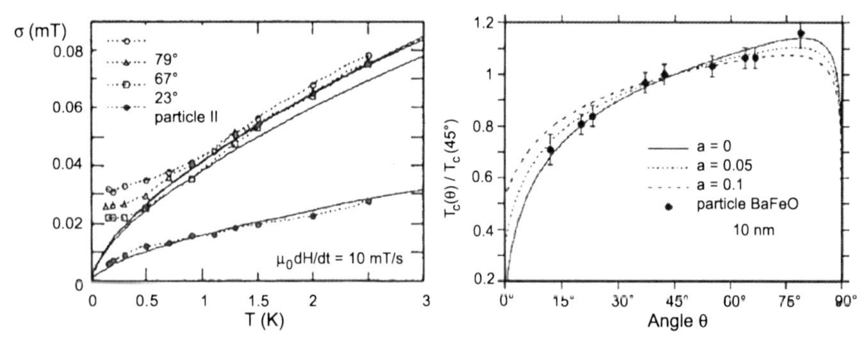

Fig. 2.4 *Left*: temperature dependence of the width of the switching field distribution for a 20 nm BaFeO nano-particle (magnetic moment $10^5 \mu_B$) and for three different orientations of the applied magnetic field. *Solid data points* were measured on a larger particle ($10^6 \mu_B$) [51, 55]. *Right*: experimental and calculated variations of the crossover temperature T_{co} with the field tilt angle. In both cases, *continuous lines* are fits to the model of Ref. [56]

2.2.1 Micro-SQUID Measurements

The triggering factor for the realization of the micro-SQUID magnetometer was a short discussion with A. Benoit from our institute (CRTBT at the time) who had developed a challenging micro-SQUID set-up to detect the persistent currents of single non-superconducting loops [50]. Our argument was simple: detecting the magnetic moment of a ferromagnetic nanoparticle (dream of generations of scientists in magnetism!) would be much easier than detecting the persistent current of a normal mesoscopic loop. Two (four) years later, we published the first measurements on micrometer (nanometer) single-particles, together with D. Mailly, K. Hasselbach, and W. Wernsdorfer [21, 28, 51–53]. This enabled us to make a series of text-book demonstrations on the foundations of nanomagnetism [54, 55] with, on the top of that, a study suggesting MQTM of a single ferrimagnetic Ba-ferrite nanoparticle (collective spin $\sim 10^5$) when the huge barrier is strongly depressed by the application of a magnetic field [51]. Above 0.4 K, the quantitative agreement with the Néel-Brown theory allowed us to identify unambiguously the dynamical aspects of uniform magnetization reversal (Fig. 2.4, left).

Below this temperature, strong deviations from this model were evidenced when the applied magnetic field was tilted (Fig. 2.4, left), which were in quantitative agreement with the predictions of the MQTM theory in the low dissipation regime [56] (Fig. 2.4, right). This unique result suggesting macroscopic quantum tunneling was nevertheless biased by the fact that the signature of resonant quantum tunnelling could not be fully identified because of the impossibility to detect hysteresis loop quantization, opening the door for other possible interpretations.

2.2.2 Mn_{12}-ac, The First Single Molecular Magnet

The study of Mn_{12}-ac was boosted by a talk given at the "scuola nazionale sui materiali nano-strutturati" on "magnetic nanocrystallized systems", Rimini (1993), in which we presented our works on MQTM (see above) and explained why nanoparticles with the narrowest possible size-distribution were important to get an unambiguous proof of this phenomenon. This talk found an immediate echo in D. Gatteschi, as he was himself working with R. Sessoli on Mn_{12}-acetate ($[Mn_{12}O_{12}(CH_3COO)_{16}(H_2O)_4]$) a system made of identical molecules with collective spin $S = 10$. That was precisely what we were looking for, except that the spin was much smaller than what we expected (we were used of nanoparticles of few thousand spins, at least). However, due to the absence of distribution and the complex character of the molecule (made of a hundreds of atoms), we did not hesitate to start a collaboration in the course of our search for MQTM, that we formalized a little later at the occasion of a short meeting at the CNRS centre of Aussois (not far from Grenoble).

At the end of the same year a first paper reported the existence of magnetic bistability in Mn_{12}-ac [57]. It took between one and two years to start our MQTM measurements. In fact, the discovery MQTM in SMMs was made in two steps, a first one showing an unambiguous relaxation plateau at low-temperatures and low magnetic fields and a second one showing a hysteresis-loop quantization in larger fields. In order to avoid the above-mentioned distribution problems, we needed a large single-crystal and a good SQUID magnetometer (some micro-SQUID experiments were done on a small crystal of Mn_{12}-ac, but they were not successful). A first study, performed in 1995 on five single-crystals of Mn_{12}-ac oriented under a magnifying lens, already claimed MQTM [28] (see Sect. 2.3.2 below). The first study performed on a single crystal (1 mm long parallelepiped, oriented under a magnifying lens) [25] appeared in 1996. The same year, two other studies had been published. One was performed on a fine powder obtained after severe pounding [26] (although it was partially oriented in a magnetic field, the residual magnetization was surprisingly smaller than the limit $M_s/2$ of a randomly oriented uniaxial powder; M_s = saturation magnetization) and the other was performed on micrometer crystals oriented in a field [27]. Those studies showed the existence of characteristic steps in the hysteresis loop of Mn_{12}-ac.

Shortly after the discovery of MQTM in SMMs these molecules became very popular and to this day constitute a very active field of research with e.g. the search of quantum tunnelling with larger spins (however one should keep in mind that the presence of quantization is as useful as the absence of distribution, for the identification of MQTM) or the possible use for magnetic memories: classical bits or quantum qubits. It has always been clear to us that the former is a hopeless project due to the impossibility to synthesize SMMs with energy barriers $\geq 10^5$ K (except for low temperature applications, or if $4f - 3d$ transition metal ions could be strongly coupled in as this is the case in inter-metallic alloys), whereas the latter is a very active field progressing rapidly (Sect. 2.6). Another and more important subject lays in the study of the inner mechanisms of quantum relaxation (Sects. 2.3 and 2.4). The unique understanding of incoherent quantum phenomena which has been reached in magnetism, is not unrelated to the fact that, in magnetism, theory can start from first principles which is not always the case for other fields such a superconductivity (for a review see e.g. [58]). Finally, a natural follow-up of the MQTM study

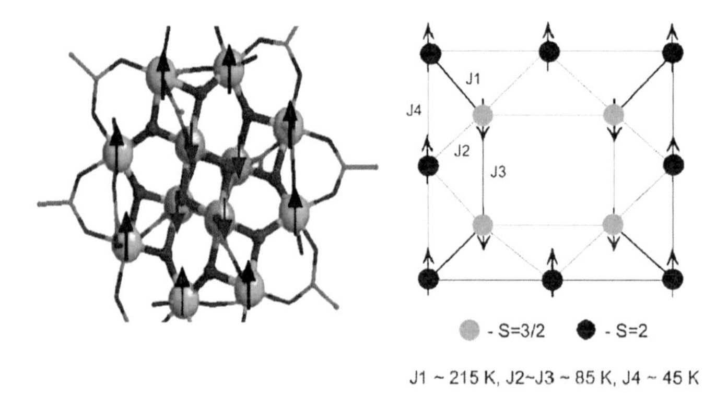

Fig. 2.5 *Left*: simplified structure of Mn_{12}-ac showing the positions and spin orientations of Mn atoms. *Right*: interaction scheme

could start a few years ago, by switching to the regime of Mesoscopic Quantum Coherence in Magnetism (MQCM) with the first observations of Rabi oscillations in SIMs [59, 60] (Sect. 2.5) and SMMs [61, 62] (Sect. 2.6). Ongoing studies on those systems are trying to identify and to study the different decoherence mechanisms in order to maximize the number of possible operations in a spin qubit, as it is done for other types of qubits.

To conclude this historical section, it is interesting to point out, that our first works in the 70's started with the same system as those with which we have been working these last years (rare-earth ions). Such a non-deliberate return to basics converged step by step and merged into the independent flow of research in other fields of physics and chemistry with now the study of coherent quantum dynamics in all kinds of qubits and the search for optimisation at the smallest possible scales.

2.3 Quantum Tunneling in Single Molecule Magnets

2.3.1 Single Molecule Magnets: Basic Properties

As explained in the historical Sect. 2.2, first evidences of MQTM were found in the SMM $[Mn_{12}O_{12}(CH_3COO)_{16}(H_2O)_4]$, hereafter referred to as Mn_{12}-ac. This molecule, synthesized by Lis more than 30 years ago [63], has a tetragonal symmetry and contains a cluster of twelve Mn ions divided into two shells with strong antiferromagnetic couplings: four $s = 3/2$ Mn^{4+} ions from the inner shell, surrounded by eight $s = 2$ Mn^{3+} ions from the outer shell, giving the collective spin $S = 10$ (Fig. 2.5).

Those molecules are chemically identical and form tetragonal crystals with an average distance between Mn_{12} molecules of the order of 1.5 nm [63]. Intermolecular exchange interactions are negligible and dipolar interactions between nearest neighbours along the **c** axis, which coincides with the magnetic anisotropy axis z,

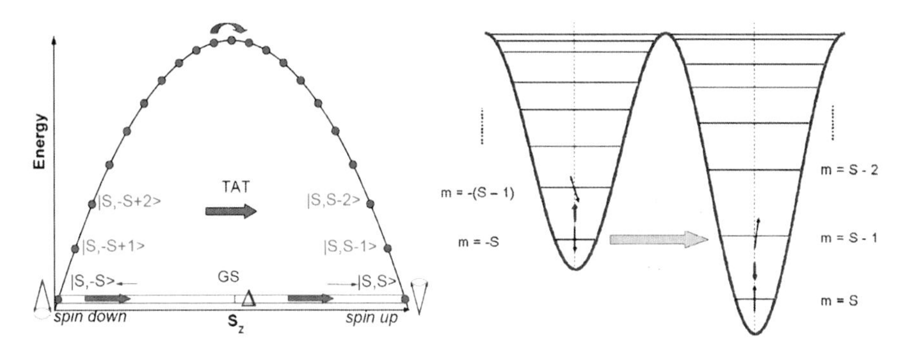

Fig. 2.6 *Left*: energy of a SMM plotted vs. S_z; the parabolic shape comes from the dominant anisotropy term DS_z^2 slightly modified by higher order longitudinal terms. Ground-state or thermally activated tunnelling take place when the spin-up and spin-down states are at resonance, provided that at least one transverse term induces a sufficient tunnel splitting Δ between them. *Right*: same energy spectrum, but plotted vs. the angle θ derived from the semi-classical expression $S_z = S\cos\theta$, in abscissa ($S \gg 1$)

are about 0.13 K [64]. This is much smaller than the magnetic anisotropy barrier of each molecule (which, as discussed in what follows, is about 60 K).

Magnetic Anisotropy: The "Giant Spin" Hamiltonian and Its Domain of Validity The magnetic response of these molecules is therefore determined mainly by the magnetic anisotropy and by the Zeeman interaction with external magnetic fields. In the case of Mn_{12}-ac, the effective spin Hamiltonian describing the energies of states with $S = 10$ (giant spin approximation), limited to fourth-order anisotropy terms [63, 65], can be approximately written as follows:

$$H = -DS_z^2 - BS_z^4 + C\left(S_+^4 + S_-^4\right) - g\mu_B \boldsymbol{H}\boldsymbol{S}, \tag{2.1}$$

with $D/k_B \approx 0.56$ K, $B/k_B \approx 1.11 \times 10^{-3}$ K and $C/k_B \approx 2.9 \times 10^{-5}$ K [57, 66]. The structure of magnetic energy levels that follows from (2.1) is shown in Fig. 2.6 as a function of S_z (left) and as a function of the semiclassical angle θ (right). Longitudinal anisotropy terms give rise to a classical barrier, of height $U_{cl}(0) \approx DS^2 + BS^4$, whereas transverse terms contribute to tunneling. The Zeeman term shifts these levels (see Fig. 2.6, right) and reduces U_{cl}. For instance, if **H** points along $\mathbf{z}U_{cl} = U_{cl}(0)(1 - H/H_A)^2$, where $H_A \approx (2S-1)D/g\mu_B \approx 10^5$ Oe is the anisotropy field.

In order to see what the limits of application of this "giant-spin" Hamiltonian (2.1) are, it is useful to analyze the thermal variation of the effective paramagnetic moment $\mu_{eff}(T) = g\sqrt{[S(S+1)]}$, determined from magnetization and susceptibility curves measured in the super-paramagnetic region between 3 and 300 K with a field parallel to the crystal **c**-axis. A$S = 10$ Ising model provides an excellent fit of data obtained below ~ 10 K, with a crossover to an Heisenberg model between 10 and 30 K [58, 67]. Above 30 K, fits based on the assumption that $S = 10$ are no longer valid. Correlatively, a fast decrease of $\mu_{eff}(T) \sim 21\mu_B$ was observed above

~ 30 K, with a broad minimum around 150–200 K. These results show that excited collective spin states with $S = 9$, $S = 8, \ldots$ cannot be neglected above 30 K, at least in regard to the equilibrium behaviour [58]. This result, in accordance with weakest intra-molecular interactions of 45–55 K (Fig. 2.5), agrees with neutron scattering experiments reporting an energy separation of about 40 K between the $S = 10$ ground state and the first excited $S = 9$ multiplets [68].

From a purely theoretical point of view one might, in principle, calculate the energy spectrum levels of a Mn_{12}-ac molecule with all the couplings between the twelve Mn atoms. However the Hilbert space dimension $(23/2+1)^4(22+1)^8 = 10^8$ is too large. Besides, due to the uncertainty in experimental determination of the coupling constants, any such attempt would not be very useful. In order to come up with a solution to this problem, at least in the region of the temperatures below 150 K, it was suggested to apply the idea of a "reduced" Hamiltonian in which the largest coupling constant $J_1 \sim 200$ K between the spins $S_1 = 2$ and $S_2 = 3/2$ "locks" them into a spin-state of $S_{12} = 1/2$ up to temperatures of the order of J_1 (similar "dimerization" was exploited in [66]). This assumption leads to an effective 8-spin Hamiltonian with a Hilbert space of 10^4 authorizing exact matrix diagonalization [58]. This "truncated" Hamiltonian includes both exchange and anisotropy (through equivalent anisotropic exchange). It should be noted that Dzyaloshinsky-Moriya (DM) interactions were ignored despite the fact that they should be, as in most low-symmetry SMMs, of the order of a few kelvins because intra-molecular couplings of are of several hundred kelvins (see [69–71] and references therein). They can induce anisotropy, remove Kramer's degeneracy, enhance tunneling or induce a coupling between multiplets without changing significantly their internal structure (energy spectrums) [65, 72–76].

Susceptibility and magnetization curves calculated from the 8-spin Hamiltonian account very well for the experimental data, with only three free parameters, thus enabling us to determine them without any ambiguity [58]. The corresponding energy level structure, given in the left-hand panel of Fig. 2.7, shows that the $S = 9$ multiplet becomes occupied above approximately 30–40 K, confirming that the $S = 10$ collective ground-state model is no longer valid above these temperatures, at least regarding equilibrium properties. However, as most experiments devoted to study MQTM in Mn_{12}-ac are done at low temperature, a description in terms of the $S = 10$ ground multiplet is quite sufficient.

In the following, I describe in detail how the discovery of MQTM in Mn_{12}-ac took place in two steps, one before and one after the 1994 NATO workshop on "Quantum tunneling of magnetization" [28].

2.3.2 *First Evidences*

Magnetization and ac-susceptibility experiments, performed in 1993–1994 on fine polycrystalline powders [57] or collections of parallelepiped Mn_{12}-ac crystals [28], showed a super-paramagnetic behaviour [3]. At high temperatures ($T > 2.5$ K),

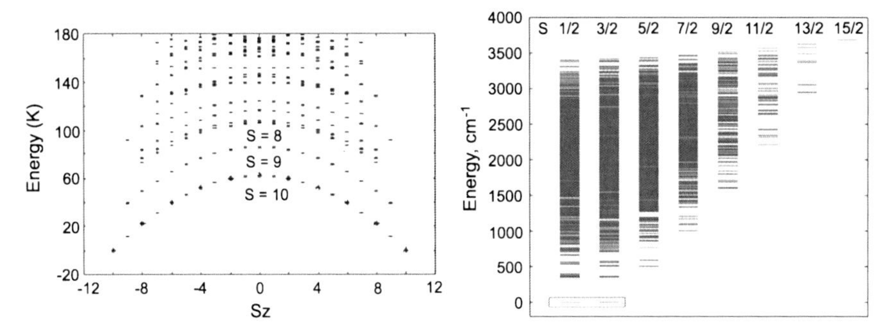

Fig. 2.7 *Left*: calculated energy spectrum of a Mn_{12}-ac cluster [58] (valid up to 180 K). *The stars* show a parabolic behavior $D(S^2 - S_z^2)$ where $-S \leq S_z \leq S$, $D/k_B = 0.627$ K, and $S = 10$. *Right*: energy spectrum of a V_{15} cluster calculated on the full Hilbert space of 2^{15}. *In a square*: the ground-state is formed of two $S = 1/2$ doublets and the first excited state is an $S = 3/2$ excited quartet (I. Tupitsyn, private communication and [76]). These 8 levels are also obtained in the 8-spins approximation

the magnetic relaxation time obeys the Arrhenius law $\tau = \tau_0 \exp(U/k_B T)$ with a prefactor $\tau_0 \sim 10^{-7}$ s and an activation energy U/k_B 61 K. This gives rise to a superparamagnetic blocking of the magnetic susceptibility below a blocking temperature $T_B \approx U/k_B \ln(t/\tau_0) \sim 3.3$ K for $t \approx 1$ h. Such an exponential behaviour is typical of extremely narrow particle size-distributions (Sect. 2.2). Above T_B a Curie-Weiss law with a very small positive paramagnetic temperature $\theta \sim 70$ mK, indicates the existence of weak dipole-dipole interactions. As temperature decreases the low-field relaxation evolves from exponential (about 10^3 sec) to non-exponential below ~ 2 K (two exponential times were observed, to be connected with a specific non-exponential behaviour discovered later, see Sect. 2.4.4). Besides, (i) a thermally-activated relaxation observed above ~ 2 K was followed at lower temperatures by a well defined plateau down to ~ 0.2 K (Fig. 2.8, left) and (ii) a minimum of relaxation was seen near $H \sim 0$ together with two less pronounced minima near $\mu_0 H \sim 0.4$ and 0.8 T (Fig. 2.8, right) [28].

Those results were attributed to MQTM between the ground-states $m = \pm S$ of Mn_{12}-ac. Indeed, in the absence of any distribution, the plateau could not have any other interpretation. Furthermore, the WKB exponent $B = 3\gamma(U_{cl} K_\perp)^{1/2}\sqrt{\varepsilon}/8k_B M_s$ (where $K_\perp$ gives the energy of transverse anisotropy terms and $\varepsilon = 1 - H/H_A$ [10] leads to a crossover temperature $T_{co} = U_{cl}/B \sim (U_{cl}/K_\perp)^{1/2} \sim 2$ K, very close to the experimental one (and later confirmed by sub-Kelvin measurements in large fields described in Sect. 2.3.2 below). The observed crossover temperature was rather important because it was measured in zero field i.e. with a large barrier $U_{cl}(0)$. At higher fields, T_{co} decreases rapidly, with the barrier $U_{cl}(0) \approx U_{cl}(0)\varepsilon$. The relaxation-time minima of Fig. 2.8 (right) were also interpreted in terms of MQTM when the level schemes of the two wells with $S = \pm 10$ are in coincidence".

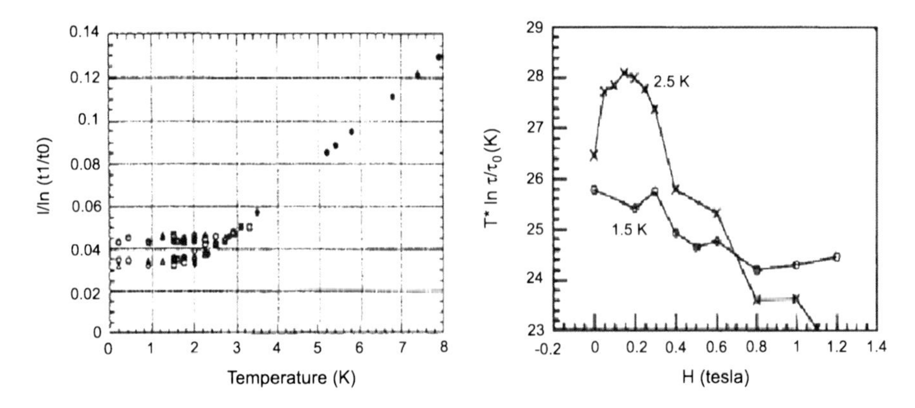

Fig. 2.8 *Left*: thermal variation of the relaxation time measured on a few large crystals of Mn_{12}-ac for fields $H = 0$, 0.4 and 0.8 T applied 45° from the easy axis. *Right*: field-dependence of the relaxation time measured just above and below T_{co} (the overall linear decrease is simply due to the Zeeman reduction of the classical barrier [28]

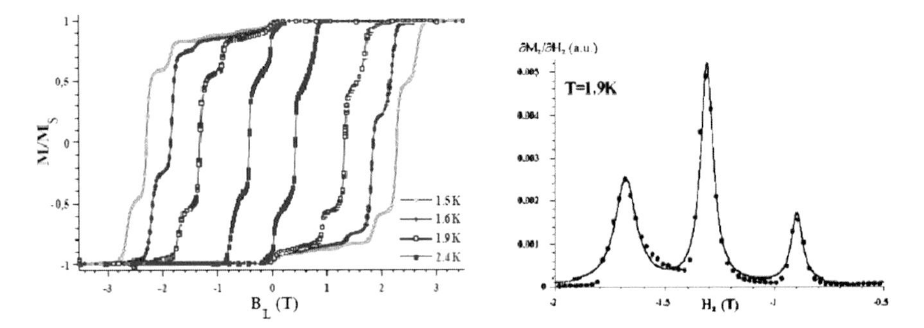

Fig. 2.9 *Left*: hysteresis loops measured on a single crystal of Mn_{12}-ac [25] with a field along **c** (easy-axis of magnetization). The steps take place when spin-up and spin-down states are in coincidence (resonant tunneling), independently of temperature. The hysteresis loop depends on temperature after the first step, when the crossover temperature decreases. *Right*: field variation of the derivative M_z/H_z taken at 1.9 K. The observed Lorentzian line-shape suggests that the system of spins is in the diluted limit, close to equilibrium

2.3.3 Main Evidences

Thermally-Activated Tunneling The above results on low-field ground-state tunneling in Mn_{12}-ac were extended a little later to the case where a longitudinal magnetic field is applied. Figure 2.9(left) shows the results of our experiments performed, at temperatures 1.5 (K) $< T < T_B$, on a single-crystal with parallelepiped shape carefully oriented along the applied field direction by the use of a magnifying lens [25]. They show a succession of plateaus (relaxation-times $\gg$ measuring-timescale) and steps (relaxation-times < measuring-timescale) [25–27].

The plot of M_z/H_z versus longitudinal field H_z (Fig. 2.9, right) gives Lorentzian peaks enabling to accurately define the magnetic fields H_n at which the magnetiza-

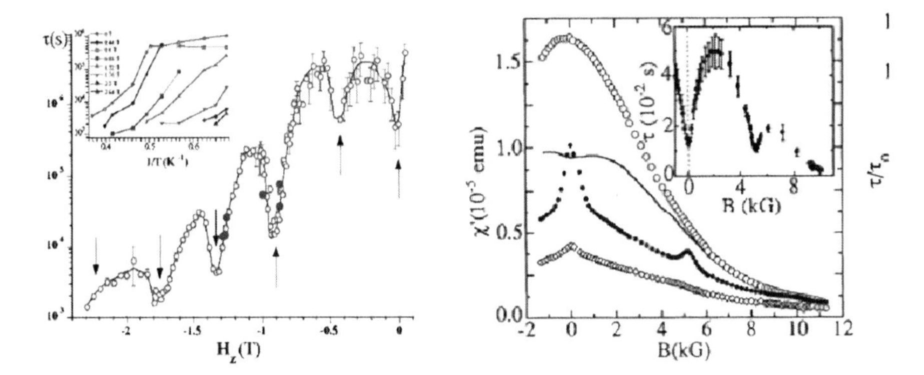

Fig. 2.10 *Left*: field dependence of the relaxation times measured on the Mn_{12}-ac hysteresis loop at 1.9 K, showing deep dips at resonant fields [25]; the relaxation times are plotted against inverse temperature in *the inset*. *Right*: real ac-susceptibility component measured at 5 K [27]; *open squares*: 980 Hz, *open ellipses* quasi-static, *solid circles*: 15 Hz. *Solid lines*: predictions for classical thermally activated relaxation. *Inset*: field dependence of the relaxation time at 5 K

tion steps occur: $H_n 0.44nT$ ($n = 0, 1, 2, \ldots$). The width is mainly associated with dipolar interactions and therefore with the value of the sample magnetization; some disorder might also contribute if the crystal quality is not good enough (more details are given in Sect. 2.4.4 below). The relaxation time shows deep minima at the field values where the steps are observed (resonances) (Fig. 2.10, left). Ac-susceptibility experiments show that the minima of relaxation time can be seen on a broader temperature range, up to 5 K (Fig. 2.10, right), with however less pronounced minima.

In those experiments the measured transition rate decreases rapidly with temperature showing that MQTM is thermally-assisted i.e. it takes place from thermally-activated excited levels [25–27]. In this case, the spin reversal probability is the product of the thermal activation probability, which promotes the spin from its ground-state $m = -S$ to an excited state $m = -S + n$, by the tunneling probability switching the spin from $m = -S + n$ to $m' = S$, giving $1/\tau_{TA} = (1/\tau_0)\exp(-B)\exp(-E_m/k_B T)$. The tunneling probability is then equivalent to changing the prefactor from $1/\tau_0$ to $1/\tau_{0Q} = (1/\tau_0)\exp(-B)$. It is well known since Néel's time that $1/\tau_0$ increases at low temperature (in some models such as $1/\sqrt{T}$). This classical effect, resulting in the slowing down motion of the "particle" at the bottom of the well, can also be attributed to a decreasing of the classical entropy when temperature decreases (associated with different paths inside the well, see e.g. [77] and references therein). However, at $T = 0$, the "particle" is at rest and the dynamics, necessarily of quantum origin, will take place only if tunneling is possible. In this case, new channels open paths extending the "particle" motion to the second well and leading to a zero-Kelvin entropy, increasing with the tunneling rate $1/\tau_{QTA} = \exp(-B_{TA})$ at finite temperature and $1/\tau_{QGS} = \exp(-B_{GS})$ at zero Kelvin. $1/\tau_{TA} \approx \exp(-E_{TA}/k_B T)$ gives $E_{TA} = E_m - k_B T S$, defining an entropy $S_{TA} = \ln(1/\tau_0) + \ln(1/\tau_{QTA})$ at low but finite temperatures and an entropy $S_{GS} = \ln(1/\tau_{QGS})$ at $T = 0$ in which quantum paths contribute by analogy with Feynman's path integrals [78].

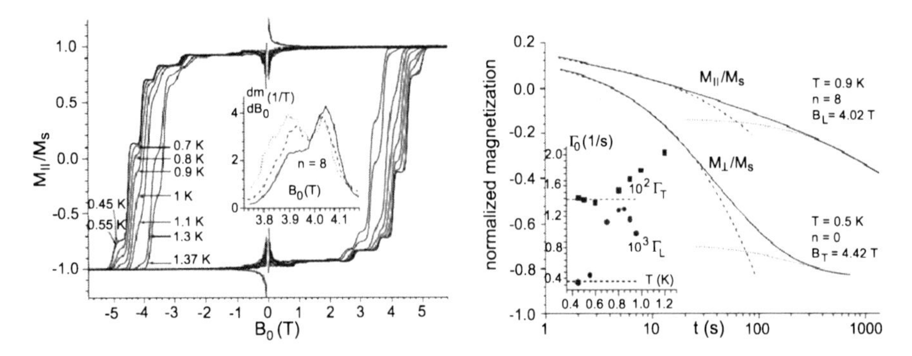

Fig. 2.11 *Left*: hysteresis loop of a Mn_{12}-ac crystal obtained from torque experiments with a magnetic field along the easy **c**-axis. The curves become independent of temperature at all fields below 0.7 K [65]. *Inset*: field-derivative of the $n = 8$ resonant transition at 0.90 (*dots*), 0.95 (*dashed*) and 1.0 K (*continuous*). The resonance splits in two: tunnelling from the ground state $m = -10$ and from the excited state $m = -9$; the other resonances show the same behaviour. *Right*: the main part of the figure gives an example of relaxation curves measured in the plateau i.e. resulting from ground-state tunneling. *Inset*: relaxation rates in a longitudinal or a transverse field vs. temperature

Ground-State Tunnelling and Relaxation Experiments performed above 1.5 K, and reported in Refs. [25–27], constituted the best proof of the thermally activated MQTM. Furthermore they confirmed the interpretation of ground-state MQTM between ~ 0.2 and 2 K, in low-field (Fig. 2.8) [28]. The observation of staircase hysteresis loops showing ground-state MQTM in Mn_{12}-ac at different fields came later, extending the low-field ground-state MQTM [28] to the case of a variable field [65]. In those sub-Kelvin experiments (Fig. 2.11, left) the field derivative of the magnetization curves near resonant transitions shows either a single peak at lowest temperature or two peaks when the temperature is high enough to excite the next upper level as shown in the inset of Fig. 2.11(left) and in Fig. 2.12, where the resonant transitions from the ground-state $m = -10$ and from the first excited state $m = -9$ are simultaneously apparent above 0.8 K. Below this temperature, only the ground-state resonance $m = -10$ is seen. Between the two, a transfer of intensity is observed as temperature increases.

This behaviour, observed for all resonances, led to the construction of Fig. 2.12 showing how MQTM evolves with temperature in this system, when the longitudinal field increases and temperature decreases. The coexistence, in a narrow temperature range, of two resonances with m and $m-1$, defines the crossover temperature $T_{co}(H_z)$ and shows that thermally-activated MQTM, taking place just above $T_{co}(H_z)$, is continuous and evolves step by step by simple spin-phonon transitions [65]. The thermal variation of $T_{co}(H_z)$, easily defined by taking the middle of the region where the first two transitions occur simultaneously, gives $T_{co} \sim 0.7$ K at ~ 5 T a value which increases when the field decreases and reaches 1.2 K at 3 T. A simple extrapolation gives, in zero longitudinal field, $T_{co}(H_z = 0) \sim 1.5$–2 K confirming the first results shown in Fig. 2.8 [28].

The above results on Mn_{12}-ac were based on hysteresis loops measured at sub-Kelvin temperatures from torque experiments and in fields up to 6 T [65]. Fig-

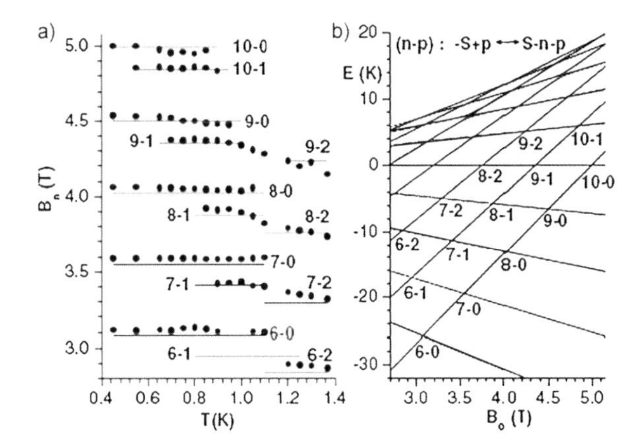

Fig. 2.12 (**a**) Peak maxima measured vs. temperature in Mn_{12}-ac between 1.4 and 0.4 K (*left*) [65]. *Horizontal lines* indicate the calculated crossing fields obtained from the energy level spectrum calculated from exact Hamiltonian diagonalisation (**b**). The shift of the crossings by the fourth order term B of (2.1) is clearly visible

ure 2.11 shows that, below 0.8 K, the different loops merge showing that tunneling takes place from the ground-state $m = 10$ at fields larger than a few Tesla (above this temperature, as in Fig. 2.9, they merge only at lower fields, in agreement with the fact that T_{co} increases with decreasing H). Similar results were obtained in the same set of experiments, but with a transverse field of about 4 T. As an example, I give in the right-hand inset of Fig. 2.11 the temperature dependence of the relaxation times measured in a transverse or a longitudinal field. The relaxation time in a transverse field of 3–4 T is clearly independent of temperature below 0.8 K, evidencing MQTM between the ground states $m = 10$ and $m = -10$ of the two symmetrical wells in the majority phase of Mn_{12}-ac [65].

Interestingly, the slow increase of the tunneling rate above 0.7–0.8 K is too smooth to be due to thermally-activated tunneling between $m = \pm 9$, ± 8 states. It rather comes from direct phonon-assisted tunneling between the ground-states $m = \pm 10$. These experiments show that, unless it is a first order transition [79], the crossover from ground-state tunneling to thermally-activated tunneling goes through an intermediate regime where tunneling takes place between non-resonant states $m = +10$ and $m = -10$ split by the magnetic field and by dipolar interactions, due to spin-phonon transitions between the two wells [65]. This phonon-assisted tunneling regime, in a large transverse field, was predicted in [80, 81] to dominate magnetic relaxation for sufficiently high transverse magnetic fields. Its existence has been confirmed by the results of EPR [82] and time-dependent heat capacity [83, 84] experiments performed under similar conditions (strong transverse magnetic fields and low temperatures).

In a longitudinal magnetic field, the relaxation rate first increases above 0.6 K, goes to a maximum near 0.8 K and then decreases (Fig. 2.12, inset). This effect was also connected with spin-phonon transitions, but within a single-well this time. This shows in particular how thermally-activated tunneling takes place from one level to the next one in relation with the results of Sect. 2.3.2. The right-hand panel of Fig. 2.11 gives an example of relaxation curves measured in the plateau, i.e., resulting from ground-state tunneling. At short time scales, relaxation follows a square root law but at longer times it becomes exponential (for details see Sect. 2.4.4).

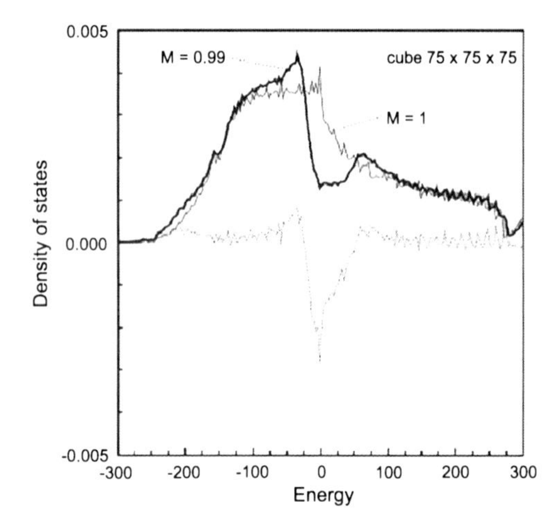

Fig. 2.13 Monte Carlo calculations predicting the effect of quantum hole digging in the initial distribution of internal fields [85]

This square root law is specific to MQTM and comes from non-equilibrated spin-state transfer from one well to the other one, by tunneling, which creates a time-dependant hole in the initial density of states. Figure 2.13 shows this hole as it was first obtained by the quantum simulations of Prokof'ev and Stamp [85].

Microwave-Activated Tunnelling After the evidences for ground-state [28] and thermally-activated [25–27] MQTM in Mn_{12}-ac, one question became rather obvious: is it possible to induce activated MQTM by microwaves? It was indeed well known that microwaves can easily promote a quantum state to an upper one if its frequency matches the level separation, as this is done in EPR [86]. The question was simply to know whether the lifetime of the excited level can be large enough in comparison with the "waiting time for tunneling". The answer was also yes as it was indeed possible to observe $m = 9m = -9$ MQTM after the application of a circularly polarized microwave of frequency $\hbar\omega = E(S) - E(S-1)(2S-1)D$ to a crystal of Fe_8 SMMs [87]. A first important remark is that, contrary to a linear polarization, a circular polarization has the advantage to distinguish between $\Delta m = +1$ (left polarization, σ^- photons) or $\Delta m = -1$ (right polarization, σ^+ photons), so that the population of only one of the two excited states $m = \pm\, 9$, is enhanced. If we start from a saturated state at $+M_s$, then tunnelling takes place essentially from S to $-S$ or $S-1$ and $-(S-1)$ implying the use of σ^+ microwave photons. The hysteresis loops of a Fe_8 single crystal measured at 60 mK under irradiation (easy axis parallel to the applied field), show that the tunneling transition near zero field is strongly enhanced for the radiation of 115 GHz, which precisely matches the $m = 10$ to $m = 9$ level separation. The observed increase of the tunneling rate at zero field, as a consequence of the absorption of photons induced by circularly polarized radiation, became evident by comparing the zero-field steps after positive $(+M_s)$, or negative $(-M_s)$ saturation. The irradiation effect was found to be much smaller for other frequencies where it resembles the thermally-activated MQTM case. A simple interpretation of the data, based on the equation of energy transfer in the bottleneck regime [86] showed that the spin temperature T_S (which is an effective temperature reflecting the occupation of excited states caused by microwave irradiation at $T \sim 0$ K), increases linearly with microwave power. This result agrees with a simple model where both the specific heat of spins and the typical spin-spin diffusion time, accounting for energy transfer by spins (also related with the level lifetime), were supposed to be independent of temperature [87]. This first study of microwave activated tunnelling in a SMM was followed by other studies [88, 89].

2.4 Theory and Comparisons with Experiments

2.4.1 Resonance Conditions

The spin $S = 10$ Hamiltonian (2.1) gives the values of the longitudinal magnetic field H_n anti-parallel to the persistent magnetization at which the intersection of energy levels occurs. The condition for the intersection of two levels with $S_z = m > 0$ and $S_z = n - m < 0$, is approximately given by [25–27]:

$$H_n \sim n D_{\text{eff}}/g\mu_B. \tag{2.2}$$

At these values of the magnetic field H_z, levels associated with $m > 0$ and $n - m < 0$ magnetic states come into resonance and tunneling channels open. The value $D_{\text{eff}}/k_B \sim 0.56$ K derived from EPR measurements [57, 66] gives, with (2.2), $H_n \sim 0.42n$ T, which differs slightly from the experimental $H_n \approx 0.44n$ T, measured at intermediate temperatures (where $m = 3$–4, Figs. 2.9–2.11). This difference can be accounted for by the fact that H_n depends on the temperature at which the experiments are done, through modifications of (2.2) by the fourth order anisotropy term:

$$H_n = nD/g\mu_B\left[1 + (B/D)\left((m-n)^2 + m^2\right)\right] \tag{2.3}$$

where $m = S_z$ and $n =$ integer number of level shifts between the two wells. D should be derived from this expression (2.3) and not from (2.2), the latter providing only an effective value D_{eff}. Assuming that the difference between $H_n \sim 0.42n$ and $H_n \sim 0.44n$ comes from that, we get $(m-n)^2 + m^2 \sim 13$, which effectively corresponds to the transitions $m \sim 3$–4 with $D_{\text{eff}}/k_B \sim 0.6$ K. Later on, a direct verification of expression (2.3) was given [69] in which a linear fit gives $D/k_B \sim 0.51 \pm 0.02$ K and $B/k_B \sim 1.0 \pm 0.3$ mK, i.e. $D_{\text{eff}}/k_B = 0.51 + 10^{-3}[(m-n)^2 + m^2] \sim 0.64$ K for the observed $H_n \sim 0.48n$. We deduce $(m-n)^2 + m^2 \sim 130$ which corresponds to transitions from the ground-state $m = S = 10$ with $n - m = 5$–6. The term B also contributes to the zero-field energy barrier $U_{\text{cl}}(0) = DS^2 + BS^4 \sim 61$ K. We shall now see how the decreasing of this barrier by application of a longitudinal or transverse field or by temperature, can be interpreted in terms of quantum fluctuations.

2.4.2 Quantum Fluctuations and Barrier Erasing

The tunneling rate between two states $-m$ and $m-n$ depends sensitively on the values of m and n. As an example, in zero longitudinal field it is extremely small with $m = \pm S$ (very long timescale) and fast with e.g. $m = \pm 1$ (very short timescale). Measurements being always performed at a given timescale, only a few sets of levels can be recorded, say $\pm m_t$. All levels above (below) those ones have larger (smaller) tunnel splitting and therefore tunnel too rapidly (slowly). If the latter do not modify the result of measurements, this is not the case for the former where the presence of tunnel splittings larger than level separations associated with diagonal anisotropy terms (cf. (2.1)) leads to a short-cut of the top of the barrier. This means that the "effective" height of the barrier in zero magnetic field is $U_{\text{eff}}(0) = U_{\text{cl}}(0)(S^2 - m_t^2)/S^2$ instead of $U_{\text{cl}}(0)$, i.e. it is reduced by quantum fluctuations resulting from large tunnel splittings near the top of the barrier. This effect was observed experimentally [69] in the high temperature relaxation regime of Mn_{12}-ac (between 2.6 and 3 K) where relaxation times follow Arrhenius law $U_{\text{eff}}(H) = T \ln[\tau(T, H)/\tau_0]$, as shown from the data points plotted in the left-hand panel of Fig. 2.14. The resonant tunneling dips reduce $U_{\text{eff}}(H)$ by about 10 % at all fields allowing one to estimate m_t. In particular the zero-field barrier shows a minimum at ~ 57 K and a maximum at ~ 64 K. The linear field decrease of the effective barrier of Fig. 2.14(left) fits the classical expression $U_{\text{eff}}(H_z) = U_{\text{eff}}(0)(1 + H_z/H_{\text{A}})^2$ although it is of pure quantum origin: when the longitudinal field increases, the barrier height and width decrease and tunneling is faster authorizing an easier barrier short-cut of the top of the barrier. This barrier short-cut takes place when the tunnel splitting associated with $\pm m$ is equal to the mean level separation between the consecutive states issued from m and $m \pm 1$. An example is given Fig. 2.14(right) for the case of a transverse field where the classical barrier reduction, $U_{\text{eff}}(H) = U_{\text{eff}}(0)(1 + H_x/H_{\text{A}})^2$, precisely follows this rule.

2.4.3 Tunnel Splittings, Spin-Parity and Observation of MQTM

Tunnel splittings are generally evaluated numerically in order to take into account the contributions of different symmetries, simultaneously and accurately. However, analytical expressions are useful for orders of magnitudes and physical discussions. Following van Hemmen and Suto [40], the tunnel splitting between the two states $-m$ and $m-n$ generated by an off-diagonal perturbation $D_p S_\pm^p$, of order p, can be written [90, 91] as $\Delta_{m,n-m}^{(\text{p})} \sim DS^2(D_p S^p/2DS^2)^{(2m-n)/p}$. Indeed, to connect two states $-m$ and $m-n$ the operator $S_\pm^p$ of order p, must be iteratively applied an integer number of times, allowing a tunnel transition between $-m$ and $m-n$ only if $(2m-n)/p =$ integer k, which leads to the above expression [91, 92]. In the case of Mn_{12}-ac we have $\Delta_{m,n-m}^{(4)} \sim DS^2(CS^2/2D)^{(2m-n)/4}$ and $\Delta_{m,n-m}^{(1)} \sim DS^2(H_x/2DS)^{(2m-n)}$ for the fourth and first order terms respectively. In samples with so-called "fast species", where isomers with tilted local

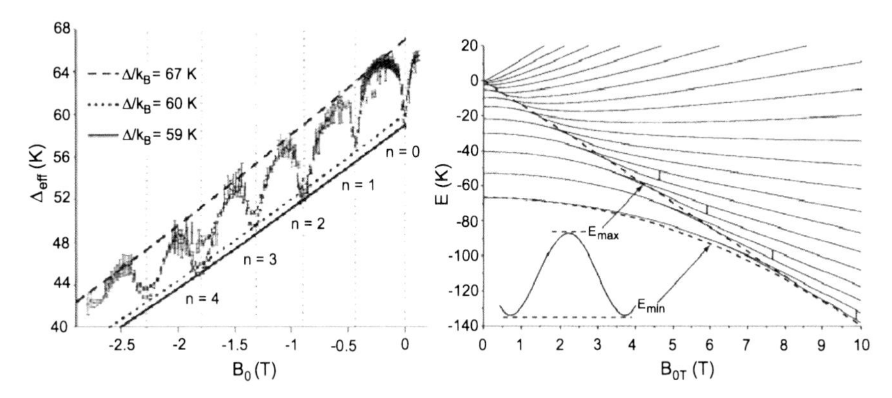

Fig. 2.14 *Left*: effective energy barrier of Mn_{12}-ac measured as a function of a magnetic field applied along the easy **c**-axis in the high temperature regime,showing a parity effect resulting from different symmetries of the crystal-field parameters [69]. *Right*: evolution of the energy spectrum vs. a transverse field, calculated from exact diagonalization of the Hamiltonian. Maxima and minima of the classical energy barrier (classical expression) are given in *dashed lines*. Interestingly these lines cross the quantum energy spectrum when tunnel splittings are equal to level separations. This is what we called the "quantum erasing of a classical barrier"

anisotropy directions induce a second order transverse term [93], the tunnel splitting $\Delta^{(2)}_{m,n-m} \sim DS^2(D_2/2D)^{(2m-n)/2}$ should also be considered.

These expressions give good orders of magnitude for tunnel splittings and allow a simple understanding of parity rules for tunneling. In Mn_{12}-ac we find $\Delta^{(4)}_{10,-10} \sim 10^{-11}$ K, a value which is about 10 times larger than the 310^{-10} K obtained experimentally from the sub-Kelvin measurements described above. However, this difference can easily be understood as due to higher-order transverse anisotropy terms [58, 69]. These terms, of unknown values, can actually make significant contributions to the tunnel splitting even if they are very small. This can be seen directly from our $\Delta^{(p)}_{m,n-m}$ expression showing that $\Delta^{(p)}_{m,n-m}$ remains of the same order of magnitude if the coefficient D_p does decrease much faster than a power-law, which is generally the case [58]. This makes any really quantitative calculation of the tunnel splitting from (2.1) rather problematic.

Having a large enough tunnel splitting is important to observe MQTM. However it is even more important that the spin-parity rules are obeyed i.e. $2m - n = kp$ for a pth order transverse term (see above). For an integer spins, elementary arithmetic shows that the contributions to tunneling are non-zero only if n is a multiple of 4 for $p = 4$, with a shift of 2 between even and odd m (4th order term $S^4_\pm$), a multiple of 2 for $p = 2$ (2nd order term $S^2_\pm$) and a simple integer if $p = 1$ (1st order term S_x). As an example, in order to connect the states $-m$ and $m - n$ with the fourth order transverse anisotropy term of Mn_{12}-ac, the S^4_- operator must be iteratively applied an integer number of times, i.e. $2m - n = 4k$, giving $n = 2(m - 2k)$, i.e. $n = 0, 4, 8, \ldots$ for even m and $n = 2, 6, 10, \ldots$ for odd m. Notice that, even if they are relatively weak, environmental magnetic fields (e.g. those arising from dipolar interactions between different molecules) can partially lift these selection rules. As was discussed in [58, 80], the combination of fourth-order anisotropy terms with such linear terms gives rise to nonzero tunnel splittings between any pair of magnetic states. More precisely, even resonances are induced by S_+^4 or S_-^4, whereas odd resonances are induced by combinations like $S_x S_+^4$ or $S_x S_-^4$ explaining why even and odd resonances have different sizes (Fig. 2.14, left, [69]). Experimentally, MQTM is observed for any integer n and not only for resonance numbers that are multiple of 2, 4, etc. In the case of Mn_{12}-ac, for instance, relaxation rates measured at odd-numbered resonances are only slightly lower than those measured at even-numbered ones [69, 94]. In addition to the combinatory effect mentioned above, one might add the likely important role of Dzyaloshinsky-Moriya interactions, which has been pointed out many times since 1997 [65, 69–71] (see Sect. 2.3.1). The low point symmetry of SMMs and the important strength of intra-molecular super-exchange interactions (several 10^2 K) does enable Dzyaloshinsky-Moriya interactions of a few Kelvin, which is quite enough to enhance tunnel splittings strongly and, more importantly, to enable tunneling for integer n, as observed experimentally. Besides, the small differences observed between odd and even resonances in Fig. 2.14(left) may come from the presence of a 2nd order term of the type $S_\pm^2$ which is related, in Mn_{12}-ac, to a minor species of isomers with local anisotropy axes tilted by a few degrees [93] (see above, the parity rule is $2m - n = 2k$). This could be a convincing evidence for the presence of a small amount of the fast species where even contributions to MQTM are larger than odd ones by $\sim 15\ \%$. Note that this percentage does not reflect the proportion of fast species precisely as it should be renormalized by the ratio of squared tunnel splittings. A rough evaluation based on published anisotropy parameter of the fast species [93] and taking this renormalization into account, gives us a few % which seems very likely.

These parity rules, required to connect two states of opposite spin components in the two potential wells, should not be mistaken for those associated with the Haldane topological phase [95] causing the Berry oscillations [96] of the tunnel splitting about $\Delta_{m,n-m}^{(p)}$ in a transverse magnetic field [97, 98] that were first observed in the system Fe_8 [99]. The simplest way to describe these oscillations analytically is to truncate the general Hamiltonian to a simple low-energy 2-level Hamiltonian [96, 98, 100–102], i.e., to consider a case identical to the classical one [103]. An extension to the quantum case has been achieved by taking into account the fourth-order anisotropy terms [58], which in particular allowed one a closer approach of the Berry phases in SMMs. However, one should not forget contributions of higher-order anisotropy terms to the tunnel splitting, which makes these calculations rather pointless if they have to be precise.

2.4.4 Quantum Tunneling and Spin-Bath

In the most naïve approach, the tunneling rate Γ is sometimes taken as equal to the quantum tunnel splitting Δ. However, this assumption is far from reality because (i) it implies that the homogeneous line width is smaller than or equal to the tunnel splitting, which is generally not the case, (ii) it neglects the dynamics i.e. the times interval during which the two levels are in coincidence (iii) it neglects environmental effects. Essential for the understanding of quantum phenomena in magnetism is the spin-bath, i.e., the ensemble of environmental spins (nuclear spins or neighboring electronic spins) interacting with the central spin [85, 97, 102, 104–106].

When both environmental and dynamical (times interval during which the two levels are in coincidence) effects are neglected, $\Gamma \sim \Delta^2/G$, where G is the level-broadening (we assume $G \gg \Delta$; if $G \to 0$, $\Gamma \sim 1 - \exp(-\Delta^2/G) \to 1$, is independent of Δ). If only environmental effects are neglected, we find the situation covered by the Landau-Zener model [107, 108], where quantum superpositions $\Uparrow \pm \Downarrow$ of the two states of a single spin $S = \frac{1}{2}$ are removed by a longitudinal magnetic bias evolving with time from $-\infty$ to $+\infty$ at a given rate $\nu = g_{z\mathrm{B}} \mathrm{d}H_z/\mathrm{d}t$. The Landau-Zener tunneling probability is given by:

$$P_{\mathrm{LZ}} = 1 - \exp\left(-^2/\hbar\nu\right) \tag{2.4}$$

At large sweeping rates, expression (2.4) gives $P_{\mathrm{LZ}} \sim \Delta^2/\hbar\nu$. This model has been applied to the tunneling problem in magnetism ever since the first MQTM results were obtained [109–111]. Note that the sweeping field term ν is equivalent to a level broadening.

In the general case where the environment is also taken into account, most important effects come from level broadenings by phonons, nuclear spins or dipolar fields which constitute the first manifestations of the phonon or the spin baths. The ways in which these level broadenings intervene are very different depending on whether they are homogeneous or inhomogeneous. The environment is able to absorb finite variations of energy and angular momentum, which is extremely important because the non-conservation of these quantities can forbid the tunneling. However, its most important "positive" effect is that it enables the observation of MQTM and this is what we will discuss now (the "negative" effects such as the suppression of coherent quantum spin dynamics will be discussed in Sect. 2.6). As we consider the spin dynamics at low temperatures, we shall mainly concentrate on the spin-bath [85, 97, 102, 104–106, 112–115] (even if it shows interesting effects (Sect. 2.3.2) the phonon-bath [79–81, 116–118] has a much smaller impact at low temperatures and low magnetic fields).

Long-range dipolar interactions spread out each resonant level $-m$, $m - n$ into frozen spin-up and spin-down distributions (inhomogeneous broadening). The degrees of coincidence of these distributions is given by the energy difference $\xi_{m,n} = E(m,n) - g_z\mu_{\mathrm{B}}(2m - n)H_z$, where $E(m,n) = D(n-m)^2 - Dm^2$. For a given pair $-m$, $m - n$ of broadened levels the writing may be simplified and the distribution of spin-down states (available for tunneling) can be written $N_{\mathrm{D}}(\xi)$ where the function N_{D} is a Lorentzian or a Gaussian (depending on the degrees of dilution/equilibration [58]). The homogeneous distributions of hyperfine or super-hyperfine interactions (they must be weak, see Sect. 2.5) create a dynamical window in this distribution allowing resonant spin-up and spin-down states to "see each-other", as this is the case with the (necessarily homogeneous) tunnel splitting of a single-spin in the Landau-Zener model. The corresponding time-scale being the nuclear coherence time T_2, the Landau-Zener probability, renormalized by the spin-bath can be written in the limit of small splitting Δ as follows [85, 97, 102, 104–106]:

$$P_{SB}(\xi) = \pi \Delta^2 e^{-|\xi|/\xi_0} \frac{N_D(\xi)}{E_N} \tag{2.5}$$

The exponential factor limits the tunnel window to the width ξ_0 of the homogeneous nuclear-spin broadening given by the hyperfine energy $\xi_0 \sim AIS$, the distribution $N_D(\xi)$ gives the number of down spins available for tunneling and E_N is the energy associated with the nuclear-spin flip timescale. Expression (2.5) can also be written as $P_{SB}(\xi) = \pi \Delta^2 e^{-|\xi|/\xi_0} N_D(\xi)\xi_0/(\hbar\xi_0 T_2^{-1})$ where T_2 is the characteristic nuclear spin flip time, which therefore defines the "sweeping field time" of the magnetic field $H_Z^N = \xi_0/gz\mu_B$ generated by nuclear spins on the electronic spin S. In this form, connections with the Landau-Zener model are clearly seen. Note that if T_2 is large enough, $T_2 \gg \hbar\xi_0/\Delta^2$, then the exponential in (2.4) cannot be expanded and the transition is adiabatic showing that the SMM spin almost always follows the nuclear spin and inversely (good entanglement). This type of "coupling" is different from the case of a large hyperfine interaction $\xi_0 \sim AIS$, where the two spins are also locked but not necessarily entangled (case of rare- earth ions, Sect. 2.5).

Resonant Tunneling Line Shapes and Observation of MQTM In the present case of a SMM crystal, each molecule is submitted to an internal field of dipolar origin varying in direction and amplitude. Longitudinal field components split the two states $\pm m$ of each molecule by a value between ~ 0 and H_{DMax} (the maximum dipolar field), generally of the order of 10–100 mT in these systems, (it can be evaluated experimentally and calculated from the crystallographic structure). In order to put states $+m$ and $-m$ of a given molecule in resonance, one has to apply a longitudinal field compensating the local dipolar field acting on it (Fig. 2.6). For states such that $\Delta_{m,n} < \xi_{m,n}(H_{\mathrm{DMax}})$, molecules have a chance to tunnel if the applied field sweeps between 0 and $\sim H_{\mathrm{DMax}}$. Under these conditions, the resonance line-width is of the order of H_{DMax}. On the other hand, resonance line widths associated to tunneling via states having $\Delta_{m,n} > \xi_{m,n}(H_{\mathrm{DMax}})$ are, as this is well known, mainly determined by $\Delta_{m,n}$ eventually corrected by phonon broadenings (see, for instance, [58, 80, 81] and below). Experimentally, resonance lines for Mn_{12} have been obtained from either the plot of $\mathrm{d}M_z/\mathrm{d}H_z$ vs. H_z or from field-dependent susceptibility data. In both cases, nearly Lorentzian line-shapes are observed (Fig. 2.7, right), the width of which (40–100 mT) depends on the value of the magnetization (i.e. of the index n) and on the shape of the sample [25, 27]. Note that Lorentzian line-shapes are expected in the limit of dilute/equilibrated static dipoles ([58] and references therein) or/and at high enough temperatures when homogeneously broadened resonance results from equilibration by spin-phonon transitions (see e.g. [25, 69, 70, 118]). An alternative explanation for the observed Lorentzian line shapes is that, at sufficiently high T, magnetic relaxation proceeds by tunneling via states close to the top of the barrier, which have tunnel splittings comparable or even larger than dipolar bias [80, 81].

The tunneling rate is maximum (center of the magnetization steps) when the applied field is such that the two distribution maxima are in resonance. The first relaxation maximum ($n = 0$) measured in Mn_{12}-ac was not exactly in zero-field but in a weakly negative field, which was interpreted as a consequence of the competition between the demagnetizing field $-NM$ (shape-dependent, where N is the demagnetizing factor) and the local Lorentz field $4\pi M/3$ [65, 69]. The same effect was also observed in Fe_8 but with a significant difference: the first resonance was observed in a small positive (and not negative) field [119]. Since the Mn_{12}-ac crystals are elongated, their demagnetizing field is smaller than the Lorentz field and the internal field is parallel to the magnetization M: one has to apply a negative field to cancel the internal field, whereas the situation is just opposite with Fe_8 where the crystals are flat.

It is also interesting to recall that ground-state MQTM could never have been detected, with the finite field resolution of existing magnetometers, in the absence of the important broadening of dipolar interactions (40–100 mT [25]). It is because in the absence of such broadenings a resonance would have required to put in coincidence two levels of width Δ, which is of the order of 10^{-10} K $\sim 10^{-11}$ T for the $m = 10$ Mn_{12}-ac ground-state resonance (and 10^{-8} K $\sim 10^{-9}$ T in Fe_8). Note that pure phonon broadening (by a factor of ~ 100) is also too small to authorize this detection at low temperatures, even if phonon-assisted tunneling is possible (see Sect. 2.3.2).

Tunneling via Thermally Activated Excited States Following the Landau-Zener description discussed above, the tunneling probability between states $-m$ and $m - n$ of a SMM, submitted to a magnetic field $H_n \sim nD$ sweeping at the rate $\nu = g_B \mathrm{d}H_z/\mathrm{d}t$, can be expressed as $P_{-m,m-n} = 1 - \exp(-{}^2_{-m,m-n}/2\nu)$, where $\Delta_{m,n-m} \sim DS^2(H_x/2DS)^{(2m-n)}$ is the tunnel splitting. This gives for the TA probability $1/\tau_{\mathrm{TA}(-m,m-n)}(1/\tau_0)[1 - \exp({}^2_{-m,m-n}/2\nu)]\exp(-E_{-m}/k_B T)$ or $1/\tau_{\mathrm{TA}(-m,m-n)} = (1/\tau_0)({}^2_{-m,m-n}/2\nu)\exp(-E_{-m}/k_B T)$ if ${}^2_{-m,m-n} \ll 2\nu$. As the population of thermal states does not involve only one state but an ensemble of states, the measured TA relaxation time is given by:

$$1/\tau_{\mathrm{TA}} = (1/\tau_0) \sum_m (\pi \Delta^2_{-m,m-n}/2\hbar\nu) \exp(-E_{-m}/k_B T) \tag{2.6}$$

According to this model, for a single spin the wave-function collapses (and coherence is thus lost) in the timescale of $\Delta_{m,n-m}/\nu$ [103]. Expression (2.6) has been extensively used to interpret MQTM experiments, derive the value of tunnel gaps, etc.

The validity of the Landau-Zener model is restricted to states with tunnel splittings larger than level broadening (typically $\Delta_{-m,m-n} > \xi_0$ for the spin-bath and $\Delta_{m,n-m} > h/\tau_0$ for the phonon-bath) and subjected to magnetic fields varying faster than environmental fluctuations (more specifically $\nu \gg \Delta_{m,n-m}/\tau_0$ for the phonon-bath and $\nu \gg \Delta_{m,n-m}/T_2$ for the spin-bath). If the latter condition is not fulfilled, the external magnetic field H_z can be considered as being quasistatic within the excited level lifetimes (τ_0). The relaxation rate is determined then by spin-phonon interactions at high temperatures and by spin-spin fluctuations at low temperatures ($k_B T \ll$ level spacing). In case $h/\tau_0 > \Delta_{m,n-m}$ [79, 116–118] the level lifetime $\tau_0 \sim 10^{-7}$–10^{-8} s is much shorter that the tunneling oscillation time $h/\Delta_{m,n-m}$ and prevents the existence of quantum coherent oscillations between states located on opposite sides of the magnetic anisotropy barrier. For excited states that fulfill the opposite condition $h/\tau_0 < \Delta_{m,n-m}$ any coherent spin dynamics is also lost, in time scales of the order of $h/\Delta_{m,n-m}$ [80, 81]. The latter prediction seems to be confirmed, for Mn_{12}-ac, by the results of nonlinear ac susceptibility experiments [67]. In any case, all these results strongly suggest that thermally activated tunneling of SMMs is a fully incoherent process. The following expression, where $hW_m = h/\tau_0$ is the phonon-line broadening associated with inter-level transitions allowed by the spin-phonon interaction, (see [58] and therein refs.)

$$1/\tau_{TA} \approx \sum_m \frac{\Delta^2_{-m,m-n} W_m}{\xi^2_{m,n} + \Delta^2_{-m,m-n} + h^2 W_m^2} \exp\left(-\frac{E_{-m}}{k_B T}\right) \tag{2.7}$$

represents both $h/\tau_0 > \Delta_{m,n-m}$ and $h/\tau_0 < \Delta_{m,n-m}$ limits. It results from the correct contribution to all orders in $\Delta_{m,n-m}$ from inelastic phonon processes.

In the situations described by (2.6) and (2.7), Berry phase oscillations of the tunnel splittings $\Delta_{-m,m-n}$ should manifest themselves also in the thermally activated regime. In this context, we should mention a recent study showing oscillations of the ac-susceptibility and blocking temperature T_B of a Fe_8 crystal as a function of the transverse magnetic field. Interestingly, a clear frequency shift was observed with respect to oscillations observed in low temperature experiments (and calculations) [77]. This shift, attributed to a mixing between the ground-state $S = 10$ and excited $S = 9$ multiplets (allowed by the presence of anti-symmetrical interactions), suggests that the traditional ground-state multiplet approximation might not be accurate enough here [77].

Mechanisms of MQTM, Square Root Relaxation, Spin-Spin Correlations and Magnetic Order Let us now see how tunneling takes place. When spin-up and spin-down state-distributions overlap, a large number of states $n - m$ and m are close to resonance, but not really at resonance because they do not "see each other", the intrinsic resonance-width (or tunnel window) $\Delta_{m,n-m}$ being extremely small ($\sim 10^{-8}$ to 10^{-9} K for SMMs ground-states). The only way to increase the number of spins at resonance in order to find a tunneling rate comparable to observed ones, would be to increase the tunnel window, but the tunnel splitting is a fixed quantity (see above) and homogeneous phonon-broadening is useless at low temperatures (it may increase the tunnel window by a factor of 100 giving $\sim 10^{-6}$–10^{-8} K which is still too small). In fact, some noise should be present to shift these levels close to resonance and put them at resonance many times per unit of time. This noise could come from the measuring tool itself, but present magnetometers are very stable and the source of fluctuations is really intrinsic as it comes from the interplay between the applied sweeping-field rate and the dynamics of electronic and nuclear spins.

After a variation of the applied magnetic field (to move from one point of the hysteresis loop to another one, inside a resonance) the nuclear spins, rather quiet (T_1 and T_2 are usually rather large), restart new flips producing hyperfine field "scans" of a few mT on the SMMs spins, bringing them across resonance at a rate of the order of $1/T_2 \sim$ some kHz. Inversely, when the spin of a given molecule flips, it produces a time-dependent magnetic field which can bring other molecules to resonance, an effect which is amplified by the long-range character of dipolar interactions. This defines an effective tunnel-window of width $\xi_0 \sim 10$ mT in which the spin-states near resonance are homogeneously broadened, can "see" each other and make states superposition (during very short periods of times), before collapsing into one or the other state (depending on the speed of the "hyperfine scan"). The initially less occupied (say spin-up) direction being favored, quantum relaxation will take place from spin-down to spin-up.

The corresponding relaxation rate has been calculated in a first model giving $\tau_N^{-1}(\xi) \approx \tau_0^{-1} \exp(-|\xi|/\xi_0)$ with $\tau_0^{-1} \approx 2\Delta^2_{-m,m-n}/\pi^{1/2}\xi_0$ [97], which accounts rather well for some of the observations made on Mn_{12}-ac or Fe_8. In these expressions, $-m$ and $n-m$ are the levels near resonance, $\Delta_{-m,m-n}$ is the tunnel splitting ($m \leq 10$) and $\xi_0 \sim \xi_{m,n-m} = \xi_m - \xi_{n-m}$ is the applied field bias ($\xi_m = E_m^0 - g_B m H_z$ and $E_m^0 = -Dm^2 - Bm^4$). In fact, this model, which assumes a time-independent density of spin-states, describes only the initial relaxation stages (very short times) for which the density of spin-states remains equal to the initial one. In a more detailed study, Prokof'ev and Stamp [85] considered both the short and long time limits, now assuming a time-dependent density of spin-states, and found again a square root law in both cases. Taking as the example the short time limit, they obtained for the inhomogeneous field-distribution of the diluted dipoles the Lorentzian function $P_\alpha(\xi) = [(1+\alpha M(t))/2]/[(\Gamma_d(t)/\pi)/\{(\xi - \alpha E(t))^2 + \Gamma_d^2(t)\}]$ where $\Gamma_d(t) = (4\pi^2 E_D/3^{5/2})[1-M(t)]$ and $E(t) = \eta V_D[1-M(t)]$. η is a sample shape-dependent constant and V_D the strength of dipolar interactions. The relaxation law was found to be a square root law [85] (see also [58]):

$$M(t)/M_s = 1 - \left(\tau_{\text{short}}^{-1} t\right)^{1/2}, \quad \tau_{\text{short}}^{-1} = \eta\Delta^2_{10,-10} P(\xi_D)/\hbar \tag{2.8}$$

It is worth mentioning here that the general validity of the square root relaxation has been questioned on theoretical grounds. For a theory that gives a time relaxation different from $t^{1/2}$, as well as relevant pro and con discussions on the subject, see Refs. [113–115].

As indicated in Sect. 2.3.2, this square root law has been verified experimentally in both Mn_{12}-ac [65, 69, 72] and Fe_8 [119, 120] for both ground-state and thermally-activated tunneling (Figs. 2.11 and 2.15) not only at saturation but also in the small initial-magnetization. However, a crossover from square root to exponential relaxation has been systematically observed in Mn_{12}-ac in the long time limit, contrary to the predicted validity of the square root relaxation in both limits [65, 69, 72]. In particular Fig. 2.15 [72] shows the scaling plot of the magnetization $M(t) = f(t/\tau(T))$ measured above 2 K, in which f is a function of $\tau(T)$ given by an Arrhenius law determined experimentally. The solid curve was calculated assuming an exponential relaxation, showing a clear deviation in the short-times/low-temperatures limit,

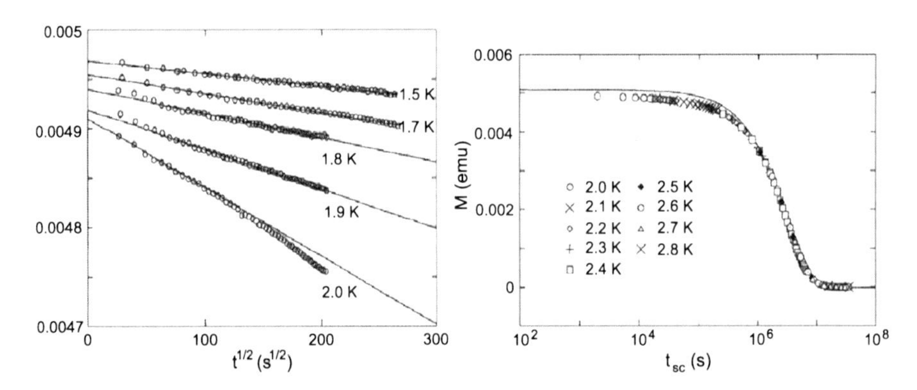

Fig. 2.15 *Left*: time decay of the remnant magnetization measured at low temperature, and plotted vs. the square root of time; *lines* show the fits. *Right*: scaling plot of the square root relaxation and its crossover to exponential at larger temperature/waiting-times (*continuous line*); for clarity, only 5 % of the data points have been plotted (they all fall in the same curve) [72]

where direct measurements showed a pure square root law. Below 1.7 K, where the scaling plot is no longer valid, the square root law becomes independent of temperature.

In Ref. [85], it was argued that the square root law is the consequence of a density-of-state transfer from the initial spin-down well (before tunneling) to the final spin-up one. The tunneling rate being very small, those densities of states are rather isolated from each other leading to the formation of a hole at resonance in the initial one. Monte-Carlo calculations (see Refs. [85, 112–114] and Fig. 2.13) clearly show such a "hole-digging" phenomenon. This prediction was verified by micro-SQUID measurements in Fe_8 and the minor species of Mn_{12}-ac [119, 121]. This is in the short-time limit, when the spins are not at equilibrium. In fact, in the long-time/high-temperature limit the spins equilibrate with the phonon and spin baths so that the hole should progressively disappear leading to the initial Lorentzian density of states. This physical picture suggests the following very simple model: being at equilibrium, the transfer of spin-states by tunneling from one well to the other does not modify the shape of the initial Lorentzian density of states $f_m(\xi) = m/(\xi^2 + m^2)$, a function of the bias field ξ, the width of which is given by the reduced magnetization $m = M/M_s$ which increases proportionally to the spins concentration as the dipolar energy does. The usual rate equation $\mathrm{d}m/\mathrm{d}t = -\Gamma m$, now writes $\mathrm{d}m/\mathrm{d}t = -\Gamma f_m(\xi) = -\Gamma m/(\xi^2 + m^2)$ which gives, after integration $\ln(m) + m^2/2\xi^2 = -(\Gamma/\xi^2)t + m_i^2/2\xi^2$ where m_i is the initial magnetization. This solution is not a stretched exponential despite the fact that the results can be fitted by such a function at high temperature. The two limits $m = m_i\sqrt{\{1-(2\Gamma/m_i)t\}}$ at short times and $m = m_i \exp(-\Gamma\xi^{-2}t)$ at long times, show the square root to exponential crossover. This model, where the relaxation rates of the exponential and square root regimes are connected (by the factor $m_i/2\xi^2$) suggests that the square root regime persists after the equilibrated hole disappears and before the exponential regime.

This simple phenomenological model should not lead us to forget that the physics of MQTM is much deeper and complex and constitutes a basis for the study of a quantum computer. The general Hamiltonian of networks of quantum gates made from solid-state qubits [104], is typically described as an ensemble of coupled 2-level systems: their formation can be speeded up by a transverse field

$$H = (\Delta_j \tau_{jx} + \varepsilon_j \tau_{jz}) + \Sigma_{ij} V_{ij} \tau_{iz} \tau_{jz} \quad (2.9)$$

where the control parameters Δ_j, ε_j and V_{ij} can be manipulated to make gate operations. This Hamiltonian mimics the full low-energy Hamiltonian of the spin-bath model (incorporating all mutual effects in a spin environment). In such a Hamiltonian, the "control parameters" are actually constituted of full expressions taking into account the coherent motion of SMM spins in interaction with nuclear spins and other environmental spins. It is important to note here that this Hamiltonian has three limiting cases, each one bringing out important aspects of the spin-bath physics: topological decoherence, orthogonality blocking, degeneracy blocking [85, 97, 102].

We shall conclude this section in an attempt to have a better qualitative understanding on how the spin-bath MQTM modifies the spin structure of a crystal. As seen above, hyperfine interactions of weak strength and long range induce MQTM through the short-lived entanglements that they favour between electronic and nuclear spins, leading to the homogeneous tunnel window. Subsequent energy transfer between molecule spins is critical for the building of multi-molecule correlations leading to the formation of SMMs ferromagnetic domains, or for triggering and propagating the avalanches described in the first part of this paper. Such ferromagnetic domains, clearly visible in the structure of $SmCo_{3.5}Cu_{1.5}$ (before and after an avalanche, Sect. 2.2, Fig. 2.1) were not observed in SMMs. One way to induce the coexistence of spin-up and spin-down domains in a SMM is to cool down the system below its Curie temperature in a field small enough to avoid saturation of the magnetization. This was done recently in a crystal in the SMM Fe_8 for which a ferromagnetic transition, expected below $T_c \sim 0.6$ K [64], was observed [122]. As with most SMMs, the Curie temperature is smaller than the blocking temperature T_B showing that, when the temperature decreases from the super-paramagnetic side, all the spins will be quenched as soon as $T < T_B$ in more or less random directions making impossible their reorientation towards a ferromagnetic state. In the absence of thermal fluctuations, this state was reached thanks to the large quantum fluctuations induced, at constant energy, by a large transverse field which was reduced before the measurements ("quantum annealing" [122]). Visualizing the spin configuration of a SMM, in the same conditions should allow the observation of magnetic domains.

Is it possible to find new systems in which the spin-bath is "simplified"? A rapid answer to this question will tentatively be given in Sect. 2.5 where we first show that quantum staircase hysteresis loops are not specific of SMMs but can also be observed with simple paramagnetic ions provided they have a large enough uniaxial anisotropy. These systems, that we called Single Ion Magnets, are easily realized with rare-earths ions. The latter generally have strong hyperfine interactions leading to a "condensation" of the nuclear degrees of freedom from the spin-bath to the central spin system. In Sect. 2.5.1, an analogy will be made with the spin-bath of SMMs, which can be considered as an ensemble of SIMs with vanishing hyperfine interactions. Section 2.6 will be devoted to the coherent regime which is a natural follow-up of the quantum relaxation regime developed above, especially with the first observations and studies of the Rabi oscillations in SIMs and SMMs.

2.5 Quantum Tunneling and Coherence in Single Ion Magnets

As explained just above, Single Ion Magnets consist of rare-earth ions diluted in an insulating matrix allowing for strong uniaxial anisotropy [29]. We started the study of these systems in order to answer a simple question: why such simple paramagnetic ions would not show MQTM, as SMMs do? Indeed (i) they carry magnetic moments as a large as the collective spins of SMMs and (ii) they can easily show anisotropy barriers at least as large as those of SMMs. We decided to start with the uniaxial system $LiY_{1-x}Ho_xF_4$ because it (i) belongs to a very well known and characterized series of which single crystals of excellent quality are made, (ii) it has an Ising doublet ground-state and (iii) the energy barrier separating the two lowest lying magnetic states is of the right order of magnitude. The choice of an insulating matrix was obviously to minimize the sources of decoherence. The main difference between SMMs and SIMs is their degree of complexity: hundred of atoms forming a collective spin with huge Hilbert space dimensions for the former and a single-atom carrying a simple paramagnetic spin with very small Hilbert space dimension for the second. In the case of rare-earth ions, another important difference comes from their much larger hyperfine interactions. It is worth mentioning that, in recent years, molecules consisting of a single lanthanide ion encapsulated by nonmagnetic ligands have been studied in detail and that they show quantum tunneling and hysteresis phenomena [30–32] resembling those found in inorganic SIMs shown above.

2.5.1 *First Evidence of MQTM in SIMs and Comparison with SMMs*

The main subject of this book being devoted to SMMs, we will pass over this section rather rapidly and show only the aspects of SIMs which are useful for a better understanding of SMMs and the possible new extensions in their study. The hysteresis loop shown Fig. 2.16(left) [29] was observed on a simple paramagnetic ensemble of spins at temperatures below a paramagnetic (and not super-paramagnetic) blocking temperature $T_B \sim 250$ mK. This system was a $LiYF_4$ single-crystal where ~ 0.2 % of Ho^{3+} was substituted for Y^{3+} (the total angular moment of Ho^{3+} is $J = L + S = 8$ and its Landé factor $g_J = 5/4$). This hysteresis loop is very

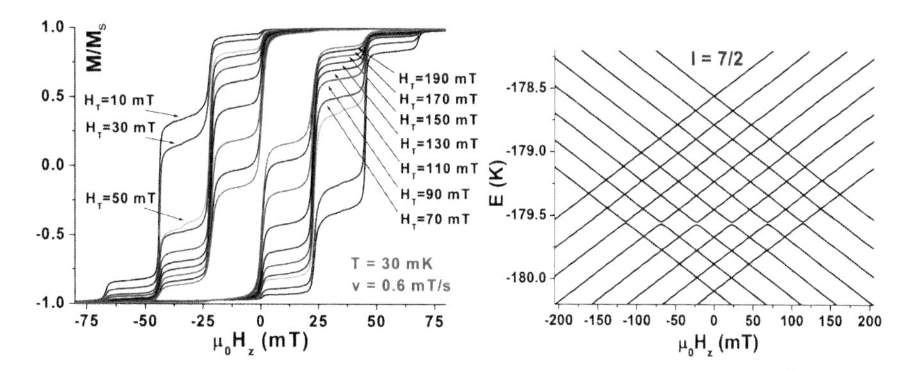

Fig. 2.16 *Left*: Hysteresis loop measured on a single crystal of $Ho_{0.002}Y{:}LiF_4$ at 30 mK and sweeping field rate of 0.6 mT/s for different transverse fields [29]. *Right*: Electro-nuclear levels scheme, obtained by exact diagonalisation of the 136×136 matrix of $H_{CF} + A_J \boldsymbol{IJ}$ on the basis J, m, I, m_I for the stable isotope ^{165}Ho. It is formed of two combs of parallel levels going up or down, with effective spins $\pm\frac{1}{2}$ and energies $E_n = \pm g_{\text{eff}}\mu_B H/2 + n\Delta E$ where $\Delta E/k_B = A_J J_z$. These levels intercept at fields $\mu_0 H_n = n\Delta E/g_{\text{eff}}\mu_B = nA_J/2g_J\mu_B(-7 \leq n \leq 7)$, giving the separation of steps of the hysteresis loop [29, 92, 124]

similar to the one of Mn_{12}-ac, with a difference however: the separation of consecutive steps is not determined by the longitudinal crystal-field term $B_0^2 O_2^0$ (equivalent to the anisotropy constant DS_z^2 of Mn_{12}-ac, giving $H_n \sim nD$) but by the hyperfine term $A_J \boldsymbol{IJ}$ with a quite large $A_J \sim 40$ mK (natural Ho has only one stable isotope with nuclear spin $I = 7/2$). As a consequence, the spin dynamics is governed by the total electro-nuclear angular momentum $\mathbf{I} + \mathbf{J}$ and not by the total angular momentum $\mathbf{J}$ (or the spin $\mathbf{S}$, as this is the case with SMMs where hyperfine couplings are weak). In particular, the moment which tunnels is not $\mathbf{J}$ or $\mathbf{S}$, but $\mathbf{I} + \mathbf{J}$. These first observations show that (i) super-paramagnetism with collective spin is not required to observe a blocking temperature T_B with hysteresis at $T < T_B$, (ii) complexity is not required to observe MQTM, (iii) nuclear spins can drive macroscopic spin-dynamics associated with MQTM.

The S_4 point symmetry group at Ho^{3+} sites is equivalent to the D_{2d} symmetry, with the crystal-field Hamiltonian:

$$H_{CF} = B_2^0 O_2^0 + B_4^0 O_4^0 + B_4^4 O_4^4 + B_6^0 O_6^0 + B_6^4 O_6^4 + B_6^{-4} O_6^{-4} + g_J \mu_B \boldsymbol{JH} \quad (2.10)$$

where O_l^m are Stevens' equivalent operators and B_l^m are crystal field parameters, initially determined by high resolution optical spectroscopy and checked many times [130]. This Hamiltonian is quite similar to (2.1), but complete and adapted to the case of rare-earths. Exact diagonalization of (2.10) leads, at zero field, to a zero-field ground-state doublet and a first excited singlet at ~ 9.5 K above it (top of the barrier) [29]. The expected weak mixing of the doublet by weak off-diagonal terms (crystal-field distribution, internal magnetic fields, Jahn-Teller effect, hyperfine interactions, …) should, in principle, lead to a single tunnel transition in zero-field (we must say that this was our initial expectation). However the hysteresis loop (Fig. 2.16) shows much more than a single step. In fact, the scheme of electro-nuclear levels obtained by exact diagonalization of the 136×136 matrix of $H_{CF} + A_J \boldsymbol{I}\boldsymbol{J}$ on the basis $|J, m, I, m_I\rangle\rangle$ is formed of two combs of parallel levels going up or down, reminiscent of the initial effective spins $\pm\frac{1}{2}$ of the Ising doublet, $E_n = \pm g_{\text{eff}}\mu_{\text{B}}H/2 + n\Delta E$ where $\Delta E/k_{\text{B}} = A_J J_z$ (Fig. 2.16, right). Levels intercept at fields $\mu_0 H_n = n\Delta E/g_{\text{eff}}\mu_{\text{B}} = nA_J/2g_J\mu_{\text{B}}(-7 \le n \le 7)$ [29, 92, 124]. The best agreement between measured and calculated resonance $H_n(\text{mT}) = 23n$ was obtained for $A_J/k_{\text{B}} = 38.6$ mK, a value comparable to the one found by NMR. This provides a new way of determining hyperfine constants in rare-earths; it is more accurate than NMR as it is simply limited by the resolution of the applied magnetic field and by the line-shape which can be made very narrow. In the present dilution of 0.2 %, inhomogeneous level broadening by dipolar interactions is ~ 20 mK. The finite avoided level crossings of the states $\frac{1}{2}, m_I \pm -\frac{1}{2}, m_{I'}$ result from off-diagonal crystal-field and hyperfine interactions $(B_4^4 O_4^4 + A_J(J^+I^- + J^-I^+)/2)$ leading to the selection rule $(m_I - m_{I'})/2 =$ odd integer. Nevertheless, the above mentioned weak but unavoidable distortions enable to observe all transitions. The used single-crystals being among the best that can be made, we believe that cell distortions are unavoidable in all types of systems, permitting tunnelling for all n-values. This should also be true with SMMs with symmetry ≥ 2, beyond the particular case of Mn_{12}-ac where symmetry lowering comes from disorder of the acetic acids of crystallization [125].

Let us now discuss shortly the tunneling of multimers of Ho^{3+} ions [29, 92, 124]: $(\mathbf{I} + \mathbf{J})$ of a single Ho^{3+} ions, $(\mathbf{I} + \mathbf{J})_1 + (\mathbf{I} + \mathbf{J})_2$ of two ions, etc and, by analogy, the case of two SMMs $(\mathbf{S}_1 + \mathbf{S}_2)$ [92]. I will show that the resonance fields of multimers do not depend on the coupling strength J_{ex} (we exclude here the case of a spin reversal "in the field of another spin" or exchange-bias tunnelling [29, 92, 124] which corresponds to a simple antiferromagnetic SMM). In the weak coupling limit (for example in the case of two distant Ho^{3+} ions), the discussion can be limited to the single-ion level scheme, where $E_p = \pm g_{\text{eff}}\mu_{\text{B}}H/2 + p\Delta E$. The resonance involving the states p, p' and $p' + 1$, entails $E_p = (E_{p'+1} + E_{p'})/2$, giving $g_{\text{eff}}\mu_{\text{B}}H_{p,p'} = (p - p' - \frac{1}{2})\Delta E$. As observed experimentally, two-ion resonances are shifted by $\frac{1}{2}$ with respect to single-ion ones. This is because the Zeeman energy is multiplied by two (two spins), while the zero-field energy $\Delta E = A_J J_z k_{\text{B}}$ is not. In the single-ion case J_z and g_{eff} cancel each other out in the expression of the resonance field, giving a direct relationship between the measured field and the hyperfine constant.

In the case of a SMM with uniaxial anisotropy, $E_n = -Dm^2 \pm g\mu_{\text{B}}mH$, a similar result can be obtained although zero-field levels are not equidistant. Co-tunneling with parallel $(\uparrow\uparrow \rightarrow \downarrow\downarrow)$ or anti-parallel $(\uparrow\downarrow \rightarrow \downarrow\uparrow)$ initial states, gives a resonance if the absolute value of the quantum number m of one of the two spins changes (e.g. from m to $m \pm 1$) while that of the other spin is unchanged (e.g. m changes to $-m$). In this case, only the first spin will contribute to change the anisotropy energy (by $\sim D$), while both spins contribute to the Zeeman energy (by $\sim 2g\mu_{\text{B}}H$), giving $g\mu_{\text{B}}H \sim D/2$. The fact that the two spins can be in different states is a consequence of weak interactions $(J_{\text{ex}} \ll D)$. Contrary to the case of equidistant levels, co-tunneling resonances are here not exactly in between single-spin resonances, unless $m \sim S$ is very large. In the strong coupling limit of, for example, two spins $\mathbf{S}_1$ and $\mathbf{S}_2$ with the same anisotropy constant D (see (2.1)) coupled by $J_{ex} \gg D$ the addition of two spins $\mathbf{S}_1$ and $\mathbf{S}_2$ should apparently give $\mathbf{S} = \mathbf{S}_1 + \mathbf{S}_2$ with a trivial resonance at $g\mu_B H_n = nD$ (cf. (2.2)). In fact, this result is wrong, the right expression being $g\mu_B H_n = nD/2$. The reason is that the ligand-field parameter D depends on a nontrivial way on the spin S. This has consequences on the way the anisotropy energy of a SMM spin S—equal to the sum of several elementary spins σ with the elementary parameter D_0—is built. It is easy to show that the expression $g\mu_B H_n = nD$ for a spin S, becomes $g\mu_B H_n = nD/2$ for a spin $2S$ or $g\mu_B H_n = nD/N$ for a spin NS. This explains the difference between the two definitions of the anisotropy energy barrier of a SMM, $U_{cl} = KV \propto KS$ (Sect. 2.2) where the extensive variable K is expressed in energy/volume-unit, and $U_{cl} = DS^2$ where D results from the choice of the Hamiltonian [92]. The expression $g\mu_B H_n = nD/N$ shows that the quantization of the hysteresis loop tends to vanish as $S \to \infty$, as expected.

We have shown that the quantum dynamics of electronuclear singlets, pairs, triplets, ... of Ho^{3+} SIMs comes from long-lived entanglements of such entities [29, 92, 124]. In order to understand the spin-bath of SMMs even better, one may imagine similar many-body SIMs electronuclear entanglements but with much weaker hyperfine and super-hyperfine interactions giving rise to very short-lived entanglements of electronic and nuclear spins within a level structure similar to the one of Fig. 2.16, but with much closer levels, the overall width being a homogeneous level broadening of hyperfine nature, which is nothing else but the tunnel window of SMMs (Sect. 2.4.4) the timescale of which is associated with decoherence by nuclear spins.

2.5.2 *First Evidence of MQCM in SIMs, Paving the Way for SMMs*

In the previous sections we have seen how coherently mixed spin-up and spin-down states collapse after tunnelling with either a spin-up or a spin-down (for a simple intuitive interpretation see [103]). In the case of SMMs, where tunnel splittings are extremely small in low fields, the time $\tau_t \sim h/\Delta$ that we can define as the tunnelling time is extremely large ($\sim 10^{-3}$ s in Mn_{12}-ac) leading to strong decoherence by the environment even if its dynamics, associated with e.g. nuclear spins, is slow. Due to such decoherence, most SMMs resonant spin-states end with a final spin-state identical to the initial one i.e. without tunnelling, whereas the rare events ending with spin reversal (tunnelling) lose their coherence immediately. The reduction of decoherence by an existing spin-bath, requires either to suppress it physically or to slow it down below the measuring timescale, the former solution being better because a frozen spin-bath produces decoherence in the presence of microwaves, see below). This is what we tried to do in our search for Mesoscopic Quantum Coherence of the Magnetization in SIMs and SMMs.

We started with Er^{3+} ions diluted in a single-crystal of $CaWO_4$ [59], a matrix isomorphic to $YLiF_4$ and containing almost no nuclear spins (only 15 % of

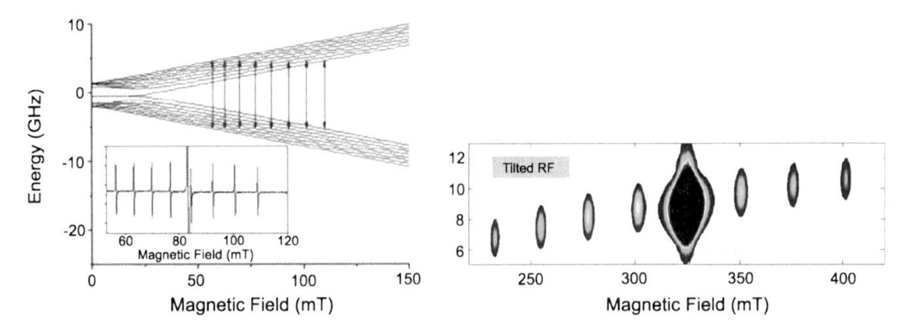

Fig. 2.17 *Left*: electro-nuclear level scheme of a ^{167}Er ion in $CaWO_4$ obtained by exact diagonalization of the 128×128 matrix of $H_{CF} + A_J \boldsymbol{I}\boldsymbol{J}$ on the basis $|J, m, I, m_I\rangle$. The zero-field 16-fold degenerated electro-nuclear ground-state is partially removed by hyperfine interactions, the remaining degeneracy being removed by a magnetic field H perpendicular to the **c**-axis [59]. *The arrows* indicate the EPR transitions with $\Delta m_I = 0$ between the two sets of 8 levels with effective spin projections $\pm\frac{1}{2}$ (each on them being labelled by its nuclear spin projection going from $-7/2$ at the centre of the figure to $+7/2$). *Inset*: measured CW EPR transitions. *Right*: measured frequencies of the associated Rabi oscillations

the W second neighbours). The total angular momentum $J = 15/2$ of the Er^{3+} ground-state multiplet and its Landé factor $g_J = 6/5$ give a magnetic moment of $9\mu_B$, comparable to that of Ho, Mn_{12}-ac or Fe_8 so that we keep in the mesoscopic regime. The crystal-field Hamiltonian of Er:$CaWO_4$ is identical to the one of Ho:$LiYF_4$ (2.10), their space group $I4_1/a$ and point symmetry S_4 being the same. Exact diagonalization of the 16×16 matrix with $H = 0$ and appropriated crystal-field parameters [59] gives an easy plane perpendicular to the **c**-axis with a doublet ground-state characterized by the g_{eff} tensor ($g_{//} = 1.247$, $g_{\perp} = 8.38$ [130]). Natural Erbium having two isotopes with nuclear spins $I = 0$ and $I = 7/2$, we extended this calculation by adding the hyperfine term $A_J \boldsymbol{I}\boldsymbol{J}$ (with $A_J = 125$ MHz and $I = 7/2$). Diagonalization of the 128×128 matrix gives an energy spectrum whose 16-fold degenerated electronuclear ground-state is partially removed by the hyperfine interactions themselves, the remaining degeneracy being removed by a magnetic field H applied perpendicular to the easy plane (Fig. 2.17, left) [59]. The eight transitions with $\Delta m_J = \pm 1$ and $\Delta m_I = 0$ have been observed in continuous and pulsed EPR. Rabi oscillations result from the coupling of two eigenstates ϕ_1 and ϕ_2 by application of a linearly polarized microwave field h_{mw}. The corresponding Hamiltonian $H_{\text{mw}} = \mu_B g_{\text{eff}} S_x h_{\text{mw}} \cos(\omega t)$ shows that, as long as its phase is preserved from the environmental fluctuations, the wave function $|\psi(t)\rangle$ of the coupled system oscillates in time between $|\phi_1\rangle$ and $|\phi_2\rangle$, according to $|\psi(t)\rangle = cos(\Omega_R t)|\phi_1\rangle - i\sin(\Omega_R t)|\phi_2\rangle$, at the Rabi frequency $\Omega_R = g_{xy}\mu_B h_{\text{mw}}/\hbar$ [126].

Pulsed EPR measurements give access to the occupation probability of, say, state $|\phi_2\rangle$ which oscillates as $\sin^2 \Omega_R t$. The EPR transitions of Er^{3+}:$CaWO_4$ were observed at ^{4}He temperatures using a Bruker X-band spectrometer at 9.7 GHz for both isotopes $I = 0$ and $I = 7/2$. An example of Rabi oscillations, measured at $T = 3.5$ K on the $I = 0$ isotope, is shown in Fig. 2.18(left) [59]. The fit to

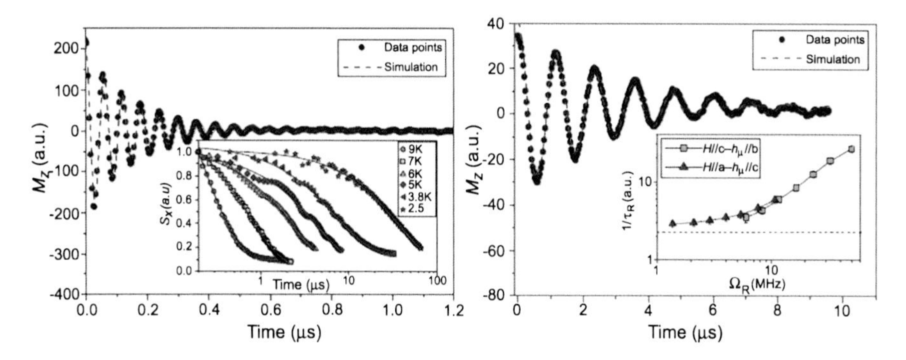

Fig. 2.18 *Left*: Rabi oscillations obtained at $T = 3.5$ K on the $I = 0$ isotope for $\mu_0 H = 0.522$ T//**c** and microwave field and frequency $h_{mw} = 0.15$ mT//**b** and 9.7 GHz [59]. The fit to $M_z = M_z(t = 0)e^{-t/\tau R} \sin(\Omega_R t)$ gives the exponential damping (or Rabi) time, $\tau_R \sim 0.2$ μs. *Right*: the same experiment performed at a microwave field 20 times smaller showing that τ_R increases, the number of Rabi oscillations $N(c)$ remaining nearly unchanged. *Inset*: oscillations damping time measured vs. Rabi frequency, showing the effect of decoherence by microwaves

$M_z = M_z(t = 0)e^{-t/\tau_R} \sin(\Omega_R t)$ gives the damping (time $\tau_R \sim 0.2$ μs of Rabi oscillations. Experiments performed at different fields showed oscillations of the second isotope $I = 7/2$: eight electronuclear oscillations separated by $\Delta H_R \sim 6$–8 mT selectively addressed by a small sweeping field (Fig. 2.17, right) [59]. These first observations of Rabi oscillations with rare-earth ions showed that coherent quantum spin dynamics can be seen at the mesoscopic scale, paving the way for the realisation of Electro-nuclear Mesoscopic Spin Qubits, which may be easily manipulated by weak applied fields at EPR and/or NMR frequencies [59]. Here, I shall skip the specific aspects of these qubits (such as, e.g. the strong crystal-field anisotropy of the Rabi frequencies when the frame of applied fields ($\mathbf{h}_{mw} \perp \mathbf{H}$) is rotated, see right hand panel of Fig. 2.17 [59, 60]). Instead, I concentrate on decoherence mechanisms and, in particular, on a basic mechanism that we found to be very general and which is also present in SMMs: the measured Rabi decay time τ_R is always smaller (or even much smaller) than T_2, the spin-spin coherence time which is often considered as the time limiting quantum calculations in an hypothetical quantum computer. Experiments performed at different microwave power (Fig. 2.18) showed that τ_R increases as power decreases, while the number of Rabi oscillations $N(c)$ remains nearly unchanged i.e. $N(c) \sim \tau_R(c)\Omega_R$ with $N(c) \sim 20$ (right-hand inset of Fig. 2.18). This increase of $\tau_R < T_2$ suggests the phenomenological expression [59]

$$1/\tau_R(c) \sim \Omega_R/N(c) + 1/T_2(c) \tag{2.11}$$

where $\tau_R(c)$, $N(c)$, and $T_2(c)$ are concentration-dependent. Expression (2.11), which was recently confirmed in quantum simulation studies [128], shows that Rabi oscillations are lost for $t \gg T_2$ in the low-power limit where $\Omega_R \to 0$, and for $t \gg N(c)/\Omega_R$ in the large power limit where $\Omega_R \gg N(c)/T_2$. In the first case, T_2 is limited by well-known spin-diffusion due to long-range dipolar interactions (as in the absence of microwaves), whereas in the second case the observed behaviour is characteristic of an inhomogeneous nutation frequency associated with weak distribution of g_{xy} due, in the case of Er^{3+} ions, to unavoidable crystal-field distributions even in these crystals of excellent quality [129, 130]. Note that the distribution of g_z contributes to the inhomogeneous line-width which remains rather small in Er:$CaWO_4$ (~ 2 mT, as in Ho:$YLiF_4$ and much smaller than in SMMs, 10–100 mT), but nevertheless gives a contribution to $1/\tau_R$ in the limit of $\Omega_R \rightarrow 0$ (but not at $\Omega_R = 0$, i.e. it does nor modify T_2).

To conclude this section, let us say that, if we consider the effect of decoherence by microwaves, the so-called "figure of merit" of a qubit $Q_{M2} = \Omega_R T_2/\pi$ becomes $Q_{MR} = \Omega_R \tau_R/\pi$. As shown above, τ_R being generally smaller (even much smaller) than T_2, the figure of merit associated with the observed number of Rabi oscillations Q_{MR} will be smaller (even much smaller) than Q_{M2}, related to the hypothetical number of oscillations based on the value of T_2. As this number corresponds to the number of quantum operations which can be made, it is important to reduce decoherence by microwaves so that $Q_{MR} \rightarrow Q_{M2}$. For that, we need to make all qubits identical to each other, as far as possible. We may also give a very simple theoretical expression for the "Rabi figure of merit" Q_{MR}. As in the presence of a Lorentzian distribution of transverse g-factors of width Γ the reciprocal Rabi time writes $1/\tau_R = \Gamma \Omega_R$ [128], it is immediate to see that the figure of merit $Q_{MR} = 1/\pi\Gamma$ is inversely proportional to the g-factor distribution-width. The condition $Q_{MR} \geq Q_{M2}$ with which disorder becomes negligible, gives $\Gamma \leq 1/\pi Q_{M2}$ i.e. a disorder weak enough so that the width of the g-factor distribution does not exceed the inverse of ~ 3 times the theoretical number of oscillations (given by T_2). Taking the example of a figure of merit of 10^4, the distribution width of the transverse g-factors of the qubit ensemble should not exceed 10^{-4}, which might be a problem in many cases.

2.6 Quantum Coherence in Single Molecule Magnets

In order to observe quantum coherent oscillations in systems of SMMs, a major problem to be addressed is the minimization of decoherence induced by dipole-dipole interactions [106]. In bulk, and despite their natural dilution (the distance between molecules is at the nm scale), dipolar interactions remain important and are of the order of 40–100 mT in Mn_{12}-ac or Fe_8 [25], which is crippling unless the ratio H/T (see below) [106]. A possible way out is diluting these molecules without modifying their architecture. Measurements performed on frozen solutions of different SMMs give values of T_2 up to a few μs [61, 127, 131, 132] while measurements performed on a single crystal gave much shorter times [133], thus suggesting that it should also be possible to observe Rabi oscillations in diluted SMMs.

In order to minimize dipolar interactions, we focused our study on a low spin SMM, the so-called V_{15} system of formula $K_6^+[V_{15}^{IV}As_6O_{42}(H_2O)]^{6-}H_2O$, and found a way to dilute the anionic clusters $[V_{15}^{IV}As_6O_{42}(H_2O)]^{6-}$ by using the

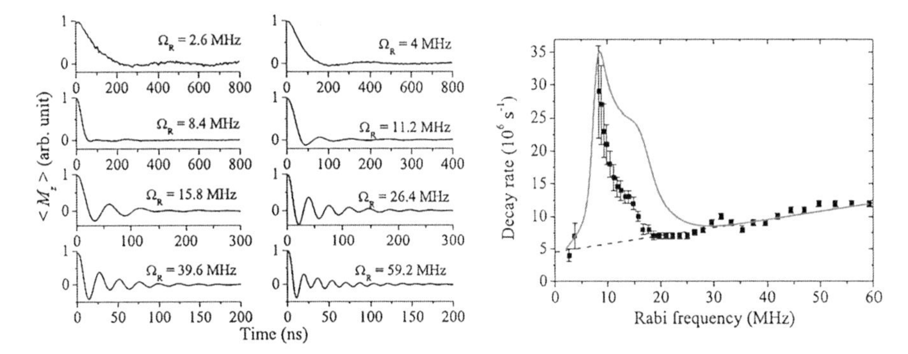

Fig. 2.19 *Left*: Rabi oscillations of the system V_{15}– DODA [135] (see also[61]). Measured $M_z(t)$ for an applied field of 0.354 T and $\mu_0 h_{mw}$ in the range 0.054 to 0.054 to 1.24 mT (or, equivalently $\Omega_R/2 = 2.6$ to 59.2 MHz, where the ratio $\Omega_R/2h_{mw} = 48$ MHz/mT was determined experimentally). *Right*: measured variation of damping rate $\tau_R^{-1} v s_R$ [135]. The broad peak in the range 8 MHz < $\Omega_R/2$ < 15 MHz implies the existence of a decoherence mechanism that is extremely sensitive to the microwave field amplitude. The calculated *continuous line* shows a shoulder at $\Omega_R \sim \omega_N$ the Larmor frequency of protons in the applied field of 0.354 T and a main peak on the left, which is shifted from the shoulder by the dipolar super-hyperfine field of protons on V_{15} clusters

cationic surfactant $[Mn_2N\{(CH_2)_{17}Me\}_2]^+$ as an embedding material (so-called DODA) [61]. In this system, each cluster of size $\sim$ 1.3 nm is made of 15 spins $\frac{1}{2}$ coupled by frustrated antiferromagnetic super-exchange interactions, so that the collective spin is $S = 1/2$. This low-spin SMM can be considered as a mesoscopic spin $\frac{1}{2}$ with a spatial extension of $\sim$ 1.3 nm. It exhibits a unique layered structure with a large central V_3^{IV} spin-triangle sandwiched between two smaller V_6^{IV} spin-hexagons [134]. Dzyaloshinsky–Moriya interactions (Sect. 2.4, [65, 69, 70]) were studied in detail in this system [73–76]. Energy spectrum calculations on the full Hilbert space of 2^{15} give two $S = 1/2$ ground-state doublets and an $S = 3/2$ excited quartet (Fig. 2.12(b)). These low-lying energy states, which are "isolated" from a quasi-continuum of states lying $\sim$ 250 K above, can also be obtained to a good accuracy using a triangular three-spin approximation coupled with an effective interaction J_0 much smaller than the frustrated intra-molecular interactions (valid below 100 K) [73–76]. The separation $3J_0/2$ between the two doublets and the quartet was determined from susceptibility measurements and high-field magnetization curves giving $J_0 \sim 2.45$ K [73]. The D-M interactions remove the degeneracy of the two low-lying doublets weakly (producing the first-order zero-field splitting $\Delta_{DM} \sim 3\sqrt{D_z}$) and of the exited quartet (producing the second-order zero-field splitting $\Delta'_{DM} \sim D_{xy}^2/8J_0 \ll \Delta_{DM}$).

EPR experiments were performed on this V_{15} hybrid material at $\sim$ 4 K using the same Bruker spectrometer operating at 9.7 GHz [61] as for SIMs (Sect. 2.5.2). The sample was characterized using different methods [61] and among them, the CW-EPR of a frozen sample at 16 K which gave the same $g_{//} \sim 1.98$ and $g_\perp \sim 1.95$ as previously obtained at the solid state. These results were also consistent with previous sub-Kelvin CW micro-SQUID experiments on a single-crystal of V_{15} [133]. A pulsed EPR experiment showed the first Rabi oscillations in a SMM [61]. Figure 2.19(left) shows the results of a new set of experiments performed later with another V_{15}-DODA sample between 2 and 20 K, for an applied field of 0.354 T and microwave field h_{mw} in the range 0.054 to 1.24 mT [135]. Each $M_z(t)$ corresponds to the superposition of oscillations between $S = 3/2$ states with slightly different Rabi frequencies (the transitions associated with the ground doublets being much less intense). Oscillations, between spin projections $m_S + 1$ and m_S occur at Rabi frequencies $\Omega_R^{(\pm 3/2 \leftrightarrow \pm 1/2)} = \sqrt{3}(g\mu_B/\hbar)h_{mw}$ and $\Omega_R^{(1/2 \leftrightarrow -1/2)} = 2(g\mu_B/\hbar)h_{mw}$. Taking $g = 1.96$, we find $\Omega_R^{(\pm 3/2 \pm 1/2)}/2\pi = 47.5h_{mw}$ MHz/mT and $\Omega_R^{(1/2-1/2)}/2\pi = 54.9h_{mw}$ MHz/mT, giving a mean Rabi frequency $\Omega_R/2\pi = (247.5 + 54.9)h_{mw}/3 \sim 50h_{mw}$ MHz/mT, very close to $\Omega_R/2h_{mw} = 48h_{mw}$ MHz/mT obtained experimentally (not shown). All curves show a fast decrease of Ω_R at short times due to the dephasing of spin-packets with different resonance frequencies in the non-homogeneous EPR line, followed by a number of damped Rabi oscillations $n_R = 7$ to 10. However, fast damping is observed, particularly in the frequency range from 8 to 15 MHz where $n_R < 3$. In order to extract the exponential damping time τ_R for different Ω_R (or h_{mw}) each curve was fitted to $j_0(\Omega_R t)e^{-t/\tau_R}$, where $j_0(z) = \int_z^\infty J_0(z)\mathrm{d}z(J_0)$ is the zero-order Bessel function) is associated with the distribution of Larmor frequencies within the EPR line [135]. The full evolution of the damping rate τ_R^{-1} vs. Ω_R is shown in the right-hand panel of Fig. 2.19, where the broad peak appearing in the range 8 MHz $< \Omega_R/2\pi <$ 15 MHz implies the existence of a new and efficient decoherence mechanism that is extremely sensitive to the microwave field amplitude. The amplitude of oscillations in that particular region does not obey a simple exponential law. The peak value $\tau_R = 36$ ns obtained at $\Omega_R/2\pi = 8$ MHz is close to an order of magnitude shorter than the coherence time $T_2 = 250$ ns measured under the same experimental conditions showing that this time, which is generally taken as a reference to certify the quality of a qubit, should at the very least be taken with caution.

We now switch to the interpretation of the observed decoherence. The slow linear variation of τ_R^{-1} with $\Omega_R/2\pi$ in the range 20–60 MHz (right-hand panel of Fig. 2.19) is a consequence of the random distribution of the Landé factor of V_{15} clusters in the frozen solution (plus intra-cluster hyperfine interactions) [135] and has the same origin as in Er:$CaWO_4$ (Sect. 2.5.2). This is not the case for the decoherence peak observed around 8 MHz, which constitutes the first observation of a decoherence window in the space of Rabi frequencies (or microwave fields). This peak is accompanied with a shoulder at $\Omega_R/2\pi \sim 15$ MHz, close to the Larmor frequency of protons (15.1 MHz in the static field of 0.354 T) suggesting a decoherence mechanism associated with resonant electronuclear cross-relaxation when the V_{15} Rabi frequency Ω_R is close to the average proton Larmor frequency ω_N. Such a mechanism of polarization transfer from the electronic to the nuclear spin-bath, is analogous to the one which takes place in the Nuclear Spin Orientation Via Electron spin Locking (NOVEL) technique of dynamic nuclear polarization that is produced under resonant microwave field and with the electronic spin nutation frequency tuned to ω_N ([135] and references therein). However, here, we are not interested in the degrees of polarization of the nuclear spin bath, but on the degrees of depolarization of the electronic spin-bath. The Hamiltonian of a "central" V_{15} spin interacting with static and microwave external magnetic fields and a large number of nuclear spins can be written [135] as:

$$H = \omega_e S_z + 2\Omega_R S_x \cos \omega t + \sum_j \omega_j I_z^j + \sum_{\substack{j \\ \alpha=x,y,z}} A_{z\alpha}^j S_z I_\alpha^j \tag{2.12}$$

where ω_e and Ω_R are the Larmor and Rabi frequencies of a cluster and ω_j are the precession frequencies of the proton spins $I = 1/2$ distributed around ω_N with half-width σ_N (average local field produced by the V_{15} spins at nuclear spins). The last term represents the super-hyperfine interaction between V_{15} and nuclear spins. The following effective Hamiltonian is obtained in the rotating reference frame [135]:

$$H' = \Omega S_x + V(t) \tag{2.13}$$

where

$$V(t) = \frac{1}{2\Omega}(\Omega_R S_z + \varepsilon S_x) \sum_j \{ [(A_{zx}^j - i A_{zy}^j) e^{i\omega_j t} I_+^j + cc] + 2 A_{zz}^j I_z^j \} \tag{2.14}$$

and $\Omega = (\varepsilon^2 + \Omega_R^2)^{1/2}$ is the distributed nutation frequency of V_{15} collective spins ($\varepsilon = \omega_e - \omega$). $V(t)$ involves two components of the local random fields induced by the nuclei at the V_{15} site (for details see [135]). Terms $\tilde{S}_x I_{+(-)}^j$ and $\tilde{S}_x I_z^j$, associated with V_{15} transverse spin components, result in a dephasing of Rabi oscillations by nuclear spins, relevant far from resonance when $\varepsilon \sim \Omega_R$ [135]. Cross-relaxation terms $\tilde{S}_z I_{+(-)}^j$ are responsible for V_{15} Larmor dephasing, inducing mutual flips of the electronic and nuclear spins, leading to energy dissipation in the applied static magnetic field. This resonant process occurs only when $\omega_j \sim \Omega_R$ i.e. when the Larmor frequency of nuclear spins is close to the Rabi frequency of V_{15} clusters. Averaging over the ensemble of spin-packets that are coherently driven by the microwave pulse, leads to simulations of the time evolution of the magnetization $\langle M_z(t) \rangle$ allowing a comparison between the calculated and the measured evolutions of the damping rate τ_R^{-1} vs. Ω_R, as shown in the right-hand panel of Fig. 2.19. In particular, the broad decoherence window and its shoulder are reproduced. A damping factor $e^{-(\beta\Omega_R + \Gamma_2)t}$ was added to the theoretical expressions in order to account for the linear background resulting from transverse g-factor distribution (Sect. 2.5.1). These results show [135] that the main peak at ~ 8 MHz is associated with a shift from the protons Larmor frequency (shoulder) created by the dipolar field of protons on the V_{15} sites. This peak comes from a mechanism of dissipative decoherence associated with a resonant process (V_{15} Rabi frequency $\sim$ protons Larmor frequency). The contribution from nuclear spins other than protons to decoherence (^{75}As, ^{14}N and ^{51}V) is negligible for $\Omega_R > 5$ MHz but can be important when $\Omega_R \to 0$, i.e. in the absence of microwaves, as in T_2 measurements [61, 135].

This first study of the decoherence of Rabi oscillations vs. variable microwave field in a SMM shows a feature that is quite general: a linear increase of Rabi decay rate τ_R^{-1} with $\Omega_R/2$, associated with the often unavoidable dispersion of the g-factors [61, 135], and a feature that seems to be specific to SMMs: the existence of a decoherence window in a certain range of frequencies associated with polarization transfer between electronic and nuclear subsystems [61, 135]. This decoherence is accompanied with energy dissipation from the electronic to the nuclear spin-bath. This seems to be a rare example, showing in which cases decoherence takes place with or without dissipation. Before ending this section we should recall that the protons which produce this decoherence window come essentially from DODA, the cationic surfactant embedding the V_{15} anionic clusters. It might therefore be of interest to measure Rabi oscillations on a single-crystal of V_{15}. Till now, the only EPR measurements made on a SMM single-crystal have been done on Fe_8, with T_2 measurements [135]. In this case, decoherence is collective due to the presence of short-lived spin-wave excitations associated with dipolar interactions in the superparamagnetic state. As mentioned above (Sect. 2.4.4), this systems orders magnetically below $T_c \sim 0.6$ K through dipole-dipole interactions [122] implying that eventual T_2 measurements below this temperature might lead to a coherence breakdown, the spin-waves lifetimes being much longer. Such a possibility may compete with another one much more attractive in which spin-waves become coherent at $T < T_c$ leading to a macroscopic coherence.

2.7 Conclusion and Perspectives

This chapter gives an overview of the subject of quantum tunneling of the magnetization in SMMs, starting from its roots in the 70's (Sect. 2.2) and ending with the coherent quantum dynamics (Sect. 2.6) which is a fast-growing subject, due in particular to its potential applications for the implementation of a quantum computer. The collective spin $S = 10$ of such molecules is large enough for both quantum and classical facets of their behavior to be observed and studied. This is why the word "mesoscopic" was linked to these studies of Quantum Tunneling of the Magnetization (Sects. 2.3 and 2.4). Last, the study of quantum coherence in magnetism (Sect. 2.6) permits to reach a degree of understanding that is for example not possible with superconducting qubits. Regarding spin qubits, SMMs might not be the best-suited for these applications in comparison with other types of qubits, but this is not proved as this type has important resources still to be explored. An example is given here, where, after the first demonstration of quantum coherence in a SMM (Rabi oscillations, Sect. 2.6), it is clearly shown in which cases decoherence is accompanied or not by dissipation, a distinction which is still not clear elsewhere. Going back in time, the study of the coherent quantum dynamics of SMMs was preceded by that of rare earths ions. These simple, but nevertheless mesoscopic, paramagnetic spins allowed us to progress towards different mechanisms of decoherence, showing for example that a very significant damping of Rabi oscillations is created by the microwaves which induce them (Sect. 2.5.2). This is quite a general phenomenon, always present if the different qubits of a system (such as a quantum computer) are not identical to each other, which is generally the case. Just before (Sect. 2.5), we extended the observation of MQTM to such rare earth ions, showing that this phenomenon is not specific to SMMs. Due to their large hyperfine interactions, MQTM and MQCM of rare earth ions also involve nuclear spins leading to “electronuclear quantum dynamics” in which the electronic and nuclear spins of each rare earth ion tunnel simultaneously. Such a two-body tunneling (extension to many-bodies tunnelling were also considered) leads to typical electro-nuclear steps in the hysteresis loop of rare-earth ions with uniaxial anisotropy (Sect. 2.5.1). If one imagines that the large hyperfine interactions of rare-earth ions decrease and become small enough, then the electronics and nuclear spins form short-living entangled pairs, as this is the case in SMMs. Such entanglements form the essential part of the dynamical spin-bath of Prokof’ev and Stamp, the quasi-static part being associated with the dipolar interactions between SMMs (Sect. 2.4). Besides, an experimental description of spin-bath effects is given explaining e.g. thermally-activated or microwave-activated MQTM, ground-state MQTM, the reasons why MQTM can be observed so easily, square root relaxation, quantum barrier-erasing and its amplification by a magnetic field, ... The physics of MQTM and MQCM in ensembles of mesoscopic spins is now rather well understood and can be used in nanotechnology e.g. for the study of small devices preparing the implementation of a quantum computer, in which each spin should be manipulated and addressed individually. An advantage of SMMs may be the possibility of elaborating supramolecular self-organized networks of nearly identical molecules (which is a crucial task to keep long coherence times, Sect. 2.6). Regarding rare earth ions, they might be embedded in a quantum-well semi-conductor layer for the implementation of a Kane-like electronic quantum computer (not published) or might be inserted in self-organized molecules containing one or several rare earths [137, 138]. However, despite the fact that the addressing of single nano-objects is currently performed with e.g., a single-atom [139] or a single-electron transistor [140], a single-shot detection of an electron [141] or of a nuclear-spin [142] (in particular of a rare-earth nuclear spin [143] if advantage is taken from the persistence of the rare-earth electronuclear steps [29], Sect. 2.5.1, at the molecular scale), it is not yet possible to manipulate and to address a large enough number of qubits and, in the case this would be possible, one would be faced with the problem of to make them, and their connections, nearly identical to prevent decoherence by microwaves (Sect. 2.6).

Acknowledgements I would like to thank my former Ph.D. students and my main collaborators who contributed to these works over so many years. Their names appear in an exhaustive list of references. Also I would like to thank Fernando Luis, Editor of this book, for his careful reading and enriching suggestions with in particular new set of references.

References

1. E. Schrödinger, Naturwissenschaften **23**, 516 (1935). Translation: J.D. Trimmer, Proc. Am. Philos. Soc. **124**, 323 (1980)

2. A.J. Leggett, J. Phys., Colloq. **39**(C6), 126 (1978)
3. L. Néel, Ann. Geophys. **5**, 99 (1949). Ibid., Cah. Phys. **12**, 1 (1942); Ibid., **13**, 1 (1943); see also, L. Néel, Oeuvre Scientifique (Editions CNRS, 1978)
4. R.F. Voss, R.A. Webb, Phys. Rev. Lett. **47**, 265 (1981)
5. A.J. Leggett, J. Phys., Colloq. **39**(C6), 1264 (1978); Ibid., in Proceedings of the 6th Int. Conf. on Noise in Physical Systems, Gaithersburg, Maryland (1981)
6. Y. Kaluzny, P. Goy, M. Gross, J.M. Raimond, S. Haroche, Phys. Rev. Lett. **51**, 1175 (1983)
7. D.P. Di Vincenzo, in *Quantum Tunneling of Magnetization—QTM'94*, ed. by L. Gunther, B. Barbara. NATO ASI Series E: Applied Science, vol. 301 (Kluwer, Dordrecht, 1995), p. 495
8. D. Loss, D.P. DiVincenzo, Phys. Rev. A **57**, 120 (1998)
9. K. Ziemelis, *Nature Milestones: Spin*. Nature Physics S, vol. S19 (2008)
10. P.C.E. Stamp, E. Chudnovsky, B. Barbara, Int. J. Mod. Phys. B **6**, 1355 (1992)
11. B. Barbara, G. Fillion, D. Gignoux, R. Lemaire, Solid State Commun. **10**, 1149 (1972)
12. B. Barbara, J. Phys. **34**, 1039 (1973). Ibid., in Symposium in Memory of Remy Lemaire: Magnetism of Rare Earth Intermetallic Alloys, Grenoble, France (1993), published in J. Magn. Magn. Mater. **129**, 79 (1994)
13. B. Barbara, C. Bècle, R. Lemaire, D. Paccard, J. de Phys. C1 **2–3**(suppl.), 299 (1971)
14. B. Barbara, M. Uehara, IEEE Trans. Magn. **12**, 997 (1976). See also, B. Barbara in Proc. of the 2nd International Symposium on Magnetic Anisotropy and Coercivity in Rare-Earth Transition Metal Alloys, July 1978, San Diego, CA, 137; see also, M. Uehara, Dissertation, Université Scientifique et Médicale de Grenoble (1975)
15. B. Barbara, M. Uehara, in *Proc. Inst. Phys. Conf.: Rare-Earth Conference*, vol. 37, (1978)
16. M. Uehara, B. Barbara, J. Phys. **47**, 2 (1986)
17. M. Uehara, B. Barbara, B. Dieny, P.C.E. Stamp, Phys. Lett. A **114**, 1 (1986)
18. B. Barbara, P.C.E. Stamp, M. Uehara, J. Phys. **49**, C–8, 529 (1988)
19. B. Barbara, J.E. Wegrowe, L.C. Sampaio, J.P. Nozières, M. Uehara, M. Novak, C. Paulsen, J.L. Tholence, Phys. Scr. **T49A**, 268 (1993)
20. B. Barbara, L.C. Sampaio, J.E. Wegrowe, J. Appl. Phys. **73**, 6703 (1993)
21. W. Wernsdorfer, K. Hasselbach, D. Mailly, A. Benoit, B. Barbara, J. Magn. Magn. Mater. **145**, 33 (1995). See also, Quantum tunneling of Magnetization—QTM'94, ed. by L. Gunther, B. Barbara, NATO ASI Series E: Applied Science, vol. 301 (Kluwer, Dordrecht, 1995), p. 227; see also, W. Wernsdorfer, Dissertation, Université Scientifique et Médicale de Grenoble (1996)
22. S. Tarucha et al., Phys. Rev. Lett. **84**, 2485 (2000)
23. M. Ciorga et al., Phys. Rev. Lett. **88**, 256804 (2002)
24. L. Kouwenhoven et al., Nature **442**(7104), 766 (2006)
25. L. Thomas, F. Lionti, R. Ballou, D. Gatteschi, R. Sessoli, B. Barbara, Nature **383**, 145 (1996). See also, L. Thomas, Dissertation, Université Scientifique et Médicale de Grenoble (1997)
26. J.R. Friedman, M.P. Sarachik, J. Tejada, R. Ziolo, Phys. Rev. Lett. **76**, 3830 (1996)
27. J.M. Hernández, X.X. Zhang, F. Luis, J. Bartolomé, J. Tejada, R. Ziolo, Europhys. Lett. **35**, 301 (1996)
28. B. Barbara, W. Wernsdorfer, L.C. Sampaio, J.G. Park, C. Paulsen, M.A. Novak, R. Ferré, D. Mailly, R. Sessoli, A. Caneschi, K. Hasselbach, A. Benoit, L. Thomas, J. Magn. Magn. Mater. **140–144**, 1825 (1995). See also, Quantum Tunneling of Magnetization—QTM'94, ed. by L. Gunther, B. Barbara, NATO ASI Series E: Applied Science, vol. 301 (Kluwer, Dordrecht, 1995)
29. R. Giraud, W. Wernsdorfer, A.M. Tkachuk, D. Mailly, B. Barbara, Phys. Rev. Lett. **87**, 5 (2001). See also, R. Giraud, A.M. Tkachuk, B. Barbara, Phys. Rev. Lett. **91**, 25 (2003); see also, R. Giraud, Dissertation, Université Scientifique et Médicale de Grenoble (2002)
30. N. Ishikawa, M. Sugita, W. Wernsdorfer, Angew. Chem., Int. Ed. Engl. **44**, 2931 (2005)
31. M.A. AlDamen, J.M. Clemente-Juan, E. Coronado, C. Martí-Gastaldo, A. Gaita-Ariño, J. Am. Chem. Soc. **130**, 8874 (2008)

32. F. Luis, M.J. Martínez-Pérez, O. Montero, E. Coronado, S. Cardona-Serra, C. Martí-Gastaldo, J.M. Clemente-Juan, J. Sesé, D. Drung, T. Schurig, Phys. Rev. B **82**, 060403(R) (2010)
33. J.M. Galligan, T. Oku, Acta Metall. **19**, 223 (1971)
34. R.C. Miller, C. Weinreich, Phys. Rev. **117**, 1460 (1960)
35. J.J. Gilman, J. Appl. Phys. **36**, 3195 (1965)
36. V. Celli et al., Phys. Rev. **131**, 58 (1963)
37. T. Egami, Phys. Status Solidi A **19**, 747 (1973). Ibid., Phys. Status Solidi **57**, 1, 1973
38. A. Haziot, A.D. Fefferman, J. Beamish, S. Balibar, Phys. Rev. B **87**, 060509(R) (2013). See also, A. Haziot, X. Rojas, A.D. Fefferman, J. Beamish, S. Balibar, Phys. Rev. Lett. **110**, 035301 (2013)
39. B. Barbara, in *Proceedings of the Int. Workshop. Studies of Magnetic Properties of Fine Particles and Their Relevance to Material Science*, ed. by J.L. Dormann, D. Fiorani (Elsevier, Amsterdam, 1991)
40. J.L. van Hemmen, S. Suto, Europhys. Lett. **1**, 481 (1986). Ibid., Physica **141B**, 37 (1986)
41. M. Henz, R. Schilling, J. Phys. C, Solid State Phys. **19**, 1765 (1986). Ibid., L711 (1986)
42. E. Chudnovsky, L. Gunther, Phys. Rev. Lett. **60**, 661 (1988). Ibid., Phys. Rev. B **37**, 9455 (1988)
43. B. Barbara, E. Chudnovsky, Phys. Lett. A **145**, 4 (1990)
44. P.C.E. Stamp, Phys. Rev. Lett. **66**, 2802 (1991)
45. R. Ferré, B. Barbara, J. Magn. Magn. Mater. **140–144**, 1861 (1995). See also, R. Ferré, Thesis of the Université Scientifique et Médicale de Grenoble (1996), Study of isolated ferromagnetic nanoparticles at submicronic scales
46. J.P. Sethna, K.A. Dahmen, C.R. Myers, Nature **410**, 242 (2001)
47. C. Paulsen, J.G. Park, B. Barbara, R. Sessoli, A. Caneschi, J. Magn. Magn. Mater. **140–144**, 1891 (1995)
48. H. Yoneda, T. Goto, Y. Fujii, B. Barbara, A. Müller, Physica B **329**, 1126 (2003)
49. A. Hernández-Mínguez, J.M. Hernández, F. Macià, A. García-Santiago, J. Tejada, P.V. Santos, Phys. Rev. Lett. **95**, 217205 (2005)
50. D. Mailly, C. Chapelier, A. Benoit, Phys. Rev. Lett. **70**, 2020 (1993)
51. W. Wernsdorfer, E.B. Orozco, K. Hasselbach, A. Benoit, D. Mailly, O. Kubo, H. Nakano, B. Barbara, Phys. Rev. Lett. **79**, 20 (1997)
52. W. Wernsdorfer, E. Bonet Orozco, K. Hasselbach, A. Benoit, B. Barbara, N. Demoncy, A. Loiseau, H. Pascard, D. Mailly, Phys. Rev. Lett. **78**, 1791 (1997)
53. E. Bonet, W. Wernsdorfer, B. Barbara, A. Benoît, D. Mailly, A. Thiaville, Phys. Rev. Lett. **83**, 4188 (1999)
54. W. Wernsdorfer, Adv. Chem. Phys. **118**, 99 (2001)
55. B. Barbara, in *Lectures Notes in Physics on Magnetism and Synchrotron Radiation*, ed. by Beaurepaire et al.(Springer, Berlin, 2001), p. 157
56. M.C. Miguel, E.M. Chudnovsky, Phys. Rev. B **54**, 389 (1996). See also, G.K. Kim, D.S. Hwang, Phys. Rev. B **55**, 6918 (1997)
57. R. Sessoli, D. Gatteschi, A. Caneschi, M.A. Novak, Nature **365**, 141 (1993)
58. I. Tupitsyn, B. Barbara, in *Magnetism: Molecules to Materials*, vol. 3, ed. by J.S. Miller, M. Drillon (Wiley-VCH, Weinheim, 2002)
59. S. Bertaina, S. Gambarelli, A. Tkachuk, I.N. Kurkin, B. Malkin, A. Stepanov, B. Barbara, Nat. Nanotechnol. **2**, 39 (2007)
60. S. Bertaina, J.H. Shim, S. Gambarelli, B.Z. Malkin, B. Barbara, Phys. Rev. Lett. **103**, 22 (2009)
61. S. Bertaina, S. Gambarelli, T. Mitra, B. Tsukerblat, A. Müller, B. Barbara, Nature **453**(7192), 203 (2008). See also, Nature **466**, 1006 (2010). Corrigendum
62. J.H. Shim, S. Bertaina, S. Gambarelli, S. Mitra, A. Müller, E. Baibekov, B.Z. Malkin, B. Tsukerblat, B. Barbara, Phys. Rev. Lett. **109**, 050401 (2012)
63. T. Lis, Acta Crystallogr. B **36**, 2042 (1980)

64. J.F. Fernández, J.J. Alonso, Phys. Rev. B **62**, 53 (2000). Ibid., Phys. Rev. B **65**, 189901(E) (2002)
65. I. Chiorescu, R. Giraud, A. Caneschi, L. Jansen, B. Barbara, Phys. Rev. Lett. **85**, 4807 (2000)
66. A.L. Barra, D. Gatteschi, R. Sessoli, Phys. Rev. B **56**, 8192 (1996)
67. R. López-Ruiz, F. Luis, V. González, A. Millán, J.L. García-Palacios, Phys. Rev. B **7**(2), 224433 (2005)
68. M. Hennion, L. Pardi, I. Mirebeau, E. Suard, R. Sessoli, A. Caneschi, Phys. Rev. B **56**, 8819 (1997)
69. B. Barbara, L. Thomas, F. Lionti, I. Chiorescu, A. Sulpice, J. Magn. Magn. Mater. **200**, 167 (1999)
70. B. Barbara, L. Thomas, F. Lionti, I. Chiorescu, A. Sulpice, J. Magn. Magn. Mater. **177–181**, 1324 (1998). See also, F. Lionti, L. Thomas, R. Ballou, B. Barbara, R. Sessoli, D. Gatteschi, J. Appl. Phys. **81**, 4608 (1997)
71. M.I. Katsnelson, V.V. Dobrovitski, B.N. Harmon, Phys. Rev. B **59**, 6919 (1999)
72. L. Thomas, A. Caneschi, B. Barbara, Phys. Rev. Lett. **83**, 12 (1999). Ibid., L. Thomas, A. Caneschi, B. Barbara, J. Low Temp. Phys. **113**, 1055 (1998)
73. B. Barbara, J. Mol. Struct. **656**, 135 (2003). See also, I. Chiorescu, W. Wernsdorfer, A. Müller, H. Bögge, B. Barbara, Phys. Rev. Lett. **84**, 3454 (2000)
74. H.D. de Raedt, S. Miyashita, K. Michielsen, M. Machida, Phys. Rev. B **70**, 064401 (2004)
75. G. Chaboussant, S.T. Ochsenbein, A. Sieber, H.U. Güdel, H. Mutka, A. Müller, B. Barbara, Europhys. Lett. **66**, 423 (2004)
76. B. Tsukerblat, A. Tarantul, A. Müller, J. Chem. Phys. **125**, 054714 (2006)
77. E. Burzurí, F. Luis, O. Montero, B. Barbara, R. Ballou, S. Maegawa, Phys. Rev. Lett. **111**(5), 057201 (2013). doi:10.1103/PhysRevLett.111.057201
78. R.P. Feynman, Selected Papers of Richard Feynman: With Commentary, in *20th Century Physics*, ed. by L.M. Brown (World Scientific, Singapore, 2000)
79. E.M. Chudnovsky, D.A. Garanin, Phys. Rev. Lett. **79**, 4469 (1997). See also, D.A. Garanin, E.M. Chudnovsky, Phys. Rev. B **56**, 11102 (1997)
80. F. Luis, J. Bartolomé, J.F. Fernández, Phys. Rev. B **57**, 505 (1998). See also, J.F. Fernández, F. Luis, J. Bartolomé, Phys. Rev. Lett. **80**, 5659 (1998)
81. F. Luis, F.L. Mettes, L.J. de Jongh, in *Magnetoscience: Molecules to Materials*, vol. 3, ed. by J.S. Miller, M. Drillon (Wiley-VCH, Weinheim, 2002), p. 169
82. G. Bellessa, N. Vernier, B. Barbara, D. Gatteschi, Phys. Rev. Lett. **83**, 2 (1999)
83. F. Luis, F.L. Mettes, J. Tejada, D. Gatteschi, L.J. de Jongh, Phys. Rev. Lett. **85**, 4377 (2000). See also, F.L. Mettes, F. Luis, L.J. de Jongh, Phys. Rev. B **64**, 174411 (2001)
84. M. Evangelisti, F. Luis, F.L. Mettes, R. Sessoli, L.J. de Jongh, Phys. Rev. Lett. **95**, 227206 (2005)
85. N.V. Prokof'ev, P.C.E. Stamp, Phys. Rev. Lett. **80**, 5794 (1998). See also, I.S. Tupitsyn, N.V. Prokof'ev, P.C.E. Stamp, Int. J. Mod. Phys. B **11**, 2901 (1997)
86. A. Abragam, B. Bleaney, *Electron Paramagnetic Resonance of Transition Ions* (Oxford University Press, Oxford, 1970)
87. L. Sorace, W. Wernsdorfer, C. Thirion, A.L. Barra, M. Pacchioni, D. Mailly, B. Barbara, Phys. Rev. B **68**, 220407 (2003)
88. J. Friedman, Y. Suzuki, K.M. Mertes, E.M. Rumberger, D.N. Hendrickson, Y. Myascodov, H. Shtrikman, N. Avraham, E. Zeldov, Phys. Rev. B **70**, 100408 (2004)
89. K. Petukhov, V. Mosser, W. Wernsdorfer, Phys. Rev. B **77**, 6 (2008)
90. D.A. Garanin, J. Phys. A, Math. Gen. **24**, L61 (1991)
91. B. Barbara, I. Chiorescu, R. Giraud, A.G.M. Hansen, A. Caneschi, Frontiers in magnetism. J. Phys. Soc. Jpn., Suppl. **69**, 383 (2000). See also, B. Barbara, Images de la Physique (Editions CNRS, 1999); See also B. Barbara, Phys.-J. **8/9**, 81 (2008)
92. B. Barbara, C. R. Phys. **6**, 934 (2005). See also, B. Barbara, Nature News & Views **421**, 32 (2003); see also, Inorg. Chim. Acta **361**, 3371 (2008)
93. R. Sessoli, H.L. Tsai, R. Shake, S. Wang, J.B. Vincent, K. Folting, D. Gatteschi, G. Christou, D.N. Hendrickson, J. Am. Chem. Soc. **115**, 1804 (1993)

94. F. Luis, J.M. Hernández, J. Bartolomé, J. Tejada, Nanotechnology **10**, 86 (1999)
95. F.D.M. Haldane, Phys. Rev. Lett. **50**, 1153 (1983)
96. M.V. Berry, Proc. R. Soc. Lond. A **392**(1802), 45 (1984)
97. N.V. Prokof'ev, P.C.E. Stamp, J. Low Temp. Phys. **104**, 143 (1996)
98. E.N. Bogachek, I.V. Krive, Phys. Rev. B **46**, 14559 (1992)
99. W. Wernsdorfer, R. Sessoli, Science **284**, 133 (1999)
100. D. Loss, D.P. DiVincenzo, G. Grinstein, Phys. Rev. Lett. **69**, 3232 (1992)
101. A. Garg, Europhys. Lett. **22**, 205 (1993)
102. I.S. Tupitsyn, N.V. Prokof'ev, P.C.E. Stamp, Int. J. Mod. Phys. B **11**, 2901 (1997). See also, I. Tupitsyn, JETP Lett., **67**, 28 (1998)
103. B. Barbara, Philos. Trans. R. Soc. Lond. A **370**, 30 (2012)
104. P.C.E. Stamp, in *Proceedings of the International Workshop Quantum and Classical Spin Manipulation*, Les Houches, France, ed. by B. Barbara (2005)
105. N.V. Prokof'ev, P.C.E. Stamp, in *Quantum Tunneling of Magnetization—QTM'94*, ed. by L. Gunther, B. Barbara. NATO ASI Series E: Applied Science, vol. 301 (Kluwer, Dordrecht, 1995), p. 347, 373. See also, N.V. Prokof'ev, P.C.E. Stamp, J. Phys. Condens. Matter **5**, L663 (1993)
106. A. Morello, P.C.E. Stamp, I. Tupitsyn, Phys. Rev. Lett. **97**, 207205 (2006)
107. C. Zener, Proc. R. Soc. Lond. Ser. A, Math. Phys. Sci. **137**, 696 (1932)
108. L.D. Landau, E.M. Lifshitz, *Quantum Mechanics* (Pergamon, Oxford, 1965)
109. S. Miyashita, J. Phys. Soc. Jpn. **64**, 3207 (1995)
110. V.V. Dobrovitski, A.K. Zvezdin, Europhys. Lett. **38**, 377 (1997)
111. L. Gunther, Europhys. Lett. **39**, 1 (1997)
112. J.J. Alonso, J.F. Fernández, Phys. Rev. Lett. **87**, 097205 (2001)
113. J.F. Fernández, J.J. Alonso, Phys. Rev. Lett. **91**, 047202 (2003). See also, comment by I.S. Tupitsyn, P.C.E. Stamp, Phys. Rev. Lett. **92**, 119701 (2004) and J.F. Fernández, J.J. Alonso's reply Phys. Rev. Lett. **92**, 119702 (2004)
114. J.F. Fernández, J.J. Alonso, Phys. Rev. B **69**, 024411 (2004)
115. I.S. Tupitsyn, P.C.E. Stamp, N.V. Prokof'ev, Phys. Rev. B **69**, 132406 (2004). See also, comment by J.J. Alonso, J.F. Fernández, Phys. Rev. B **72**, 026401 (2005) and reply by I.S. Tupitsyn, P.C.E. Stamp, N.V. Prokof'ev, Phys. Rev. B **72**, 026402 (2005)
116. A. Würger, Phys. Rev. Lett. **81**, 212 (1998). Ibid., J. Phys. Condens. Matter **10**, 44 (1998)
117. J. Villain, F. Hartmann-Bourtron, R. Sessoli, A. Rettori, Europhys. Lett. **27**, 159 (1994). See also, P. Politi, A. Rettori, F. Hartmann-Boutron, J. Villain, Phys. Rev. Lett. **75**, 537 (1995); See also, F. Hartmann-Boutron, P. Politi, J. Villain, Int. J. Mod. Phys. **10**, 2577 (1996); See also, A. Fort, A. Rettori, J. Villain, D. Gatteschi, R. Sessoli, Phys. Rev. Lett. **80**, 612 (1998)
118. M.N. Leuenberger, D. Loss, Europhys. Lett. **46**, 692 (1999)
119. W. Wernsdorfer, T. Ohm, C. Sangregorio, R. Sessoli, D. Mailly, C. Paulsen, Phys. Rev. Lett. **82**, 3903 (1999)
120. T. Ohm, C. Sangregorio, C. Paulsen, Eur. Phys. J. B **6**, 195 (1998). Ibid., J. Low Temp. Phys. **113**, 1141 (1998)
121. W. Wernsdorfer, R. Sessoli, D. Gatteschi, Europhys. Lett. **47**(2), 254 (1999)
122. E. Burzurí, F. Luis, B. Barbara, R. Ballou, E.L. Ressouche, O. Montero, J. Campo, S. Maegawa, Phys. Rev. Lett. **107**, 097201 (2011)
123. J. Magariño, J. Tuchendler, P. Beauvillain, I. Laursen, Phys. Rev. B **13**, 2805 (1976)
124. B. Barbara, R. Giraud, W. Wernsdorfer, D. Mailly, A.M. Tkachuk, P. Lejay, H. Susuki, J. Magn. Magn. Mater. **272–276**, 1024 (2004)
125. A. Cornia, R. Sessoli, L. Sorace, D. Gatteschi, A.L. Barra, C. Daiguebonne, Phys. Rev. Lett. **89**, 257201 (2002)
126. I.I. Rabi, Phys. Rev. **51**, 652 (1937)
127. A. Ardavan, O. Rival, J.J.L. Morton, S.J. Blundell, A.M. Tyryshkin, G.A. Timco, R.E.P. Winpenny, Phys. Rev. Lett. **98**, 057201 (2007)
128. H. de Raedt, B. Barbara, S. Miyashita, K. Michielsen, S. Bertaina, S. Gambarelli, Phys. Rev. B **85**, 014408 (2012)

129. Y. Zhang, N.A.W. Holzwarth, R.T. Williams, Phys. Rev. B **57**, 12738 (1998)
130. A.A. Antipin et al., Phys. Solid State **10**, 468 (1968). See also, I. N. Kurkin, L.Ya. Shekun, Fiz. Tverd. Tela **9**, 444 (1967)
131. C. Schlegel, J. van Slageren, M. Manoli, E.K. Brechin, M. Dressel, Phys. Rev. Lett. **101**, 147203 (2008)
132. C.J. Wedge, G.A. Timco, E.T. Spielberg, R.E. George, F. Tuna, S. Rigby, E.J.L. McInnes, R.E.P. Winpenny, S.J. Blundell, A. Ardavan, Phys. Rev. Lett. **108**, 107204 (2012)
133. W. Wernsdorfer, A. Müller, D. Mailly, B. Barbara, Europhys. Lett. **66**(6), 861 (2004)
134. A. Müller, J. Döring, Angew. Chem., Int. Ed. Engl. **27**, 1721 (1988). See also, D. Gatteschi, L. Pardi, A.L. Barra, A. Müller, J. Döring, Nature **354**, 463 (1991)
135. J.H. Shim, S. Bertaina, S. Gambarelli, T. Mitra, A. Müller, E.I. Baibekov, B.Z. Malkin, B. Tsukerblat, B. Barbara, Phys. Rev. Lett. **109**, 050401 (2012)
136. S. Takahashi, I.S. Tupitsyn, J. van Tol, C.C. Beedle, D.N. Hendrickson, P.C.E. Stamp, Nature **476**, 76 (2011)
137. F. Luis, A. Repollés, M.J. Martínez-Pérez, D. Aguilà, O. Roubeau, D. Zueco, P.J. Alonso, M. Evangelisti, A. Camón, J. Sesé, L.A. Barrios, G. Aromí, Phys. Rev. Lett. **107**, 117203 (2011)
138. G. Aromí, D. Aguilà, P. Gamez, F. Luis, O. Roubeau, Chem. Soc. Rev. **41**, 537 (2012)
139. F.Q. Xie, Ch. Obermair, Th. Schimmel, Solid State Commun. **132**, 437 (2004)
140. D. Goldhaber-Gordon, H. Shtrikman, D. Mahalu, D. Abusch-Magder, U. Meirav, M.A. Kastner, Nature **6**, 156 (1998). Ibid., Phys. Rev. Lett. **81**, 5225 (1998)
141. J. Köhler, A.J.M. Disselhorst, M.C.J.M. Donckers, E.J.J. Groenen, J. Schmidt, W.E. Moerner, Nature **363**, 242 (1993)
142. P. Neumann, J. Beck, M. Steiner, F. Rempp, H. Fedder, P.R. Hemmer, J. Wrachtrup, F. Jelezko, Science **329**(599), 542–544 (2010)
143. R. Vincent, S. Klyatskaya, M. Ruben, W. Wernsdorfer, F. Balestro, Nature **488**, 357 (2012)

Chapter 3
Spin Tunneling in Magnetic Molecules That Have Full or Partial Mechanical Freedom

Eugene M. Chudnovsky

Abstract While most of the research on spin tunneling in molecules has focused on crystals of molecular magnets, future experiments may involve molecules loosely attached to a substrate, as well as free magnetic molecules. What would be the effect of the mechanical freedom on spin tunneling? Exact solutions for this set of problems have been recently obtained. They involve anomalous commutation relations for spin and rotational angular momentum in the rotating frame of reference. Application of these findings to magnetic molecules points towards important effect of the mechanical freedom on spin tunneling. In a free molecule the tunneling is prohibited unless the molecule is sufficiently heavy and the tunnel splitting is large.

3.1 Introduction

North and south magnetic poles of a small particle or a molecule can interchange via quantum tunneling [1]. Due to this effect, crystals of magnetic molecules have been shown to exhibit stepwise magnetization curve [2]. The tunneling implies quantum superposition of states characterized by a definite orientation of the magnetic moment. The hope has been expressed that isolated magnetic molecules may one day become elements of quantum computers. Long coherence time of spin states is required for quantum computation. This made researchers think about ways to isolate magnetic molecules from the dissipative environment [3]. Efforts have been made to suspend a single magnetic molecule between conducting leads and deposit magnetic molecules on surfaces [4]. Quantum states of molecules that are partially isolated from the environment must be easier to manipulate. Such molecules, however, may have some degree of mechanical freedom that must be taken into account in the quantum problem.

E.M. Chudnovsky (✉)
Department of Physics and Astronomy, Herbert H. Lehman College, The City University of New York, Bronx, NY 10468-1589, USA
e-mail: Eugene.Chudnovsky@Lehman.CUNY.edu

J. Bartolomé et al. (eds.), *Molecular Magnets*, NanoScience and Technology,
DOI 10.1007/978-3-642-40609-6_3,

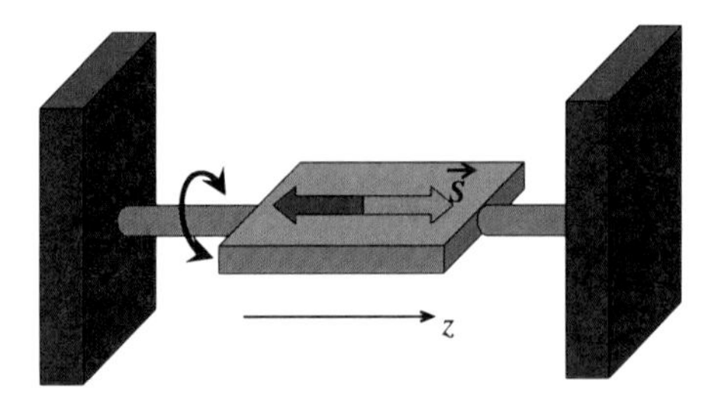

Fig. 3.1 Torsional nano-oscillator with a two-state macrospin

To simplify notations, we shall treat this problem as if magnetism were of spin origin. This is not necessarily true for small atomic clusters or molecules containing rare-earth atoms. In this case the spin **S** in all our expressions must be replaced with the internal angular momentum (due to both spin and orbital electronic states) that is responsible for the magnetic moment of the nanomagnet. An important distinction, however, should be made between that internal angular momentum and the mechanical angular momentum **L** that corresponds to the rotation of the nanomagnet as a whole. The latter, unlike the internal angular momentum, does not contribute to the magnetic moment of the electrically neutral particle. For a free particle one should find the entangled quantum states of the spin and mechanical angular momentum. For a nanomagnet having partial mechanical freedom the quantities of interest are the tunnel splitting Δ and the spin decoherence rate Γ.

The problem is clearly related to the conservation of the total angular momentum [5]. Even before it is applied to a free molecule, one can ask what happens to the total angular momentum when spin 10 of a Mn-12 molecule tunnels between opposite orientations in a crystal. The answer one usually gets is that the spin Hamiltonian for this problem, e.g., $H = -DS_z^2 + dS_y^2$, does not possess the full rotational symmetry and, therefore, it does not conserve the total angular momentum, $\mathbf{J} = \mathbf{S} + \mathbf{L}$. This, however, simply sweeps the problem under the rug because the full Hamiltonian that describes the spin and quantized phonon modes in the solid containing that spin must conserve the total angular momentum. Conservation of angular momentum is responsible for the parameter-independent lower bound [6, 7], $\Gamma = S^2\Delta^5\rho^{3/2}/(12\pi\hbar^4 G^{5/2})$, on the spin decoherence rate in a solid of mass density ρ and shear modulus G. The latter is a measure of the rigidity of the solid with respect to elastic twists, required for the transfer of the angular momentum between the spin and the solid. It can be shown [8] that when $G \to 0$ the tunnel splitting Δ disappears as $\exp(-\mathrm{const}/\sqrt{G})$. Similar effect occurs in a mechanical nanoresonator with a magnetic molecule [9, 10]. This problem, depicted in Fig. 3.1, will be discussed in Sect. 3.2. It turns out that the spin tunnelling is suppressed by the zero-point oscillations of the nanoresonator [11]. The effect becomes more pronounced with decreasing the size and the spring constant of the resonator. Coupling to a light mechanical resonator may also lead to strong decoherence of quantum oscillations of the spin [12].

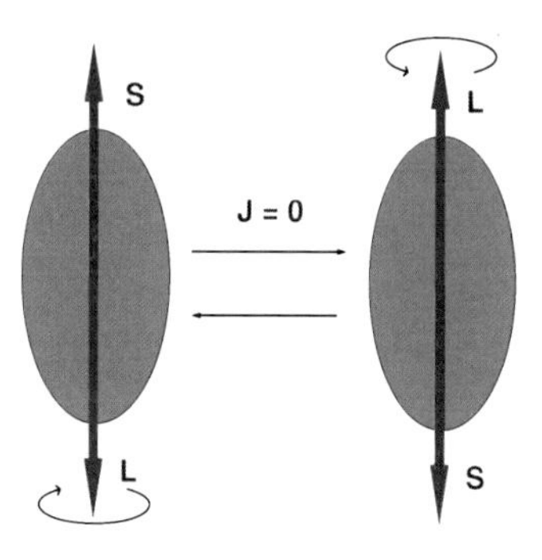

Fig. 3.2 Tunneling between states with zero total angular momentum

For a free magnetic molecule some insight into the importance of the conservation of angular momentum can be obtained by considering a spin **S** embedded in a rigid body that is free to rotate. While large magnetic molecules can hardly be treated as mechanically rigid, this approximation catches the essence of the effect of mechanical rotations. Spin tunneling leads to the ground state that is a superposition of spin-up and spin-down states. It is separated from the first excited state by the tunnel splitting gap Δ. Delocalization in the spin space due to tunneling lowers the energy of the degenerate non-tunnel states by $\Delta/2$. The only problem is that the tunneling of the spin alone violates conservation of the total angular momentum. It may occur, however, in a rotational state with $J = 0$, as is illustrated in Fig. 3.2. In this case, spin tunneling is accompanied by the tunneling of the mechanical angular momentum $L = S$. The corresponding ground state entangles the spin with mechanical rotations. It has a mechanical energy $\hbar^2 L^2/(2I_z) = \hbar^2 S^2/(2I_z)$ (we consider dimensionless S and L), where I_z is the moment of inertia for rotation about the spin quantization (easy magnetization) axis. Whether this state has advantage in energy over the non-rotating state with a frozen direction of the magnetic moment depends on whether the energy gain, $\Delta/2$, due to spin tunneling is greater than the energy loss, $\hbar^2 S^2/(2I_z)$, due to mechanical rotation. In the limit of a macroscopic rigid body, when $I_z \to \infty$, this effect becomes irrelevant and spin tunneling always lowers the energy. For a small moment of inertia, however, the mechanical energy needed to sustain spin tunneling is large and the ground state is a non-rotating state with a frozen orientation of the magnetic moment. In Section 3 we will present a rigorous solution of this problem that shows that the above energy argument misses the critical value of I_z only by a factor of 2.

The question of conservation of angular momentum in spin tunneling was more academic than practical until experimentalists began the effort of isolating magnetic molecules from the environment. Rigorous quantum-mechanical solution for a rotating two-state spin system was obtained only recently, first for the case when rotations were allowed about a fixed axis [13, 14], and then for arbitrary rotations of a symmetric quantum rotator [15]. They provided a framework for treating situations when nanomagnets have partial or full mechanical freedom. We shall consider these problems in the following sections. At the end we will discuss implications of our findings for experiments.

3.2 Nanomechanics of a Two-State Spin System Rotating About a Fixed Axis

3.2.1 *Quantum Mechanics of a Two-State Spin System*

The general form of a spin Hamiltonian is

$$\hat{H}_S = \hat{H}_{\parallel} + \hat{H}_{\perp}, \tag{3.1}$$

where $\hat{H}_{\parallel}$ commutes with S_z and $\hat{H}_{\perp}$ does not. The states $|\pm S\rangle$ are degenerate ground states of $\hat{H}_{\parallel}$, where S is the total spin of the nanomagnet. We are interested in a situation when $\hat{H}_{\perp}$ only slightly perturbs these states, adding to them small contributions from other $|m_S\rangle$ states. We will call these degenerate perturbed states $|\psi_{\pm S}\rangle$. Physically they describe the magnetic moment aligned in one of the two directions along the anisotropy axis. Full perturbation theory that accounts for degeneracy of $\hat{H}_S$ provides quantum tunneling between the $|\psi_{\pm S}\rangle$ states for integer S. The ground state and first excited state are symmetric and antisymmetric combinations of $|\psi_{\pm S}\rangle$, respectively,

$$\begin{aligned} \Psi_+ &= \frac{1}{\sqrt{2}}\big(|\psi_S\rangle + |\psi_{-S}\rangle\big) \\ \Psi_- &= \frac{1}{\sqrt{2}}\big(|\psi_S\rangle - |\psi_{-S}\rangle\big), \end{aligned} \tag{3.2}$$

which satisfy

$$\hat{H}_S \Psi_\pm = E_\mp \Psi_\pm, \tag{3.3}$$

where

$$E_+ - E_- \equiv \Delta. \tag{3.4}$$

The tunnel splitting Δ is generally very small compared to the distance to other spin energy levels, which makes the two-state approximation very accurate at low energies.

It is convenient to describe these lowest energy spin states $\Psi_\pm$ with a pseudospin-1/2. The components of the corresponding Pauli operator $\boldsymbol{\sigma}$ are

$$\begin{aligned} \sigma_x &= |\psi_{-S}\rangle\langle\psi_S| + |\psi_S\rangle\langle\psi_{-S}| \\ \sigma_y &= i|\psi_{-S}\rangle\langle\psi_S| - i|\psi_S\rangle\langle\psi_{-S}| \\ \sigma_z &= |\psi_S\rangle\langle\psi_S| - |\psi_{-S}\rangle\langle\psi_{-S}|. \end{aligned} \tag{3.5}$$

The projection of $\hat{H}_S$ onto $|\psi_{\pm S}\rangle$ states is

$$\hat{H}_\sigma = \sum_{m,n=\psi_{\pm S}} \langle m|\hat{H}_S|n\rangle |m\rangle\langle n|. \tag{3.6}$$

Expressing $|\psi_{\pm S}\rangle$ in terms of $\Psi_{\pm}$ one obtains

$$\langle \psi_{\pm S}|\hat{H}_S|\psi_{\pm S}\rangle = 0, \qquad \langle \psi_{\pm S}|\hat{H}_S|\psi_{\mp S}\rangle = -\frac{\Delta}{2}, \tag{3.7}$$

which gives the two-state Hamiltonian

$$\hat{H}_\sigma = -\frac{\Delta}{2}\sigma_x \tag{3.8}$$

having eigenvalues $\pm\Delta/2$. In the absence of tunneling a classical magnetic moment is localized in the up or down state. It is clear that delocalization of the magnetic moment due to spin tunneling reduces the energy by $\Delta/2$.

3.2.2 Renormalization of the Spin Tunnel Splitting in a Nano-oscillator

We now place the nanomagnet considered in the previous subsection in a torsional oscillator shown in Fig. 3.1, with the quantization (easy magnetization) axis parallel to the axis of mechanical rotations. The full Hamiltonian of such system,

$$\hat{H} = \hat{H}'_S + \hat{H}_{\mathrm{rot}}, \tag{3.9}$$

consists of the spin part, $\hat{H}'_S$, and the mechanical part

$$\hat{H}_{\mathrm{rot}} = \frac{1}{2I_z}\left(\hbar^2 L_z^2 + I_z^2\omega_r^2\phi^2\right). \tag{3.10}$$

Here I_z is the moment of inertia of the oscillator, ω_r is its resonant frequency, and ϕ is the angle of rotation. Operator of the mechanical angular momentum, $L_z = -i\partial/\partial\phi$, satisfies commutation relation

$$[\phi, L_z] = i. \tag{3.11}$$

The subtle point is that $\hat{H}'_S$ in (3.9) is different from $\hat{H}_S$ in (3.1). The latter was written in the coordinate frame rigidly coupled with the axes of the nanomagnet. As the nanomagnet is now allowed to rotate together with the mechanical oscillator, its spin Hamiltonian is given by

$$\hat{H}'_S = \hat{R}\hat{H}_S\hat{R}^{-1}, \tag{3.12}$$

where

$$\hat{R} = e^{-iS_z\phi}. \tag{3.13}$$

This gives

$$\langle \psi_{\pm S} | \hat{H}'_S | \psi_{\pm S} \rangle = 0, \qquad \langle \psi_{\mp S} | \hat{H}'_S | \psi_{\pm S} \rangle = -\frac{\Delta}{2} e^{\pm 2iS\phi} \tag{3.14}$$

for the matrix elements of $\hat{H}'_S$. Noticing that

$$S_z |\psi_{\pm S}\rangle \cong S_z |\pm S\rangle = \pm S |\psi_{\pm S}\rangle, \tag{3.15}$$

it is easy to project Hamiltonian (3.12) onto $\psi_{\pm S}$. A simple calculation yields [13]

$$\hat{H}'_\sigma = -\frac{\Delta}{2}\left[\cos(2S\phi)\sigma_x + \sin(2S\phi)\sigma_y\right] = -\frac{\Delta}{2}\left(e^{-2iS\phi}\sigma_+ + e^{2iS\phi}\sigma_-\right), \tag{3.16}$$

where $\sigma_\pm = \frac{1}{2}(\sigma_x \pm i\sigma_y)$. This expression generalizes (3.8) for the case of $\phi \neq 0$.

Standard quantization of mechanical rotations of the oscillator gives

$$\phi = \sqrt{\frac{\hbar}{2I\omega_r}}\left(a + a^\dagger\right). \tag{3.17}$$

The full Hamiltonian of the system then becomes

$$H = \hbar\omega_r\left(a^\dagger a + \frac{1}{2}\right) - \frac{\Delta}{2}\left[e^{-i\beta(a+a^\dagger)}\sigma_+ + e^{i\beta(a+a^\dagger)}\sigma_-\right], \tag{3.18}$$

where

$$\beta = \sqrt{\frac{2\hbar S^2}{I_z\omega_r}}. \tag{3.19}$$

The simplest case corresponds to $\omega_r \gg \Delta/\hbar$, when the excited states of the mechanical oscillator are separated by the large energy gap from the lowest energy spin states. In this case one can simply average $\hat{H}'_\sigma$ of (3.16) over the ground state of the oscillator to obtain the effective spin Hamiltonian. Noticing that

$$\langle 0|e^{-2iS\phi}|0\rangle = \langle 0|e^{2iS\phi}|0\rangle = \langle 0|e^{i\beta(a+a^\dagger)}|0\rangle = 1 - \frac{1}{2}\beta^2 + \cdots = e^{-\beta^2/2}, \tag{3.20}$$

one obtains

$$\hat{H}^{\text{eff}}_\sigma = -\frac{\Delta}{2}e^{-\beta^2/2}(\sigma_+ + \sigma_-) = -\frac{\Delta_{\text{eff}}}{2}\sigma_x \tag{3.21}$$

with

$$\Delta_{\text{eff}} = \Delta e^{-\beta^2/2}, \tag{3.22}$$

as compared to the Hamiltonian $\hat{H}_\sigma = -\frac{\Delta}{2}\sigma_x$ of (3.8) unperturbed by mechanical rotations.

To have a sense of the renormalization of the spin tunnel splitting by the coupling to the oscillator, one should express ω_r in terms of the spring constant (torsional rigidity) k of the oscillator and its moment of inertia I_z. Writing ω_r as $\sqrt{k/I_z}$ gives

$$\beta = \sqrt{\frac{2\hbar S^2}{\sqrt{kI_z}}}. \tag{3.23}$$

Together with (3.22) this shows that coupling to a macroscopic oscillator with large k and I_z has little effect on spin tunneling. The reduction of I_z at a constant k eventually leads to the exponential freezing of the tunnel splitting, $\Delta_{\text{eff}} \propto \exp(-\text{const}/I_z^{1/2})$.

In-depth study [12] shows that the behavior of the system depends on two dimensionless parameters:

$$\alpha = \frac{2(\hbar S)^2}{I_z \Delta}, \qquad r = \frac{\hbar \omega_r}{\Delta}. \tag{3.24}$$

At large r, the spin once prepared in the state up, oscillates between up and down at a frequency $\Delta_{\text{eff}}/\hbar$. This is easy to see by writing this state as

$$\begin{aligned} \Psi(t) &= \frac{1}{\sqrt{2}}\left(\Psi_+ e^{i\Delta t/(2\hbar)} + \Psi_- e^{-i\Delta t/(2\hbar)}\right) \\ &= \cos\left(\frac{\Delta t}{2\hbar}\right)|\psi_S\rangle + \sin\left(\frac{\Delta t}{2\hbar}\right)|\psi_{-S}\rangle \end{aligned} \tag{3.25}$$

and computing the corresponding expectation value of σ_z:

$$\langle \Psi_t|\sigma_z|\Psi_t\rangle = \cos\left(\frac{t\Delta}{\hbar}\right). \tag{3.26}$$

Numerical solution [12] of the Schrödinger equation with the Hamiltonian (3.18) shows that at small r the coupling to mechanical oscillations produces strong decohering effect on the quantum oscillations of the spin. The typical small r behavior of the spin is shown in Fig. 3.3. At large α the spin tunneling disappears altogether. In the next Section we shall see that this is also true for a free nanomagnet.

3.3 Free Quantum Rotator with a Two-State Macrospin

3.3.1 Anomalous Commutation Relations

The problem of a free quantum rotator with a spin has a natural solution in the rotating coordinate frame that is rigidly coupled with the rotator. In this Subsection, we re-derive some known but largely forgotten facts about commutation relations of the operators of angular momentum in the rotating frame. This treatment applies to

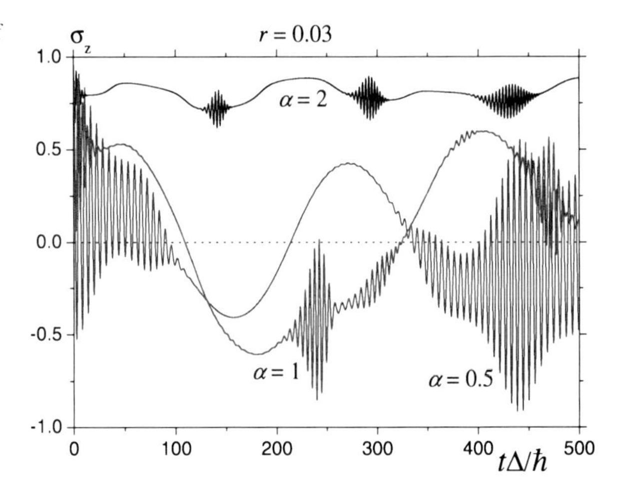

Fig. 3.3 Time dependence of the expectation value of σ_z at different values of α for $r = 0.03$ [12]. At $\alpha = 2$ the spin tunneling is frozen and the spin points in one direction

the general case of mechanical rotations of a quantum system with internal angular momentum degrees of freedom [16]. In our case this internal degree of freedom is the spin, **S**. Starting with the usual commutation relations for the components of **S**, components of the angular momentum of mechanical rotations **L**, and total angular momentum $\mathbf{J} = \mathbf{S} + \mathbf{L}$ in the laboratory frame, we derive commutation relations in the rotating (body) frame. We show that in the body frame all components of **J** and **S** commute, and therefore the corresponding quantum numbers provide good description of the quantum states of the system.

Let the X, Y, Z axes make up the laboratory frame that is fixed in space. The x, y, z axes define the coordinate system that is rigidly coupled to the body, and are directed along its principle moments of inertia. Initially, these two frames coincide. At any other instant the orientation of the rotating xyz frame relative to the fixed XYZ frame is specified by the Euler angles [17] ϕ, θ, ψ. A vector **R** with components R_A (we use uppercase Roman letters to indicate the laboratory frame components) as measured in the fixed XYZ frame can be projected onto the rotating xyz coordinate frame. In the rotating frame, this vector is described by **r** and has components r_α (lowercase Greek indices denote components in the rotating frame). The transformation $\mathbf{r} = \mathbf{CR}$ is given by the rotation matrix **C**,

$$\mathbf{C} = \begin{pmatrix} \lambda_{xX} & \lambda_{xY} & \lambda_{xZ} \\ \lambda_{yX} & \lambda_{yY} & \lambda_{yZ} \\ \lambda_{zX} & \lambda_{zY} & \lambda_{zZ} \end{pmatrix} \tag{3.27}$$

where the $\lambda_{\alpha A}$ are direction cosines between laboratory frame and rotating frame axes.

The total angular momentum **J** obeys the usual commutation relations in the laboratory XYZ frame,

$$[J_A, J_B] = i\epsilon_{ABC} J_C, \tag{3.28}$$

where ϵ_{ABC} is the fully antisymmetric Levi-Civita tensor (summation over repeated indices is implicit throughout this chapter). In the rotating xyz frame, the total angular momentum has components

$$J_\alpha = \lambda_{\alpha A} J_A \tag{3.29}$$

and the sign of i in the commutation relation is reversed,

$$[J_\alpha, J_\beta] = -i\epsilon_{\alpha\beta\gamma} J_\gamma. \tag{3.30}$$

The components of the mechanical angular momentum can be resolved in either frame, or in terms of the Euler angles ϕ, θ, ψ. The operator forms of the corresponding angular momenta are

$$p_\phi = -i\hbar\frac{\partial}{\partial\phi}, \qquad p_\theta = -i\hbar\frac{\partial}{\partial\theta}, \qquad p_\psi = -i\hbar\frac{\partial}{\partial\psi} \tag{3.31}$$

which mutually commute. The rotational angular momentum operators can be projected onto the laboratory frame coordinate system,

$$\begin{aligned} L_X &= -\cot\theta\cos\phi p_\phi - \sin\phi p_\theta + \csc\theta\cos\phi p_\psi \\ L_Y &= -\cot\theta\sin\phi p_\phi + \cos\phi p_\theta + \csc\theta\sin\phi p_\psi \\ L_Z &= p_\phi, \end{aligned} \tag{3.32}$$

or the body frame coordinate system,

$$\begin{aligned} L_x &= -\csc\theta\cos\psi p_\phi + \sin\psi p_\theta + \cot\theta\cos\psi p_\psi \\ L_y &= \csc\theta\sin\psi p_\phi + \cos\psi p_\theta - \cot\theta\sin\psi p_\psi \\ L_z &= p_\psi. \end{aligned} \tag{3.33}$$

The commutation relations can be obtained by direct calculation, with the laboratory frame components satisfying

$$[L_A, L_B] = i\epsilon_{ABC} L_C, \tag{3.34}$$

while the rotating frame components obey

$$[L_\alpha, L_\beta] = -i\epsilon_{\alpha\beta\gamma} L_\gamma. \tag{3.35}$$

The spin obeys the same regular commutation relations in either frame. To show this, we define the spin components in the laboratory frame where

$$[S_A, S_B] = i\epsilon_{ABC} S_C. \tag{3.36}$$

Using the fact that $[S_A, L_B] = 0$, it is easy to see that the components of $\mathbf{J} = \mathbf{L} + \mathbf{S}$ in the laboratory frame satisfy the commutation relation given by (3.28). Projecting this spin onto the rotating axes, $S_\alpha = \lambda_{\alpha A} S_A$, and noticing that $[S_A, \lambda_{\beta B}] = 0$, we obtain

$$[S_\alpha, S_\beta] = \lambda_{\alpha A}\lambda_{\beta B}[S_A, S_B] = i\epsilon_{ABC}\lambda_{\alpha A}\lambda_{\beta B} S_C = i\epsilon_{\alpha\beta\gamma}\lambda_{\gamma C} S_C = i\epsilon_{\alpha\beta\gamma} S_\gamma. \tag{3.37}$$

The relation

$$\epsilon_{ABC}\lambda_{\alpha A}\lambda_{\beta B} = \epsilon_{\alpha\beta\gamma}\lambda_{\gamma C} \tag{3.38}$$

follows from the fact that for a special orthogonal matrix any element is equal to its cofactor. In order to obtain the same sign for all angular momenta in the rotating frame, we define reversed spin $\mathbf{S} \to \tilde{S} = -S$, giving

$$[\tilde{S}_\alpha, \tilde{S}_\beta] = -i\epsilon_{\alpha\beta\gamma}\tilde{S}_\gamma. \tag{3.39}$$

Now we may write $\mathbf{J} = \mathbf{L} - \tilde{\mathbf{S}}$, and the components of $\mathbf{J}$ satisfy the anomalous commutation relations, (3.30).

Alternatively, the commutation relation for the total angular momentum in the lab frame can be calculated directly. Using the fact that the directional cosines transform according to

$$[\lambda_{\alpha A}, J_B] = [\lambda_{\alpha A}, L_B] = i\epsilon_{ABC}\lambda_{\alpha C}, \tag{3.40}$$

and that the spin and rotational angular momenta do not commute in the rotating frame

$$[\tilde{S}_\alpha, L_\beta] = -\lambda_{\beta B}[\lambda_{\alpha A}, L_B] S_A = -i\epsilon_{\alpha\beta\gamma}\tilde{S}_\gamma \tag{3.41}$$

gives

$$\begin{aligned} [J_\alpha, J_\beta] &= [L_\alpha, L_\beta] - [L_\alpha, \tilde{S}_\beta] - [\tilde{S}_\alpha, L_\beta] + [\tilde{S}_\alpha, \tilde{S}_\beta] \\ &= -i\epsilon_{\alpha\beta\gamma}(L_\gamma + \tilde{S}_\gamma) = -i\epsilon_{\alpha\beta\gamma} J_\gamma. \end{aligned} \tag{3.42}$$

Similarly, we can show that

$$[J_\alpha, \tilde{S}_\beta] = [L_\alpha, \tilde{S}_\beta] - [\tilde{S}_\alpha, \tilde{S}_\beta] = 0, \tag{3.43}$$

allowing us to simultaneously choose quantum numbers corresponding to both $\mathbf{J}$ and $\tilde{\mathbf{S}}$.

3.3.2 Rotating Two-State Spin System

The full Hamiltonian of a rotating nanomagnet is given by the sum of the rotational energy and magnetic anisotropy energy

$$\hat{H} = \frac{\hbar^2 L_x^2}{2I_x} + \frac{\hbar^2 L_y^2}{2I_y} + \frac{\hbar^2 L_z^2}{2I_z} + \hat{H}_{\tilde{S}}. \tag{3.44}$$

Note that the mechanical part and the spin part of this Hamiltonian are not independent because in the body frame the operators $\mathbf{L}$ and $\tilde{\mathbf{S}}$ do not commute, $[L_i, \tilde{S}_j] = -i\epsilon_{ijk}\tilde{S}_k$. It, therefore, makes sense to express the above Hamiltonian in terms of the commuting body-frame operators of the total angular momentum $\mathbf{J}$ and reversed spin $\tilde{\mathbf{S}}$:

$$\hat{H} = \frac{\hbar^2}{2}\left(\frac{J_x^2}{I_x} + \frac{J_y^2}{I_y} + \frac{J_z^2}{I_z}\right) + \frac{\hbar^2}{2}\left(\frac{\tilde{S}_x^2}{I_x} + \frac{\tilde{S}_y^2}{I_y} + \frac{\tilde{S}_z^2}{I_z}\right)$$
$$+ \hbar^2\left(\frac{J_x\tilde{S}_x}{I_x} + \frac{J_y\tilde{S}_y}{I_y} + \frac{J_z\tilde{S}_z}{I_z}\right) + \hat{H}_{\tilde{S}}. \tag{3.45}$$

For a symmetric rigid rotor with $I_x = I_y$ this Hamiltonian reduces to

$$\hat{H} = \frac{\hbar^2\mathbf{J}^2}{2I_x} + \frac{\hbar^2 J_z^2}{2}\left(\frac{1}{I_z} - \frac{1}{I_x}\right) + \hbar^2\left(\frac{J_x\tilde{S}_x + J_y\tilde{S}_y}{I_x} + \frac{J_z\tilde{S}_z}{I_z}\right) + \hat{\hat{H}}_{\tilde{S}}, \tag{3.46}$$

where

$$\hat{\hat{H}}_{\tilde{S}} = \hat{H}_{\tilde{S}} + \frac{\hbar^2}{2}\left(\frac{1}{I_z} - \frac{1}{I_x}\right)\tilde{S}_z^2 + \frac{\hbar^2\tilde{\mathbf{S}}^2}{2I_x}. \tag{3.47}$$

The last term in $\hat{\hat{H}}_{\tilde{S}}$ is an unessential constant, $\hbar^2 S(S+1)/(2I_x)$.

The second term in (3.47) provides renormalization of the crystal field in a freely rotating particle. For, e.g., the biaxial spin Hamiltonian, $\hat{H}_S = -DS_z^2 + dS_y^2$, one has

$$D \to D_{\text{eff}} = D - \frac{\hbar^2}{2}\left(\frac{1}{I_z} - \frac{1}{I_x}\right). \tag{3.48}$$

Since Δ/D scales as $(d/D)^S$, this leads to the renormalization of Δ:

$$\Delta_{\text{eff}} = \Delta\left[1 - \frac{\hbar^2}{2I_z D}(1-\lambda)\right]^{1-S}, \tag{3.49}$$

where we have introduced the aspect ratio for the moments of inertia,

$$\lambda = \frac{I_z}{I_x}. \tag{3.50}$$

The range of λ for a symmetric rotator is $0 \le \lambda \le 2$. (For, e.g., a symmetric ellipsoid with semiaxes $a = b \neq c$, one has $\lambda = 2a^2/(a^2 + c^2)$.) Depending on the shape of the rotator, the tunnel splitting can therefore increase or decrease. This effect is typically small [14].

Projection of (3.46) on the two spin states along the lines of the previous Section gives

$$\hat{H} = \frac{\hbar^2\mathbf{J}^2}{2I_x} + \frac{\hbar^2 J_z^2}{2}\left(\frac{1}{I_z} - \frac{1}{I_x}\right) - \frac{\Delta}{2}\sigma_x - \frac{\hbar^2 S}{I_z}J_z\sigma_z. \tag{3.51}$$

where we have used

$$\langle\psi_{\pm S}|S_z|\psi_{\pm S}\rangle = \pm S, \qquad \langle\psi_{\pm S}|S_{x,y}|\psi_{\pm S}\rangle = 0. \tag{3.52}$$

We construct eigenstates of this Hamiltonian according to

$$|\Psi_{JK}\rangle = \frac{1}{\sqrt{2}}\left(C_{\pm S}|\psi_S\rangle \pm C_{\mp S}|\psi_{-S}\rangle\right)|JK\rangle \tag{3.53}$$

where

$$\begin{aligned} \mathbf{J}^2|JK\rangle &= J(J+1)|JK\rangle, \quad J = 0, 1, 2, \ldots \\ J_z|JK\rangle &= K|JK\rangle, \quad K = -J, \ldots, J. \end{aligned} \tag{3.54}$$

Solution of $\hat{H}|\Psi_{JK}\rangle = E|\Psi_{JK}\rangle$ gives energy levels as

$$E_{JK}^{(\pm)} = \frac{\hbar^2 J(J+1)}{2I_x} + \frac{\hbar^2 K^2}{2}\left(\frac{1}{I_z} - \frac{1}{I_x}\right) \pm \sqrt{\left(\frac{\Delta}{2}\right)^2 + \left(\frac{\hbar^2 KS}{I_z}\right)^2}, \tag{3.55}$$

The upper (lower) sign in (3.55) corresponds to the lower (upper) sign in (3.53). For $K \neq 0$ each state is degenerate with respect to the sign of K. For $K = 0, 1, 2, \ldots$ the coefficients in (3.53) are given by

$$C_{\pm} = \sqrt{1 \pm \alpha K/\sqrt{S^2 + (\alpha K)^2}}, \tag{3.56}$$

with α given by (3.24).

3.3.3 Ground State

Minimization of the energy in (3.55) with respect to J, taking into account the fact that J cannot be smaller than K, immediately yields $J = K$, that is, the ground state always corresponds to the maximal projection of the total angular momentum onto the spin quantization axis. In semiclassical terms this means that the minimal energy states in the presence of spin tunneling always correspond to mechanical rotations about the magnetic anisotropy axis. This is easy to understand by noticing that the sole reason for mechanical rotation is the necessity to conserve the total angular momentum while allowing spin tunneling to lower the energy. To accomplish this the magnetic particle needs to oscillate between clockwise and counterclockwise rotations about the spin quantization axis in unison with the tunneling spin. If such mechanical oscillation costs more energy than the energy gain from spin tunneling, then both spin tunneling and mechanical motion must be frozen in the ground state as, indeed, happens in very light particles (see below). Rotations about axes other than the spin quantization axis can only increase the energy and, thus, should be absent in the ground state.

For further analysis it is convenient to write (3.55) in the dimensionless form,

$$\frac{E_{JK}^{(\pm)}}{\Delta} = \frac{\alpha}{4}\left[\frac{J(J+1)-K^2}{S^2}\lambda + \frac{K^2}{S^2}\right] \pm \frac{1}{2}\sqrt{1+\frac{K^2}{S^2}\alpha^2}. \tag{3.57}$$

For a given λ, as α increases the ground state switches from $J=0$ to higher J when

$$E_{00}^{(-)}\left[\alpha_J^0(\lambda)\right] = E_{JJ}^{(-)}\left[\alpha_J^0(\lambda)\right]. \tag{3.58}$$

Solution of this equation for $\alpha_J^0(\lambda)$ gives

$$\alpha_J^0 = \frac{(2S)^2(J+\lambda)}{J[(2S)^2-(J+\lambda)^2]}. \tag{3.59}$$

This first transition occurs for the smallest value of $\alpha_J^0(\lambda)$ and the transition is from $J=0$ to the corresponding critical value, J_c. For $\alpha < \alpha_{J_c}^0$ the ground state corresponds to $J=0$ and $C_{\pm S}=1$. After the first transition from $J=0$ to $J=J_c$, the ground state switches to sequentially higher J at values of α which satisfy

$$E_{J-1\,J-1}^{(-)}\left[\alpha_J(\lambda)\right] = E_{JJ}^{(-)}\left[\alpha_J(\lambda)\right]. \tag{3.60}$$

Solution of this equation for $\alpha_J(\lambda)$ gives

$$\alpha_J = \frac{(2S)^2 T(J,\lambda)}{\sqrt{(2S)^2(2J-1)^2 - T(J,\lambda)^2}\sqrt{(2S)^2 - T(J,\lambda)^2}}, \tag{3.61}$$

with

$$T(J,\lambda) = 2J - 1 + \lambda. \tag{3.62}$$

The critical α_J has poles at $\lambda = 2(S-J)+1$. For $\lambda \geq 1$ there is no longer a ground state transition to $J=S$, even for very large values of α.

Because the ground state is completely determined by the parameters α and λ, we can depict the ground state behavior in a quantum phase diagram shown in Fig. 3.4. The curves separate areas in the (α, λ) plane that correspond to different values of J and different values of the magnetic moment. The latter is due entirely to the spin of the nanomagnet (or the internal angular momentum in place of the spin), as L_z represents mechanical motion of the nanomagnet as a whole, and not electronic orbital angular momentum. Thus,

$$\mu = -g\mu_B\langle\Psi_{JK}|S_z|\Psi_{JK}\rangle = -g\mu_B S\frac{\alpha K}{\sqrt{S^2+(\alpha K)^2}}. \tag{3.63}$$

Here g is the spin gyromagnetic factor, and the minus sign reflects the negative gyromagnetic ratio $\gamma = -g\mu_B/\hbar$.

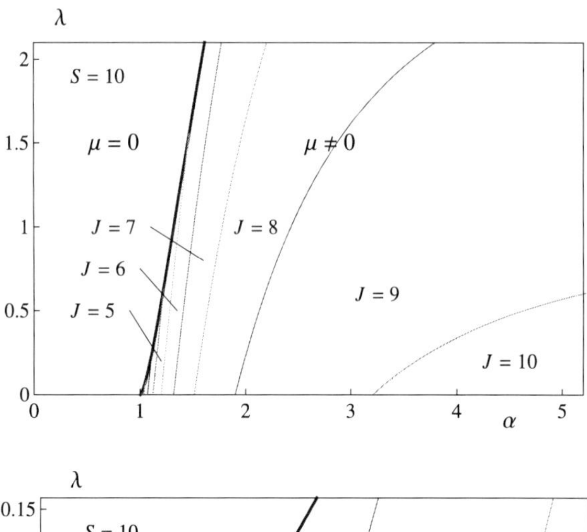

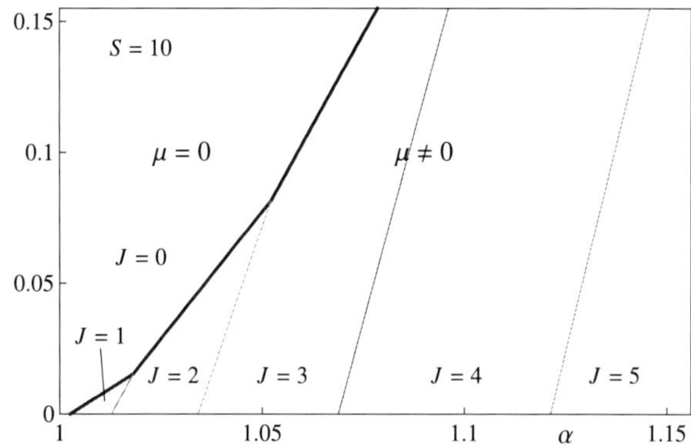

Fig. 3.4 Quantum phase diagram for the total angular momentum J and magnetic moment μ of a symmetric rotator with spin 10, ratio of the principal moments of inertia λ, and magneto-mechanical ratio α [15]. *The lower picture* shows the fine structure of the diagram near the first critical α

3.4 Conclusions

In free magnetic molecules like Mn-12 or Fe-8 spin tunneling should be completely frozen. Indeed, according to (3.24) the spin tunneling in a free rotor can occur only if $\alpha = 2\hbar^2 S^2/(I_z \Delta)$ is not too large compared to one. For the moment of inertia $I_z \sim 10^{-42}$ kg m^2, which is in the right ball park for Mn-12 and Fe-8 spin-10 molecules, this requires Δ of order 0.1 K. The natural tunnel splitting in Mn-12 and Fe-8 is much smaller. Of course, these molecules can hardly be treated as rigid rotators. It is, obvious, however, that their "softness" can only further impede the transfer of the angular momentum between spin and mechanical degrees of freedom that is needed for spin tunneling.

For a magnetic molecule attached to a surface by some kind of a molecular leg, another relevant parameter is the torsional rigidity of the leg k, see (3.22) and (3.23). For $I_z \sim 10^{-42}$ kg m^2 the value of $\beta^2/2$ in $\Delta_{\text{eff}} = \Delta \exp(-\beta^2/2)$ becomes of order unity at $k < 10^{-22}$ N m ($\omega_r < 10^{10}$ s^{-1}). At such values of k, that correspond to a loose connection with the surface, spin tunneling will be strongly suppressed. At $\omega_r < \Delta/\hbar$ significant decoherence of spin states will occur as well. For comparison, the renormalization of Δ in a Mn-12 molecule attached to a carbon nanotube [3]

($k \sim 10^{-18}$ N m) must be very small. These effects should be taken into account when designing qubits based upon molecular magnets.

Acknowledgements Research on spin-rotation coupling reviewed in this Chapter was conducted jointly with Professor Dmitry Garanin. Our students Reem Jaafar and Michael O'Keeffe participated in obtaining some of the results. This work has been supported by the research grant No. DMR-1161571 from the US National Science Foundation.

References

1. E.M. Chudnovsky, J. Tejada, *Macroscopic Quantum Tunneling of the Magnetic Moment* (Cambridge University Press, Cambridge, 1998)
2. J.R. Freedman, M.P. Sarachik, J. Tejada, R. Ziolo, Phys. Rev. Lett. **76**, 3830 (1996)
3. L. Bogani, W. Wernsdorfer, Nat. Mater. **7**, 179 (2008)
4. M. Mannini, F. Pineider, C. Danieli, F. Totti, L. Sorace, Ph. Sainctavit, M.-A. Arrio, E. Otero, L. Joly, J.C. Cezar, A. Cornia, R. Sessoli, Nature **468**, 417 (2010)
5. E.M. Chudnovsky, Phys. Rev. Lett. **72**, 3433 (1994)
6. E.M. Chudnovsky, Phys. Rev. Lett. **92**, 120405 (2004)
7. E.M. Chudnovsky, D.A. Garanin, R. Schilling, Phys. Rev. B **72**, 094426 (2005)
8. E.M. Chudnovsky, M.F. O'Keeffe, D.A. Garanin, J. Supercond. Nov. Magn. **25**, 1007 (2012)
9. R. Jaafar, E.M. Chudnovsky, Phys. Rev. Lett. **102**, 227202 (2009)
10. R. Jaafar, E.M. Chudnovsky, D.A. Garanin, Europhys. Lett. **89**, 27001 (2010)
11. A.A. Kovalev, L.X. Hayden, G.E.W. Bauer, Y. Tserkovnyak, Phys. Rev. Lett. **106**, 147203 (2011)
12. D.A. Garanin, E.M. Chudnovsky, Phys. Rev. X **1**, 011005 (2011)
13. E.M. Chudnovsky, D.A. Garanin, Phys. Rev. B **81**, 214423 (2010)
14. M.F. O'Keeffe, E.M. Chudnovsky, Phys. Rev. B **83**, 092402 (2011)
15. M.F. O'Keeffe, E.M. Chudnovsky, D.A. Garanin, J. Magn. Magn. Mater. **324**, 2871 (2012)
16. J.H. Van Vleck, Rev. Mod. Phys. **23**, 213 (1951)
17. H. Goldstein, C.P. Poole, J.L. Safko, *Classical Mechanics* (Addison-Wesley, Reading, 2001)

Chapter 4
A Microscopic and Spectroscopic View of Quantum Tunneling of Magnetization

Junjie Liu, Enrique del Barco, and Stephen Hill

Abstract This chapter takes a microscopic view of quantum tunneling of magnetization (QTM) in single-molecule magnets (SMMs), focusing on the interplay between exchange and anisotropy. Careful consideration is given to the relationship between molecular symmetry and the symmetry of the spin Hamiltonian that dictates QTM selection rules. Higher order interactions that can modify the usual selection rules are shown to be very sensitive to the exchange strength. In the strong coupling limit, the spin Hamiltonian possesses rigorous D_{2h} symmetry (or C_∞ in high-symmetry cases). In the case of weaker exchange, additional symmetries may emerge through mixing of excited spin states into the ground state. Group theoretic arguments are introduced to support these ideas, as are extensive results of magnetization hysteresis and electron paramagnetic resonance measurements.

4.1 Spin Hamiltonian

The concept of an effective spin-Hamiltonian involving only spin variables has been employed in the study of paramagnetic species for well over half a century. This formalism is particularly suited to the study of transition metal complexes in which the ground state is very often an orbital singlet that is well isolated from excited orbital states due to the strong influence of the ligand field [1]. The ground state multiplicity is determined entirely by the spin state of the ion in this situation, even though the ground wave functions are not exact eigenstates of $\hat{\mathbf{S}}^2$ because of residual spin-orbit (SO) coupling. It is this coupling that gives rise to the familiar anisotropic zero-field

J. Liu
Department of Physics, University of Florida, Gainesville, FL 32611, USA
e-mail: jliu@magnet.fsu.edu

E. del Barco (✉)
Department of Physics, University of Central Florida, Orlando, FL 32816, USA
e-mail: delbarco@physics.ucf.edu

S. Hill
Department of Physics and National High Magnetic Field Laboratory, Florida State University, Tallahassee, FL 32310, USA
e-mail: shill@magnet.fsu.edu

J. Bartolomé et al. (eds.), *Molecular Magnets*, NanoScience and Technology,
DOI 10.1007/978-3-642-40609-6_4, © Springer-Verlag Berlin Heidelberg 2014

splitting (zfs) and Zeeman terms in the resultant spin Hamiltonian, parameterized by the zfs $\overleftrightarrow{D}$ and Lande $\overleftrightarrow{g}$-tensors.

4.1.1 Giant-Spin Approximation Hamiltonian

The magnetic moment of a typical polynuclear transition metal cluster is determined by the exchange interactions between the spins associated with the constituent ions. As detailed in this chapter, there are a number of ways to extend the spin Hamiltonian formalism to this multi-ion situation. By far the simplest is the so-called Giant Spin Approximation (GSA), in which one assigns a total (giant) spin quantum number, S, to the lowest-lying (m_s) magnetic levels [2]; for a ferromagnetic molecule, S is obtained from the algebraic sum of the spin values associated with each of the ions. If the exchange coupling within the molecule is large in comparison to the single-ion zfs interactions, then this ground spin multiplet will be well separated from excited spin states. One may then employ a GSA Hamiltonian to describe the magnetic properties of the molecule, provided that the temperature is sufficiently low that excited spin states are not thermally populated.

A series expansion in terms of the spin component operators $\hat{S}_x$, $\hat{S}_y$, and $\hat{S}_z$, employing so-called Extended Stevens operators, results in the following effective zfs Hamiltonian [3–5]:

$$\hat{H}_{\mathrm{zfs}} = \sum_{p}^{2S} \sum_{q=0}^{p} B_p^q \hat{O}_p^q, \tag{4.1}$$

where $\hat{O}_p^q(\hat{S}_x, \hat{S}_y, \hat{S}_z)$ represent the operators, and B_p^q the associated phenomenological (or effective) zfs parameters. The subscript, p, denotes the order of the operator, which must be even due to the time reversal invariance of the SO interaction; the order is also limited by the total spin, S, of the molecule such that $p \leq 2S$. The superscript, q ($\leq p$), denotes the rotational symmetry of the operator about the z-axis. Equation (4.1) has been employed with great success in the study of single-molecule magnets (SMMs), particularly in terms of describing low-temperature quantum tunneling of magnetization (QTM) behavior and electron paramagnetic resonance (EPR) data [2]. In fact, (4.1) has even been applied quite successfully in cases where the ground spin multiplet is not so well isolated from excited spin states [6–9]. Fourth and higher order operators are often found to be important in these cases. This chapter examines the microscopic origin of these terms, which are often negligible for the constituent ions. However, the usually dominant 2nd-order zfs interaction is first considered.

4.1.1.1 Second Order Anisotropy

Although fourth and higher order Stevens operators ($p \geq 4$) are allowed for a molecule with $S \geq 2$, it is not always necessary to include all of them (up to

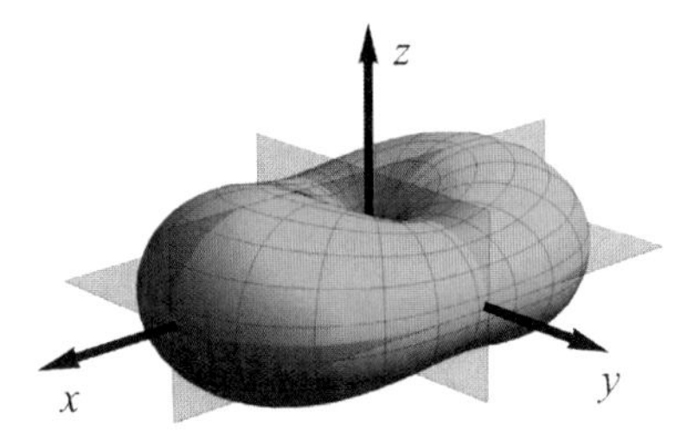

Fig. 4.1 Potential energy surface corresponding to the 2nd-order anisotropy tensor. The surface is generated employing (4.2) with $|E/D| = 1/5$ and $D < 0$. The radial distance to the surface represents the energy of a spin as a function of its orientation

$p = 2S$). For SMMs comprised of transition metal ions, the low-energy physics is usually dominated by 2nd-order SO anisotropy. Therefore, the simplest zero-field GSA Hamiltonian used in characterizing SMMs is often written as:

$$\hat{H}_{\text{zfs}} = D\hat{S}_z^2 + E\left(\hat{S}_x^2 - \hat{S}_y^2\right). \tag{4.2}$$

Equation (4.2) includes only 2nd-order terms, where D $(= 3B_2^0)$ parameterizes the uniaxial anisotropy and E $(= B_2^2)$ the rhombicity. For an approximately uniaxial system, $D\hat{S}_z^2$ is the dominant anisotropy, with z chosen as the quantization axis. In biaxial cases, the ratio between E and D is usually restricted such that $|E/D| < 1/3$; one can always perform a rotation of the coordinate system such that this criterion is satisfied.

One of the main goals of this section is to understand the influence of molecular symmetry on the QTM properties of SMMs. Hence, it is important to examine the symmetry of (4.2) since, strictly speaking, the symmetry of the Hamiltonian should be compatible with the symmetry of the molecule under investigation. In addition, the nature of SO coupling ensures that the spin Hamiltonian be invariant under time-reversal (p is even), i.e., the spin Hamiltonian naturally possesses C_i symmetry. As such, the physics is invariant to inversion of either the total spin moment or the applied field. A classical representation of (4.2) is shown in Fig. 4.1, with $|E| = |D|/5$ and $D < 0$. This graphical representation is obtained by substituting the spin operators in (4.2) by their classical equivalents, as follows:

$$\begin{aligned} \hat{S}_x &\rightarrow S\sin\theta\cos\phi; \\ \hat{S}_y &\rightarrow S\sin\theta\sin\phi; \\ \hat{S}_z &\rightarrow S\cos\theta; \end{aligned} \tag{4.3}$$

where θ and ϕ are the inclination and azimuthal angles in spherical coordinates, respectively. In this representation, the spin is treated as a macroscopic magnetic moment for which all the three components (S_x, S_y and S_z) can be determined simultaneously. The surface shown in Fig. 4.1 represents the energy of the spin as a function of its orientation, where the radial distance to the surface corresponds to its energy.

As can be seen, (4.2) contains the following symmetry elements: (a) three orthogonal C_2 axes, corresponding to x, y and z, and (b) three orthogonal mirror planes, corresponding to the xy, yz, and zx-planes. These symmetry elements, together with the C_i symmetry, give rise to a D_{2h} symmetry for (4.2), which is obviously a much higher symmetry than most real molecules. Consequently, even though (4.2) can often account very well for low-temperature thermodynamic measurements performed on SMMs (e.g. ac susceptibility and magnetization), it may nevertheless fail to explain symmetry-sensitive quantum mechanical (spectroscopic) observables such as QTM steps, Berry-phase interference (BPI) patterns and EPR/neutron spectra.

It should be noted from the preceding discussion that the $B_2^1\hat{O}_2^1$ ($\hat{O}_2^1 \equiv \frac{1}{2}[\hat{S}_x\hat{S}_z + \hat{S}_z\hat{S}_x]$) term was neglected in (4.2), even though it is perfectly allowed within the GSA. With the exception of C_i, the $\hat{O}_2^1$ operator satisfies none of the symmetry operations described in the preceding paragraph. However, such a term is unnecessary, as can be seen when writing the 2nd-order GSA Hamiltonian in the more compact form:

$$\hat{H}_{\text{zfs}} = \hat{\mathbf{S}} \cdot \overleftrightarrow{D} \cdot \hat{\mathbf{S}}, \tag{4.4}$$

where $\overleftrightarrow{D}$ is a 3×3 matrix corresponding to the full 2nd-order anisotropy tensor. In (4.4), D and E are related to the diagonal elements of $\overleftrightarrow{D}$ (see below) while B_2^1 appears as off-diagonal elements. The only restriction on $\overleftrightarrow{D}$ is that it must be Hermitian in order to guarantee the Hamiltonian be Hermitian; indeed, $D_{xz} = D_{zx} = \frac{1}{2}B_2^1$. Consequently, $\overleftrightarrow{D}$ can always be diagonalized by rotating the original Cartesian coordinate frame. Upon doing so, all of the off-diagonal elements of the rotated matrix vanish, i.e., $B_2^1 = 0$ in the new Cartesian coordinate frame. Finally, one may adjust the absolute values of the resultant eigenvalues without altering the symmetry of the Hamiltonian simply by subtracting $\frac{1}{2}(D_{xx} + D_{yy})\overleftrightarrow{I}$ from $\overleftrightarrow{D}$ ($\overleftrightarrow{I}$ is the identity matrix). The zfs Hamiltonian can then be rewritten as (4.2) with

$$D = D_{zz} - \frac{1}{2}(D_{xx} + D_{yy}) \quad \text{and} \quad E = \frac{1}{2}(D_{xx} - D_{yy}), \tag{4.5}$$

where D_{ii} ($i = x, y, z$) refer to components of the diagonalized (rotated) $\overleftrightarrow{D}$ tensor. In other words, (4.4) is equivalent to (4.2), requiring just two parameters, D and E, to completely describe the effective 2nd-order anisotropy within the GSA. Inclusion of $\hat{O}_2^1$ results simply in a rotation of the surface depicted in Fig. 4.1. Consequently, the 2nd-order GSA Hamiltonian necessarily possesses at least D_{2h} symmetry.

The preceding discussion demonstrates something very important: even though 2nd-order terms typically represent the dominant interactions within the GSA description of a SMM, the resultant $\overleftrightarrow{D}$ tensor possesses an artificially high (D_{2h}) symmetry which may not be compatible with the structural symmetry of a particular molecule under investigation. The consequences of this property of the GSA in terms of the resultant QTM will be discussed in detail in the following sections.

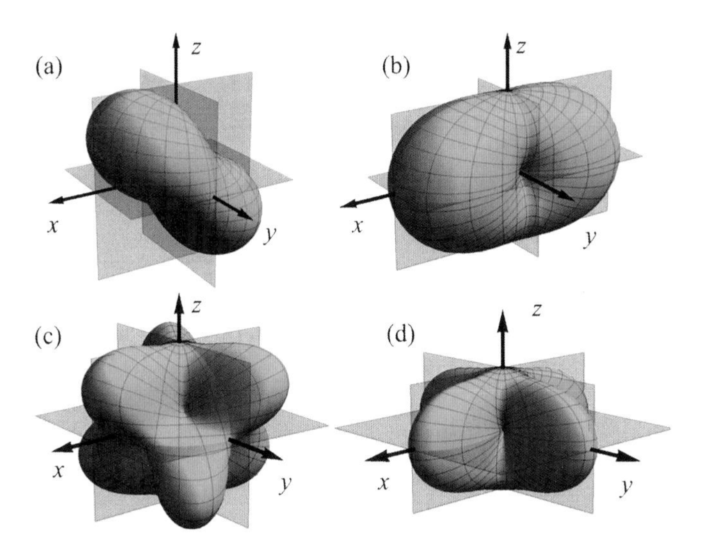

Fig. 4.2 Potential energy surfaces corresponding to the fourth order Stevens operators $\hat{O}_4^1$ (**a**), $\hat{O}_4^2$ (**b**), $\hat{O}_4^3$ (**c**) and $\hat{O}_4^4$ (**d**). As can be seen, the q-even operators include the xy-plane as an extra symmetry element while the q-odd operators include improper rotations (see main text)

4.1.1.2 Higher-Order Anisotropies

For a SMM with $S \geq 2$, Stevens operators of order four (and higher) are allowed in the GSA Hamiltonian—note that p can take on any even value from 2 to $2S$. The values of the 4th-order parameters are often deceptively small, especially for SMMs with large spin values. For example, $|B_4^0/D| \sim 5 \times 10^{-5}$ for the Mn_{12} SMMs, yet the $B_4^0\hat{O}_4^0$ GSA term contributes $\sim 20\,\%$ to the energy barrier. This is due not only to the higher order of $\hat{S}_z$ in $\hat{O}_4^0$, but also because of the way in which the $\hat{O}_4^0$ operator is defined—a multiplier of 35 is associated with S_z^4. In general, the contribution of higher-order terms to the energies of spin states may be expected to be smaller than those of the 2nd-order terms. However, this rule of thumb breaks down in the weak exchange limit (or for particularly high-symmetry molecules [10]); indeed, it is in this limit that one may call into question the validity and/or usefulness of the GSA. Axial ($q = 0$) 4th-order terms lead to a non-parabolic energy barrier, which gives rise to non-even spacings between EPR and QTM resonance fields [11–14]. More importantly, the higher-order transverse ($q \neq 0$) terms introduce additional symmetries into the GSA Hamiltonian, enabling a more precise description of the quantum properties of SMMs.

Figure 4.2 displays the classical energy surfaces corresponding to the 4th-order Stevens operators; the $\hat{O}_4^0$ surface is not shown since it commutes with $\hat{S}_z$ and possesses C_∞ (cylindrical) rotational symmetry. All of the surfaces, and hence the operators, exhibit rotational symmetries which are compatible with the superscript q. However, one may note a systematic difference between the q-odd and q-even operators: the q-even operators have the xy-plane as an additional mirror plane; this is not the case for the q-odd operators. In particular, the lobes of the $\hat{O}_4^2$ and $\hat{O}_4^4$ operators lie within the xy-plane, whereas those of $\hat{O}_4^1$ and $\hat{O}_4^3$ alternately lie above and below this plane. The symmetries of the 4th-order operators can be understood in terms of combinations of rotational symmetries and the intrinsic C_i symmetry of the Hamiltonian. For the q-even operators, the direct product of the rotational and inversion groups leads to: $C_2 \times C_i = C_{2h}$ symmetry for $\hat{O}_4^2$; and $C_4 \times C_i = C_{4h}$ for $\hat{O}_4^4$. Thus, the xy-plane is introduced as a new symmetry element. In contrast, for the q-odd operators, $C_1 \times C_i = C_i$ for $\hat{O}_4^1$ and $C_3 \times C_i = S_6$ for $\hat{O}_4^3$. The resultant symmetry groups corresponding to these operators include an improper rotation (C_i can be treated as the improper rotation S_2). The absence of the xy-mirror plane for the q-odd operators suggests that the molecular hard plane may not coincide with the xy-plane, which leads to several intriguing phenomena described later in this chapter.

The inclusion of $p \geq 4$ Stevens operators in the GSA has a significant influence on the interpretation of QTM measurements. When limited to 2nd-order anisotropy, the zero-field Hamiltonian can only mix spin projection states that differ in m_s by an even number, i.e., $\Delta m_s = |m_{s1} - m_{s2}| = 2n$ ($n =$ integer). This means that only k-even ($k = m_{s1} + m_{s2}$) QTM resonances should be observable for parallel applied fields ($H \parallel z$). Moreover, in molecules for which rhombicity is symmetry forbidden ($E = 0$), a purely 2nd-order Hamiltonian would be cylindrically symmetric. Consequently, for $H \parallel z$, m_s remains an exact quantum number and QTM should be completely forbidden. However, plenty of such high-symmetry molecules exist and are known to exhibit clear QTM behavior, including Mn_{12} and several other SMMs discussed in this chapter. In these situations, it is necessary to include higher order anisotropies in the GSA. The corresponding operators introduce new spin-mixing rules which can lead, for example, to k-odd QTM resonances.

The advantage of the GSA lies in the fact that one can usually restrict the total number of zfs parameters involved in data analysis to just a few by considering the overall symmetry of the molecule under study. Furthermore, the GSA Hilbert space includes only the $2S + 1$ states that belong to the ground spin multiplet, such that the Hamiltonian matrix has dimension $(2S + 1) \times (2S + 1)$. This makes data analysis for large clusters computationally possible. However, the GSA completely ignores the internal degrees of freedom within a molecule, thus completely failing to capture the underlying physics in cases where the total spin fluctuates [15–17]. Moreover, when a molecule possesses very little symmetry (e.g. C_i), the number of GSA zfs parameters cannot be restricted on the basis of symmetry and, in principle, all possible terms (up to $p = 2S$) should be taken into account. In these cases, it may be advantageous to employ a multi-spin Hamiltonian, particularly in situations where microscopic insights are desired.

4.1.2 Multi-Spin Hamiltonian

In the multi-spin (MS) model, a molecule is treated as a cluster of magnetic ions (spins) which are coupled to each other via pairwise exchange interactions. The corresponding zero-field Hamiltonian is:

$$\hat{H}_{\text{zfs}} = \sum_i \hat{\mathbf{s}}_i \cdot \overleftrightarrow{R}_i^T \cdot \overleftrightarrow{d}_i \cdot \overleftrightarrow{R}_i \cdot \hat{\mathbf{s}}_i + \sum_{i<j} \hat{\mathbf{s}}_i \cdot \overleftrightarrow{J}_{i,j} \cdot \hat{\mathbf{s}}_j, \tag{4.6}$$

where $\hat{\mathbf{s}}_i$ represents the spin operator of the ith ion, and $\overleftrightarrow{d}_i$ is the 2nd-order zfs tensor associated with this same ion; lowercase symbols are used here to differentiate parameters/variables employed in both models, i.e., lowercase $\equiv$ MS and uppercase $\equiv$ GSA. For the sake of simplicity, $\overleftrightarrow{d}_i$ is written in the diagonal form:

$$\overleftrightarrow{d} = \begin{bmatrix} e_i & 0 & 0 \\ 0 & -e_i & 0 \\ 0 & 0 & d_i \end{bmatrix}, \tag{4.7}$$

where the local coordinate frame of $\overleftrightarrow{d}_i$ is chosen to match the local principal anisotropy axes of the ith ion. $\overleftrightarrow{R}_i$ is the Euler matrix, specified by the Euler angles θ_i, ϕ_i and ψ_i, which transforms the local coordinate frame of the ith ion into the molecular coordinate frame. The matrix $\overleftrightarrow{J}_{i,j}$ specifies the exchange interaction between the ith and jth ions. It should be emphasized that all of the parameters in (4.6) should be constrained by the structure of the molecule under study, i.e., the overall symmetry of the Hamiltonian must be compatible with the molecular symmetry.

The MS model captures physics associated with internal molecular degrees of freedom that are not easily understood within the GSA framework. First and foremost, the MS model is capable of describing phenomena in which the total spin of a molecule fluctuates, i.e., it gives the energies of excited spin states in addition to the ground state, and includes the mixing between these states [18]. Secondly, the parameters in the MS Hamiltonian have clear physical significance, i.e., they describe the magnetic properties of the constituent ions and the coupling between them. Moreover, many of these parameters can be independently verified through measurements of related compounds [19]. In contrast, the parameters deduced on the basis of a GSA are purely phenomenological. For example, comparisons between the two models have shown that higher order anisotropies in the GSA arise from the interplay between the local 2nd-order single-ion anisotropy and the magnetic interactions between the ions, leading to mixing of excited spin states into the ground spin multiplet [16, 18, 20]. In other words, the phenomenological $p \geq 4$ zfs parameters are a direct manifestation of physics that goes beyond the GSA. The following section deals with this issue in detail, with a focus on the correlation between QTM behavior and the structural symmetries of real molecules.

4.2 Quantum Tunneling of Magnetization in High-Symmetry Mn_3 Single-Molecule Magnets

The first clear observation of QTM selection rules, i.e., a complete absence of symmetry forbidden resonances [21], was reported for the trigonal SMM

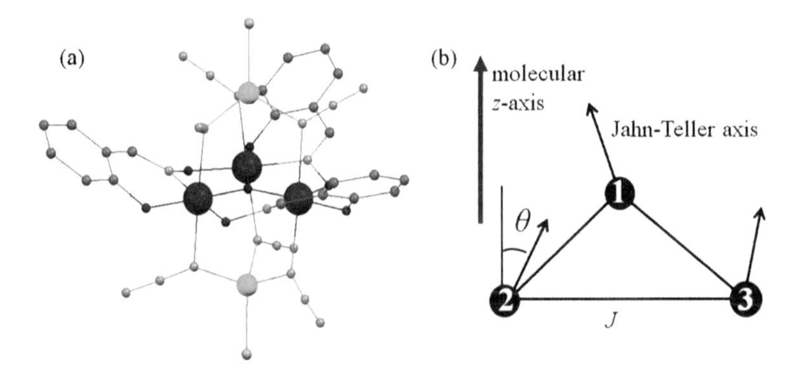

Fig. 4.3 The molecular structure (**a**) and schematic representation of the magnetic core (**b**) of the Mn_3 SMM. Color code: Mn = purple, Zn = green, O = red, N = blue, C = black and Cl = dark gold. H-atoms have been omitted for clarity

$[NE_4]_3[Mn_3Zn_2(salox)_3O(N_3)_6Cl_2]$ (henceforth Mn_3) [22, 23]. This section focuses on QTM in SMMs with trigonal symmetry, emphasizing (i) symmetry-enforced selection rules that allow quantum relaxation in k-odd resonances (Sect. 4.2.2), (ii) the role of disorder (Sect. 4.2.3), and (iii) the microscopic origin of the $B_4^3\hat{O}_4^3$ GSA interaction which is predicted to give rise to unusual BPI patterns (Sect. 4.2.4).

4.2.1 The Mn_3 Single-Molecule Magnet

Several Mn_3 SMMs are known to crystallize in the trigonal space group $R3c$ with racemic mixtures of C_3 symmetric chiral molecules [18, 22, 23]. The structure of the $[NE_4]_3[Mn_3Zn_2(salox)_3O(N_3)_6Cl_2]$ molecule is shown in Fig. 4.3(a). The magnetic core consists of three ferromagnetically coupled Mn^{III} ($s = 2$) ions, which form an equilateral triangle with two Zn^{II} ions located above and below the Mn_3 plane, thus forming a trigonal bipyramidal structure. The C_3 axis of the molecule is perpendicular to the Mn_3 plane, while the local easy-axes of the individual spins are defined by the Jahn-Teller (JT) elongation axes of Mn^{III} ions, which are tilted slightly with respect to the C_3 axis. At low temperatures, the spin $S = 6$ ground state multiplet can be described with the following GSA Hamiltonian:

$$\hat{H} = D\hat{S}_z^2 + B_4^0\hat{O}_4^0 + B_4^3\hat{O}_4^3 + B_6^6\hat{O}_6^6 + \mu_B\vec{B}\cdot\overleftrightarrow{g}\cdot\hat{\mathbf{S}} \tag{4.8}$$

Due to the C_3 symmetry of the molecule, the 2nd order transverse anisotropy term, $E(\hat{S}_x^2 - \hat{S}_y^2)$, is rigorously forbidden. Hence, the leading trigonal ($\hat{O}_4^3$) and hexagonal ($\hat{O}_6^6$) transverse zfs terms are instead included in (4.8).

Mn_3 is highly attractive in the context of understanding the origin of QTM at a microscopic level. The dimension of the MS Hamiltonian matrix for three $s = 2$

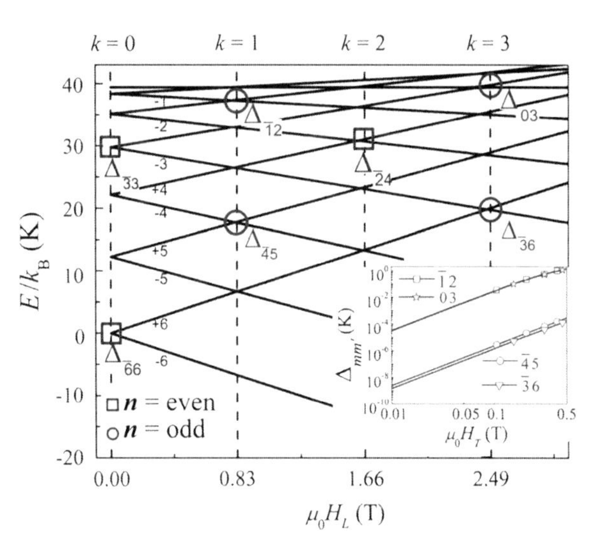

Fig. 4.4 Zeeman diagram for a spin $S = 6$ multiplet with easy-axis anisotropy ($D < 0$ in (4.8)) and $H//z$. All possible non-zero tunneling gaps for C_3 symmetry are labeled according to the scheme discussed in the main text. *The inset* shows the H_T dependence of the odd-n tunneling gaps

spins is just $[(2s + 1)^3]^2 = 125 \times 125$. The C_3 symmetry reduces the number of interaction parameters to just a single exchange constant, J, and identical d and e values for each ion; it also guarantees identical θ_i Euler angles ($= 8.5°$) for the three spins, with $\phi_i = (i - 1) \times 120°$. The remaining parameters have then been determined from fits to EPR and magnetization hysteresis measurements [18, 21–23]. Lastly, the structure contains no solvent molecules. This is rare among SMMs and removes a major source of disorder [24]. Consequently, exceptional spectroscopic data (QTM and EPR) are available against which one can test theoretical models.

4.2.2 QTM Selection Rules in Mn_3

For large spin systems, the effects of $q \neq 0$ zfs terms typically manifest themselves at energy scales that are orders of magnitude smaller than those of the axial ($q = 0$) terms. One must therefore focus on avoided level crossings, where the tunneling gaps are governed by the transverse terms in (4.8). Figure 4.4 displays the Zeeman diagram corresponding to the nominal spin $S = 6$ ground state multiplet of the Mn_3 molecule. Due to symmetry restrictions ($q = 3n$ for C_3 symmetry, where n is an integer), non-zero tunneling gaps are limited to level crossings with $\Delta m_s = 3n$, where m_s is the projection of the total spin onto the molecular C_3 (z-) axis. All such gaps, $\Delta_{m_s,m_s'}$, have been labeled in Fig. 4.4 for QTM resonances $k \leq 3$, where k ($= m_s + m_s'$) denotes an avoided crossing between pairs of levels with spin projections m_s and m_s' ($\bar{m}_s$ denotes $-|m_s|$).

By performing a mapping of the energy diagram obtained via exact diagonalization of (4.6) onto that of the GSA Hamiltonian (4.8) one can obtain microscopic insights into the emergence of $p \geq 4$ transverse terms in the latter approximation. Published zfs parameters were employed for simulations involving (4.6), i.e.,

Table 4.1 Comparison of tunneling gaps obtained for Mn_3 from the MS and GSA models for resonances $k = 0, 1, 2$ and 3, for the two cases $\theta = 0$ (*top*) and $\theta = 8.5°$ (*bottom*)

k	n	Δ	GSA-gap (K)	MS-gap (K)	Ratio
JT-axes parallel to the molecular z-axis					
0	2	$\Delta_{\bar{3},3}$	2.60×10^{-2}	2.66×10^{-2}	0.98
0	4	$\Delta_{\bar{6},6}$	1.10×10^{-6}	1.05×10^{-6}	1.05
2	2	$\Delta_{\bar{2},4}$	2.37×10^{-2}	2.35×10^{-2}	1.01
JT-axes tilted $\theta = 8.5°$ away from the molecular z-axis					
0	2	$\Delta_{\bar{3},3}$	2.76×10^{-2}	2.91×10^{-2}	0.95
0	4	$\Delta_{\bar{6},6}$	1.26×10^{-6}	1.25×10^{-6}	1.01
1	3	$\Delta_{\bar{4},5}$	4.68×10^{-5}	4.19×10^{-5}	1.12
1	1	$\Delta_{\bar{1},2}$	6.33×10^{-2}	6.31×10^{-2}	1.00
2	2	$\Delta_{\bar{2},4}$	2.45×10^{-2}	2.61×10^{-2}	0.94
3	3	$\Delta_{\bar{3},6}$	8.66×10^{-5}	7.53×10^{-5}	1.15
3	1	$\Delta_{0,3}$	1.76×10^{-1}	1.76×10^{-1}	1.00

$d = -4.2$ K and $e = 0.9$ K [21]. An isotropic exchange constant J $(= -10$ K) was employed, set to a value that is artificially high in order to isolate the ground state from excited multiplets, thereby simplifying analysis of higher-lying QTM gaps (see Fig. 4.4). The Euler angles were set to $\phi_1 = 0$, $\phi_2 = 120°$ and $\phi_3 = 240°$ (all $\psi_i = 0$) to preserve C_3 symmetry, while θ_i $(= \theta)$ was allowed to vary in order to examine its influence on QTM selection rules.

The situation in which the JT axes of the three Mn^{III} ions are parallel to the C_3 axis is first considered, i.e., $\theta = 0$. The top section of Table 4.1 lists the magnitudes of even-n QTM gaps involving pairs of levels with $\Delta m_s = 3n$, deduced via diagonalization of (4.6) in the absence of a transverse field, H_T $(\perp z)$. The odd-n, $H_T = 0$ gaps are identically zero, as can be seen from their dependence on H_T (Fig. 4.4 inset): the power-law behavior indicates no contribution from zfs interactions (at $H_T = 0$). Consequently, one expects only even-n zfs terms of the form $B_p^{3n}\hat{O}_p^{3n}$ in the GSA: those satisfying this requirement have six-fold rotational symmetry about the C_3 axis, i.e., a higher symmetry than the real molecule (further explanation is given below). For comparison, these QTM gaps are simulated employing (4.8) with $B_4^3 = 0$ and $B_6^6 = 4.3 \times 10^{-7}$ K. As seen in Table 4.1, an excellent overall agreement between the two models is obtained. Small differences may be attributed to higher-order six-fold terms such as $B_8^6\hat{O}_8^6$, $B_{10}^6\hat{O}_{10}^6$, *etc.*, which have been neglected in this analysis.

A more realistic situation involves a tilting of the JT axes away from the C_3 axis by $\theta = 8.5°$, as is the case for Mn_3 [23]. Both even- and odd-n, $H_T = 0$ QTM gaps are generated in this situation, i.e., k-odd QTM resonances become allowed. This may be understood within the framework of the GSA as being due to the emergence of zfs interactions possessing three-fold rotational symmetry about the molecular C_3 axis, i.e., $B_p^{3n}\hat{O}_p^{3n}$ with $n = 1$ and $p \geq 4$; the leading such term is $B_4^3\hat{O}_4^3$. Table 4.1

(a)

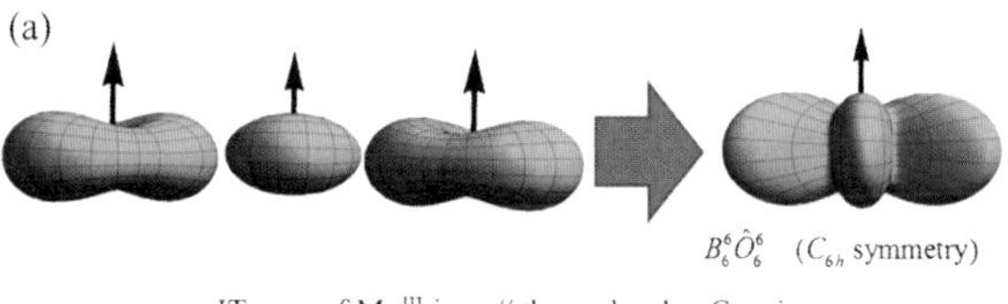

(b)

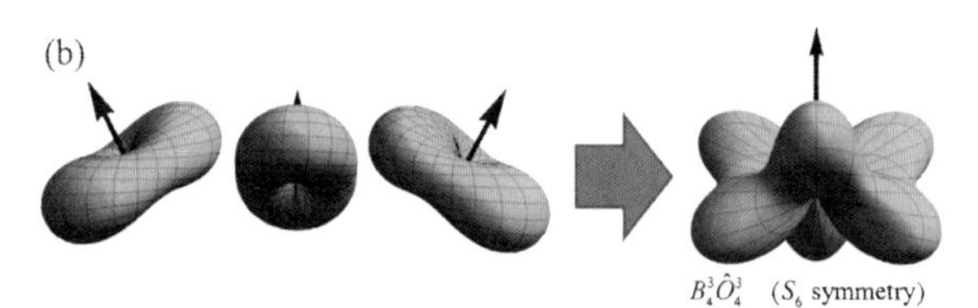

Fig. 4.5 The influence of the orientations of the JT-axes of the Mn^{III} ions on the zero-field magneto symmetry of the Mn$_3$ SMM. In (**a**), the JT-axes of the Mn^{III} ions (*left*) are parallel to the molecular C_3 axis; consequently, the resultant Hamiltonian of the molecule (*right*) possesses C_{6h} symmetry. In (**b**), the JT-axes of the Mn^{III} ions (*left*) are tilted away from the molecular C_3 axis; consequently, the resultant Hamiltonian of the molecule possesses S_6 symmetry

lists the QTM gaps evaluated via diagonalization of (4.8) using $B_6^6 = 4.3 \times 10^{-7}$ K and $B_4^3 = 4.77 \times 10^{-4}$ K. Excellent agreement is once again achieved between the GSA and MS Hamiltonians. Minor deviations may, in principle, be corrected by introducing higher-order transverse terms such as $B_6^3\hat{O}_6^3$.

The emergence of the $B_4^3\hat{O}_4^3$ interaction in the GSA description of Mn$_3$ clearly indicates a lowering of the symmetry of the spin Hamiltonian upon tilting the JT axes. To understand this one must consider both the symmetry of the molecule and the intrinsic symmetry of the zfs tensors of the individual ions. Considering only 2nd order SO anisotropy, the Hamiltonian of a single Mn^{III} ion possesses D_{2h} symmetry (as noted in Sect. 4.1.1), with three mutually orthogonal C_2 axes. When the JT axes are parallel ($\theta = 0$), the local z-axis of each Mn^{III} center coincides with the molecular C_3 axis. The resultant Hamiltonian should then possess $C_3 \times C_2 \times C_i = C_{6h}$ symmetry (see Fig. 4.5(a)), requiring $B_4^3 = 0$; the additional C_i symmetry arises from the time-reversal invariance of the SO interaction. In contrast, when the JT axes are tilted, the C_2 and C_3 axes do not coincide. In addition, the xy-mirror symmetry of the molecule is broken, as is that of the spin Hamiltonian. The rotational symmetry then reduces to three-fold and, hence, $B_4^3\hat{O}_4^3$ is allowed; the symmetry in this case is $C_3 \times C_i = S_6$ (Fig. 4.5(b)).

It is possible to reinforce the preceding discussion via group theoretic arguments, without the need to write down an exact expression for the Hamiltonian. When the external magnetic field is applied parallel to the molecular C_3-axis, the C_{6h} symmetry reduces to C_6, and the 13 basis functions of the $S = 6$ ground state fall into six distinct one-dimensional irreducible representations [25]. These functions can be grouped according to their behavior under a C_6 rotation: $|-6\rangle, |0\rangle, |+6\rangle \in \Gamma_1$; $|-2\rangle, |+4\rangle \in \Gamma_2$; $|+2\rangle, |-4\rangle \in \Gamma_3$; $|-3\rangle, |+3\rangle \in \Gamma_4$; $|+1\rangle, |-5\rangle \in \Gamma_5$; $|-1\rangle$,

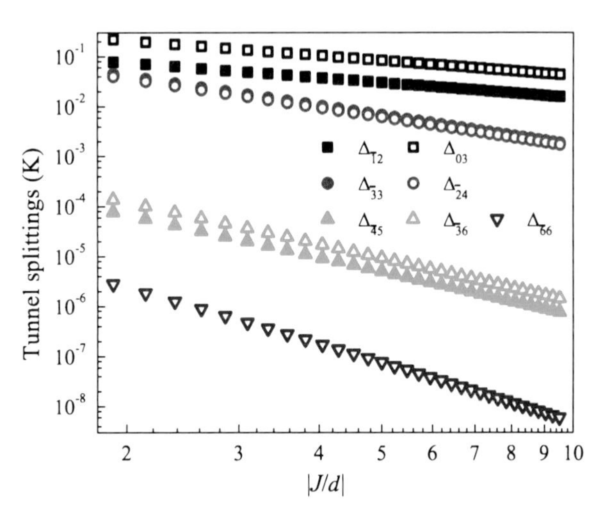

Fig. 4.6 Calculated QTM gaps for the Mn_3 SMM as a function of the coupling constant J. Simulations were performed with the JT-axes tilted 8.5° away from the molecular C_3-axis. The QTM gaps associated with same $|\Delta m|$ value are rendered in the same color. Note that the results are plotted on a logarithmic scale

$|+5\rangle \in \Gamma_6$, where $\Gamma_{1...6}$ are the six irreducible representations following the Bethe notation [25]. Because the Hamiltonian operator belongs to the totally symmetric representation, $\langle m_s|\hat{H}|m_s'\rangle$ is non-zero only when $|m_s\rangle$ and $|m_s'\rangle$ belong to the same representation [26]. As can be seen, such states have $\Delta m_s = 3n$, with n even, which is the criterion for state mixing in C_6 symmetry. When the symmetry of the Hamiltonian is reduced to S_6 (C_3 upon application of $H//z$) the basis functions may be grouped into three different irreducible representations: $|0\rangle, |\pm3\rangle, |\pm6\rangle \in \Gamma_1$; $|+4\rangle, |+1\rangle, |-2\rangle, |-5\rangle \in \Gamma_2$; $|-4\rangle, |-1\rangle, |+2\rangle, |+5\rangle \in \Gamma_3$. Here, the selection rule for mixing is $\Delta m_s = 3n$, with n being integer, again in agreement with the preceding calculations.

Before concluding this section, the influence of the exchange coupling, J, on the QTM observed in Mn_3 deserves further consideration. The J dependence of higher-order ($p \geq 4$) coefficients in the GSA has been discussed previously for several other high-symmetry SMMs [16, 18, 20, 27, 28]. In these cases, the 2nd-order transverse anisotropy ($q > 0$) cancels exactly, emerging at higher orders as a consequence of the mixing of spin states. This is illustrated for Mn_3 in Fig. 4.6, which plots the power law dependence of several QTM gaps as a function of the ratio of J/d; the single-ion zfs parameters given above were employed in these calculations. It is found that the QTM gaps are proportional to $|J|^{-n}$, i.e., $B_4^3 \propto |J|^{-1}$ and $B_6^6 \propto |J|^{-2}$ [18]. Note that this implies a complete suppression of QTM in the strong coupling limit ($|J| \gg |d|$).

4.2.3 *The Influence of Disorder on QTM*

An important conclusion of the preceding analysis is the demonstration of the existence of k-odd QTM resonances, i.e., a quite realistic parameterization of (4.6) generates zfs terms in the GSA containing odd powers of $\hat{S}_+$ and $\hat{S}_-$. These ideas should apply quite generally. For example, the disorder potential associated with the distortion of a symmetric molecule likely contains zfs terms (e.g. $\hat{O}_4^1$ or $\hat{O}_4^3$) that unfreeze k-odd QTM resonances (as explicitly demonstrated in Sect. 4.3.3), contrary to the belief that odd QTM resonances *cannot* be generated in this way [29]. However, it remains to be seen whether this can account for the absence of selection rules in SMMs such as Mn_{12}. We note that these arguments do not apply to zero-field ($k = 0$) QTM in half-integer spin systems, which is strictly forbidden according to Kramers' theorem [30].

This revives a partly unresolved and somewhat controversial issue concerning the influence of disorder on the QTM characteristics of SMMs. Disorder became a focus of attention in some of the early spectroscopic investigations of the Mn_{12}-acetate and Fe_8Br SMMs, revealing significant distributions (or strains) in the measured GSA D parameters [31–34]. Around the same time, Chudnovsky and Garanin argued that long-range strains nucleated by line dislocations could give rise to a broad distribution of transverse 2nd-order anisotropies in otherwise high-symmetry crystals of SMMs such as Mn_{12}-acetate, i.e., a broad distribution (on a logarithmic scale) in E centered about an average value of zero [35, 36]. Importantly, the Chudnovsky/Garanin theory pointed out that disorder would lead to local variations in molecular symmetry away from the ideal (S_4 for Mn_{12}-acetate), and that this could modify the selection rules governing QTM. This motivated intense efforts aimed at carefully studying QTM in Mn_{12}-acetate, including selective hole-burning experiments targeted at subsets of molecules belonging to different parts of the relaxation time distribution [37–40]. A breakthrough was achieved as a result of crystallographic investigations by Cornia et al. [41], that revealed a form of intrinsic disorder associated with the acetic acid solvent that co-crystallizes with the standard form of Mn_{12}-acetate (we refer the reader to Refs. [13, 39–43] for detailed discussion). The acetic acid forms a hydrogen-bond to the Mn_{12} core, resulting in a non-trivial distortion of the molecule. However, while each solvent molecule occupies a position between two Mn_{12}'s, it can only hydrogen-bond to one of them, with $50:50$ probability. Hence, real Mn_{12}-acetate crystals contain a statistical distribution of several different solvent isomers, some of which maintain approximate four-fold symmetry, while more than 50 % have a lower (rhombic) symmetry [41]. EPR, inelastic neutron scattering and magnetic hysteresis measurements subsequently yielded excellent qualitative and quantitative agreement with the model proposed by Cornia, thus demonstrating for the first time that solvent disorder can have a profound influence on QTM relaxation [13, 39, 40, 42, 43].

Many more recent studies have reinforced the idea that solvent disorder can significantly influence QTM relaxation in SMMs. First of all, magnetization and EPR studies have shown that the anomalous distributions in zfs parameters found for Mn_{12}-acetate are absent in several newer high-symmetry (S_4) Mn_{12} SMMs that do not suffer from the intrinsic solvent disorder (or for which the interaction between the solvent and the SMM core is far weaker than in the original acetate) [24, 28, 44–47]. Interestingly, the deliberate removal of solvent from the newer Mn_{12}'s (by pumping on the samples at room temperature) has been shown to accelerate the low temperature magnetization relaxation, without affecting the height of the classical relaxation barrier [24]. Meanwhile, EPR studies demonstrate that the solvent loss induces disorder that looks very similar to the intrinsic disorder in Mn_{12}-acetate [24]. This again suggests that the induced (extrinsic) disorder causes the faster relaxation, presumably as a result of quantum tunneling processes. This leads to known sample handling problems, i.e., crystals containing volatile solvent (e.g. $Mn_{12}BrAc{\cdot}CH_2Cl_2$) can change beyond recognition as far as their QTM and EPR characteristics are concerned if they are cooled under vacuum [24, 47, 48].

The reason why the ideal Mn_{12} SMM is so susceptible to disorder is because it has such a high symmetry; the nominally forbidden 2nd-order transverse anisotropy rapidly reemerges upon the introduction of weak disorder, either through solvent loss or otherwise. This is not the case for lower symmetry molecules that already possess a 2nd-order rhombic zfs interaction [49]. This has caused some confusion in the literature. As an aside, we note that internal transverse dipolar/hyperfine fields can, in principle, also affect QTM selection rules in high-spin SMMs [50]. Indeed, early work demonstrated that a combination of allowed transverse zfs interactions, together with transverse dipolar/hyperfine fields, may explain the observed absence of QTM selection rules in Mn_{12}-acetate in the thermally activated regime [51, 52]. However, this explanation fails in the pure QTM regime, where tunneling couples low lying spin states at a high order of perturbation theory ($\Delta m_s \gg 1$). More recent studies claim that differences in QTM relaxation observed for the various high-symmetry Mn_{12} SMMs are due entirely to differences in the widths of the dipolar field distributions, which obviously depend on the crystal structures [50]. However, the recent preparation of two versions of Mn_{12}-acetate that are identical in almost all respects (including the lattice constants), apart from the co-crystallizing solvent (acetic acid in one case, methanol in the other), seem to rule out this assertion [24]. While dipolar fields undoubtedly play a crucial role in the collective QTM relaxation in SMM crystals [53], the marked differences in relaxation rates found for the two Mn_{12}-acetates appear to be related to the disorder associated with the hydrogen bonding acetic acid solvent, which is not present in the methanol variant.

The large dimension of the Mn_{12} MS Hamiltonian presents a considerable challenge in terms of gaining theoretical insights into the effects of disorder. However, there exist many smaller molecules with equally high symmetry which are, thus, more amenable to this type of study. Indeed, this issue is revisited in Sect. 4.3.3, which deals explicitly with the distortion of a Ni_4 SMM that possesses the same intrinsic S_4 symmetry as the ideal Mn_{12}'s [16]. Aside from the obvious computational advantages, several smaller SMMs are also known to crystallize with no lattice solvent molecules [18, 22, 54]. More importantly, there exist families of low-nuclearity SMMs for which some members co-crystallize with solvent, while others do not. These include Mn_3, Mn_4 and Ni_4, which represent the focus of the remainder of this chapter. The spectroscopic differences between solvated and solvent-free SMMs are quite dramatic. For instance, D-strain is almost absent in the latter, giving rise to remarkably sharp EPR spectra. This again implicates solvent molecules as a major source of disorder in SMM crystals. The key finding involved a solvent-free Mn_3 compound, which is the only SMM to display a complete absence of a symmetry-forbidden QTM resonance [21]. When combined with the observation of

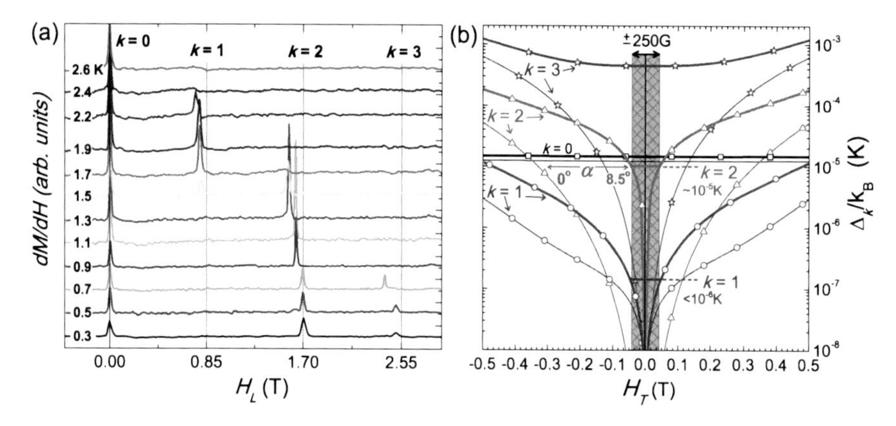

Fig. 4.7 (**a**) Field derivative of the magnetization curves obtained for a Mn_3 single crystal at different temperatures, with $B \parallel z$. (**b**) Ground-state tunnel splittings associated with resonances $k = 0$ (*black squares*), $k = 1$ (*red circles*), $k = 2$ (*green triangles*), and $k = 3$ (*blue stars*) as a function of the transverse field H_T, with the JT-axes aligned along the C_3 axis (*thin lines*) and tilted by 8.5 degrees away from the C_3 axis (*thick lines*). The strength of the dipolar magnetic field in the sample is represented by the central gray area, with the corresponding splitting values achieved for such dipolar field values for resonances $k = 1$ and $k = 2$ (*dashed horizontal lines*)

uniquely sharp EPR spectra [22], this result suggests that it is the absence of solvent that unmasks the intrinsic QTM selection rules, again hinting at the connection between solvent, disorder and the absence of QTM selection rules in other SMMs.

Figure 4.7(a) shows derivatives of magnetization hysteresis curves for Mn_3, recorded at different temperatures from 0.3 K to 2.6 K, with $H \parallel z$. At low temperatures, the $k = 1$ resonance is completely absent. It eventually appears for temperatures above 1.3 K as a result of a symmetry allowed thermally activated QTM process. As discussed above, the trigonal symmetry of this molecule enforces the $|\Delta m| = 3n$ selection rule when taking into account the 8.5 degrees misalignment of the JT axes from the molecular C_3 axis (S_6 reduced to C_3 when a longitudinal field is applied). The effect can be seen in Fig. 4.7(b), which shows the tunnel splittings for the four lowest resonances $k = 0 - 3$, calculated by diagonalization of the MS Hamiltonian of (4.6) with the parameters given in Ref. [21]. In the absence of a transverse field ($H_T = 0$), the ground state tunnel splitting is always absent for resonances $k = 1$ and $k = 2$, while the degeneracy is only broken in resonances $k = 0$ and $k = 3$. Consequently, one expects steps in the hysteresis curves (peaks in the derivatives) appearing only at $k = 0$ and $k = 3$. The absence of the $k = 1$ resonance at low temperatures constitutes direct evidence for the expected QTM selection rule, an observation made possible because of the highly ordered solvent-free crystal structure. Following the same reasoning, resonance $k = 2$ should also be absent at low temperatures, since the ground tunnel splitting couples spin states differing by $|\Delta m| \neq 3n$. However, as can be seen in Fig. 4.7(b), the $k = 2$ splitting grows quickly as a function of the transverse field, reaching observable magnitudes for field values provided by internal dipolar fields. This result shows that internal Zeeman fields can indeed unfreeze some QTM resonances, but not all of them. In-

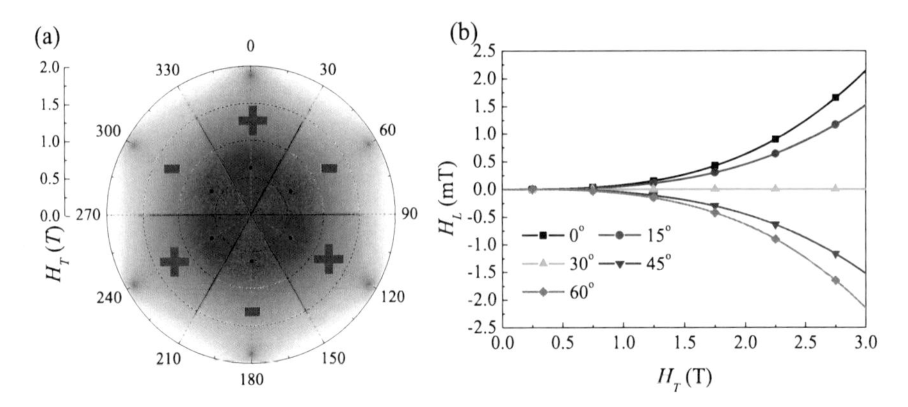

Fig. 4.8 (**a**) The calculated BPI patterns associated with the ground state $k = 0$ QTM resonance for the Mn_3 SMM: the color contour plot shows $\Delta_{\bar{6},6}$ as function of H_T (with B_6^6 set to zero); a compensating H_L field is required that alternates between positive (*red*) and negative (*blue*) values. (**b**) The compensating H_L field for $\Delta_{\bar{6},6}$, as a function of the magnitude of H_T; note the curvature (except for the 30° trace, for which $H_L = 0$)

deed, the ground state tunnel splitting associated with the $k = 1$ resonance remains almost two orders of magnitude smaller than that of $k = 2$ for the same transverse field. One would expect the influence of dipolar fields to diminish further still in the pure QTM regime for SMMs with larger S values.

The Mn_3 SMM illustrates perfectly how crystalline disorder can mask the fundamental QTM behavior in SMMs; in this particular case, it is the absence of disorder that unmasks intrinsic symmetry-enforced quantum properties. This, in turn, allows fundamental insights into the influence of the internal molecular degrees of freedom on the QTM phenomenon. The low-nuclearity of the Mn_3 SMM proved particularly helpful by making this a computationally tractable problem. The following section digs deeper into the unusual BPI patterns predicted for trigonal SMMs.

4.2.4 Berry Phase Interference in Trigonal Symmetry

This section focuses explicitly on the BPI patterns generated by the $\hat{O}_4^3$ operator. In contrast to all of the even-q GSA terms, the xy-plane does not correspond to a symmetry element for the odd-q interactions, as discussed in Sect. 4.1.1.2. Hence, the $\hat{O}_4^3$ operator is expected to result in BPI patterns which have not been observed in previous studies of SMMs, essentially all of which possess even rotational symmetry with only even-q zfs interactions [55–57].

The influence of $B_4^3\hat{O}_4^3$ is quite fascinating. In order to simplify discussion, Fig. 4.8 was generated with $B_6^6 = 0$; details of the calculations can be found in Ref. [58]. The $\Delta_{\bar{6},6}$ ($k = 0$) QTM resonance exhibits the most intriguing new features. One might expect a six-fold behavior due to the intrinsic C_i symmetry of the

Hamiltonian, i.e., the spectrum should be invariant under inversion of H_T. However, this assumes that $H_L = 0$. In fact, application of a transverse field causes a shift of the $\Delta_{\bar{6},6}$ minimum away from $H_L = 0$, as illustrated in Fig. 4.8(b), i.e., the $k = 0$ QTM resonance shifts away from $H_L = 0$ in the presence of a finite transverse field. The resultant transverse-field BPI patterns appear to exhibit a hexagonal symmetry in Fig. 4.8(a). However, the color coding represents the polarity of the required compensating longitudinal field, H_L. Thus, on the basis of the sign of H_L, one sees that the BPI minima in fact exhibit a three-fold rotational symmetry, which is consistent with the symmetry of the $B_4^3\hat{O}_4^3$ interaction. One way to interpret this result is to view the $\hat{O}_4^3$ operator as generating an effective internal longitudinal field, H_L^*, under the action of an applied transverse field; H_L^* is then responsible for the shift of the $k = 0$ resonance from $H_L = 0$. This can be seen from the expression of the $\hat{O}_4^3 = \frac{1}{2}[\hat{S}_z, \hat{S}_-^3 + \hat{S}_+^3]$ operator, which, unlike the q-even operators, contains an odd power of $\hat{S}_z$, akin to the Zeeman interaction with $H \parallel z$. An alternative view may be derived from the S_6 surface depicted in Fig. 4.2(c), where one sees that the hard/medium directions do not lie within the xy-plane, contrary to the case for the q-even operators. In other words, the classical hard plane is not flat, but corrugated with a 120° periodicity. Consequently, application of a longitudinal field is required in order to insure that the total applied field lies within the hard plane when rotating H_T. Note that the predicted BPI patterns nevertheless exhibit the required C_i symmetry, i.e., they are invariant with respect to inversion of the total field.

Figure 4.8(b) plots the shift of the $k = 0$ resonance ($\Delta_{\bar{6},6}$ minimum) away from $H_L = 0$ upon applying a transverse field, H_T, for several orientations within the xy-plane. The shift is positive for 0° and 15°, and negative for 45° and 60°, with no shift at 30° (i.e. the 30° resonance occurs at $H_L = 0$). In other words, the quantum molecular hard plane is not flat, but rather corrugated, with a 120° periodicity. This is consistent with the classical energy surface shown in Fig. 4.2(c). It is also notable that the H_L shift displays a non-linear dependence on H_T, which indicates that the exact locations of the hard directions depend on the magnitude of H_T. Finally, it should be emphasized that these phenomena, especially the shift of the $k = 0$ resonance from $H_L = 0$, cannot be generated by any of the even-q operators [57], where the xy plane necessarily corresponds to the hard plane because of the additional mirror symmetry about this plane (see the discussion in Sect. 4.1.1.2).

4.3 Quantum Tunneling of Magnetization in the High-Symmetry Ni_4 Single-Molecule Magnet

4.3.1 The Ni_4 Single-Molecule Magnet

The $[Ni(hmp)(dmb)Cl]_4$ SMM (henceforth Ni_4) possesses q-even rotational symmetry [14, 16, 54, 59, 60]. The complex crystallizes in an $I41/a$ space group without any lattice solvent molecules. The structure of the Ni_4 molecule is shown in

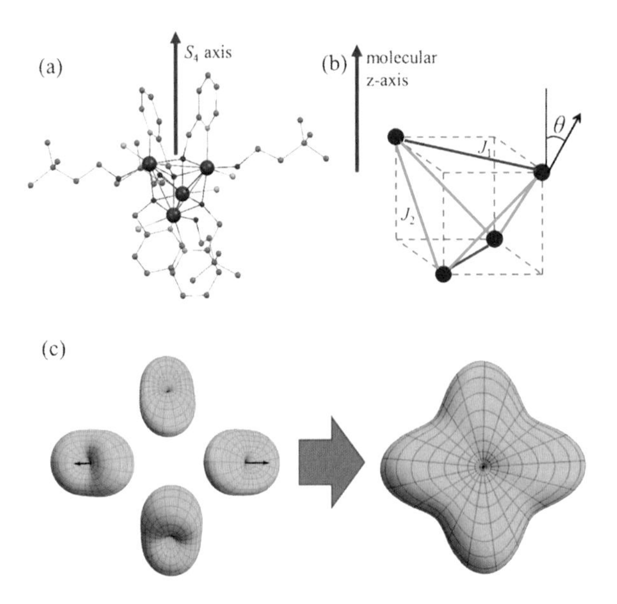

Fig. 4.9 The structure (**a**) and schematic representation of the magnetic core (**b**) of the Ni_4 SMM. Color code: Ni = olive, O = red, N = blue, C = black and Cl = dark gold. H-atoms have been omitted for clarity. (**c**) Representation of the zero-field magneto symmetry of the Ni_4 SMM resulting from the situation in which the 2nd-order single-ion zfs tensors have their C_2 axes tilted away from the molecular S_4 axis. Once added, the time reversal symmetry of the SO interaction guarantees that the resultant zero-field Hamiltonian of the molecule possesses C_{4h} symmetry (see text for details)

Fig. 4.9(a). The magnetic Ni_4O_4 core is a slightly distorted cube with the Ni^{II} ions ($s = 1$) located on opposite corners, as sketched in Fig. 4.9(b). The distorted cube retains S_4 symmetry, with the S_4-axis indicated in Fig. 4.9(a). The four Ni^{II} ions are ferromagnetically coupled, leading to a spin $S = 4$ molecular ground state. The Ni_4 SMM exhibits extremely fast zero-field QTM, which significantly reduces the effective relaxation barrier. Nevertheless, it does display a small magnetic hysteresis [60]. However, the fast relaxation unfortunately precludes the observation of $k > 0$ QTM resonances. Nevertheless, a theoretical study of Ni_4 proves enlightening. The molecule can be described with the following spin Hamiltonian:

$$\hat{H}_{\text{zfs}} = \sum_{i=1}^{4} \hat{s}_i \cdot \left(\overleftrightarrow{n}^T\right)^i \cdot \overleftrightarrow{d} \cdot \left(\overleftrightarrow{n}\right)^i \cdot \hat{s}_i + \sum_{i<j} \hat{s}_i \cdot \overleftrightarrow{J}_{ij} \cdot \hat{s}_i. \tag{4.9}$$

This Hamiltonian differs from (4.6) in that the individual rotation matrices, $\overleftrightarrow{R}_i$, are replaced by a single matrix, $\overleftrightarrow{n}$, which explicitly takes the rotational symmetry of the molecule into account, including cases involving improper rotations. The zero-field anisotropy can then be parameterized by a single $\overleftrightarrow{d}$-tensor (corresponding to one of the ions), specified with respect to the molecular coordinate frame. If the

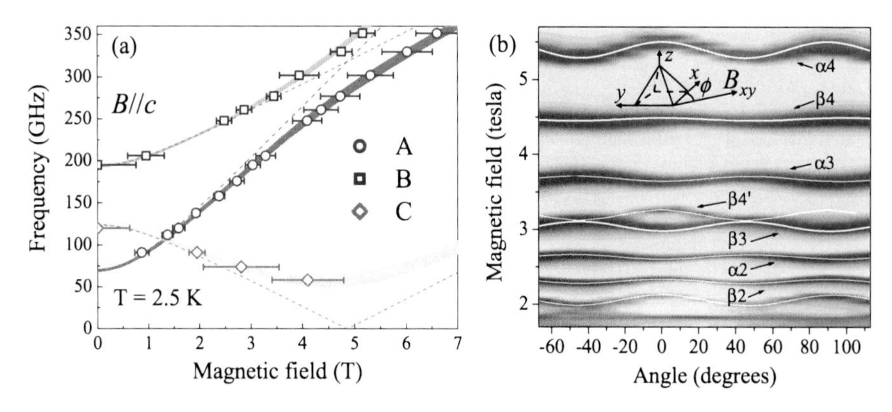

Fig. 4.10 (**a**) Frequency dependence of the positions (in field) of the three EPR transitions associated with the isolated Ni^{II} triplet ($S = 1$) state in a diluted single crystal of the compound $[Zn_{3.91}Ni_{0.09}(hmp)_4(dmb)_4Cl_4]$ (see Ref. [19] for assignments of the A, B and C peaks). *The colored curves* correspond to best fits to the data employing the following single-ion zfs parameters: $d = -5.30(5)\ cm^{-1}$, $e = \pm 1.20(2)\ cm^{-1}$, $g_z = 2.30(5)$, and a tilting of the local z-axes of 15° away from the symmetry (c-) axis of the crystal. The energy splittings around 5 T provide a direct measure of the tilting of the local $\overleftrightarrow{d}$ tensors; the dashed curves correspond to the non-tilted case, for which these splittings are zero. The widths of the colored curves reflect the uncertainty in the orientations of the local x- and y-axes, which were subsequently deduced from two axis rotation studies [19]. (**b**) 2D color map of the EPR absorption intensity as a function of the magnetic field strength and its orientation within the hard plane of a single crystals of the $S = 4$ SMM $[Ni_4(hmp)(dmb)Cl]_4$ (see Ref. [61] for explanation of the peak labeling). Superimposed on the absorption maxima (darker red regions) are fits (white curves) to the data that involve just a single adjustable zfs parameter, $B_4^4 = 4 \times 10^{-4}\ cm^{-1}$ (over and above those deduced from easy-axis measurements [16])

local coordinate frames of the individual ions are related by a series of proper rotations (C_2, C_4, etc.) within the molecular coordinate frame, then $\overleftrightarrow{n}$ may be replaced by a single rotation matrix $\overleftrightarrow{R}$ (corresponding, e.g., to a 90° rotation about z for a molecule with C_4 symmetry). On the other hand, if the local coordinate frame of the ith ion is related to the molecular frame by an improper rotation (S_4 or S_4^3), then $\overleftrightarrow{n} = \sigma \overleftrightarrow{R}$, where σ represents a reflection in the plane perpendicular to the S_4 axis. Note that, for S_4 symmetry, $\overleftrightarrow{n}^2$ is equivalent to a C_2 rotation, and $\overleftrightarrow{n}^4$ is equivalent to the identity matrix.

The Ni_4 SMM is a particularly ideal platform for comparison with Mn_3. The molecule possesses a well separated $S = 4$ ground state with the $S = 3$ excited spin multiplets located roughly 30 K above in energy. The 3×3 Hamiltonian matrix associated with a single Ni^{II} ion contains only two 2nd-order zfs parameters, d and e, i.e., higher order single-ion anisotropies ($p \geq 4$) are strictly forbidden. The zfs of the individual Ni^{II} ions, as well as their orientations, have been directly measured through EPR studies on an isostructural diluted $Zn_{4-x}Ni_x$ compound (see Fig. 4.10(a) and Ref. [19]). Due to the restriction of S_4 symmetry, only two independent Heisenberg interaction parameters, J_1 and J_2, are allowed; these interactions

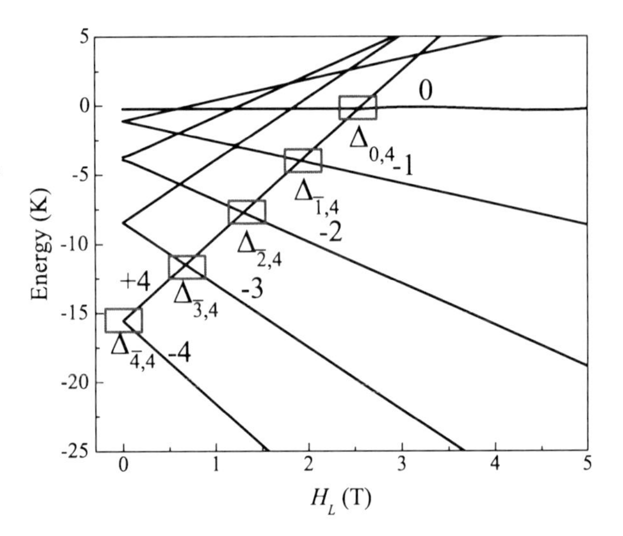

Fig. 4.11 Zeeman diagram for the ground state $S = 4$ multiplet associated with the Ni_4 SMM, simulated employing (4.9). The $k = 0$ to 4 ground state QTM splittings are labeled in the figure

can be determined by dc susceptibility measurements [60]. Therefore, all of the parameters in (4.9) are known independently. Meanwhile, the molecule possesses the same S_4 symmetry as Mn_{12}, which prohibits the rhombic anisotropy term in the GSA Hamiltonian. The high symmetry of the molecule has been confirmed by single-crystal EPR measurements, where exceptionally sharp resonances are again observed, with a four-fold modulation pattern upon rotating H_T (see Fig. 4.10(b) and Ref. [61]). This clearly illustrates the presence of a 4th- (or higher-) order transverse GSA anisotropy which is responsible for the fast QTM.

4.3.2 Quantum Tunneling of Magnetization in the Ni_4 SMM

In analogy with Mn_3, the transverse GSA anisotropy in Ni_4 is assessed by calculating the QTM gaps, focusing on the $k = 0, 1, \ldots, 4$ ground state resonances, as shown in Fig. 4.11; the $|\Delta m|$ values associated with these resonances equal $8, 7, \ldots, 4$, respectively. The simulations were performed using the published zfs parameters $d = -7.6$ K, $e = 1.73$ K and $J_1 = J_2 = -10$ K [14, 16, 19]. Previous EPR studies also show that the local easy-axes of the Ni^{II} ions are tilted away from the molecular z-axis by $\theta = 15°$ (See Fig. 4.9(c)). However, the $\theta = 0°$ case is also examined in order to further explore the influence of easy-axis tilting on the symmetry of the molecular Hamiltonian.

Figure 4.12 shows the ground state QTM gaps as a function of transverse field (H_T), deduced via exact diagonalization of (4.9). As seen in the figure, $\Delta_{\bar{4},4}$ ($k = 0$) and $\Delta_{0,4}$ ($k = 4$) retain non-zero values in the absence of a transverse field, while all other tunnel splittings vanish at $H_T = 0$. This result is not surprising based on the S_4 molecular symmetry, where only $|\Delta m| = 4n$ (n is an integer) QTM resonances are allowed. However, unlike the Mn_3 SMM, the QTM selection rules corresponding

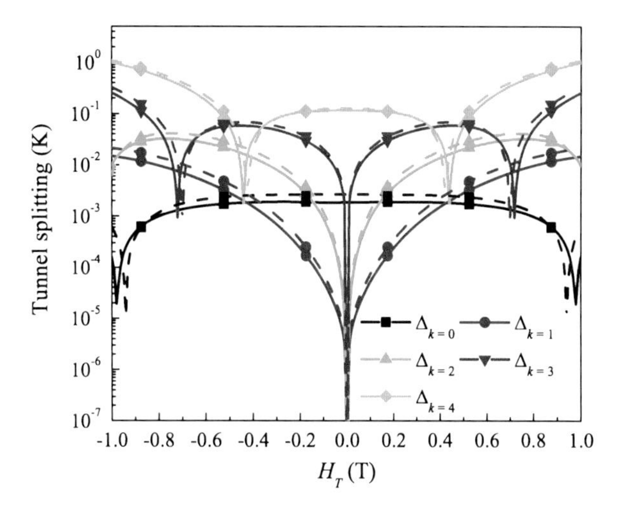

Fig. 4.12 The ground state QTM gaps for the Ni_4 SMM as a function of H_T. The simulations were performed employing (4.9) with the parameters given in the main text. *The solid lines* were generated with $\theta = 0$ and *the dash lines* were generated with $\theta = 15°$

to the $\theta = 15°$ and $0°$ situations are exactly the same. In both scenarios, only the $\Delta_{\bar{4},4}$ $(k = 0)$ and $\Delta_{\ 0,4}$ $(k = 4)$ gaps are non-zero, while the other k-even QTM gap, $\Delta_{\bar{2},4}$ $(k = 2)$, vanishes when $H_T = 0$. These results imply that the easy-axis tilting does not affect the symmetry of the Hamiltonian, contrary to the case for the Mn_3 SMM. This can be understood in terms of the different symmetry properties associated with q-even and q-odd cases. In the even case, the molecular z-axis must also be a C_2 axis. Consequently, forcing the local C_2 axes of the individual ions to be parallel to the molecular z-axis $(\theta = 0°)$ does not introduce an extra C_2 symmetry to the molecular Hamiltonian. In contrast, the molecular z-axis is not a C_2 axis in a molecule with odd rotational symmetry. Therefore, the symmetry of the molecular Hamiltonian changes when $\theta = 0°$.

In the preceding discussions of Mn_3, the QTM selection rules can be simply understood in terms of the rotational symmetry of the molecule (C_6 or C_3). In contrast, the selection rules for Ni_4 cannot be fully explained by the S_4 molecular symmetry; one must additionally take into account the intrinsic C_i symmetry of the spin Hamiltonian. Upon application of a magnetic field parallel to the molecular z-axis, the S_4 symmetry group reduces to C_2, for which the $\Delta_{\bar{2},4}$ $(k = 2)$ QTM resonance should be allowed. This clearly contradicts the simulation in Fig. 4.12, which suggests a higher symmetry. However, one must also consider the C_i symmetry associated with the SO interaction. The consequential zero-field spin Hamiltonian then possesses $S_4 \times C_i = C_{4h}$ symmetry, which corresponds to the symmetry of the $\hat{O}_4^4$ interaction, as seen in Fig. 4.2(d). Upon application of a longitudinal field, the C_{4h} group reduces to the C_4 group, for which the expected QTM selection rule $|\Delta m| = 4n$ is recovered. The C_i symmetry is guaranteed by the nature of the SO interaction. This property is not limited to spin Hamiltonians, i.e., it applies to any Hamiltonian dictated by crystal field and/or SO physics, where the C_i symmetry should apply regardless of whether the orbital angular momentum is quenched or not. In other words, it is always necessary to consider the C_i symmetry in addition to the structural symmetry, especially when improper rotations are involved. Unfor-

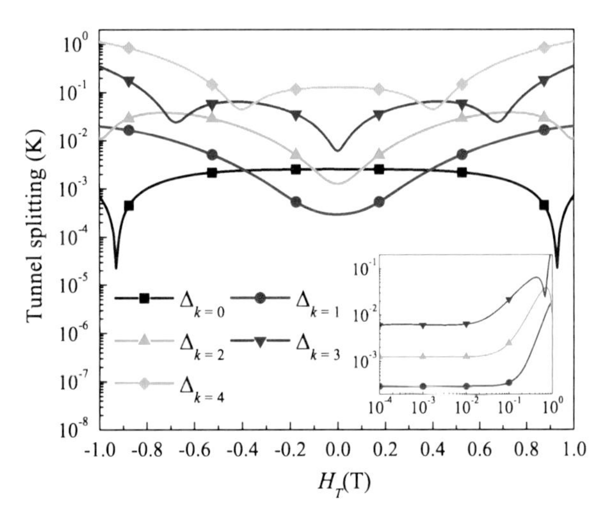

Fig. 4.13 The effect of disorder on the ground state QTM gaps for the Ni_4 SMM. The simulations were performed employing (4.9) by misaligning the zfs tensor of one of the Ni^{II} ions with respect to the unperturbed molecular z-axis (see details in the main text)

tunately, observation of $k > 0$ QTM steps in Ni_4 is impractical due to the extremely fast tunneling at $k = 0$. This tunneling should be greatly suppressed if the ground spin state of the molecule is increased, as is the case for Mn_{12}. However, it would be interesting to obtain a four-fold symmetric SMM constituted of four $s = 2$ Mn^{III} ions, for which it would be possible to study the $k > 0$ QTM steps. Moreover, the Hamiltonian dimension of just 625×625 would be quite manageable.

4.3.3 Disorder

In the presence of random disorder, one would expect the symmetry of most molecules to be lowered, leading to an absence of QTM selection rules. The Ni_4 molecule provides an excellent platform to study this issue. Figure 4.13 was generated by adjusting the orientation of the zfs tensor of one of the Ni^{II} ions in the molecule, i.e., the zfs tensors of three of the Ni ions are tilted 15° from the molecular z-axis, while the other is tilted 10°. It should be emphasized that it is not trivial to find the orientation of the molecular easy-axis in this situation, i.e., it no longer coincides with the molecular z-axis. For each resonance ($k = 0$ to 4), a search was performed for the minimum QTM gap by varying the orientation of the applied field. As seen in the figure, all resonances posses a non-zero QTM gap at $H_T = 0$. The inset to Fig. 4.13 plots $\Delta_{\bar{3},4}$ ($k = 1$), $\Delta_{\bar{2},4}$ ($k = 2$) and $\Delta_{\bar{1},4}$ ($k = 3$) on a log-log scale, clearly demonstrating that these QTM gaps, which are forbidden for S_4 symmetry, also saturate at non-zero values when $H_T \to 0$. These results show that a small disorder effectively unfreezes all QTM steps without the assistance of a transverse field. This argument can be reinforced by group theoretic considerations. With random disorder, the symmetry of a molecule is lowered to C_1, resulting in a spin Hamiltonian with C_i symmetry. Upon applying a longitudinal field, the C_i group reduces

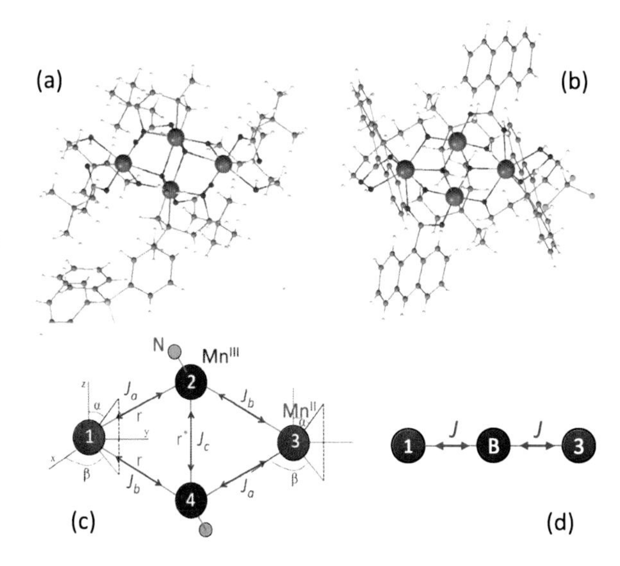

Fig. 4.14 (**a**) The Mn_4-anca molecule. (**b**) The Mn_4-Bet molecule. Color code: Mn = purple, O = red, N = blue, C = grey, H = white, B = pink and Cl = green. (**c**) Sketch showing the different exchange interactions used to solve the four spin MS Hamiltonian for these molecules. (**d**) Trimer model representing the Mn_4 molecules assuming an infinite J coupling between the two central Mn^{III} ions

to C_1, where all of the states necessarily belong to the same one-dimensional irreducible representation [25]. Therefore, mixing between all states is allowed. We note that this kind of disorder can be introduced by small crystallographic defects, which always exist to some degree in real samples. Thus, exceptionally clean crystals are required in order to observe symmetry imposed QTM selection rules. Importantly, the preceding discussion clearly demonstrates that disorder can be responsible for the observation of k-odd QTM steps in SMMs with even rotational symmetries.

4.4 Quantum Tunneling of Magnetization in Low-Symmetry Mn_4 Single-Molecule Magnets

In order to contrast results presented in previous sections, EPR and QTM/BPI results are presented here for two related Mn_4 SMMs that possess almost no symmetry. Both molecules crystallize in the triclinic $P\bar{1}$ space group. One of the structures co-crystallizes with solvent, while the other does not. Consequently, significant differences are observed in terms of the widths of EPR and QTM resonances due to the different degrees of disorder in the two crystals. In addition, small structural differences associated with the Mn_4 cores result in different coupling strengths between the Mn ions within the two molecules which, in turn, result in different QTM behavior.

4.4.1 The Mn_4 Single-Molecule Magnets

The Mn_4 molecules (Figs. 4.14(a) and (b)) possess mixed-valent butterfly-type structures, with two central Mn^{III} ions ($s_2 = s_4 = 2$) in the body positions and two Mn^{II} ions ($s_1 = s_3 = 5/2$) in the wing positions (see Fig. 4.14(c)). Magnetic superexchange is mediated through oxygen bridges. The Mn_4 molecule that co-crystallizes with solvent is $[Mn_4(anca)_4(Hedea)_2(edea)_2]\cdot 2CHCl_3\cdot 2EtOH$, henceforth Mn_4-anca (see Fig. 4.14(a) and Ref. [8] for more details). The solvent-free molecule is $[Mn_4(Bet)_4(mdea)_2(mdeaH)_2](BPh_4)_4$, henceforth Mn_4-Bet (see Fig. 4.14(b) and Ref. [9] for more details). Both molecules crystallize in the centrosymmetric triclinic space group $P\bar{1}$. The asymmetric unit therefore consists of half the molecule ($Mn^{III}Mn^{II}$), with the other half generated via an inversion. This also ensures that all four Mn ions lie in a plane. Temperature dependent susceptibility data recorded at different magnetic fields indicate that both molecules possess a spin $S = 9$ ground state at low temperatures, implying overall ferromagnetic coupling within the molecules (note that this does not rule out the possibility that one of the exchange paths could be antiferromagnetic).

4.4.2 EPR and QTM Spectroscopy in Mn_4 SMMs with and Without Solvent

A selection of EPR and QTM measurements from Refs. [8, 9] are presented in Fig. 4.15: (a) displays 165 GHz EPR spectra recorded at different temperatures for a single crystal of Mn_4-anca, with the magnetic field applied close to the molecular easy-axis; (b) shows equivalent spectra obtained for Mn_4-Bet at a frequency of 139.5 GHz and similar temperatures. The first thing to note are the obvious differences in the EPR linewidths in the two figures. This again provides a clear illustration of the inferior quality of samples in which the SMMs co-crystallize with interstitial solvent molecules. In the present example, the Mn_4-anca sample is the more disordered, resulting in a broader distribution of GSA zfs parameters. A series of nine absorption peaks can clearly be seen for the Mn_4-anca SMM in Fig. 4.15(a), which have been labeled 1 through 9. These correspond to transitions from consecutive spin projection (m_S) states belonging to the $S = 9$ ground state multiplet, where the numbering denotes the absolute m_S value associated with the level from which the EPR transition was excited, e.g., resonance $\alpha = 9$ corresponds to the $m_S = -9 \rightarrow -8$ transition. The fact that all of the spectral weight transfers to the $\alpha = 9$ resonance as $T \rightarrow 0$ indicates uniaxial anisotropy, i.e., $D < 0$ according to the GSA Hamiltonian of (4.2). The uneven spacing between the labeled ground state resonances is indicative of 4th- (and higher-) order anisotropy within the GSA (or weak exchange within the MS picture). Finally, as the temperature is increased, a few weaker resonances (not labeled) can be seen to appear in Fig. 4.15(a) between the labeled transitions. These additional peaks are associated with the population of higher-lying, excited spin states, e.g., $S = 8$.

The EPR spectra obtained for Mn_4-Bet (Fig. 4.15(b)) are not so simple to interpret. First and foremost, many more peaks are observed, suggesting the population of many more spin states. Based upon the knowledge gained from Mn_4-anca, and the results of subsequent simulations, the nine resonances corresponding to transitions

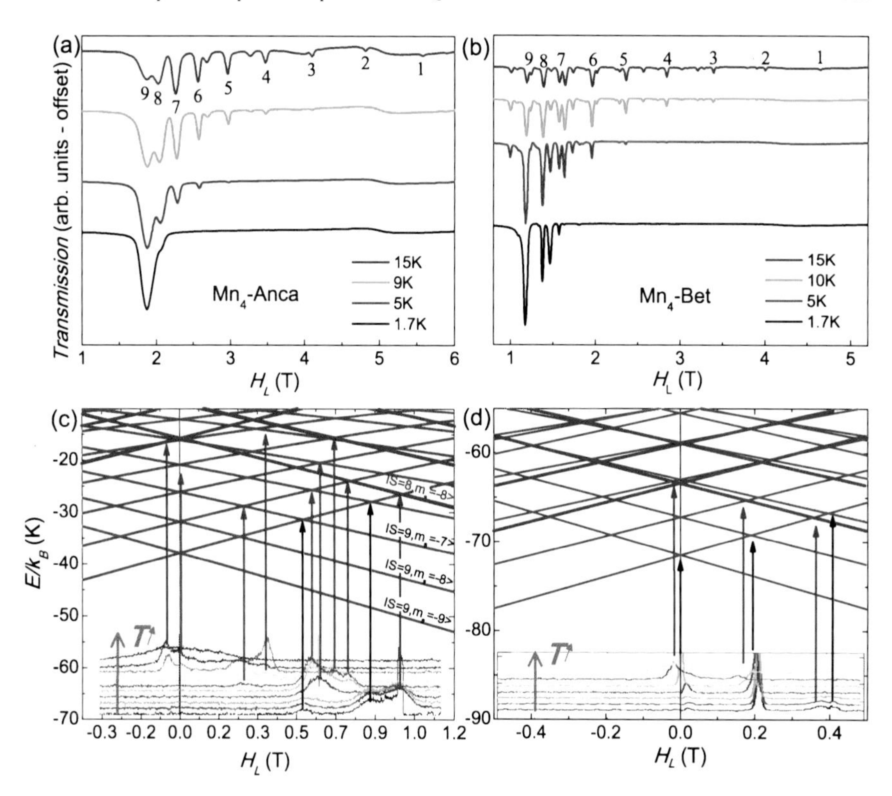

Fig. 4.15 EPR spectra obtained at different temperatures with the field along the easy-axis at: 165 GHz for Mn_4-anca (**a**); and 139.5 GHz for Mn_4-Bet (**b**). Zeeman diagrams depicting the low–lying energy levels for Mn_4-anca (**c**) and Mn_4-Bet (**d**), obtained by diagonalization of the MS Hamiltonian (4.6) using the trimer model depicted in Fig. 4.14(d). *Arrows* relate the QTM peaks observed in the field derivatives of the magnetization curves obtained at different temperatures (*bottom* of the graphics) with the corresponding crossings between spin levels; with *black arrows* indicating crossings of the ground state $|S = 9, m_s = 9\rangle$ (*red* for excited states) with other states within the same spin multiplet ($S = 9$), and *blue arrows* signaling both ground and excited crossings involving levels of different spin length ($S = 9$ and $S = 8$)

within the $S = 9$ ground state multiplet have been labeled in the figure. Meanwhile, the peaks that are not labeled correspond to transitions within low-lying excited spin multiplets. Clearly, the emergence of excited state EPR transitions at much lower temperatures indicates significantly weaker exchange coupling in the Mn_4-Bet molecule; note that excited state intensity is seen even at the lowest temperature, whereas this is not the case until ~ 9 K in the Mn_4-anca sample. The weaker exchange and higher crystal quality associated with the Mn_4-Bet sample lead to the observation of unusual MQT/BPI behavior, as detailed in the following section.

Although one can reproduce the peak positions of the nine labeled EPR transitions in both Figs. 4.15(a) and (b) using the GSA (including $p \geq 4$ terms), a MS Hamiltonian becomes essential to account for transitions within excited spin states. In other words, one starts to see the limitations of the GSA in these two cases—

especially for Mn_4-Bet. Diagonalization of the exact MS Hamiltonian that considers all four Mn ions and the couplings between them (as indicated in Fig. 4.14(c)) is manageable on a standard computer. However, a convenient and reasonable approximation replaces the ferromagnetically coupled central Mn^{III} ions with a single $s_B = 4$ spin, resulting in a linear trimer consisting of the central spin, s_B, and the two outer $s_A = 5/2$ Mn^{II} spins, as depicted in Fig. 4.14(d). This approximation, which contains elements of both the GSA and MS models, has been compared with the more exact four-spin model for the case of Mn_4-Bet in Ref. [7]. The two models give good agreement in terms of the lowest-lying portions of the energy-level diagrams that dictate the low-temperature QTM and EPR properties, provided that the ferromagnetic coupling between the Mn^{III} ions is not too weak. Although it involves a level of approximation, the trimer model captures all of the important physics associated with these low-symmetry SMMs. Moreover, the smaller Hamiltonian matrix dimension enables much faster fitting (hours instead of days), and employs fewer parameters—just two d tensors and a single exchange coupling constant, J. Finally, working with a single J parameter to identify the internal exchange energy becomes particularly useful in relating the distinct behavior of the two molecules to different intramolecular couplings which result from slight structural disparities between the two compounds. In the following, we diagonalize the MS Hamiltonian (4.6) using the trimer model (Fig. 4.14(c)) to account for the energy landscape associated with the lowest lying molecular spin multiplets, which result from the main anisotropy terms in the Hamiltonian (i.e. axial terms). The full MS Hamiltonian (4.6) including the four manganese ions (Fig. 4.14(d)) is used to account for the behavior of the tunnel splittings, which result from the smaller anisotropy terms in the Hamiltonian (i.e. transverse terms) and are more sensitive to small variations of the internal degrees of freedom of the molecules.

In fitting the data in Figs. 4.15(a) and (b), as well as other EPR data obtained at different temperatures and applied field orientations (see Refs. [8, 9]), one finds that the exchange coupling constant J has a strong influence on the positions of the EPR peaks (particularly the relative spacings between peaks). This again highlights the fact that one cannot use a GSA to realistically describe these results, especially those of Mn_4-Bet, i.e., there is no exchange parameter in the GSA (all energy splittings are parameterized in terms of the SO anisotropy). It is thus preferable to use the MS approach whenever computational resources will allow, as is the case for all of the low-nuclearity SMMs described in this chapter. Indeed, there are many other interesting features associated with the magnetic behavior of SMMs that cannot be explained with the GSA, regardless of how complex the corresponding Hamiltonian is, e.g., QTM resonances involving level crossings between different spin states.

QTM spectroscopy also facilitates comparisons between the two Mn_4 molecules. Figures 4.15(c) and (d) show the association between the observed QTM resonance positions and the corresponding energy-level diagrams for the Mn_4-anca and Mn_4-Bet molecules, respectively. The QTM resonances are determined from the positions of the peaks in the derivatives of the magnetization versus field curves obtained at different temperatures, as shown in the lower portions of Figs. 4.15(c) and (d). Note that the effects of the solvent disorder can again be seen, causing broader QTM resonances for Mn_4-anca in comparison to Mn_4-Bet. The energy level diagrams are obtained via exact diagonalization of the MS Hamiltonian of (4.6) using the trimer model depicted in Fig. 4.14(c) with the following parameters: $J = 5.42$ K, $d_1 = d_2 = d_A = -0.115$ K, $d_B = -2.22$ K in Mn_4-anca (Fig. 4.15(c)); and $J = 1.90$ K, $d_A = -0.115$ K, $d_B = -2.00$ K in Mn_4-Bet (Fig. 4.15(d)), with isotropic $g = 2.0$ for all ions in both cases. The correspondence between QTM resonances and their associated level crossings are indicated by vertical arrows. Crossings involving the ground state $|S = 9, m_s = 9\rangle$ with other levels within the same multiplet ($S = 9$) are indicated by black arrows and expected to appear at the lowest temperatures, for which transitions involving all excited states (red arrows) should vanish. Interestingly, some of the resonances (blue arrows) correspond to crossings between levels associated with different spin states, i.e., $S = 9$ and $S = 8$. The main difference between the two molecules resides in the value of J, being more than double for Mn_4-anca. This results in the lowest spin projection states ($m_S = \pm 8$) associated with the first excited state ($S = 8$) being much closer to the $m_S = \pm 9$ ground states in Mn_4-Bet (~ 8 K separation) than in Mn_4-anca (~ 22 K separation). These findings are consistent with the temperature dependence of the EPR spectra, which suggested the excited states to be appreciably lower in energy in the Mn_4-Bet molecule in comparison to Mn_4-anca. The differences in J values can be reconciled with the minor structural differences between the two molecules. It is well known that the superexchange coupling between two transition metal ions is very sensitive to the bridging angle, to the extent that the sign of the interaction can switch from ferromagnetic to antiferromagnetic within a small range of angles [18, 62, 63]. Indeed, there are measurable differences in the bond angles and distances associated with these two Mn_4 molecules.

4.4.3 *Berry Phase Interference in Mn_4-Bet*

The spectroscopic results presented in the previous section illustrate how small structural perturbations can lead to significant changes in the exchange coupling within a molecule. Crucially, the Mn_4-Bet SMM resides in a particularly interesting region of the 'anisotropy' versus 'exchange' parameter space in which excited spin states exert a significant influence on the low-energy/low-temperature quantum properties. First and foremost, it can be seen that some of the QTM resonances involve level crossings between different spin multiplets. More importantly, the QTM properties within this intermediate exchange regime ($d \sim J$) are extraordinarily sensitive to the internal magnetic structure of the molecule. As noted in previous sections, the physics associated with the strong exchange limit ($J \gg d$) is dominated by the 2nd-order GSA anisotropy. Consequently, any observable BPI patterns should display a high degree of symmetry (D_{2h}), regardless of the molecular symmetry. However, in the intermediate exchange regime, one may expect any BPI effects to mimic the symmetry of the molecule under investigation much more closely. This symmetry can be expressed exactly using a four spin MS model in the case of Mn_4-Bet, although we note that one can also reproduce most of the features of these

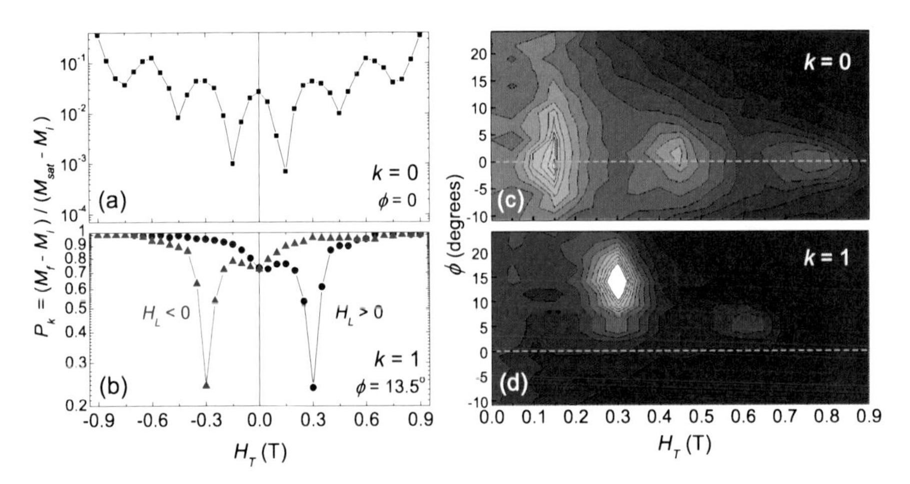

Fig. 4.16 Modulation of the QTM probabilities for resonances $k = 0$ (**a**) and $k = 1$ (**b**) as a function of H_T applied at different angles, ϕ, within the xy-plane of the Mn_4-Bet SMM. The asymmetry of the BPI pattern of oscillations in resonance $k = 1$ is inverted upon reversal of H_L. (**c**) and (**d**) show the modulation of the tunnel splitings of resonances $k = 0$ and $k = 1$, respectively, for different directions of the transverse field

experiments using a GSA that includes appropriate 4th- (and higher-) order terms. The virtual lack of any symmetry associated with Mn_4-Bet leads to some remarkable BPI patterns, which also shed light on a previous mystery surrounding another high-nuclearity Mn_{12} SMM [64].

Low temperature QTM measurements performed in the presence of a transverse magnetic field, H_T, enable exploration of the dominant symmetries associated with the Hamiltonian describing a SMM. Figure 4.16(a) shows the modulation of the QTM probability for the $k = 0$ resonance for Mn_4-Bet as a function of the magnitude of H_T applied parallel to the hard axis of the molecule ($\phi = 0°$, see Ref. [7] for details). As explained in Sect. 4.2.4, the oscillations correspond to BPI (constructive/destructive interference associated with equivalent tunneling trajectories on the Bloch sphere), with minima occurring at regularly spaced field values ($\Delta H_T = 0.3$ T). Experiments designed to probe the modulation of the $k = 0$ QTM gap as a function of the orientation of a fixed transverse field within the hard plane (see Ref. [7]) reveal a two-fold behavior. One may be tempted to ascribe this to a rhombic anisotropy. However, the molecule possesses a much lower symmetry ($P\bar{1}$). In fact, the two-fold pattern is a direct manifestation of the C_i symmetry associated with the SO interaction. Since no longitudinal field (H_L) is present for the $k = 0$ resonance, the Hamiltonian must be invariant with respect to inversion of H_T—hence the apparent two-fold behavior. Note that the $k = 0$ BPI oscillations do, indeed, respect the symmetry of the Hamiltonian, i.e., they are invariant under inversion of H_T.

Due to the absence of H_L, $k = 0$ turns out *not* to be the most interesting QTM resonance, because the C_i symmetry guarantees symmetric BPI patterns about $H_T = 0$. In contrast, this is clearly not the case for the BPI pattern associated with the $k = 1$

resonance, as can be seen in Fig. 4.16(b). In this case, a single interference minimum is observed at $H_T = 0.3$ T for only one polarity of the transverse field, i.e., the corresponding BPI minimum is completely absent under inversion of H_T. Such a result is not so surprising when one recognizes that there are no mirror symmetries within $P\bar{1}$. Hence, there is no reason why the BPI patterns should be invariant under inversion of just one component of the applied field. However, the Hamiltonian, and therefore the BPI patterns, must be invariant under a full inversion of the applied field, i.e., inversion of both H_T and H_L together. This indeed turns out to be the case for the $k = 1$ resonance, as clearly seen in Fig. 4.16(b).

Another interesting feature observed in the BPI patterns of Mn_4-Bet is the motion of the minima associated with the $k = 1$ resonance within the xy-plane. This can be observed in Fig. 4.16(d), which shows a color contour plot of the QTM probability in $k = 1$, as a function of the magnitude and the orientation of H_T. Two minima can clearly be observed; they are again spaced by ~ 0.3 T, and are located half way between the $k = 0$ minima, as is usually the case for even/odd resonances. However, the $k = 1$ minima do not appear at the same orientations within the xy-plane as those of $k = 0$. Moreover, the orientations of the two observed $k = 1$ minima do not even coincide: $\phi = 13.5°$ for the first minimum and $\phi = 6°$ for the second one. Note that, in contrast to $k = 1$, all of the $k = 0$ minima lie along the nominal hard anisotropy axis ($\phi = 0$), as seen in Fig. 4.16(c). In essence, the hard directions associated with the $k = 1$ resonance (for which both H_L and H_T are finite) do not occur along a fixed axis, as would be the case for a rhombic molecule. This property, which is analogous to the behavior seen in Fig. 4.7(b), is a direct consequence of the absence of any mirror symmetries in the $P\bar{1}$ space group. It is impossible to simultaneously satisfy both C_i symmetry and a mirror symmetry if the BPI minima do not reside on a fixed axis in space. However, if the mirror symmetry is broken, then the BPI minima may in principle occur anywhere on the Bloch sphere, so long as they occur in pairs related by inversion. The results displayed in Fig. 4.16(d) were obtained by rotating H_T while keeping H_L fixed; the space above and below the xy-plane was not explored due to experimental constraints. One cannot rule out further minima above or below the xy-plane. Indeed, this may explain why the 2nd $k = 1$ minimum is so weak, i.e., its real location may be above or below the xy-plane. In fact, this could also be true for the $k = 0$ minima, as again emphasized in Fig. 4.7(b) for the case of the Mn_3 molecule, which lacks xy mirror symmetry. More detailed (and time consuming) experiments are clearly required to further explore this issue in Mn_4-Bet.

In order to simulate the observed BPI patterns, one must obviously break some or all of the mirror symmetries within the Mn_4-Bet spin Hamiltonian, whilst also respecting the inversion symmetry of the real molecule. There really is only one way to achieve this, involving misalignment of the 2nd-order zfs tensors associated with the Mn^{III} and the Mn^{II} ions. The molecular inversion symmetry guarantees that the JT axes associated with the Mn^{III} ions be parallel to each other; likewise the zfs tensors associated with the Mn^{II} ions must be co-linear. However, there is no requirement that the tensors associated with the two types of ion be co-linear. Indeed, all of the results in Fig. 4.16 have been reproduced in Ref. [7] following exactly this approach. Although the trimer model (Fig. 4.14(c)) can reproduce the observed behavior quite well, the four spin Hamiltonian (4.6) was employed in order to describe the geometry in Fig. 4.14(d), since it gives a better quantitative agreement and allows for a more physical interpretation of the observations (e.g. the real dipolar coupling between the four Mn ions can be employed, which involves no fitting parameters). Using this approach, we find the optimal parameter set to be as follows: (central Mn^{III}'s) $d_2 = d_4 = -4.99$ K and $e_2 = e_4 = 0.82$ K, with the easy and hard axes along z ($\alpha_2 = 0$) and x ($\beta_2 = 0$), respectively; (outer Mn^{II}'s) $d_1 = d_3 = -0.67$ K and $e_1 = e_3 = 0$, with the axes rotated with respect to the central spin by identical Euler angles $\alpha_{1,3} = 45°$, $\beta_{1,3} = 0°$ (as required by inversion symmetry); γ being zero for all ions; finally, isotropic ferromagnetic exchange constants $J_a = -3.84$ K, $J_b = -1.20$ K and $J_c = -3.36$ K were employed. It should be stressed that these parameters were additionally constrained by the EPR and QTM data in Fig. 4.15. They do not necessarily constitute the correct parameterization, but they account for all experimental observations. Note that the zfs tensors of the two Mn^{II} ions are tilted by 45 degrees with respect to the Mn^{III} tensors, thereby breaking the molecular xy mirror symmetry. This results in a breaking of the xy mirror symmetry of the corresponding spin Hamiltonian. Note that this would not be the case for a molecule with even rotational symmetry, because of the C_i symmetry associated with the SO interaction. However, in $P\bar{1}$ ($q = 1$), the xy mirror symmetry of the Hamiltonian is broken, as was also the case for the trigonal ($q = 3$) Mn_3 SMM. One could, in principle, also explain these results using a GSA by introducing 4th- (and higher-) order terms. However, a more natural and satisfying account of the results is obtained by diagonalizing the four-spin Hamiltonian, which easily allows for a tilting of the zfs tensors of the four Mn ions.

Interestingly, there again exists a connection to Mn_{12}, albeit a wheel molecule that bears no resemblance to the well studied, high-symmetry Mn_{12}'s discussed in Sect. 4.2.3. The Mn_{12} wheel molecule [64], which possesses the exact same $P\bar{1}$ symmetry as the Mn_4 molecules considered in this section, attracted considerable controversy on account of the observation of asymmetric $k > 0$ BPI patterns [17, 65]. Initial attempts to account for this behavior involved treating the molecule as a dimer, including an unphysical Dzyaloshinskii-Moriya coupling between the two halves of the dimer (this interaction is forbidden on account of the molecular inversion symmetry [66]). The present studies have shown that asymmetric $k > 0$ BPI patterns are, in fact, quite natural in low-symmetry molecules. As in the case of Mn_3, a detailed understanding of the QTM characteristics in the Mn_4 molecules is made possible due to the low nuclearity of the system, which enables the employment of a MS Hamiltonian. This, in turn, provides fundamental insights that are much harder to achieve when studying larger molecules using a GSA.

4.5 Summary and Outlook

This chapter focuses on the microscopic factors that dictate the QTM behavior observed in polynuclear transition-metal SMMs, with particular focus on molecular symmetry. The examples provided involve relatively simple, low-nuclearity clusters (Mn_3, Mn_4 and Ni_4) which display essentially the same physics as the original Mn_{12} and Fe_8 SMMs that have occupied chemists and physicists working in this field for nearly two decades. The simpler systems are amenable to analysis using a microscopic spin Hamiltonian that incorporates both the single-ion physics, and isotropic exchange coupling between the constituent ions, and relies on relatively few parameters. One can therefore systematically investigate the role of internal spin degrees of freedom within a molecule, in contrast to the approximate GSA approach employed for most studies of Mn_{12} and Fe_8. Comparisons between theory and experiment are presented for a range of cluster symmetries, with remarkable quantitative agreement achieved in all cases.

Until fairly recently, most SMM research was directed towards polynuclear $3d$ transition metal clusters, with the synthetic goal of maximizing both the molecular spin state and the magneto-anisotropy. However, a number of fundamental factors have limited progress based on this strategy, with the record blocking temperature for a Mn_6 cluster [67] only just surpassing that of the original Mn_{12} SMM [24]. Limiting factors include: (i) a tendency for superexchange interactions between constituent transition metal spins to be both weak (few cm^{-1}) and often antiferromagnetic; (ii) the fact that strong crystal-field effects typically quench the orbital momentum associated with $3d$ elements, thus significantly suppressing the magneto-anisotropy; and (iii) the difficulties associated with maximally projecting any remaining (2nd order SO) anisotropy onto the ground spin state. In fact, careful studies of this issue suggest that one is unlikely to achieve anisotropy barriers that significantly exceed those of the constituent ions [18]. This is perhaps best illustrated by the optimum Mn_3, Mn_6 and Mn_{12} SMMs, which possess similar barriers (to within a factor of < 2 [18]). This is because the molecular anisotropy, D, is given by a weighted sum of the anisotropies of the constituent ions (d_i), where the weighting is inversely proportional to the total molecular spin, S [18, 68]. Thus, D decreases as S increases, and the theoretical best that one can hope to achieve is an anisotropy barrier ($\sim DS^2$) that scales linearly with S (or N, the number of magnetic ions in a ferromagnetic molecule). Even then, many challenges remain—some fundamental (quantum tunneling, spin state mixing, etc. [6]), some synthetic. The synthetic challenges, in particular, become more complex with increasing molecule size, e.g., maintaining ferromagnetic couplings, maintaining parallel arrangements of the individual anisotropy tensors, etc. Thus, it is perhaps no surprise that the optimum $[Mn^{III}]_N$ SMM has a nuclearity of just six [67]!

Given the aforementioned situation, it has become clear that a more direct route to magnetic molecules that might one day be used in practical devices involves the use of ions that exhibit considerably stronger magnetic anisotropies than those that have traditionally been employed in the synthesis of large polynuclear clusters, i.e., ions for which the orbital momentum is not quenched, and/or heavier elements in which strong SO effects are manifest. Examples include certain high-symmetry and low-coordinate $3d$ transition metal complexes (Fe^{II} [69], Co^{II} [70], even Ni^{II} [71]), as well as elements further down the periodic table such as the $4f$ and $5f$ elements. Indeed, over the past few years, a number of mononuclear complexes have been shown to exhibit magnetization blocking of pure molecular origin [69, 70, 72–74]. However, the quantum magnetization dynamics of these single-ion molecular nanomagnets has yet to be studied in detail, and much remains to be learned theoretically. Obviously, much of the physics related to exchange which is discussed in this chapter does not apply in these cases. Nevertheless, the spin Hamiltonian formalism remains applicable, as does the crucial importance of molecular and crystallographic symmetry. In particular, 4th and higher-order crystal-field interactions may be expected to play a crucial role in the quantum dynamics of mononuclear lanthanide SMMs [75]. Thus, one would expect similar combinations of theory and spectroscopy to contribute in future to this evolving field of research.

Acknowledgements This work was supported by the US National Science Foundation, grant numbers DMR0804408 (SH), CHE0924374 (SH), and DMR0747587 (EdB). Work performed at the National High Magnetic Field Laboratory is supported by the National Science Foundation (grant number DMR1157490), the State of Florida and the Department of Energy.

References

1. C. Rudowicz, S.K. Misra, Appl. Spectrosc. Rev. **36**(1), 11–63 (2001)
2. D. Gatteschi, R. Sessoli, J. Villain, *Molecular Nanomagnets* (Oxford University Press, Oxford, 2006)
3. A. Abragam, B. Bleaney, *Electron Paramagnetic Resonance of Transition Ions* (Dover, New York, 1986)
4. C. Rudowicz, C.Y. Chung, J. Phys. Condens. Matter **16**(32), 5825 (2004)
5. S. Stoll, A. Schweiger, J. Magn. Reson. **178**(1), 42–55 (2006)
6. S. Hill, M. Murugesu, G. Christou, Phys. Rev. B **80**, 174416 (2009)
7. H.M. Quddusi, J. Liu, S. Singh, K.J. Heroux, E. del Barco, S. Hill, D.N. Hendrickson, Phys. Rev. Lett. **106**, 227201 (2011)
8. J. Liu, C.C. Beedle, H.M. Quddusi, E. del Barco, D.N. Hendrickson, S. Hill, Polyhedron **30**, 2965–2968 (2011)
9. K.J. Heroux, H.M. Quddusi, J. Liu, J.R. O'Brien, M. Nakano, E. del Barco, S. Hill, D.N. Hendrickson, Inorg. Chem. **50**, 7367–7369 (2011)
10. C.C. Beedle, W.-G. Wang, C. Koo, J. Liu, A.-J. Zhou, M. Nakano, J.R. O'Brien, W. Wernsdorfer, S. Hill, M.-L. Tong, X.-M. Chen, D.N. Hendrickson, Inorganic Chemistry (2013). Under review
11. S. Hill, J.A.A.J. Perenboom, N.S. Dalal, T. Hathaway, T. Stalcup, J.S. Brooks, Phys. Rev. Lett. **80**, 2453–2456 (1998)
12. A.L. Barra, D. Gatteschi, R. Sessoli, Chemistry **6**(9), 1608–1614 (2000)
13. S. Takahashi, R.S. Edwards, J.M. North, S. Hill, N.S. Dalal, Phys. Rev. B **70**(9), 094429 (2004)
14. C. Kirman, J. Lawrence, S. Hill, E.-C. Yang, D.N. Hendrickson, J. Appl. Phys. **97**(10), 10M501 (2005)
15. S. Carretta, E. Liviotti, N. Magnani, P. Santini, G. Amoretti, Phys. Rev. Lett. **92**(20), 207205 (2004)
16. A. Wilson, J. Lawrence, E.C. Yang, M. Nakano, D.N. Hendrickson, S. Hill, Phys. Rev. B **74**(14), 140403 (2006)
17. C.M. Ramsey, E. del Barco, S. Hill, S.J. Shah, C.C. Beedle, D.N. Hendrickson, Nat. Phys. **4**, 277–281 (2008)
18. S. Hill, S. Datta, J. Liu, R. Inglis, C.J. Milios, P.L. Feng, J.J. Henderson, E. del Barco, E.K. Brechin, D.N. Hendrickson, Dalton Trans. **39**(20), 4693–4707 (2010)

19. E.-C. Yang, C. Kirman, J. Lawrence, L.N. Zakharov, A.L. Rheingold, S. Hill, D.N. Hendrickson, Inorg. Chem. **44**(11), 3827–3836 (2005)
20. R. Maurice, C. deGraaf, N. Guihéry, Phys. Rev. B **81**, 214427 (2010)
21. J.J. Henderson, C. Koo, P.L. Feng, E. del Barco, S. Hill, I.S. Tupitsyn, P.C.E. Stamp, D.N. Hendrickson, Phys. Rev. Lett. **103**, 017202 (2009)
22. P.L. Feng, C. Koo, J.J. Henderson, M. Nakano, S. Hill, E. del Barco, D.N. Hendrickson, Inorg. Chem. **47**(19), 8610–8612 (2008)
23. P.L. Feng, C. Koo, J.J. Henderson, P. Manning, M. Nakano, E. del Barco, S. Hill, D.N. Hendrickson, Inorg. Chem. **48**(8), 3480–3492 (2009)
24. G. Redler, C. Lampropoulos, S. Datta, C. Koo, T.C. Stamatatos, N.E. Chakov, G. Christou, S. Hill, Phys. Rev. B **80**(9), 094408 (2009)
25. G.F. Koster, J.O. Dimmock, P.G. Wheeler, H. Statz, *Properties of the Thirty-Two Point Groups* (MIT Press, Cambridge, 1963)
26. F.A. Cotton, *Chemical Applications of Group Theory*, 3rd edn. (Wiley-Interscience, New York, 1990)
27. E. Liviotti, S. Carretta, G. Amoretti, J. Chem. Phys. **117**(7), 3361–3368 (2002)
28. A.-L. Barra, A. Caneschi, A. Cornia, D. Gatteschi, L. Gorini, L.-P. Heiniger, R. Sessoli, L. Sorace, J. Am. Chem. Soc. **129**(35), 10754–10762 (2007)
29. J. van Slageren, S. Vongtragool, B. Gorshunov, A. Mukhin, M. Dressel, Phys. Rev. B **79**, 224406 (2009)
30. W. Wernsdorfer, S. Bhaduri, C. Boskovic, G. Christou, D.N. Hendrickson, Phys. Rev. B **65**, 180403 (2002)
31. S. Maccagnano, R. Achey, E. Negusse, A. Lussier, M.M. Mola, S. Hill, N.S. Dalal, Polyhedron **20**, 1441 (2001)
32. K. Park, M.A. Novotny, N. Dalal, S. Hill, P. Rikvold, Phys. Rev. B **65**, 014426 (2001)
33. S. Hill, S. Maccagnano, K. Park, R.M. Achey, J.M. North, N.S. Dalal, Phys. Rev. B **65**, 224410 (2002)
34. K. Park, M.A. Novotny, N.S. Dalal, S. Hill, P. Rikvold, Phys. Rev. B **66**, 144409 (2002)
35. E.M. Chudnovsky, D.A. Garanin, Phys. Rev. Lett. **87**(18), 187203 (2001)
36. E.M. Chudnovsky, D.A. Garanin, Phys. Rev. B **65**, 094423 (2002)
37. K.M. Mertes, Y. Suzuki, M.P. Sarachik, Y. Paltiel, H. Shtrikman, E. Zeldov, E.M. Rumberger, D.N. Hendrickson, G. Christou, Phys. Rev. Lett. **87**(22), 227205 (2001)
38. E. del Barco, A.D. Kent, E.M. Rumberger, D.N. Hendrickson, G. Christou, Europhys. Lett. **60**, 768–774 (2002)
39. E. del Barco, A.D. Kent, E.M. Rumberger, D.N. Hendrickson, G. Christou, Phys. Rev. Lett. **91**, 047203 (2003)
40. E. del Barco, A.D. Kent, S. Hill, J.M. North, N.S. Dalal, E.M. Rumberger, D.N. Hendrickson, N. Chakov, G. Christou, J. Low Temp. Phys. **140**, 119–174 (2005)
41. A. Cornia, R. Sessoli, L. Sorace, D. Gatteschi, A.L. Barra, C. Daiguebonne, Phys. Rev. Lett. **89**, 257201 (2002)
42. S. Hill, R.S. Edwards, S.I. Jones, J.M. North, N.S. Dalal, Phys. Rev. Lett. **90**, 217204 (2003)
43. R. Bircher, G. Chaboussant, A. Sieber, H.U. Güdel, H. Mutka, Phys. Rev. B **70**, 212413 (2004)
44. S. Hill, N. Anderson, A. Wilson, S. Takahashi, K. Petukhov, N.E. Chakov, M. Murugesu, J.M. North, E. del Barco, A.D. Kent, N.S. Dalal, G. Christou, Polyhedron **24**, 2284–2292 (2005)
45. N.E. Chakov, S.-C. Lee, A.G. Harter, P.L. Kuhns, A.P. Reyes, S.O. Hill, N.S. Dalal, W. Wernsdorfer, K.A. Abboud, G. Christou, J. Am. Chem. Soc. **128**, 6975–6989 (2006)
46. P. Subedi, A.D. Kent, B. Wen, M.P. Sarachik, Y. Yeshurun, A.J. Millis, S. Mukherjee, G. Christou, Phys. Rev. B **85**, 134441 (2012)
47. C. Lampropoulos, M. Murugesu, A.G. Harter, W. Wernsdofer, S. Hill, N.S. Dalal, K.A. Abboud, G. Christou, Inorg. Chem. **52**, 258–272 (2013)
48. E. del Barco, A.D. Kent, N.E. Chakov, L.N. Zakharov, A.L. Rheingold, D.N. Hendrickson, G. Christou, Phys. Rev. B **69**, 020411 (2004)

49. C. Carbonera, F. Luis, J. Campo, J. Sánchez-Marcos, A. Camón, J. Chaboy, D. Ruiz-Molina, I. Imaz, J. vanSlageren, S. Dengler, M. González, Phys. Rev. B **81**, 014427 (2010)
50. E. Burzurí, C. Carbonera, F. Luis, D. Ruiz-Molina, C. Lampropoulos, G. Christou, Phys. Rev. B **80**, 224428 (2009)
51. J.F. Fernandez, F. Luis, J. Bartolome, Phys. Rev. Lett. **80**, 5659–5662 (1998)
52. F. Luis, J. Bartolome, J.F. Fernandez, Phys. Rev. B **57**, 505–513 (1998)
53. N.V. Prokof'ev, P.C.E. Stamp, Phys. Rev. Lett. **80**, 5794–5797 (1998)
54. J. Lawrence, E.-C. Yang, R. Edwards, M.M. Olmstead, C. Ramsey, N.S. Dalal, P.K. Gantzel, S. Hill, D.N. Hendrickson, Inorg. Chem. **47**, 1965–1974 (2008)
55. A. Garg, Europhys. Lett. **22**(3), 205 (1993)
56. C.-S. Park, A. Garg, Phys. Rev. B **65**, 064411 (2002)
57. F. Li, A. Garg, Phys. Rev. B **83**(13), 132401 (2011)
58. J. Liu, E. del Barco, S. Hill, Phys. Rev. B **85**(1), 012406 (2012)
59. E.-C. Yang, W. Wernsdorfer, S. Hill, R.S. Edwards, M. Nakano, S. Maccagnano, L.N. Zakharov, A.L. Rheingold, G. Christou, D.N. Hendrickson, Polyhedron **22**, 1727–1733 (2003)
60. E.-C. Yang, W. Wernsdorfer, L.N. Zakharov, Y. Karaki, A. Yamaguchi, R.M. Isidro, G.-D. Lu, S.A. Wilson, A.L. Rheingold, H. Ishimoto, D.N. Hendrickson, Inorg. Chem. **45**, 529–546 (2006)
61. J. Lawrence, S. Hill, E.-C. Yang, D.N. Hendrickson, Phys. Chem. Chem. Phys. **2009**(11), 6743–6749 (2009)
62. R. Inglis, S.M. Taylor, L.F. Jones, G.S. Papaefstathiou, S.P. Perlepes, S. Datta, S. Hill, W. Wernsdorfer, E.K. Brechin, in *Dalton Transactions*, (2009), pp. 9157–9168
63. R. Inglis, L.F. Jones, C.J. Milios, S. Datta, A. Collins, S. Parsons, W. Wernsdorfer, S. Hill, S.P. Perlepes, S. Piligkos, E.K. Brechin, in *Dalton Transactions*, (2009), pp. 3403–3412
64. E. del Barco, S. Hill, C.C. Beedle, D.N. Hendrickson, I.S. Tupitsyn, P.C.E. Stamp, Phys. Rev. B **82**, 104426 (2010)
65. W. Wernsdorfer, T.C. Stamatatos, G. Christou, Phys. Rev. Lett. **101**, 237204 (2008)
66. E. del Barco, S. Hill, D.N. Hendrickson, Phys. Rev. Lett. **103**, 059701 (2009)
67. C.J. Milios, A. Vinslava, W. Wernsdorfer, S. Moggach, S. Parsons, S.P. Perlepes, G. Christou, E.K. Brechin, J. Am. Chem. Soc. **129**(10), 2754–2755 (2007)
68. O. Waldmann, Inorg. Chem. **46**, 10035–10037 (2007)
69. W.H. Harman, T.D. Harris, D.E. Freedman, H. Fong, A. Chang, J.D. Rinehart, A. Ozarowski, M.T. Sougrati, F. Grandjean, G.J. Long, J.R. Long, C.J. Chang, J. Am. Chem. Soc. **132**, 18115–18126 (2010)
70. J.M. Zadrozny, J. Liu, N.A. Piro, C.J. Chang, S. Hill, J.R. Long, Chem. Commun. **48**, 3927–3929 (2012)
71. R. Ruamps, R. Maurice, M. Boggio-Pasqual, N. Guihery, L. Batchelor, J. Liu, S. Hill, T. Mallah, A.-L. Barra, J. Am. Chem. Soc. **135**, 3017–3026 (2013)
72. N. Ishikawa, M. Sugita, T. Ishikawa, S. Koshihara, Y. Kaizu, J. Am. Chem. Soc. **125**, 8694–8695 (2003)
73. M. AlDamen, J.M. Clemente-Juan, E. Coronado, C. Martì-Gastaldo, A. Gaita-Ariño, J. Am. Chem. Soc. **130**, 8874–8875 (2008)
74. J.D. Rinehart, J.R. Long, J. Am. Chem. Soc. **131**, 12558–12559 (2009)
75. S. Ghosh, S. Datta, L. Friend, S. Cardona-Serra, E. Coronado, S. Hill, Dalton Trans. **41**, 13697 (2012)

Part II
Beyond Single Molecules

Chapter 5
Magnetic Avalanches in Molecular Magnets

Myriam P. Sarachik

Abstract The reversal of the magnetization of crystals of molecular magnets that have a large spin and high anisotropy barrier generally proceeds below the blocking temperature by quantum tunneling. This is manifested as a series of controlled steps in the hysteresis loops at resonant values of the magnetic field where energy levels on opposite sides of the barrier cross. An abrupt reversal of the magnetic moment of the entire crystal can occur instead by a process commonly referred to as a magnetic avalanche, where the molecular spins reverse along a deflagration front that travels through the sample at subsonic speed. In this chapter, we review experimental results obtained to date for magnetic deflagration in molecular nanomagnets.

5.1 Background

First reported by Paulsen and Park [1], magnetic avalanches occur in many different molecular magnets. Systematic experimental studies of avalanches have focussed largely on crystals of Mn_{12}-ac [$Mn_{12}O_{12}(CH_3COO)_{16}(H_2O)_4$] a particularly simple, high-symmetry prototype of this class of materials.

As shown in the left panel of Fig. 5.1, the magnetic core of Mn_{12}-ac has four Mn^{4+} ($S = 3/2$) ions in a central tetrahedron surrounded by eight Mn^{3+} ($S = 2$) ions. The ions are coupled by superexchange through oxygen bridges with the net result that the four inner and eight outer ions point in opposite directions, yielding a total spin $S = 10$ [2]. The magnetic core is surrounded by acetate ligands, which serve to isolate each core from its neighbors in a body-centered tetragonal lattice. A crystalline sample typically contains $\sim 10^{17}$ or more (nearly) identical, weakly interacting single molecule nanomagnets in (nearly) identical crystalline environments.

The interesting physics and potential applications of Mn_{12}-ac and similar materials derive from the fact that: (i) the exchange between ions within the magnetic core is very strong, resulting in a sizable, rigid spin-10 magnetization per molecule

M.P. Sarachik (✉)
City College of New York, CUNY, New York, NY 10031, USA
e-mail: sarachik@sci.ccny.cuny.edu

J. Bartolomé et al. (eds.), *Molecular Magnets*, NanoScience and Technology,
DOI 10.1007/978-3-642-40609-6_5, © Springer-Verlag Berlin Heidelberg 2014

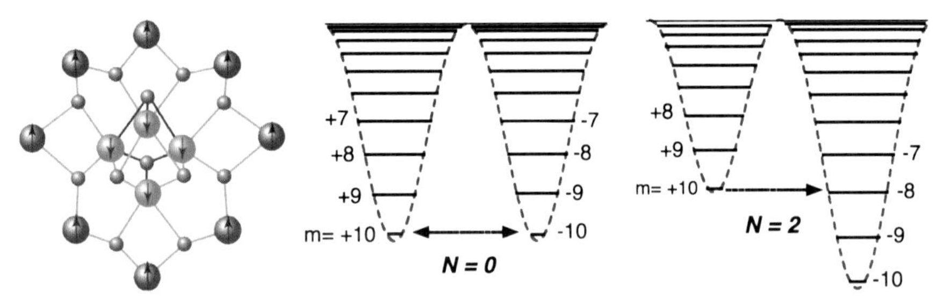

Fig. 5.1 *Left panel*: Chemical structure of the core of the Mn_{12} molecule. The four inner spin-down Mn^{3+} ions each have spin $S = 3/2$; the eight outer spin-up Mn^{4+} ions each have spin $S = 2$, yielding a net spin $S = 10$ for the magnetic cluster; the small grey spheres are O bridges; the arrows denote spin direction. Acetate ligands and water molecules have been removed for clarity; *Middle panel*: Double-well potential in the absence of magnetic field showing spin-up and spin-down levels separated by the anisotropy barrier. Different spin projection states $|m\rangle$ are indicated. *The arrows* denote quantum tunneling. *Right panel*: Double-well potential for the $N = 2$ step in a magnetic field applied along the easy axis

with no internal spin degrees of freedom at low temperatures, and (ii) the anisotropy is exceptionally large, so that the spins are bistable at low temperature, exhibiting slow relaxation and hysteresis below a blocking temperature T_B. To lowest order, the spin Hamiltonian is given by:

$$\mathcal{H} = -DS_z^2 - g_z \mu_B H_z S_z + \cdots + \mathcal{H}', \tag{5.1}$$

where the first term denotes the anisotropy barrier, the second is the Zeeman energy that splits the spin-up and spin-down states in a magnetic field, and the last term, $\mathcal{H}'$, contains all symmetry-breaking operators that do not commute with S_z, thereby allowing quantum tunneling. For Mn_{12}-ac, $D = 0.548$ K, $g_z = 1.94$; μ_B is the Bohr magneton.

As illustrated in the middle and right-hand panel of Fig. 5.1, the energy is modeled as a double-well potential, with one well corresponding to the spin pointing along the easy axis in one direction and the other to the spin pointing in the opposite direction. In zero field, there is a set of discrete, doubly degenerate energy levels corresponding to $(2S + 1)$ projections, $m = +10, +9, \ldots, 0, \ldots, -9, -10$, of the total spin along the easy (c-axis) of the crystal. Applying a magnetic field along the easy axis lowers the energy of the potential well with spins pointing in the direction of the field relative to the potential well for spins opposite to the field.

The relaxation rate decreases as the temperature is reduced, and below a (sweep rate-dependent) blocking temperature ($T_B \sim 3$ K), the large anisotropy barrier gives rise to slow relaxation and hysteresis loops that display steps [3] as a function of magnetic field H_z as the magnetic field is swept from full magnetization in one direction to full magnetization in the other [3–5]. The left panel of Fig. 5.2 shows the magnetization M as a function of magnetic field $\mu_0 H_z$. These steps, characteristic of molecular magnets, can be understood with reference to the double well potential of Fig. 5.1: faster relaxation occurs by spin-tunneling at the "resonant" values of the magnetic field that correspond to alignment of levels on opposite sides of

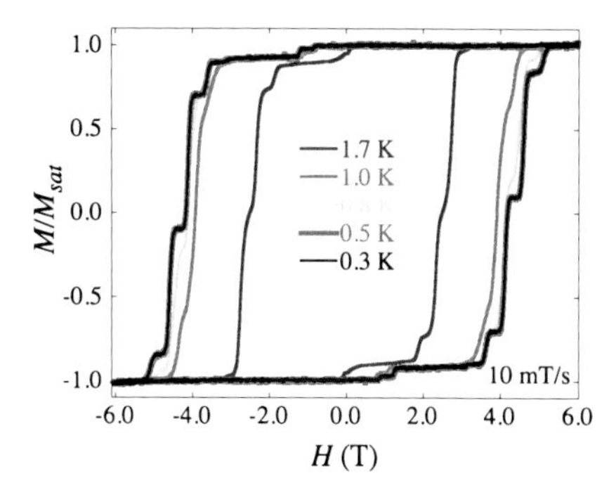

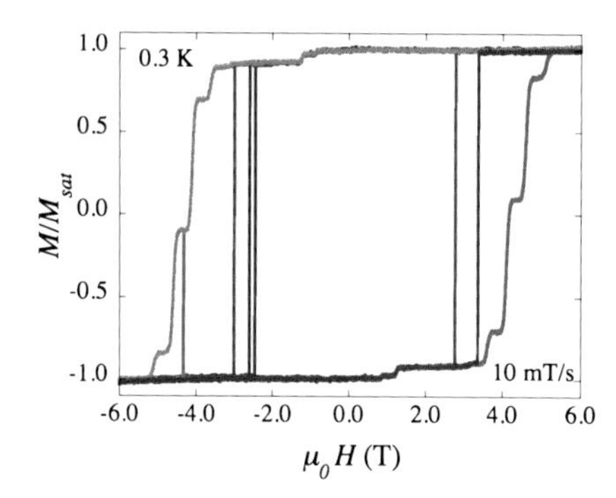

Fig. 5.2 *Left*: Hysteresis loops of a Mn_{12}-ac crystal for magnetic field applied along the uniaxial *c*-axis direction at different temperatures below the blocking temperature; the magnetization is normalized by its saturation value; magnetic field was swept at 10 mT/s. *Right*: Hysteresis loops at 0.25 K interrupted by magnetic avalanches (*vertical lines*)

the anisotropy barrier. Full saturation of the magnetization is thereby reached in a stepwise fashion, where the detailed form of the steps depend on sweep-rate and temperature. For reviews, see Refs. [6–13] and articles in the current volume.

By contrast, a magnetic avalanche signals a sudden reversal of the full magnetization of the crystal, as shown in the right panel of Fig. 5.2. This process has been attributed to a thermal runaway which can be understood again with reference to the right panel of Fig. 5.1: when tunneling of a molecular spin occurs from the lowest state of the metastable (left-hand) well to an excited state in the stable (right-hand) well, the subsequent decay to the ground state results in the release of heat that, under appropriate conditions, can further accelerate the magnetic relaxation. Direct measurements of the heat emitted have confirmed the thermal nature of these avalanches.

From time-resolved measurements of the local magnetization using an array of micron-sized Hall sensors placed on the surface of Mn_{12}-ac crystals, Suzuki et al. [14] discovered that a magnetic avalanche propagates through the crystal at subsonic speed in the form of a thin interface between regions of opposite spin magnetization. Figure 5.3(a) shows traces recorded during an avalanche by sensors placed sequentially along the easy axis near the center of a Mn_{12}-ac sample. Figure 5.3(b) is a plot of the sensor position versus the time of arrival of the peak. The inset is a schematic that illustrates the bunching of field lines at the propagating front that gives rise to the observed peaks. From these measurements one deduces that the front separating up- and down-spins travels with a constant (field-dependent) speed on the order of 1 to 30 m/s, two to three orders of magnitude slower than the speed of sound.

From a thermodynamic point of view, a crystal of Mn_{12} molecules placed in a magnetic field opposite to the magnetic moment is equivalent to a metastable (flammable) chemical substance. The release of energy by a metastable chemical substance is combustion or slow burning, technically referred to as deflagration [15]. It occurs as a flame front of finite width propagates at a constant speed small compared to the speed of sound. For "magnetic deflagration" in Mn_{12}-ac, the role of

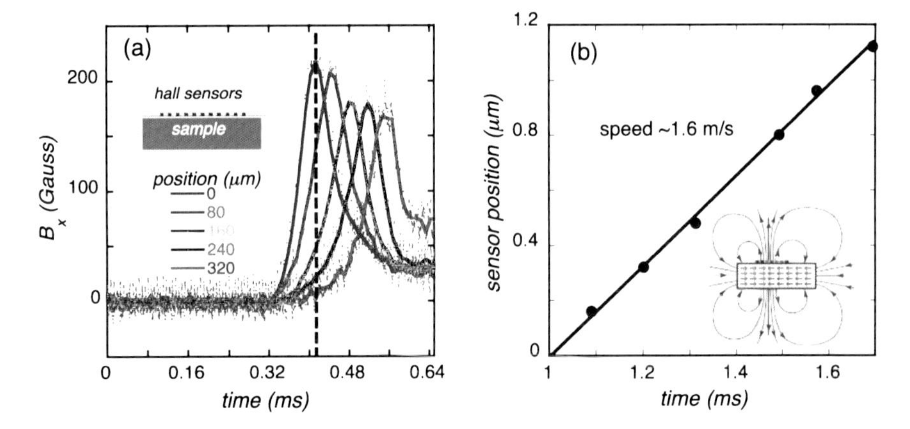

Fig. 5.3 (**a**) The local magnetization measured as a function of time by an array of micron-sized Hall sensors placed along the surface of the sample. *The inset* shows the placement of the Hall sensors on the crystal; (**b**) The sensor position as a function of the time at which the sensor registered the peak. The propagation speed for this avalanche is 2.2 m/s, approximately three orders of magnitude below the speed of sound. *The inset* illustrates the "bunching" of magnetic field lines as the deflagration front travels past a given Hall sensor

the chemical energy stored in a molecule is played by the difference in the Zeeman energy, $\Delta E = 2g\mu_B HS$, for states of the Mn_{12}-ac molecule that correspond to **S** parallel and antiparallel to **H**.

The avalanches that have been studied experimentally to date are driven predominantly by the increase in temperature associated with an input of energy. As further discussed below, Chudnovsky and Garanin have proposed a comprehensive theory to account for this process [16]. In a subsequent series of papers, the same authors have pointed out that the decay rate is also affected near spin-tunneling resonances by dipolar fields that can block or unblock the tunneling [17–20]. They found that the magnetization adjusts self-consistently in such a way that the system is on resonance over a broad spatial extent, with the consequence that there can be propagating spin reversal fronts that are driven by dipolar interactions. In general, both dipolar field and temperature are expected to control the propagation of quantum deflagration [19, 20].

This review provides an overview in Sect. 5.2 of the work done to date on avalanches where temperature is the dominant driver of the deflagration front. Section 5.3 briefly considers the possibility of tunneling fronts driven by dipolar interactions. Section 5.4 ends the review with a brief summary and suggestions for future research.

5.2 Temperature-Driven Magnetic Deflagration

Although the probability of a spontaneous avalanche has been shown to be higher at resonant magnetic fields than off-resonance [21], avalanche ignition is unpredictable in a swept external magnetic field, the experimental protocol that has generally been used to study the steps in the hysteresis loops. Avalanche ignition under these conditions is a stochastic process that depends on factors such as the sweep rate, the temperature, the quality of the crystal, and perhaps other factors. In order to carry out systematic studies of avalanche characteristics one needs to trigger avalanches in a controlled manner. This has been achieved using a heater [22], and by using surface acoustic waves (which serve to heat the sample) [23]. Recent studies [24] have used current pulses. Control of the location as well as the time of ignition could be accomplished using optical methods.

5.2.1 Avalanche Ignition

McHugh et al. [22] used a resistive wire element as a simple electric heater to trigger avalanches in a manner similar to the work of Paulsen and Park [1]. In these experiments, an external magnetic field is ramped to, and held at a fixed value. The heater is then turned on to slowly heat the sample until an avalanche is triggered at a temperature measured by a small thermometer. Avalanches launched by this method occur at well-defined, reproducible ignition temperatures. Figure 5.4(a) shows a typical temperature profile: starting at the base temperature of 300 mK, the temperature gradually rises until an abrupt sharp increase in the temperature signals the ignition of an avalanche. For this particular avalanche triggered at $\mu_0 H_z = 0.83$ T, the ignition temperature is about 1 K.

Single crystals of Mn_{12}-ac are known to contain two types of molecules. In addition to the primary or "major" species described at the beginning of this review, as-grown crystals contain a second "minor" species at a level of ≈ 5 percent with lower (rhombohedral) symmetry [25–27]. These faster-relaxing molecules can be modeled by the same effective spin Hamiltonian, (5.1), with a lower anisotropy barrier of 0.49 K. Avalanches of each species can be studied in the absence of the other through an appropriate magnetic protocol described in Ref. [28].

Interestingly, avalanches are separately triggered by the two species in low magnetic field. As shown in Fig. 5.4(b), at low fields the minor species relaxes prior to and independently of the major species, while above ~ 0.7 T the major and minor species ignite together and propagate as a single front. It is analogous to grass and trees that can sustain separate burn fronts that abruptly merge into a single front when the grass becomes sufficiently hot to ignite the trees.

Despite the turbulent conditions that one might expect for deflagration (as in chemical combustion), quantum mechanical tunneling clearly plays a role, as demonstrated in Fig. 5.5, where the temperature of ignition is plotted as a function of a preset, constant magnetic field [22]. The temperature required to ignite avalanches exhibits an overall decrease with applied magnetic field, reflecting the fact that larger fields reduce the barrier (see the double-well potential in Fig. 5.1). The role of quantum mechanics is clearly evidenced by the minima observed in the ignition temperature at the resonant magnetic fields due to tunneling when levels cross on opposite sides of the anisotropy barrier.

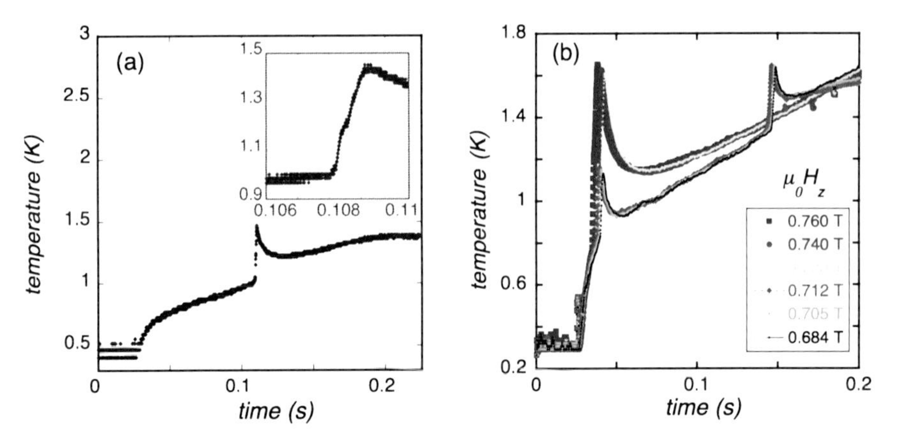

Fig. 5.4 (**a**) Temperature recorded by a thermometer in contact with a Mn_{12} crystal during the triggering of an avalanche at 0.83 T. The heater is turned on at $\sim$ 0.03 s, the temperature then increases slowly until an abrupt rise in temperature at 0.11 s signals the ignition of an avalanche. *The inset* shows data taken near ignition with higher resolution. The noise at low temperatures derives from digitizing the analog output of the thermometer, which depends weakly on temperature below 0.4 K; (**b**) Temperature profiles for avalanches of major and minor species triggered at low fields in a Mn_{12} crystal. The two types of avalanches are triggered separately below a sample-dependent magnetic field, while at higher fields ignition of the minor species triggers the ignition of the major species

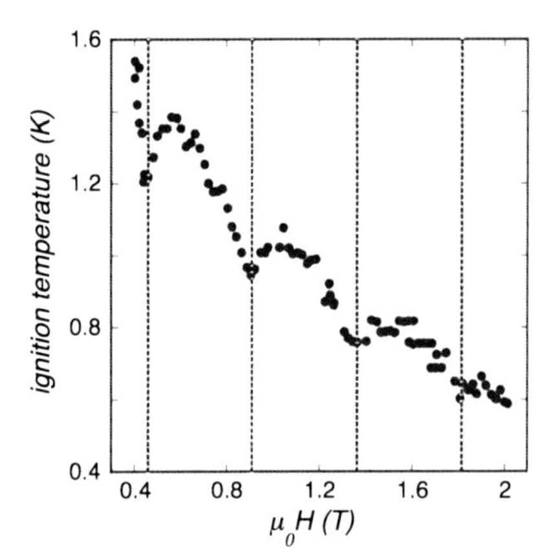

Fig. 5.5 Temperature required to ignite avalanches plotted as a function of magnetic field. *The vertical lines* denote the magnetic fields where sharp minima occur in the ignition temperature corresponding to tunneling near the top of the anisotropy barrier. The overall decrease in ignition temperature is due to the reduction of the anisotropy barrier as the field is increased

In the ignition studies described above, the barrier against spin reversal was lowered by applying a longitudinal magnetic field, H_z, along the uniaxial c-direction, which serves to unbalance the potential wells and lower the barrier against spin reversal. Tunneling can also be promoted by applying a transverse field H_x which reduces the anisotropy barrier by introducing a symmetry-breaking term, $g\mu_B H_x S_x$, to the Hamiltonian, (5.1). Macià et al. [21] investigated the threshold for avalanche ignition in Mn_{12}-ac as a function of the magnitude and direction of a magnetic field applied at various angles with respect to the anisotropy axis and as a function of tem-

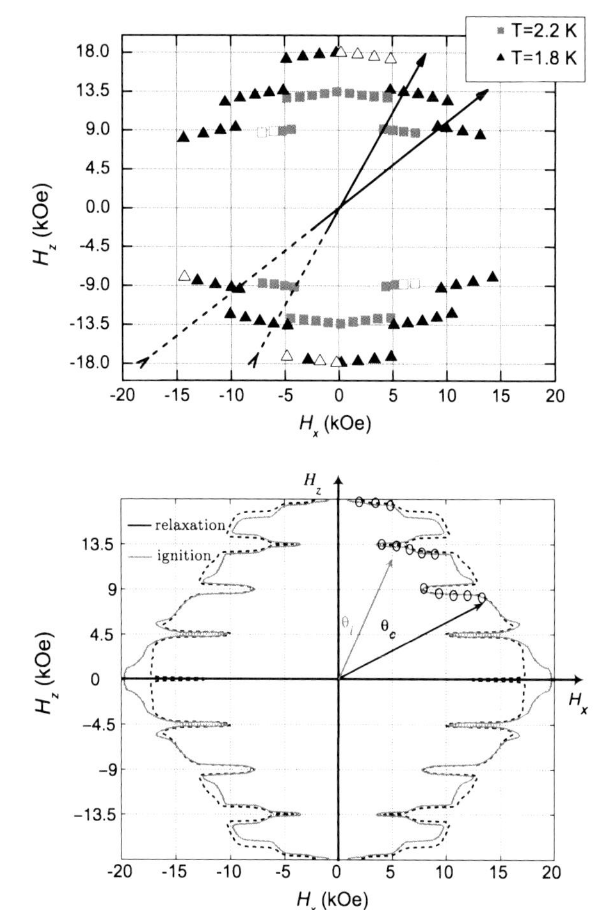

Fig. 5.6 *Top*: Angle dependence of metastability measured through the occurrence of avalanches. *Squares* (*triangles*) denote parameter values where deflagration occurs for initial temperature 2.2 K (1.8 K); *Bottom*: Theoretical calculation for the area of stability against ignition of avalanches (*solid curve*) and against slow relaxation (*dashed curve*) (as is true for the determination of blocking temperatures, the position of *the dashed line* depends on the typical experimental time scale). *Circles* denote points where avalanches are predicted to occur at a given angle θ_i within the first quadrant. The angle θ_c denotes the crossing point between areas of slow relaxation and avalanche stability. These results were obtained with T_f as a parameter varying from 6.8 K for $H = 4600$ Oe to 10.9 K for $H = 9200$ Oe. From Macià et al. [21]

perature. As the external field is increased at a constant rate from negative saturation to positive values, both H_z and H_x increase, tracing a trajectory in the (H_z, H_x) parameter space. Examples of sweeps starting from zero are shown by the arrows in Fig. 5.6. An avalanche was recorded for each pair (H_x, H_z) denoted by a square (for $T = 2.2$ K) or a triangle (for $T = 1.8$) K.

A theory of magnetic deflagration developed by Garanin and Chudnovsky [16] that considers only thermal effects (no dipolar interactions) explains the main features of the ignition experiments of McHugh et al. in which the critical relaxation rate was reached by varying T_0 with a heater, and the experiments of Macià et al., where the ignition threshold was reached by controlling the barrier U using H_x and H_z.

A deflagration front is expected to develop when the rate at which energy is released by the relaxing metastable spins exceeds the rate of energy loss through the boundaries of the crystal. This condition can be expressed in terms of a critical relaxation rate, [16]

$$\Gamma_c = \frac{8k(T_0)k_B T_0^2}{U\langle E\rangle l^2}, \tag{5.2}$$

where $\Gamma_c = \Gamma_0 \exp[-U/k_B T_0]$, T_0 is the initial temperature, k is the thermal conductivity, l^2 is the characteristic cross section of the crystal, and $\langle E\rangle$ is the average amount of heat released per molecule when its spin relaxes to the stable state. The energy released when a single molecule relaxes is the Zeeman energy $\Delta E = 2g\mu_B S B_z$. To obtain the average energy per molecule, one has to consider the fraction of molecules that relax:

$$\langle E\rangle = 2g\mu_B S\left(\frac{\Delta M}{2M_s}\right)B_z, \tag{5.3}$$

where M_s is the saturation magnetization and $\Delta M = |M_z - M_s|$ is the change from initial to final magnetization.

Calculations based on (5.2) yield the curves shown in Fig. 5.6(b). Two areas are defined in the (H_z, H_x) parameter space where the spins are expected to be metastable against relaxation: the solid line denotes the region of metastability against relaxation by triggering avalanches while the dashed curve delineates the region of metastability against slow, stepwise relaxation.[1] If the experimental trajectory, denoted by the arrows, crosses the grey solid line first, an avalanche will ignite. If the dashed line is crossed first, the metastable spins will relax slowly without triggering an avalanche. This defines a critical angle θ_c, above which an avalanche cannot occur.

Macià et al. measured the ignition threshold by applying an increasing external field at an angle with respect to the crystal. The relaxation rate increases as the field grows until Γ_c is reached and deflagration ignites, as shown in Fig. 5.6(a). For sufficiently large values of H_x, they found that the slow relaxation of the metastable spins occurs before deflagration can ignite. This defines a line in parameter space separating regions where one or the other mode of relaxation occurs, as shown in Fig. 5.6(b). The theory predicts that the transverse field should result in a significant decrease in the magnetization metastability at the resonant fields of H_z. The data recorded in Fig. 5.6(a) confirm this and are consistent with the ignition temperatures of Fig. 5.5. In addition, ignition thresholds were measured at two different temperatures. The area of stability is clearly reduced by the increased initial temperature, as expected.

5.2.2 Avalanche Speed

Hernández-Mínguez et al. [23, 29] carried out a systematic investigation of avalanche speeds as a function of a preset, constant magnetic field $\mu_0 H_z$ for

---

[1] As is true for the determination of blocking temperatures, the position of the dashed line depends on the typical experimental time scale.

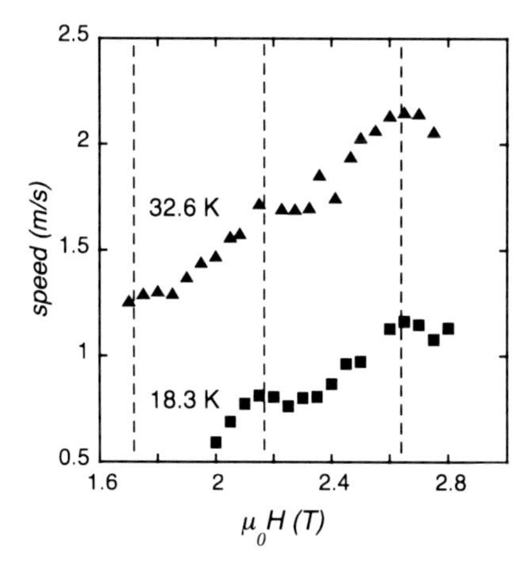

Fig. 5.7 The speed of propagation of the magnetic avalanche deflagration front as a function of the (fixed) field at which the avalanche is triggered. Data are shown for category C avalanches for which the average energy released, $\langle E \rangle$, is held constant at 18.3 K and 32.6 K (see text). Note the enhancement of propagation velocity at magnetic fields corresponding to quantum tunneling (denoted by *vertical dotted lines*). From McHugh thesis [30]

avalanches triggered by surface acoustic waves. From SQUID-based measurements of the total magnetization of a crystal of known dimensions, and the realization that the avalanche propagates as an interface between regions of opposite magnetization [14], they deduced that the speed of propagation of the avalanches is enhanced at the resonant fields where tunneling occurs, confirming the important role of quantum mechanics and prompting the authors to name the phenomenon "quantum magnetic deflagration". Similar results were obtained from local, time-resolved magnetization measurements using micron-sized Hall sensors [30], as shown in Fig. 5.7.

McHugh et al. [31] reported a detailed, systematic investigation of the speed of magnetic avalanches for various experimental conditions. The speed of propagation of an avalanche is described approximately [14] by the expression, $v \sim (\kappa/\tau_0)^{1/2} \exp[-U(H)/2k_B T_f]$, where U is the barrier against spin reversal, T_f is the flame temperature at or near the propagating front where energy is released by the reversing spins, κ is the thermal diffusivity, and τ_0 is an attempt time. We note that the energy barrier U and the flame temperature T_f appear only as the ratio U/T_f in the above expression for the velocity. It is therefore convenient to plot the speed of the avalanche as a function of U/T_f.

In the studies of McHugh et al. [31], avalanches were controllably triggered using three different protocols, as follows:

(A) From fixed (maximum) initial magnetization in various external fields; there is full (maximum) magnetization reversal, $\Delta M/2M_s = 1$; both U and T_f vary;

(B) In fixed external field starting from different initial magnetization; here the amount of "fuel" $\Delta M/2M$ is varied for a fixed magnetic field (thus U is held constant); the avalanches differ primarily through the amount of energy released—the flame temperature T_f varies;

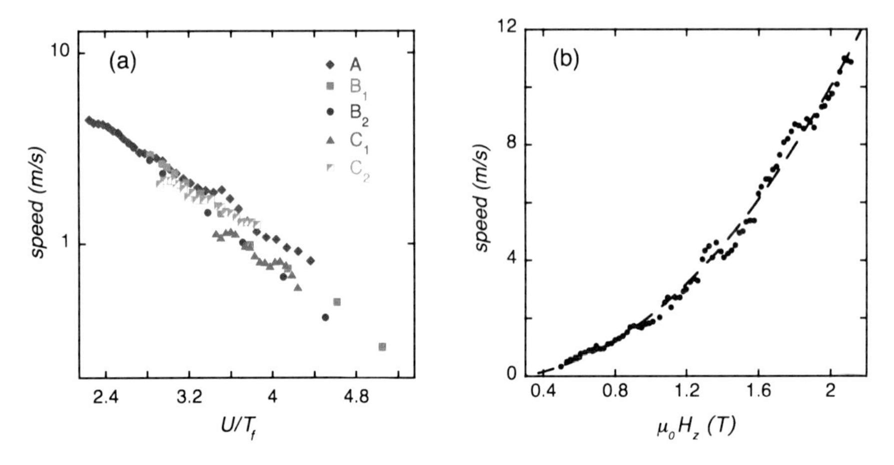

Fig. 5.8 (**a**) Avalanche speeds for a single crystal with various initial magnetic preparations. A denotes avalanches with $\Delta M/2M_s = 1$; B_1 and B_2 denote data taken at $\mu_0 H_z = 2.2$ T and 2.5 T, respectively; C_1 and C_2 denote avalanches with estimated flame temperatures $T_f \approx 10$ K and 12 K, respectively. (**b**) Avalanche speeds for different crystal with $\Delta M/2M_s = 1$. The fit requires an unphysical temperature dependence for the thermal diffusivity, $\kappa \propto T^{3.5}$

(C) for fixed energy release, thus fixed T_f, by adjusting external magnetic fields and initial magnetization.

The theory of magnetic deflagration [16] provides the following theoretical expression for the speed of the deflagration front:

$$v = \sqrt{\frac{3k_B T_f \kappa \Gamma(B, T_f)}{U(B)}}. \tag{5.4}$$

If one assumes the thermal diffusivity κ is approximately independent of temperature, or that its temperature dependence is unimportant compared to that of other parameters in the problem, then all measured avalanche velocities should collapse onto a single curve when plotted as a function of (U/T_f).

Figure 5.8(a) [31] shows the measured avalanche velocity as a function of (U/T_f) obtained using the three different protocols described above. Although an approximate collapse is obtained, there are clear and systematic deviations. That these different experimental protocols introduce systematic variations, albeit small, suggests that the theory is incomplete.

Shown in Fig. 5.8(b), an attempt to fit to the theory by allowing the thermal diffusivity to vary as a power law of the temperature for avalanches of type (A) that involve full magnetization reversal yields $\kappa \sim T^{3.5}$. This is a distinctly unphysical result, as the thermal diffusivity is generally a strongly decreasing function of temperature [32] for these materials. We note that experimental measurements of the thermal diffusivity of Mn_{12} are not available.

In brief, the Chudnovsky-Garanin theory of deflagration captures the main features found in the experiments. However, although the ignition experiments have yielded results that agree with it in detail, the theory does not provide a fully satisfactory description of the speed of propagation of the deflagration fronts. It is possible that dipolar interactions (discussed in the next section) play a sufficiently important role to account for the discrepancies.

5.3 Cold Deflagration

Although small compared to other energies, dipolar interactions are now recognized as playing an important role in many molecular magnets. This is confirmed by reports of ferromagnetism mediated by dipolar interactions below 1 K in Mn_{12}-ac [33, 34], as well as in other molecular magnets [35–37]. Even in the paramagnetic phase, where long range order is not realized, the change of the spin state of a molecule results in a change of the long-range dipolar field acting on other nearby spins [38, 39]. Thus, dipolar interactions can tune spins in and out of resonance and can thereby have a profound influence on the spin dynamics.

D.A. Garanin and E.M. Chudnovksy [17–20] have proposed that propagating fronts of spin reversal (avalanches) can occur that are driven by the dipolar interactions between magnetic molecules. Their numerical simulations show that dipole-dipole forces establish spatially inhomogeneous states in molecular magnets[2] such that there is a self-consistent adjustment of the metastable population acting to create a dipolar field that is constant over a sizable region of the sample, thereby bringing the system to resonance over an extended region where all the spins can relax collectively by tunneling. This, in turn, can lead to propagating fronts of spin reversal, which they have dubbed "cold deflagration".

Interestingly, Garanin and Chudnovsky have noted that such collective traveling spin reversal fronts are potential sources of Dicke superradiance [41–47] at frequencies in the teraHertz range, a particularly interesting region of the electromagnetic spectrum where few sources are available. If self-organization does result in a uniform dipolar field within the tunneling front, the resonant condition is fulfilled for a macroscopic number of magnetic molecules inside the front, and it is indeed plausible that these avalanches could emit a superradiant electromagnetic signal. Intense bursts of radiation have indeed been detected experimentally during magnetic avalanches. There has been much speculation that this could be Dicke superradiance, but experiments have been inconclusive on this very interesting issue [48–53].

The avalanches that have been studied experimentally to date have been triggered in large longitudinal bias fields near the higher-number field resonances. In these circumstances, the spins tunnel from a metastable state and decay to a ground state of opposite spin that is much lower in energy, releasing Zeeman energy to the phonon system and generating heat. This triggers thermal avalanches, as confirmed by a measured increase in the temperature of the crystal. It will be of great interest

[2] Experimental evidence of such inhomogeneous states has been obtained by local measurements in Mn_{12}-ac that show oscillating magnetization near a tunneling resonance; see [40].

to find magnetic avalanches that are driven (or partially driven) by dipole-dipole interactions, and to study the relative roles of cold and "hot" deflagration for different parameters (temperature, parallel and perpendicular magnetic field, sweep rate, and so on). The possibility that superradiance will be emitted in the process is particularly exciting.

Although hints of cold deflagration may have been found, there are no definitive reports of this process to date. We note that a particularly large effect is expected in the presence of a strong transverse field which promotes tunneling and lowers the anisotropy barrier, so that relaxation toward equilibrium can proceed by tunneling at zero longitudinal bias field without thermal assistance and without releasing Zeeman energy into the system. However, recent experiments [24] show that in a strong transverse magnetic field the relaxation near tunneling resonances becomes so rapid that it is difficult to create an initial state with a sizable out-of-equilibrium population sufficient to trigger a tunneling front. This is a major experimental challenge for realizing dipole-driven spin-reversal fronts.

5.4 Summary and Outlook for the Future

Once considered events to be avoided, as they interfere with a detailed study of the stepwise process of magnetization via spin-tunneling, magnetic avalanches have recently been the focus of attention and renewed interest, partly stimulated by the theoretical suggestion that the radiation emitted during an avalanche may be in the form of coherent (Dicke) superradiance [41]. Although the issue of coherence of the radiation has yet to be resolved, recent studies have clarified the nature of the avalanche process itself.

Magnetic avalanches proceed as traveling fronts along which the molecules reverse their spin, releasing Zeeman energy which drives the spins to reverse throughout the crystal. These spin-reversal fronts propagate at subsonic speeds, and are analogous to the process of chemical combustion, technically known as chemical deflagration: here a chemical reaction propagates along a front where energy is released that drives the reaction front at subsonic speed. A burning sheet of paper is a clear example of chemical deflagration. A great advantage of the magnetic analog is that, unlike burning paper, it is non-destructive, fully reversible and continuously tunable using an external magnetic field. Magnetic deflagration is thus amenable to carefully controlled study.

In this chapter, we have reviewed experiments on the ignition and the speed of propagation of a magnetic avalanche driven by the release of Zeeman energy at the deflagration front. The conditions for ignition and the speed of propagations both show clear effects of quantum mechanics at the resonant fields that allow tunneling across the anisotropy barrier. The theory of magnetic deflagration of Chudnovsky and Garanin is in excellent agreement with the parameters determined experimentally for ignition. The theory also provides a good qualitative fit to the observed avalanche velocity, but there are detailed discrepancies that suggest that additional factors need to be included to obtain good quantitative agreement. The effect of dipolar interactions must clearly be included in a full theory [20].

Dipole-dipole interactions are sufficiently strong in some molecular magnets that they lead to long-range ordering at low temperatures. A particularly interesting consequence of dipolar interactions is "cold deflagration" proposed by Garanin and Chudnvosky, a process in which spin-reversal fronts are driven predominantly by dipolar effects. These have not yet been realized (or perhaps recognized) experimentally. Garanin and Chudnovsky suggest that self-organization of the internal dipolar fields brings molecules into resonance over a broad front that may serve as a source of coherent teraHertz radiation. In addition to the intrinsic interest, it would be interesting to search for cold deflagration as a potential source of Dicke superradiance in this difficult and important range of the electromagnetic spectrum, where few sources are available.

More experimental work is clearly needed. Currently underway, a detailed investigation of avalanche ignition in fixed transverse and fixed longitudinal (bias) field is yielding interesting, new results [24]. Measurements of the thermal diffusivity would provide an important constraint on the theory, as would a reliable determination of the (local) temperature of the deflagration front. Investigations of the influence of sample shape, size and quality would also be illuminating. Spatial control of the avalanche ignition points, possibly by optical means, could provide important information. Studies of the shape of the deflagration front, and its character (turbulent or laminar) would be particularly interesting.

The possibility of observing a transition to detonation is intriguing [20, 54]. Deflagration is but one type of combustion process. Another, more violent type, is detonation, where heat spreads from the reaction front as a shock wave rather than by diffusion. It is natural to ask whether crystals of molecular magnets can support the magnetic analog of chemical detonation. Decelle et al. [55] have reported results hinting at this possibility using high external field sweep rates (4 kT/s). The interpretation of these experiments is not entirely clear, and much work remains to be done.

We close by noting once more that, to the degree that magnetic deflagration resembles chemical deflagration, the magnetic manifestation of this process offers some clear advantages for the study of chemical combustion. The magnetic analog is non-destructive and reversible, enabling a broad range of controlled studies on a single sample. Unlike the chemical process, it is a particularly interesting realization of deflagration in which quantum mechanical tunneling plays an important role.

Acknowledgements The author thanks the students whose experiments made this review possible, most particularly Yoko Suzuki and Sean McHugh; and Eugene Chudnovsky and Dmitry Garanin for their careful reading of this manuscript. Support was provided by NSF Grant No. DMR-00451605.

References

1. C. Paulsen, J.G. Park, in *Quantum Tunneling of Magnetization—QTM'94*, ed. by L. Gunther, B. Barbara (Kluwer, Dordrecht, 1995), pp. 189–207

2. R. Sessoli, D. Gatteschi, A. Caneschi, M.A. Novak, Nature (London) **365**, 141 (1993)
3. J.R. Friedman, M.P. Sarachik, J. Tejada, R. Ziolo, Phys. Rev. Lett. **76**, 3830 (1996)
4. J.M. Hernandez, X.X. Zhang, F. Luis, J. Bartholome, J. Tejada, R. Ziolo, Europhys. Lett. **35**, 301 (1996)
5. J.M. Hernandez, X.X. Zhang, F. Luis, J. Tejada, J.R. Friedman, M.P. Sarachik, R. Ziolo, Phys. Rev. B **55**, 5858 (1997)
6. B. Barbara, L. Thomas, F. Lionti, I. Chiorescu, A. Sulpice, J. Magn. Magn. Mater. **200**, 167–181 (1999)
7. J.R. Friedman, in *Exploring the Quantum/Classical Frontier: Recent Advances in Macroscopic Quantum Phenomena*, ed. by J.R. Friedman, S. Han (Nova Science, Hauppauge, 2002), pp. 219–249
8. K.M. Mertes, Y. Suzuki, M.P. Sarachik, Y. Myasoedov, H. Shtrikman, E. Zeldov, E.M. Rumberger, D.N. Hendrickson, G. Christou, Solid State Commun. **127**, 131–139 (2003)
9. D. Gatteschi, R. Sessoli, Angew. Chem., Int. Ed. Engl. **42**, 268 (2003)
10. D. Gatteschi, R. Sessoli, J. Villain, *Molecular Nanomagnets* (Oxford University Press, London, 2006)
11. B. Barbara, Inorg. Chim. Acta **361**, 3371–3379 (2008)
12. R. Bagai, G. Christou, Chem. Soc. Rev. **38**, 1011 (2009)
13. J.R. Friedman, M.P. Sarachik, Annu. Rev. Condens. Matter Phys. **1**, 109–128 (2010)
14. Y. Suzuki, M.P. Sarachik, E.M. Chudnovsky, S. McHugh, R. Gonzalez-Rubio, N. Avraham, Y. Myasoedov, E. Zeldov, H. Shtrikman, N.E. Chakov, G. Christou, Phys. Rev. Lett. **95**, 147201 (2005)
15. L.D. Landau, E.M. Lifshitz, *Fluid Dynamics* (Pergamon, Elmsford, 1987)
16. D.A. Garanin, E.M. Chudnovsky, Phys. Rev. B **76**, 054410 (2007)
17. D.A. Garanin, E.M. Chudnovsky, Phys. Rev. Lett. **102**, 097206 (2009)
18. D.A. Garanin, Phys. Rev. B **80**, 014406 (2009)
19. D.A. Garanin, R. Jaafar, Phys. Rev. B **81**, 180401 (2010)
20. D.A. Garanin, S. Shoyeb, Phys. Rev. B **85**, 094403 (2012)
21. F. Macià, J.M. Hernandez, J. Tejada, S. Datta, S. Hill, C. Lampropoulos, G. Christou, Phys. Rev. B **79**, 092403 (2009)
22. S. McHugh, R. Jaafar, M.P. Sarachik, Y. Myasoedov, A. Finkler, H. Shtrikman, E. Zeldov, R. Bagai, G. Christou, Phys. Rev. B **76**, 172410 (2007)
23. A. Hernández-Mínguez, J.M. Hernández, F. Macià, A. García-Santiago, J. Tejada, P.V. Santos, Phys. Rev. Lett. **95**, 217205 (2005)
24. P. Subedi, S. Vélez, F. Macià, S. Li, M.P. Sarachik, J. Tejada, S. Mukherjee, G. Christou, A.D. Kent, Phys. Rev. Lett. **110**, 207203 (2013)
25. A. Caneschi, T. Ohm, C. Paulsen, D. Royal, C. Sangregorio, R. Sessoli, J. Magn. Magn. Mater. **177**, 1330 (1998)
26. Z. Sun, D. Ruiz, N.R. Dilley, M. Soler, J. Ribas, K. Folting, M.B. Maple, G. Christou, D.N. Hendrickson, Chem. Commun. **19**, 1973 (1999)
27. W. Wernsdorfer, R. Sessoli, D. Gatteschi, Europhys. Lett. **47**, 254 (1999)
28. S. McHugh, R. Jaafar, M.P. Sarachik, Y. Myasoedov, A. Finkler, E. Zeldov, R. Bagai, G. Christou, Phys. Rev. B **80**, 024403 (2009)
29. A. Hernández-Mínguez, F. Macià, J.M. Hernández, J. Tejada, P.V. Santos, J. Magn. Magn. Mater. **320**, 1457 (2008)
30. S. McHugh, Thesis, CUNY (2009)
31. S. McHugh, B. Wen, X. Ma, M.P. Sarachik, Y. Myasoedov, E. Zeldov, R. Bagai, G. Christou, Phys. Rev. B **79**, 174413 (2009)
32. C. Enss, S. Hunklinger, *Low-temperature Physics* (Springer, Berlin, 2005)
33. F. Luis, J. Campo, J. Gómez, G.J. McIntyre, J. Luzón, D. Ruiz-Molina, Phys. Rev. Lett. **95**, 227202 (2005)
34. B. Wen, P. Subedi, L. Bo, Y. Yeshurun, M.P. Sarachik, A.D. Kent, A.J. Millis, C. Lampropoupos, G. Christou, Phys. Rev. B **82**, 014406 (2010)

35. A. Morello, F.L. Mettes, F. Luis, J.F. Fernández, J. Krzystek, G. Aromí, G. Christou, L.J. de Jongh, Phys. Rev. Lett. **90**, 017206 (2003)
36. M. Evangelisti, A. Candini, A. Ghirri, M. Affronte, G.W. Powell, I.A. Gass, P.A. wood, S. Parsons, E.K. Brechin, D. Cllison, S.L. Leath, Phys. Rev. Lett. **97**, 167202 (2006)
37. E. Burzurí, F. Luis, B. Barbara, R. Ballou, E. Ressouche, O. Montero, J. Campo, S. Maegawa, Phys. Rev. Lett. **107**, 097203 (2011)
38. D.A. Garanin, Eur. Phys. J. B **85**, 107 (2012)
39. J. Liu, B. Wu, L. Fur, B. Diener, Q. Niu, Phys. Rev. B **65**, 224401 (2002)
40. N. Avraham, A. Stern, Y. Suzuki, K.M. Mertes, M.P. Sarachik, E. Zeldov, Y. Myasoedov, H. Shtrikman, E.M. Rumberger, D.N. Hendrickson, N.E. Chakov, G. Christou, Phys. Rev. B **72**, 144428 (2005)
41. E.M. Chudnovsky, D.M. Garanin, Phys. Rev. Lett. **89**, 157201 (2002)
42. C.L. Joseph, C. Calero, E.M. Chudnovsky, Phys. Rev. B **70**, 174416 (2004)
43. V.K. Henner, I.V. Kaganov, Phys. Rev. B **68**, 144420 (2003)
44. V.I. Yukalov, Laser Phys. Lett. **2**, 356 (2005)
45. V.I. Yukalov, E.P. Yukalova, Europhys. Lett. **70**, 306 (2005)
46. M.G. Benedict, P. Foldi, F.M. Peeters, Phys. Rev. B **72**, 214430 (2005)
47. I.D. Tokman, V.I. Pozdnjakova, G.A. Vugalter, A.V. Shvetsov, Phys. Rev. B **77**, 094414 (2008)
48. J. Tejada, R. Amigo, J.M. Hernández, E.M. Chudnovsky, Phys. Rev. B **66**, 014431 (2003)
49. M. Bal, J.R. Friedman, K.M. Mertes, W. Chen, E.M. Rumberger, D.N. Hendrickson, N. Avraham, Y. Myasoedov, H. Shtrikman, E. Zeldov, Phys. Rev. B **70**, 140403 (2004)
50. J. Tejada, E.M. Chudnovsky, J.M. Hernández, R. Amigo, Appl. Phys. Lett. **84**, 2373 (2004)
51. A. Hernández-Mínguez, M. Jordi, R. Amigo, A. García-Santiago, J.M. Hernández, J. Tejada, Europhys. Lett. **69**, 270 (2005)
52. M. Bal, J.R. Friedman, W. Chen, M. Tuominen, C.C. Beedle, E.M. Rumberger, D.N. Hendrickson, Europhys. Lett. **82**, 17005 (2008)
53. O. Shafir, A. Keren, Phys. Rev. B **79**, 180404 (2009)
54. M. Modestov, V. Bychkov, M. Marklund, Phys. Rev. Lett. **107**, 207208 (2011)
55. W. Decelle, J. Vanacken, V.V. Moshchalkov, J. Tejada, J.M. Hernández, F. Macià, Phys. Rev. Lett. **102**, 027203 (2009)

Chapter 6
Theory of Deflagration and Fronts of Tunneling in Molecular Magnets

D.A. Garanin

Abstract Decay of metastable states in molecular magnets leads to energy release that results in temperature increase that boosts the decay rate. This is the mechanism of the recently discovered magnetic deflagration that is similar to regular chemical burning and can propagate in a form of burning fronts. Near spin-tunneling resonances the decay rate is also affected by the dipolar field that can block or unblock tunneling. There are non-thermal fronts of tunneling in which the magnetization adjusts in such a way that the system is on resonance within the front core. Both dipolar field and temperature control fronts of quantum deflagration. The front speed can reach sonic values if a strong transverse field is applied to boost tunneling.

6.1 Introduction

Deflagration or burning is the decay of metastable states accelerated by a temperature rise due to energy released in this process [1, 2]. In most cases the decay rate has the Arrhenius temperature dependence, $\Gamma = \Gamma_0 \exp[-U/(k_B T)]$, where U is the energy barrier. Because of very strong positive feedback, burning can have a form of a thermal runaway: almost undetectably slow relaxation at the beginning followed by an explosion-like relaxation at the end (explosions at ammunition-storage sites, Bhopal disaster, etc.). In other cases there is a burning front propagating with a constant speed away from the ignition point. These fronts are driven by the heat conduction from the hot burned region to the cold unburned region before the front. Burning of a sheet of paper is a good example of a deflagration front.

Molecular magnets (MM), of which the most famous is $Mn_{12}Ac$ [3], are burnable materials because of their bistability resulting from a strong uniaxial anisotropy that creates an energy barrier [4]. One can make magnetic state metastable by applying a magnetic field along the anisotropy axis. Burning, of course, should lead to a much faster relaxation than a regular relaxation at fixed low temperatures. Indeed, in early

D.A. Garanin (✉)
Department of Physics and Astronomy, Lehman College, City University of New York, 250 Bedford Park Boulevard West, Bronx, New York 10468-1589, USA
e-mail: dmitry.garanin@lehman.cuny.edu

J. Bartolomé et al. (eds.), *Molecular Magnets*, NanoScience and Technology, DOI 10.1007/978-3-642-40609-6_6, © Springer-Verlag Berlin Heidelberg 2014

experiments on relaxation of large specimens of MM [5–7] an abrupt and nearly total relaxation of the metastable magnetization has been detected but not explained. The 2005 space-resolved experiments of the Sarachik group [8] on long crystals of $Mn_{12}Ac$ have shown propagating fronts of relaxation that were interpreted as deflagration fronts. In this experiments, regularly-spaced Hall probes at the sides of the crystal detected the transverse magnetic field created by the non-uniformity of the magnetization [9]. Measurements of the time dependence of the total magnetization by the Tejada group, inspired by the above experiment, have shown a linear time dependence that was attributed to a deflagration front travelling through a Mn_{12} crystal [10]. Here quantum maxima of the front speed vs the bias field have been detected, Fig. 4 of Ref. [10]. Discovery of magnetic deflagration, mainly on $Mn_{12}Ac$ [11–14], opened an active field of experimental research. Experiments at high sweep rates [15, 16] have shown spin avalanches propagating at a fast speed. In this region, deflagration can go over into detonation [17]. Magnetic deflagration (coupled to a structural phase transition) has also been observed on manganites [18] and intermetallic compounds [19, 20]. By contrast, it is problematic to observe deflagration fronts on another popular MM Fe_8 because of the pyramidal shape of its crystals.

One can ask if deflagration can exist in traditional magnetic systems, many having a strong uniaxial anisotropy. Unfortunately, the energy release in magnetic systems is much weaker than in the case of a regular (chemical) deflagration. Thus, at room temperatures, the ensuing temperature increase is too small to change the relaxation rate and support burning. Only at low temperatures the increase of the relaxation rate becomes large. A hallmark of magnetic deflagration is its non-destructive character. "Burned" MM can be recycled (put again into the metastable state) by simply reversing the longitudinal magnetic field.

A comprehensive theory of magnetic deflagration given in Ref. [21] includes calculations of the stationary speed of the burning front, ignition time due to local increase of temperature or change of the magnetic field, as well as the analysis of stability of the low-temperature state with respect to deflagration that depends on the heat contact of the MM crystal with the environment. However, up to now there is no complete accordance between the theory and experiment for several reasons. First, thermal diffusivity κ of Mn_{12} that plays a crucial role in deflagration has not been measured up to now. Second, there is no completely satisfactory theory of relaxation in molecular magnets that takes into account important collective effects such as superradiance and phonon bottleneck.

Because of their not too large spin ($S = 10$ for Mn_{12} and Fe_8), molecular magnets are famous exponents of spin tunneling [22–25] which has a resonance character and leads to the steps in dynamic hysteresis curves at the values of the longitudinal magnetic field where quantum levels of the spin at the two sides of the potential barrier match [26–28]. Since the discovery of magnetic deflagration there has been a quest for quantum effects in it. The simplest approach [10, 21] uses the fact that usually spin tunneling occurs via pairs of quantum levels just below the classical barrier. This tunneling is thermally assisted and can be described by an effective lowering of the energy barrier at resonance values of the bias field (Fig. 2 of Ref. [7]). Thus using the Arrhenius relaxation rate with such an effective barrier does incorporate spin tunneling. Experimentally it was found that spin tunneling strongly affects ignition of deflagration (Fig. 5 of Ref. [11]) and to a smaller extent the front speed (Fig. 5 of Ref. [11] and Fig. 4 of Ref. [10]).

Quantum effects in deflagration should be sensitive to the dipolar field created by the sample. In a long uniformly magnetized crystal of $Mn_{12}Ac$ the dipolar field is $B^{(D)} = 52.6$ mT, as calculated microscopically in Ref. [29], while the measured value [14] is very close to it. This creates a dipolar energy bias $W^{(D)} = g\mu_B B^{(D)}(m' - m)$ between the pair of resonant quantum levels m and m' (quantum numbers for S_z in the two energy wells). This energy bias typically largely exceeds the tunnel splitting Δ that contributes to the resonance width. In the deflagration front the dipolar field typically changes between $+B^{(D)}$ and $-B^{(D)}$ and so does the energy bias. As the result, spin tunneling in the deflagration front does not occur at a fixed resonance condition. This can explain why the observed quantum maxima in the front speed can be not as strong as expected, compared to the effect of tunneling on the ignition of deflagration.

Further theoretical research led to the idea of the dipole-dipole interaction (DDI) playing an active role in deflagration by controlling the relaxation rate, as temperature does in regular deflagration. Adding to the external bias field, the dipolar field can set particular magnetic molecules on or off resonance, facilitating or blocking their tunneling relaxation. The problem is self-consistent since tunneling of one magnetic molecule changes dipolar fields on the other ones. A numerical solution of this problem in a form of a moving front of tunneling at zero temperature (sometimes called "cold deflagration") has been found in Ref. [30]. An analytical solution for the front of tunneling in the realistic strong-DDI case has been obtained in Ref. [31].

Pure non-thermal fronts of tunneling can occur in the case of a very good thermal contact of the MM crystal with the environment, so that its temperature does not increase and remains so low that tunneling takes place directly from the metastable ground state into a matching excited state on the other side of the barrier. This process can be efficient only if a strong transverse field is applied and the corresponding tunnel splitting Δ is large enough. In this case the speed of fronts of tunneling can theoretically exceed the speed of a regular deflagration by a large margin. Indeed, the dipolar field in the crystal changes instantaneously, in contrast to the temperature changing via heat conduction. In addition, the relaxation rate due to tunneling directly from the ground state can be much higher than the relaxation rate due to barrier-climbing processes in the regular deflagration.

If an MM crystal is thermally insulated, its temperature is increasing as a result of a decay of the metastable state, so that there can be a mixture of both mechanisms of deflagration considered above [32]. Whereas far from resonances a regular deflagration takes place, near resonances tunneling leads to a great increase of the front speed. A more detailed treatment of the quantum-thermal deflagration for a realistic model of $Mn_{12}Ac$ with S_z^4 terms in the effective Hamiltonian was recently given in Ref. [33].

Theories of fronts of tunneling mentioned above are based on the model simplification considering it as one dimensional. In the regular deflagration, there is

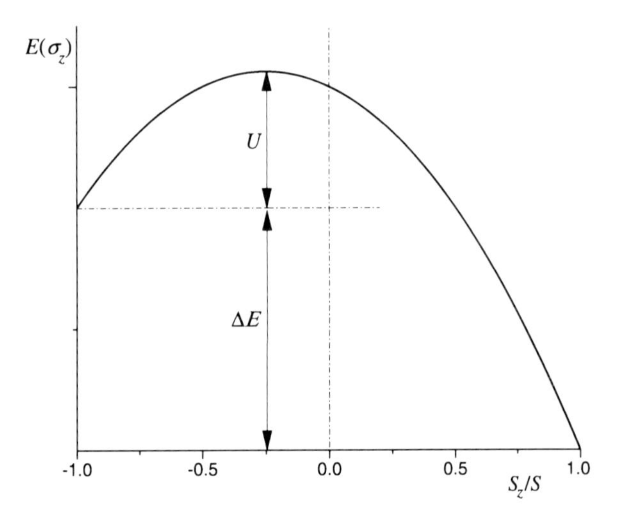

Fig. 6.1 Energy barrier of a biased molecular magnet

a mechanism that makes fronts flat and smooth (laminar), so that the deflagration problem in long crystals indeed becomes $1d$. In the case of dipolar-driven fronts of tunneling, it is not immediately clear whether fronts are flat or not, and, moreover, there is a mechanism that favors non-laminar fronts. The full $3d$ theory of fronts of tunneling that will be presented below, numerically yields non-flat and non-laminar fronts. The latter slows down the front speed in comparison to the simplified $1d$ theory but, nevertheless, the speed can reach values comparable with the speed of sound in MM near tunneling resonances in strong transverse fields.

In the main part of this contribution, the regular (thermal) magnetic deflagration will first be considered. Then calculation of the dipolar field in molecular magnets will be explained. The final part is devoted to the theory of fronts of tunneling.

6.2 Magnetic Deflagration

For the generic model of a molecular magnet the energy has the form

$$\mathcal{H} = -DS_z^2 - g\mu_B B_z S_z + \mathcal{H}', \tag{6.1}$$

where $D > 0$ is the uniaxial anisotropy constant and $\mathcal{H}'$ stands for all terms that do not commute with S_z and thus cause spin tunneling. In $Mn_{12}Ac$ there is an additional smaller longitudinal term $-AS_z^4$, the implications of which will be discussed later. In the biased case $B_z > 0$, the dependence of the energy on S_z is sketched in Fig. 6.1. The energy barrier U shown in Fig. 6.1 has the form

$$U = (1-h)^2 U_0, \qquad U_0 = DS^2, \qquad h \equiv g\mu_B B_z/(2DS). \tag{6.2}$$

With $S = 10$ the zero-field energy barrier U_0 has a large value of 67 K in $Mn_{12}Ac$. The energy of the metastable state is given by $\Delta E = 2Sg\mu_B B_z$.

In the absence of spin tunneling at low temperatures, $U/(k_B T) \gg 1$, the rate equation describing relaxation of the metastable population n (the fraction of magnetic molecules in the left well) has the form

$$\dot{n} = -\Gamma \left(n - n^{(\mathrm{eq})} \right), \tag{6.3}$$

where the relaxation rate is given by

$$\Gamma = \Gamma_0 \exp\left(-\frac{U}{k_B T} \right) \left[1 + \exp\left(-\frac{\Delta E}{k_B T} \right) \right]. \tag{6.4}$$

Here the second term in the square brackets describes back transitions from the stable well to the metastable well. In the strong-bias case, $\Delta E \gg k_B T$, this term can be omitted. The equilibrium metastable population $n^{(\mathrm{eq})}$ is given by

$$n^{(\mathrm{eq})} = 1 \Big/ \left[\exp\left(\frac{\Delta E}{k_B T} \right) + 1 \right]. \tag{6.5}$$

In the strong-bias case it can be neglected.

The second equation describing deflagration is the heat conduction equation

$$C\dot{T} = \nabla \cdot k \nabla T - \dot{n} \Delta E, \tag{6.6}$$

where k is thermal conductivity and C is heat capacity. The second term on the right is the energy release due to decay of the metastable state. The heat capacity is mainly due to phonons, whereas the magnetic contribution is relatively small. At low temperatures only acoustic phonons are excited, whereas high-energy optical phonons are frozen out, thus C has the form [34]

$$C = A k_B (T/\Theta_D)^{\alpha}, \tag{6.7}$$

where $\alpha = 3$ in three dimensions, $A = 12\pi^4/5 \simeq 234$ is a numerical factor and Θ_D is the Debye temperature, $\Theta_D \simeq 40$ K for $Mn_{12}Ac$. Although at low temperatures this expression is in a reasonable accordance with measurements on $Mn_{12}Ac$ [35], its applicability range is very narrow, $T \lesssim 5$ K. On the other hand, the temperature generated in deflagration (the so-called flame temperature) is typically above 10 K. The heat capacity of $Mn_{12}Ac$ can be well described within a broad temperature range with the help of the extended Debye model (EDM) [36] that comprises three different acoustic phonon modes as well as optical modes. Practically, one can use measured values of C [35].

It is convenient to use the relation $C = d\mathcal{E}/dT$ to rewrite (6.6) in terms of the energy $\mathcal{E}$ (here due to phonons) as

$$\dot{\mathcal{E}} = \nabla \cdot \kappa \nabla \mathcal{E} - \dot{n} \Delta E, \tag{6.8}$$

where $\kappa = k/C$ is thermal diffusivity. The latter has not yet been measured, although a crude estimate $\kappa \simeq 10^{-5}$ m^2/s was deduced from experiments [8, 13]. This value is comparable with that of metals. Temperature dependence of κ that could be substantial at low temperatures remains unknown.

Equations (6.3) and (6.8), together with (6.4) and the relation

$$\mathcal{E}(T) = \int_0^T C(T')dT', \tag{6.9}$$

is a strongly-nonlinear system of equations. It is easy to solve these equations numerically but it costs efforts to do it analytically. The two main problems to solve are (i) stability of the low-temperature state with respect to thermal runaway or ignition of a deflagration front and (ii) the shape and speed of the stationary deflagration front in long crystal.

6.2.1 Ignition of Deflagration

If the sample is perfectly thermally insulated, the whole released energy remains inside and the temperature monotonically increases. This leads to a thermal instability that can take a considerable time to develop, the ignition time. If there is a thermal contact with the environment, maintained at a constant low temperature T_0, there are two possible cases. In the subcritical case, the temperature rise in the sample due to slow decay leads to temperature gradients and heat flow toward the sample boundaries that ensures a stationary low-temperature state (proper conditions of explosives' storage). In the supercritical case, heat loss through the boundaries is insufficient to balance the increase of released heat owing to the rising temperature. This leads to ignition of a self-supporting burning process. In small crystals of MM, temperature gradients are higher and heat loss to the environment is more efficient. In larger crystals, temperature gradients are lower and thermal instability is more likely. This is why deflagration was observed in larger crystals.

Thermal instability occurs because of a stronger temperature dependence of the relaxation rate, (6.4), than that of the heat exchange with the environment. The essence of the problem is contained in the old Semenov model of explosive instability [37] described by a single equation

$$\dot{T} = Q_{\text{reaction}} - Q_{\text{cooling}}, \tag{6.10}$$

where $Q_{\text{reaction}} \sim \Gamma(T)$ and $Q_{\text{cooling}} = \alpha(T - T_0)$. In the case B in Fig. 6.2, the thermal contact to the bath is too weak, $Q_{\text{cooling}} < Q_{\text{reaction}}$ at all T, so that the system is absolutely unstable. In case A, the thermal contact is stronger and there is a stability range $T < T_2$, where the stationary state $T = T_1$ is an attractor. However, heating the system above T_2 leads to a thermal explosion.

Semenov's model is zero-dimensional, whereas in MM crystals the problem is at least one-dimensional and more complicated. There are different cases of thermal instability, mainly instability of a large crystal initially at uniform temperature (that begins at the center), instability due to heating one end of a long crystal, and the

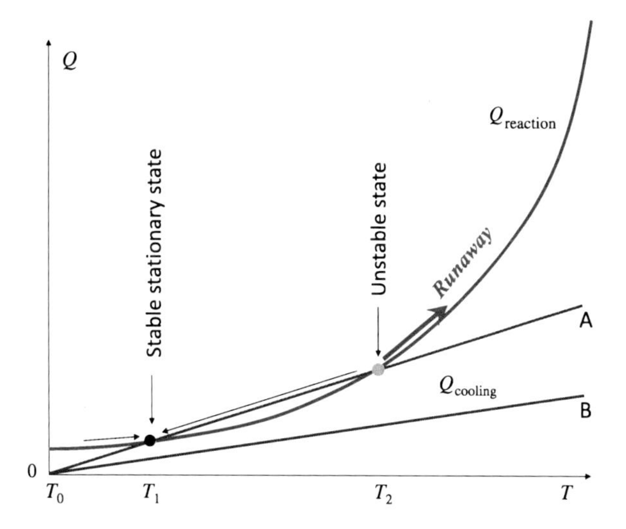

Fig. 6.2 Semenov's mechanism of a thermal runaway, (6.10)

instability due a magnetic field gradient that makes the barrier lower at one side of the crystal. Analysis of all these cases has been done in Ref. [21]. In particular, in simplest case of the uniform energy barrier and constant temperature T_0 maintained at the boundaries, the crystal loses stability against formation and propagation of the flame (magnetic avalanche) when the rate of the spin flip for an individual molecule, $\Gamma(H, T_0)$, exceeds

$$\Gamma_c = \frac{k_B T_0}{U} \frac{8 k T_0}{l^2 n_i \Delta E}. \tag{6.11}$$

Here k is thermal conductivity at T_0 and the length parameter l is uniquely determined by geometry, being of the order of the smallest dimension of the crystal, whereas n_i is the metastable population in the initial state.

Experimentally magnetic deflagration can be initiated either by heating one end of the crystal [11–13] or by sweeping the magnetic field in the positive direction, which reduces the energy barrier and makes the condition in (6.11) satisfied [8]. In Ref. [10] deflagration was ignited by surface acoustic waves (SAW), instead of by heating.

6.2.2 *Deflagration Fronts*

Fronts of magnetic burning propagating in long crystals of molecular magnets are flat and smooth, i.e., the problem of deflagration is one-dimensional. The stability of flat fronts can be immediately seen. Indeed, if a fraction of a front gets ahead of neighboring fractions, the heat released at this place will be propagating not exactly straight ahead (as in a flat front) but also sideways. This will slow down this leading fraction of the front and speed up the lagging fractions surrounding it. Thus any local deviation from a flat front will disappear with time.

In a stationary-moving front, all physical quantities depend only on the combined variable that can be chosen, e.g., in the time-like form $u \equiv t - z/v$, where v is the front speed. In terms of u the deflagration equations have the form

$$\begin{aligned} \frac{dn}{du} &= -\Gamma(T)\big(n - n^{(\mathrm{eq})}(T)\big) \\ \frac{d\mathcal{E}}{du} &= \frac{1}{v^2}\frac{d}{du}\kappa\frac{d\mathcal{E}}{du} - \frac{dn}{du}\Delta E \end{aligned} \tag{6.12}$$

plus (6.9). Integrating the energy equation one obtains

$$\mathcal{E} + n\Delta E - \frac{\kappa}{v^2}\frac{d\mathcal{E}}{du} = \mathrm{const.} \tag{6.13}$$

Far before and far behind the front, the term with the derivative vanishes. Thus one obtains the energy conservation law in the form

$$\mathcal{E}_i + n_i\Delta E = \mathcal{E}_f + n^{(\mathrm{eq})}(T_f)\Delta E, \tag{6.14}$$

where i stands for "initial" (before the front) and f stands for "final" or "flame". This is a transcendental equation for the flame temperature T_f that has to be solved together with (6.9). If $\mathcal{E}_i \approx 0$ (low initial temperature) and $n^{(\mathrm{eq})}(T_f)$ is negligible (full-burning case realised at a strong bias, see (74) of Ref. [21]) one immediately finds the flame energy from $n_i\Delta E = \mathcal{E}_f$, and then T_f follows by inverting (6.9). In the incomplete-burning regime at small bias, a pulsating instability of stationary deflagration fronts [38] was found. The operations above assume that the heat is not exchanged via the sides of the crystal. In the opposite case, the energy conservation becomes invalid and the theory has to be extended.

One can immediately get an idea of the front speed by rewriting the deflagration equations (6.12) in the dimensionless form. In terms of the reduced variables

$$\tilde{n} \equiv n/n_i, \qquad \tilde{\mathcal{E}} \equiv \mathcal{E}/(n_i\Delta E), \qquad \tilde{u} \equiv u\Gamma_f \tag{6.15}$$

and parameters

$$\tilde{\Gamma} \equiv \Gamma/\Gamma_f, \qquad \tilde{\kappa} \equiv \kappa/\kappa_f \tag{6.16}$$

(6.12) become

$$\begin{aligned} \frac{d\tilde{n}}{d\tilde{u}} &= -\Gamma\big(\tilde{n} - \tilde{n}^{(\mathrm{eq})}\big) \\ \frac{d\tilde{\mathcal{E}}}{d\tilde{u}} &= \frac{1}{\tilde{v}^2}\frac{d}{d\tilde{u}}\tilde{\kappa}\frac{d\tilde{\mathcal{E}}}{d\tilde{u}} - \frac{d\tilde{n}}{d\tilde{u}}, \end{aligned} \tag{6.17}$$

where the reduced front speed $\tilde{v}$ is related to the actual front speed v by

$$v = \tilde{v}\sqrt{\kappa_f\Gamma_f}. \tag{6.18}$$

References [2, 8] give the expression above without $\tilde{v}$ for the front speed.

It turns out that $\tilde{v}$ in (6.18) is not merely a number but rather it is a function of dimensionless parameters such as

$$W_f \equiv U/(k_B T_f). \tag{6.19}$$

Because of the non-linearity of the first of (6.17), their general analytical solution that defines $\tilde{v}$ does not exist. There are two parameter ranges in the problem: Slow-burning high-barrier range $W_f \gg 1$ and fast-burning low-barrier range $W_f \lesssim 1$.

In the former, burning occurs in the front region where the temperature is already close to T_f. Assuming that κ is temperature independent, $\tilde{\kappa} = 1$, and linearizing $\Gamma(T)$ near T_f, one can solve the problem analytically. Within the full-burning approximation ($n^{(\mathrm{eq})} \Rightarrow 0$) the reduced front speed is given by [21]

$$\tilde{v} = \sqrt{\frac{C_f T_f}{n_i \Delta E} \frac{k_B T_f}{U}}. \tag{6.20}$$

With the help of (6.7) (that is not accurate, however!) this result simplifies to

$$\tilde{v} = \sqrt{(\alpha + 1)/W_f}. \tag{6.21}$$

The applicability range of these expressions is $\tilde{v} \ll 1$.

The corresponding profile of the metastable population n in the front has the form

$$\tilde{n} = \frac{1}{1 + e^u} = \frac{1}{2}\left(1 - \tanh\frac{\tilde{u}}{2}\right) \tag{6.22}$$

that corresponds to the symmetric tanh magnetization profile $\sigma_z = 1 - 2n = \tanh(\tilde{u}/2)$. In real units the result reads

$$n = \frac{n_i}{2}\left[1 + \tanh\left(\frac{z}{2\tilde{v} l_d} - \frac{\Gamma_f t}{2}\right)\right], \tag{6.23}$$

where $l_d = \sqrt{\kappa_f/\Gamma_f}$ is the *a-priori* with of the deflagration front. Magnetization profile of this kind can be seen in Fig. 11 of Ref. [21] and in the upper panel of Fig. 10 of Ref. [33]. The reduced energy in the front is given by

$$\tilde{\mathcal{E}} = \left(1 - e^{-u}\right)^{-\tilde{v}^2} = (1 - \tilde{n})^{\tilde{v}^2}. \tag{6.24}$$

Since in the high-barrier approximation $\tilde{v} \ll 1$, the formula above yields $\tilde{\mathcal{E}} \approx 1$ in the active burning region and actually everywhere except for the region far ahead of the front where $\tilde{n}$ is very close to 1. This justifies the approximation made.

It should be noted that the full-burning approximation used above requires a bias high enough thus the barrier low enough, $W_f \lesssim 6$, according to (79) of Ref. [21]. Thus the applicability range of the slow-burning high-barrier approximation is rather limited. The theory can be improved by taking into account incomplete burning. However, this makes analytics cumbersome because of the transcendental equation (6.13) defining T_f. Numerical solution for the deflagration front poses no problems, nevertheless. Because of incomplete burning, T_f and thus the front speed decrease below the values given above.

In the low-barrier fast-burning regime $W_f \lesssim 1$ there is no rigorous analytical solution to the problem. Additionally, the Arrhenius form of the relaxation rate, (6.4), becomes invalid. In this regime the magnetization profile is asymmetric, as can be seen in the upper panel of Fig. 12 of Ref. [33].

Making the simplifying approximation for the relaxation rate

$$\tilde{\Gamma}(\tilde{\mathcal{E}}) = \begin{cases} 0, & \tilde{\mathcal{E}} < \tilde{\mathcal{E}}_0 \\ 1, & \tilde{\mathcal{E}} > \tilde{\mathcal{E}}_0, \end{cases} \tag{6.25}$$

where $\tilde{\mathcal{E}}_0$ will be defined below, one can solve the problem of a stationary deflagration front in the whole parameter range. Let us search for the front in which $\tilde{\mathcal{E}} = \tilde{\mathcal{E}}_0$ at $u = 0$. In the reduced form of the energy equation (6.13),

$$\frac{d\tilde{\mathcal{E}}}{d\tilde{u}} = \tilde{v}^2(\tilde{\mathcal{E}} + \tilde{n} - 1), \tag{6.26}$$

one has $\tilde{n} = 1$ before the front, $u < 0$. Thus the energy equation solves here to

$$\tilde{\mathcal{E}} = \tilde{\mathcal{E}}_0 e^{\tilde{v}^2 \tilde{u}}. \tag{6.27}$$

On the other hand, for $u > 0$ the solution of the population equation $d\tilde{n}/d\tilde{u} = -\tilde{\Gamma}\tilde{n} = -\tilde{n}$ reads $\tilde{n} = e^{-\tilde{u}}$. Inserting this into (6.26), and integrating the differential equation, one obtains the solution

$$\tilde{\mathcal{E}} = \left(\tilde{\mathcal{E}}_0 - \frac{1}{1+\tilde{v}^2}\right) e^{\tilde{v}^2 \tilde{u}} + 1 - \frac{\tilde{v}^2}{1+\tilde{v}^2} e^{-\tilde{u}}. \tag{6.28}$$

The first term of this expression must vanish because of the boundary condition $\tilde{\mathcal{E}}(\infty) = 1$. This defines the reduced front speed,

$$\tilde{v} = \sqrt{\frac{1}{\tilde{\mathcal{E}}_0} - 1}. \tag{6.29}$$

To define $\tilde{\mathcal{E}}_0$, consider the reduced Arrhenius relaxation rate

$$\tilde{\Gamma} = \exp\left[W_f\left(1 - \frac{1}{\tilde{T}}\right)\right] \tag{6.30}$$

and require

$$W_f\left(1 - \frac{1}{\tilde{T}_0}\right) = -1 \tag{6.31}$$

as the switching point between $\tilde{\Gamma} = 0$ and $\tilde{\Gamma} = 1$. This yields

$$\tilde{T}_0 = \frac{W_f}{1 + W_f}. \qquad (6.32)$$

Using (6.7), one obtains

$$\tilde{\mathcal{E}}_0 = \tilde{T}_0^{\alpha+1} = \left(\frac{W_f}{1 + W_f}\right)^{\alpha+1}. \qquad (6.33)$$

Substituting this into (6.29), one finally obtains

$$\tilde{v} = \sqrt{\left(\frac{1 + W_f}{W_f}\right)^{\alpha+1} - 1}. \qquad (6.34)$$

Limiting cases of this formula are

$$\tilde{v} \cong \begin{cases} \sqrt{(\alpha + 1)/W_f}, & W_f \gg 1 \\ 1/W_f^{(\alpha+1)/2}, & W_f \ll 1. \end{cases} \qquad (6.35)$$

It is remarkable that the rigorously obtained high-barrier slow-burning result of (6.21) is recovered exactly. In the low-barrier fast-burning case the reduced front speed becomes large, as well as the actual front speed of (6.18). One can see that (6.34) is in good agreement with the numerical solution shown in Fig. 6.3.

The high-speed regime of the deflagration should be superseded by detonation when the front speed approaches the speed of sound. In detonation, thermal expansion resulting from burning sends a shock wave into the cold region before the front. As a consequence, the temperature before the front rises as a result of compression, initiating burning. Such a mechanism was recently considered for $Mn_{12}Ac$ in Ref. [17].

6.3 Fronts of Tunneling

6.3.1 Tunneling Effects in the Relaxation Rate

The relaxation rate Γ including spin tunneling is at the foundation of the quantum theory of deflagration in molecular magnets. In the generic model of MM, (6.1), tunneling resonances occur at the values of the total bias field $B_{\mathrm{tot},z}$ (including the self-produced dipolar field) equal to

$$B_k = kD/(g\mu_B), \quad k = 0, \pm 1, \pm 2, \ldots \qquad (6.36)$$

for all the resonances. Spin tunneling leads to the famous steps in the dynamic hysteresis curves [26–28]. In the real $Mn_{12}Ac$ there is an additional term $-AS_z^4$ that

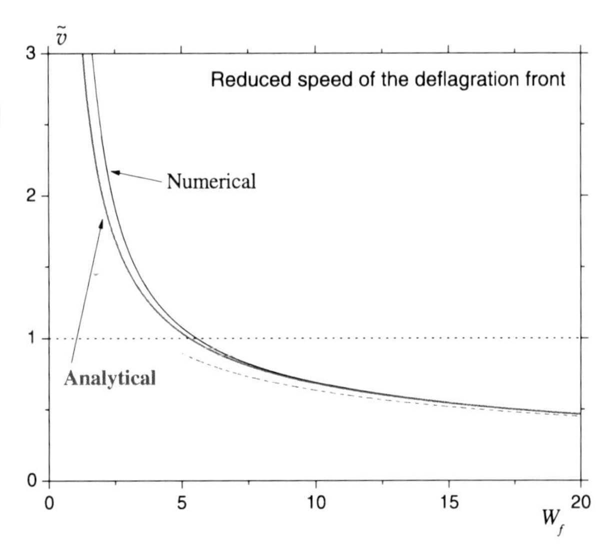

Fig. 6.3 Reduced speed of a deflagration front defined by (6.18). The numerical result has been obtained in Ref. [21] within the full-burning approximation using the low-temperature form of the heat capacity, (6.7). Analytical result is (6.34). *The dotted line* is the high-W_f asymptote, (6.21)

makes higher-energy resonances be achieved at smaller B_z than low-energy resonances. The resulting tunneling multiplets

$$g\mu_B B_{km} = k\left[D + \left(m^2 + (m+k)^2\right)A\right] \tag{6.37}$$

were used to experimentally monitor [39, 40] the transition between thermally assisted and ground-state tunneling [41] in $Mn_{12}Ac$. Below B_k will stand for the resonance field B_{km}, for simplicity of notations.

In the case of an isolated magnetic molecule, the probability of a spin to be in one of the resonant quantum states is oscillating with time with the frequency $\Delta/\hbar$, where Δ is the tunnel splitting. However, coupling to the environment, e.g., to phonons, introduces damping to these oscillations. If the decay rate of at least one of the resonance states, Γ_m or $\Gamma_{m'}$, exceeds $\Delta/\hbar$, tunneling oscillations of the spin are overdamped. This can be illustrated in the case of a resonance between the metastable ground state $|-S\rangle$ and the matching excited state at the other side of the barrier $|m'\rangle$ of a biased MM at zero temperature. Ignoring all other levels, that is justified at $T = 0$, one can write down the Schrödinger equation in the form [31]

$$\begin{aligned} \dot{c}_{-S} &= -\frac{i}{2}\frac{\Delta}{\hbar}c_{m'} \\ \dot{c}_{m'} &= \left(\frac{iW}{\hbar} - \frac{1}{2}\Gamma_{m'}\right)c_{m'} - \frac{i}{2}\frac{\Delta}{\hbar}c_{-S}, \end{aligned} \tag{6.38}$$

where

$$W \equiv \varepsilon_{-S} - \varepsilon_{m'} = \left(S + m'\right)g\mu_B(B_{\text{tot},z} - B_k) \tag{6.39}$$

is the energy bias between the two levels. Whereas the level $|-S\rangle$ is undamped, the level $|m'\rangle$ can decay into lower-lying levels in the same well via phonon-emission processes. At $T = 0$ there are no incoming relaxation processes for $|m'\rangle$. In this case the damped Schrödinger equation above is accurate, as it can be shown to follow from the density matrix equation. In the underdamped case $\Gamma_{m'} \lesssim \Delta/\hbar$ the solution of these equations is oscillating. The first choice for studying tunneling dynamics in molecular magnets is the overdamped case $\Gamma_{m'} \gg \Delta/\hbar$, since for not too strong transverse fields $B_\perp$ the tunnel splittingΔ is a high power of $B_\perp$ (Ref. [42]) and typically it is much smaller than $\Gamma_{m'}$. In the overdamped case the variable $c_{m'}$ in (6.38) adiabatically adjusts to the instantaneous value of c_{-S} and the solution greatly simplifies. Setting $\dot{c}_{m'} = 0$ in the second of these equations, one obtains

$$c_{m'} = \frac{\Delta}{2\hbar} \frac{c_{-S}}{W/\hbar + i\Gamma_{m'}/2}. \tag{6.40}$$

Inserting this into the first of (6.38) yields a closed differential equation for c_{-S}. Using $n = |c_{-S}|^2$ for the metastable occupation number, one arrives at the rate equation

$$\dot{n} = -\Gamma n, \tag{6.41}$$

where the dissipative resonance-tunneling rate Γ is given by [43]

$$\Gamma = \frac{\Delta^2}{2\hbar^2} \frac{\Gamma_{m'}/2}{(W/\hbar)^2 + (\Gamma_{m'}/2)^2}. \tag{6.42}$$

This is a Lorentzian function with the maximum at the resonance, $W = 0$. (6.41) and (6.42) were used in Refs. [30, 31] to study dipolar-controlled fronts of tunneling at $T = 0$, or "cold deflagration". The full system of (6.38) could also be used to this purpose but nothing had been published up to date.

At nonzero temperatures, tunneling transitions via higher energy level pairs become possible (thermally-assisted tunneling) and one has to take into account nonresonant thermal transitions over the top of the barrier. This makes the problem more complicated, and one needs to use the density matrix equation (DME) taking into account spin-phonon interactions explicitly. One of the first works using DME for $Mn_{12}Ac$ was Ref. [43] in which spin tunneling was considered with the help of the high-order perturbation theory [42] for a small transverse field $B_\perp$. The spin-phonon processes taken into account were due to dynamic tilting of the anisotropy axis by transverse phonons. Ref. [43] could qualitatively explain thermally-assisted tunneling via the level pairs just below the classical barrier. However, tunneling via low-lying resonant level pairs or tunneling directly out of the metastable ground state are inaccessible by this method because large enough splitting requires nonperturbatively large transverse field that can only be dealt with numerically.

Further work on spin-phonon relaxation in MM lead to elucidation of the universal relaxation mechanism [44, 45]. This mechanism consists in distortionless rotation of the crystal field acting on a magnetic molecule, actually the same mechanism as used in Ref. [43]. It was, however, understood that this mechanism does not require any poorly-known spin-lattice coupling constants and everything can be expressed through much easier accessible crystal-field parameters. This mechanism

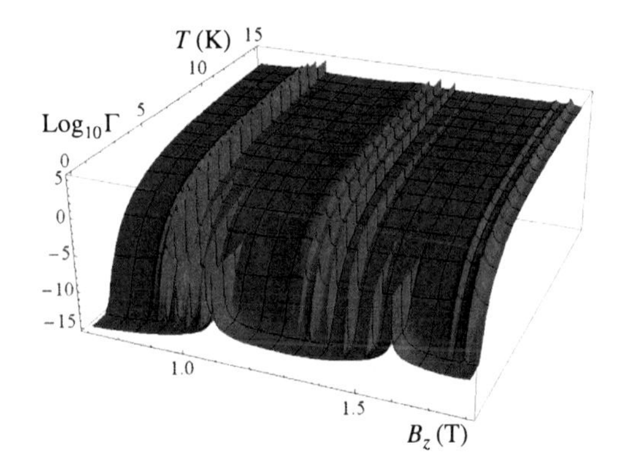

Fig. 6.4 Relaxation rate of $Mn_{12}Ac$ vs temperature and longitudinal magnetic field in a small transverse field. Resonance multiplets with $k = 2, 3$ are seen

was overlooked in older theories of spin-lattice relaxation based on energy contributions responsible for the coupling. Rotations, to the contrary, cost no energy and the effect has a purely inertial origin.

The universal relaxation mechanism allows a general numerical implementation of the DME fully based on the crystal field parameters, recently accomplished in Ref. [46] that summarizes the current state of the problem. Another important feature of Ref. [46] is using the so-called semi-secular approach capable of dealing with resonant pairs of levels and thus describe spin tunneling. Conventional implementations of the DME (see, e.g., Ref. [47, 48]) use the secular approximation that crashes on tunneling resonances. In Ref. [46] the relaxation rate Γ is extracted from the time-dependent numerical solution of the DME (expressed in terms of eigenvalues and eigenfunctions of the density matrix) as the inverse of the integral relaxation time [49, 50]. Unlike using the lowest eigenvalue of the density matrix, this method also works at elevated temperatures.

The temperature and field dependence of Γ in $Mn_{12}Ac$ at a small transverse field ($B_\perp = 0.04$T that typically arises due to a 1° misalignment of the easy axis and the applied longitudinal field) is shown in Fig. 6.4. One can see very narrow and high maxima of Γ (note that $\log \Gamma$ is plotted!) due to spin tunneling. Maxima corresponding to the ground-state tunneling, for which the maximum in Γ does not disappear at $T = 0$, correspond to the highest value of B_z in the multiplet. There are $k = 2$ and $k = 3$ tunneling multiplets seen in Fig. 6.4. Note that tunneling via low-lying resonances is relatively weak and it is eclipsed by the thermal activation contribution at higher temperatures.

At stronger transverse field such as $B_\perp = 3.5$ T in Fig. 6.5, the barrier is strongly reduced and high-lying tunneling resonances are broadened away. Here, one can see the ground-state resonance ($B_z = 0.522$ T) and the first-excited-state resonance ($B_z = 0.490$ T) for $k = 1$ multiplet. The ground-state resonance does not disappear at the highest temperature that has an important implication in the dynamics of fronts of tunneling. Note the much higher tunneling rate at $T = 0$, in comparison to the previous figure.

A puzzle in the theory of relaxation of molecular magnets is the prefactor Γ_0 in the Arrhenius relaxation rate, (6.4), being by two orders of magnitude too small.

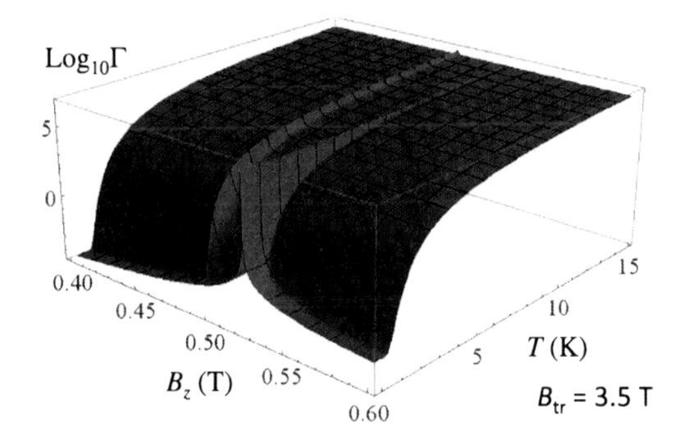

Fig. 6.5 Relaxation rate of $Mn_{12}Ac$ vs temperature and longitudinal magnetic field in the transverse field $B_\perp = 3.5$ T. One can see the ground-state resonance at $B_z = 0.522$T and the first-excited-state resonance at $B_z = 0.490$T for $k = 1$ multiplet

This was already recognized in the early work [43]. Using the standard spin-lattice relaxation model considering one spin in an infinite elastic matrix, it is impossible to arrive at $\Gamma_0 \simeq 10^7$ s^{-1} observed in experiments [35, 51] without introducing artificially strong spin-phonon interactions [52]. For a strongly diluted molecular magnet, considering a single spin could be justified, but in the regular case it can not. High density of magnetic molecules should lead to such collective effects as superradiance [53–55] and phonon bottleneck [56–58]. Possibility of superradiance in fast avalanches triggered by a fast field sweep has been discussed in Ref. [15]. References [59, 60] report microwave emission from MM that can be interpreted as superradiance. However, it would be difficult to address these complicated issues while dealing with the quantum deflagration problem, so that the calculated relaxation rate will be simply multiplied by 100 to approximately match the experiment, as was done in Ref. [33].

6.3.2 Dipolar Field in Molecular Magnets

Very sharp resonance peaks in the relaxation rate Γ seen in Figs. 6.4 and 6.5 require an accurate calculation of the dipolar field in the crystal that can self-consistently control tunneling by setting individual molecules on or off resonance. The equations describing this are the same relaxational equations (6.3) and thermal equation (6.8), as before, only with Γ depending on the total magnetic field

$$B_{\mathrm{tot},z}(\mathbf{r}) = B_z + B_z^{(D)}(\mathbf{r}), \tag{6.43}$$

where B_z is the external bias field and $B_z^{(D)}$ is the self-consistently calculated dipolar field. In the case of cold deflagration, the thermal equation can be discarded and one has to solve only the relaxational equation (6.41). Since the dipolar field depends on the magnetization everywhere in the crystal, the equations of quantum deflagration are integro-differential equations. Note that the transverse component of the dipolar field can be discarded because its effect is small.

For the purpose of calculating the dipolar field, conventional magnetostatics (see, e.g., Ref. [61]) is unsuitable because it provides an irrelevant magnetic field formally averaged over the microscopic scale that ignores the lattice structure. The physically relevant dipolar field is the field created at positions of magnetic molecules by all other molecules. It is a microscopic quantity that depends on the lattice structure. To illustrate this point, magnetostatic field in a uniformly magnetized long sample is $\mathcal{B}^{(D)} = 4\pi M$, where M is the magnetization. However, the microscopically calculated dipolar field in a long uniformly magnetized crystal of $Mn_{12}Ac$ is much smaller, $B_z^{(D)} = 5.26M$.

It is convenient to express the z component of the dipolar field at site i (i.e., at a particular magnetic molecule) in the form

$$B_z^{(D)} = (Sg\mu_B/v_0)D_{zz}, \tag{6.44}$$

where v_0 is the unit-cell volume. For $Mn_{12}Ac$ one has $Sg\mu_B/v_0 = 5.0$ mT. The reduced dipolar field D_{zz}, created by all other molecular spins polarized along the z axis is given by

$$D_{zz}(\mathbf{r}_i) = \sum_j \phi(\mathbf{r}_j - \mathbf{r}_i)\sigma_z(\mathbf{r}_j), \qquad \phi(\mathbf{r}) = v_0 \frac{3(\mathbf{e}_z \cdot \mathbf{n})^2 - 1}{r^3}, \qquad \mathbf{n} \equiv \frac{\mathbf{r}}{r}, \tag{6.45}$$

where $\sigma_z \equiv S_z/S$. To calculate the sum over the lattice for the site i, one can introduce a small sphere of radius r_0 around i satisfying $v_0^{1/3} \ll r_0 \ll L$, where L is the (macroscopic) size of the sample. The field from the spins at sites j inside this sphere can be calculated by direct summation over the lattice, whereas the field from the spins outside the sphere can be obtained by integration. The sum of the two contributions does not depend of r_0. If the magnetization in the crystal depends only on the coordinate z along the symmetry axis of the crystal that coincides with the magnetic easy axis z (that is the case for a flat deflagration front), the integral over the volume can be expressed via the integral over the crystal surfaces. The corresponding contribution can be interpreted as that of molecular currents flowing on the surface. The details are given in the Appendix to Ref. [29].

In particular, for a uniformly magnetized ellipsoid the total result has the form

$$D_{zz} \equiv \sigma_z \sum_j \phi(\mathbf{r}_j) = \bar{D}_{zz}\sigma_z, \tag{6.46}$$

independently of i, where

$$\bar{D}_{zz} = \bar{D}_{zz}^{(\mathrm{sph})} + 4\pi\nu\left(1/3 - n^{(z)}\right) \tag{6.47}$$

and ν is the number of molecules per unit cell ($\nu = 2$ for $Mn_{12}Ac$ having a body-centered tetragonal lattice). Here $\bar{D}_{zz}^{(\mathrm{sph})}$ comes from the summation over a small sphere and the remaining terms come from the integration. For the demagnetizing factor one has $n^{(z)} = 0$, $1/3$, and 1 for a cylinder, sphere, and disc, respectively. One obtains $\bar{D}_{zz}^{(\mathrm{sph})} = 0$ for a simple cubic lattice, $\bar{D}_{zz}^{(\mathrm{sph})} < 0$ for a tetragonal lattice with $a = b < c$, and $\bar{D}_{zz}^{(\mathrm{sph})} > 0$ for that with $a = b > c$. The latter is the case for $Mn_{12}Ac$

having $\bar{D}_{zz}^{(\mathrm{sph})} = 2.155$. For a long cylinder this results in $\bar{D}_{zz}^{(\mathrm{cyl})} = 10.53$ or, in real units [14, 29],

$$B_z^{(D)} = 52.6\ \mathrm{mT}. \tag{6.48}$$

The dipolar energy per magnetic molecule can be written in the form $E_0 = -(1/2)\bar{D}_{zz}E_D$, where

$$E_D \equiv (Sg\mu_B)^2/v_0 \tag{6.49}$$

is the characteristic dipolar energy, $E_D/k_B = 0.0671$ K for $Mn_{12}Ac$. The role of the DDI in spin tunneling is defined by the ratio of the typical dipolar bias $W^{(D)} = 2Sg\mu_B B_z^{(D)} = 2E_D\bar{D}_{zz}^{(\mathrm{cyl})}$ to the width of the overdamped tunneling resonance $\Gamma_{m'}$ in (6.42). It is thus convenient to introduce the parameter

$$\tilde{E}_D \equiv 2E_D/(\hbar\Gamma_{m'}) \tag{6.50}$$

that is always large. For instance, using the experimental Arrhenius prefactor $\Gamma_0 \simeq 10^7\ \mathrm{s}^{-1}$ for $\Gamma_{m'}$, one obtains $\tilde{E}_D \simeq 10^3$.

For a cylinder of length L and radius R with the symmetry axis z along the easy axis, magnetized with $\sigma_z = \sigma_z(z)$, the reduced dipolar field along the symmetry axis has the form [29]

$$D_{zz}(z) = \int_{-L/2}^{L/2} dz' \frac{2\pi\nu R^2\sigma_z(z')}{[(z'-z)^2 + R^2]^{3/2}} - k_D\sigma_z(z), \tag{6.51}$$

where $\sigma_z = 1 - 2n$ is polarization of pseudospins representing spins of magnetic molecules ($\sigma_z = \pm 1$ in the ground and metastable states, respectively) and

$$k_D \equiv 8\pi\nu/3 - \bar{D}_{zz}^{(\mathrm{sph})} = 4\pi\nu - \bar{D}_{zz}^{(\mathrm{cyl})} > 0, \tag{6.52}$$

$k_D = 14.6$ for $Mn_{12}Ac$. In (6.51), the integral term is the contribution of the crystal surfaces, while the lattice-dependent local term is the contribution obtained by direct summation over lattice site within the small sphere r_0 minus the integral over this sphere that must be subtracted from the integral over the whole crystal's volume. For other shapes such as elongated rectangular, one obtains qualitatively similar expressions [31].

A striking feature of (6.51) is that the integral and local terms have different signs. The integral term changes at the scale of R while the local term can change faster, that creates a non-monotonic dependence of $D_{zz}(z)$. In the case of a regular magnetic deflagration, the spatial magnetization profile in the slow-burning limit is of the type $\sigma_z(z) = -\tanh[(z - z_0)/l_d]$, where l_d is the width of the deflagration front that satisfies $l_d \ll R$, c.f. (6.23). The resulting dipolar field is shown in Fig. 6.6, where the line is the result of (6.51) and points represent the dipolar field along the symmetry axis of a long cylindrical crystal calculated by direct summation of microscopic dipolar fields over the $Mn_{12}Ac$ lattice. One can see that (6.51) is pretty accurate, small discrepancies resulting from l_d being not large enough in comparison to the lattice spacing a. The central region with the large positive slope

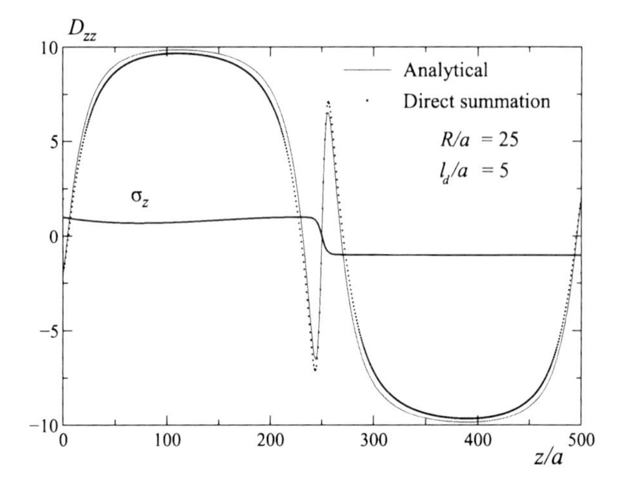

Fig. 6.6 Reduced dipolar field in a deflagration front in the slow-burning limit, created by the magnetization profile $\sigma_z(z) = -\tanh[(z - z_0)/l_d]$. Analytical result: (6.51); *Points*: Direct summation of dipolar fields over $Mn_{12}Ac$ lattice

is dominated by the local term of (6.51) that changes in the direction opposite to that of the magnetization. For $R \ggg l_d$, D_{zz} reaches the values ± 14.6 due to the local term before it begins to slowly change in the opposite direction. In real units the dipolar field at the local maximum and minimum is $\pm B_z^{(k_D)}$, where

$$B_z^{(k_D)} = 72.9 \text{ mT}, \tag{6.53}$$

exceeding the dipolar field of the uniformly magnetized long cylinder (6.48). Also one can see from Fig. 6.6 that the dipolar field becomes opposite to the magnetization at the ends of the cylinder, that should lead to an instability of the uniformly-magnetized state in zero external field.

The $1d$ theory of fronts of tunneling [30–33] is based on the simplifying assumption that the deflagration front is flat, $\sigma_z = \sigma_z(z)$, and the dipolar field is given by (6.51) everywhere. Since, in fact, the dipolar field also depends on the distance from the crystal's symmetry axis, it is likely that such a more complicated structure of B_z will self-consistently affect the front structure, making it non-flat.

There is also a question of stability of a smooth front at a small scale. Whereas flat and smooth fronts of regular burning are stable, there is an instability mechanism for a flat front in the presence of tunneling controlled by dipolar fields that will be explained below. This is why it is important to develop a full $3d$ theory of fronts of tunneling.

If the magnetization σ_z of a MM crystal depends on all the coordinates x, y, z but this dependence still has a macroscopic scale, one can again use the method of calculating the dipolar field that combines summation over a small sphere (where σ_z does not change) and integration over the remaining volume of the crystal. In this case the integral over the volume does not reduce to an integral over the surface and it has to be done numerically, i.e.

$$D_{zz}(\mathbf{r}) = \frac{\nu}{v_0} \int_{|\mathbf{r}'-\mathbf{r}|>r_0} d\mathbf{r}' \phi(\mathbf{r}' - \mathbf{r}) \sigma_z(\mathbf{r}') + \sigma_z(\mathbf{r}) \bar{D}_{zz}^{(\mathrm{sph})}, \tag{6.54}$$

where ϕ is defined in (6.45). A problem with this integral is that the contribution of the excluded region is comparable with the total result because of the singularity of the DDI. The solution to this problem is, for any point $\mathbf{r}$, to add and subtract the dipolar field in a uniformly magnetized crystal with $\sigma_z = \sigma_z(\mathbf{r})$. The total reduced dipolar field can be thus represented as

$$D_{zz}(\mathbf{r}) = \frac{\nu}{v_0} \int d\mathbf{r}' \phi(\mathbf{r}' - \mathbf{r}) \big(\sigma_z(\mathbf{r}') - \sigma_z(\mathbf{r})\big) + \sigma_z(\mathbf{r}) \big(\nu \bar{\mathcal{D}}_{zz}(\mathbf{r}) - k_D\big), \qquad (6.55)$$

Because of the terms subtraction at $\mathbf{r}' \to \mathbf{r}$, the contribution of the excluded small sphere in the integral is negligible and the integral can be extended to the whole volume of the crystal.

In the solution of the deflagration problem, it is convenient to discretize the volume of the crystal and use the same grid to sample the magnetization variables and to calculate the dipolar field. Then the values of the integral for all points of a rectangular grid can be computed via a summation method based on the fast Fourier transform (FFT) that takes $\sim N \log(N)$ operations, where N is the number of grid points. Straightforward calculation of the integral costs $\sim N^2$ operations and it has to be avoided.

The remainder of (6.55) corresponds to a uniformly magnetized crystal and its structure is similar to (6.51). Again, the term with k_D is the local contribution, while $\bar{\mathcal{D}}_{zz}(\mathbf{r})$ is the contribution of surface molecular currents, the result of conventional magnetostatics. For a crystal of a rectangular shape with dimensions $2L_x \times 2L_y \times 2L_z$ and $-L_x \leq x \leq L_x$ etc. the result can be obtained as a particular case of (88) of Ref. [62] and it has the form

$$\bar{\mathcal{D}}_{zz}(\mathbf{r}) = \sum_{\eta_x, \eta_y, \eta_z = \pm 1} \arctan \frac{(L_x + \eta_x x)^{-1} (L_y + \eta_y y)(L_z + \eta_z z)}{\sqrt{(L_x + \eta_x x)^2 + (L_y + \eta_y y)^2 + (L_z + \eta_z z)^2}} + (x \Rightarrow y), \qquad (6.56)$$

in total 16 different arctan terms. A transformation of this formula yields the needed result for box-shape samples with dimensions $L_x \times L_y \times L_z$ and $0 \leq x \leq L_x$ etc.

6.3.3 *Fronts of Tunneling at $T = 0$*

The theory of dipolar-controlled fronts of tunneling at $T = 0$ ("cold deflagration") [30, 31] uses the relaxational equation (6.41) with the resonance tunneling rate of (6.42), in which the energy bias W is given by (6.39) with $B_{\mathrm{tot},z}$ of (6.43). Within the $1d$ approximation [30, 31], the dipolar field is given by (6.44) and (6.51) for a cylinder. The problem is thus an integro-differential equation.

It is convenient to use the reduced energy bias $\tilde{W} \equiv W/(2E_D)$ that has the form

$$\tilde{W} = \tilde{W}_{\mathrm{ext}} + D_{zz}, \qquad \tilde{W}_{\mathrm{ext}} = \frac{(S + m')g\mu_B}{2E_D}(B_z - B_k), \qquad (6.57)$$

where $m' = S - k$ is close to S for not too strong bias. Propagating dipolar-controlled fronts of tunneling have been found numerically [30, 31] and analytically [31] within the dipolar window near the resonance

$$0 \leq \tilde{W}_{\text{ext}} \leq \bar{D}_{zz}^{(\text{cyl})}, \tag{6.58}$$

where $\bar{D}_{zz}^{(\text{cyl})} = 10.53$. In real units this yields the dipolar window

$$B_k \leq B_z \leq B_k + B_z^{(D)}, \tag{6.59}$$

where $B_z^{(D)}$ is given by (6.48) for Mn_{12}Ac.

The solution for the front of tunneling depends on several parameters such as the transverse size of the crystal R and the resonant value of the relaxation rate of (6.42), $\Gamma_{\text{res}} = \Delta^2/(\hbar^2 \Gamma_{m'})$. Rewriting the equations in a reduced form [31], one immediately finds that the front speed is of order $\Gamma_{\text{res}} R$. The only non-trivial parameter is $\tilde{E}_D$, (6.50). An analytical solution of the problem is possible because of the large value of $\tilde{E}_D$. The front speed is given by [31]

$$v = v^* \Gamma_{\text{res}} R, \qquad v^* \simeq \frac{B_z - B_k}{B_k + B_z^{(D)} - B_z}, \tag{6.60}$$

within the dipolar window, independently of $\tilde{E}_D$. Above $B_k + B_z^{(D)}$ the front speed is zero. The reason for this is that for the external field above $B_k + B_z^{(D)}$, the total field well before the front (where all spins are directed in the metastable negative direction and produce the dipolar field $-B_z^{(D)}$) is above its resonance value B_k (and spin tunneling would even increase the total field). Thus in this case resonance tunneling transitions cannot occur. To the contrast, just below $B_k + B_z^{(D)}$ the field well before the front is a little bit below the resonance and increases closer to the front where the magnetization is switching. In this case, there is a wide region where the system is close to the resonance, and the front speed becomes very high. Thus as B_z crosses the value $B_k + B_z^{(D)}$ from below, the front speed diverges and then drops abruptly to the value corresponding to the regular deflagration.

Let us compare the speed of fronts of tunneling $v \simeq \Gamma_{\text{res}} R$ with the speed of regular deflagration, (6.18). With a sufficiently strong transverse field applied, one can have $\Delta/\hbar \sim \Gamma_{m'}$ at the applicability limit of the overdamped approximation, and then $\Gamma_{\text{res}} \sim \Gamma_{m'} \gg \Gamma_f$ because thermal activation goes over high levels of the magnetic molecule where the distances between the levels and thus the energies of phonons involved are much smaller than for the low-lying levels, and also because Γ_f is exponentially small since $T_f \lesssim U$. Additionally, estimation of l_d with $\kappa_f = 10^{-5}\text{m}^2/\text{s}$ and the experimental value $\Gamma_0 = 10^7\text{s}^{-1}$ yield $l_d \sim 3 \times 10^{-4}$ mm for B_z near the first tunneling resonance and even smaller for larger bias. As in the experiment the width of the crystal was much larger than l_d (0.3 mm in Ref. [8], 0.2 mm in Ref. [11], and 1 mm in Ref. [10]), one can see that $\Gamma_{\text{res}} R \gg \Gamma_f l_d$ is quite possible in a strong transverse field, and then the front of spin tunneling is much faster than the front of spin burning. A very conservative estimation with

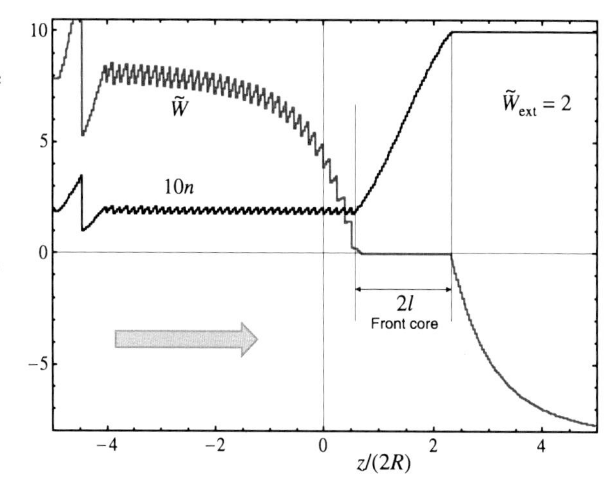

Fig. 6.7 Spatial profiles of the metastable population n and the reduced bias $\tilde{W}$ in the front for $\tilde{W}_{\text{ext}} = 2$ and $\tilde{E}_D = 20$. Everywhere in the front the system is near the resonance, $\tilde{W} \approx 0$. At this value of $\tilde{W}_{\text{ext}}$ the solution begins to lose stability and periodic structures behind the front begin to emerge

$\Gamma_{\text{res}} \Rightarrow \Gamma_0 = 10^7 \text{s}^{-1}$ and $v^* \Rightarrow 1$ for the crystal 0.2 mm thick yields $v \sim 1000$ m/s. As said above, in a strong transverse field one can have $\Gamma_{\text{res}} \gg \Gamma_0$, so that the speed of a spin-tunneling front can easily surpass the speed of sound that is about 2000 m/s in molecular magnets (see analysis in Ref. [36]).

A hallmark of the cold deflagration is residual metastable population behind the front [31] that can be rewritten as

$$n_f = (B_z - B_k)/B_z^{(D)} \tag{6.61}$$

(here $n = n_i = 1$ before the front). One can see that the change of n across the front $\Delta n = 1 - n_f$ goes to zero at the right border of the dipolar window, $B_z = B_k + B_z^{(D)}$. This reconciles the situation with the general requirement that the rate of change of the magnetization of the crystal $\dot{M}$, limited by the tunneling parameter Δ, remains finite. Indeed,

$$\dot{M} \propto (1 - n_f)v = \Gamma_{\text{res}} R(B_z - B_k)/B_z^{(D)} \tag{6.62}$$

reaches only a finite value $\dot{M} \propto \Gamma_{\text{res}} R$ at the right border of the dipolar window before it drops to zero.

To obtain a numerical solution for the cold deflagration, the integro-differential equation was discretized to make the integral in (6.51) a sum and the whole problem a set of coupled non-linear first-order differential equations. The program was written in Wolfram Mathematica. A typical result for spatial profiles of the metastable population n and total energy bias $\tilde{W}$ are shown in Fig. 6.7. In the cold deflagration front, magnetization and dipolar field are self-consistently adjusting in such a way that inside the front core of the width R the spins are on resonance and can tunnel. To the contrary, before and after the front magnetic molecules are off resonance and tunneling is blocked. One of the reasons why fronts of tunneling can be so fast is that their width R entering the expression for the front speed, (6.60), is much larger than the width of the deflagration front l_d, c.f. (6.18). The solution shown in Fig. 6.8

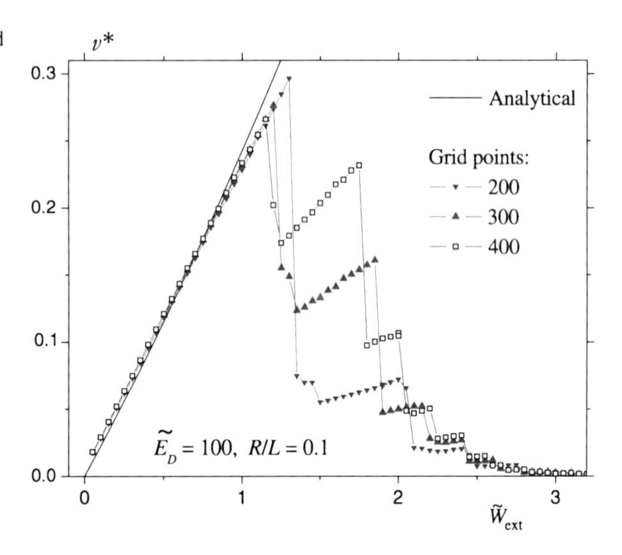

Fig. 6.8 Reduced front speed v^* of (6.60) vs the reduced bias $\tilde{W}_{ext}$ of (6.57) for different number of grid points. For $\tilde{W}_{ext} \lesssim 1$ (the laminar regime) the numerical results are in a good accordance with (6.60) (*straight line*)

is an example of the laminar solution for the cold deflagration front that is realized for a not too strong bias, $\tilde{W}_{ext} \lesssim 1$–2 or $B_z - B_k \lesssim 5$–10 mT.

For a stronger bias, the laminar solution becomes unstable. The front of tunneling is moving with a non-constant speed, leaving spatially-nonuniform distribution of the unburned metastable population behind. The spatial dependence of the dipolar field becomes discontinuous and the resonance condition in the front is not fulfilled (see Fig. 6 of Ref. [31]). As a result, the front speed begins to decrease as the instability develops with the increase of the bias, Fig. 6.8. The instability of the solution is manifesting itself in the dependence on the discretization, absent in the laminar regime.

The only experimentally feasible method to ignite cold deflagration is the sweep of the bias field B_z. When B_z is swept in the positive direction in a negatively magnetized MM crystal, the resonance condition is first achieved at the ends of the crystal where the (negative) dipolar field is weaker (see, e.g., the right side of Fig. 6.6). Spin tunneling at the ends of the crystal caused by field sweep leads to change of the dipolar field in this region that brings the system closer to the resonance in a region of the depth of order R, the transverse size of the crystal. At some moment, a spatial structure close to a stationary front of tunneling is formed and it begins to propagate into the depth of the crystal, the field sweep playing no role anymore. This mechanism is illustrated in Fig. 9 of Ref. [31]. Numerical calculations show that front of tunneling is ignited at the "magic" value of the reduced bias $\tilde{W}_{ext} \simeq 5$, weakly dependent on $\tilde{E}_D$. For this value of the bias, the front of tunneling is non-laminar.

Fronts propagating at other values of the bias, including laminar fronts, can be ignited by a modified procedure proposed in Ref. [31]. First, a global bias is being changed, as before, by a uniform field sweep until the desired value of $\tilde{W}_{ext}$ is reached. After that, front of tunneling can be ignited by a local increase of the bias near the crystal's end using a small coil producing a local magnetic field. This

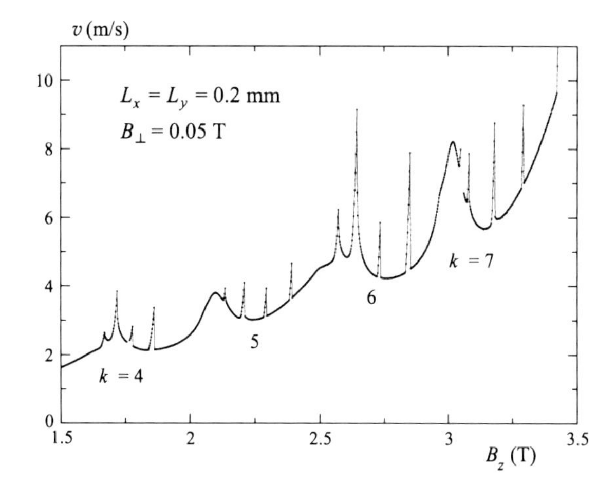

Fig. 6.9 Numerically calculated speed of the deflagration front in a long $Mn_{12}Ac$ crystal for a weak transverse field

method works well in the numerical solution of the cold deflagration problem. However, such kind of experiment has not been performed yet.

Cold deflagration can be most likely observed on thinner crystals having a good thermal contact to the environment, so that the heat released inside the crystal gets quickly removed and the temperature does not increase. As said above, the effect only exists within dipolar windows near tunneling resonances.

It was shown that disorder in resonance fields of individual magnetic molecules is compensated for by adjustment of the dipolar field in the front, so that fronts of tunneling survive [30].

6.3.4 1*d* Theory of Quantum Deflagration

Here we consider a more general situation in which the temperature of the crystal is increasing as the result of the decay of the metastable state, the case when the crystal is thermally insulated. The decay process is controlled by both the temperature (for any bias) and by the dipolar field (near tunneling resonances). The theory of the general quantum-thermal deflagration includes the relaxation equation (6.3) and the heat conduction (energy diffusion) equation (6.8), as well as the expression for the dipolar field (6.51) in the 1*d* approximation. The relaxation rate $\Gamma(T, B_z)$ was calculated for the generic $Mn_{12}Ac$ model (6.1) in Ref. [32] and for the realistic model of $Mn_{12}Ac$ containing the $-AS_z^4$ term that splits tunneling resonances in Ref. [33].

Whereas an analytical solution of this problem has not been found, its qualitative features can be well understood and the numerical solution based on discretization is available. In the case of a zero or weak transverse field, that was the case in all experiments up to date, spin tunneling is thermally assisted and it only modifies the main effect of regular deflagration, resulting in tunneling peaks in the field dependence of the front speed $v(B_z)$. As in the case of regular deflagration, ignition can be achieved by raising the temperature at an end of the crystal.

Figure 6.9 shows the front speed calculated for the bias and crystal size corresponding to the experiments in Refs. [11–13] and using the relaxation rate Γ shown

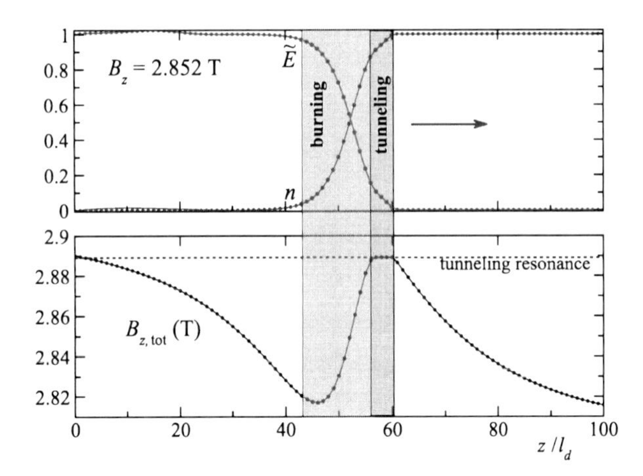

Fig. 6.10 Spatial profiles of the deflagration front in a small transverse field, $B_\perp = 0.05$ T at the peak of the front speed at $B_z = 2.852$ T. There is a resonance spin tunneling at the face of the front and burning in its central and rear parts. In the tunneling region, the total field $B_{z,\mathrm{tot}}$sticks to its resonance value

in Fig. 3 of Ref. [33]. The tunneling peaks are quite pronounced, at variance with the results of these experiments. The latter can be due to a large ligand disorder in $Mn_{12}Ac$ that leads to a substantial scatter of the anisotropy constant D and thus of the positions of the resonances of individual molecules [63–65], especially for the bias as strong as here. Just above 3 T and just below 3.5 T there are regions where the speed is too high to be measured in this calculation, an effect of ground-state tunneling.

Spatial profiles of the magnetization, energy, and the total bias field in the deflagration front give an idea of the role played by spin tunneling. Figure 6.10 shows the spatial profiles at the asymmetric peak of v at $B_z = 2.852$ T in Fig. 6.9. Here the front speed is high because of tunneling at the face of the front, where in the lower panel the total bias field is flat at the level of the tunneling resonance at $B_{z,\mathrm{tot}} = 2.889$ T. Magnetization distribution adjusts so that the dipolar field ensures resonance for a sizable group of spins that can tunnel. Tunneling of these spins results in energy release, the temperature and relaxation rate increase, and tunneling gives way to burning in the central and rear areas of the front.

Formation of the asymmetric maxima of the front speed can be explained as follows. When B_z increases, the peak of $B_{z,\mathrm{tot}}$ that arises due to the local dipolar field (central part of Fig. 6.6) reaches the resonant value. In thick crystals ($R \gg l_d$) this happens if $B_z + B_z^{k_D} = B_k$, where $B_z^{k_D}$ is given by (6.53). This defines the left border of the dipolar window $B_z = B_k - B_z^{k_D}$ (that differs from $B_z = B_k$ for the cold deflagration). At the left border of the dipolar window, a strong increase of $v(B_z)$ begins. The maximum of $B_{z,\mathrm{tot}}$ sticks to the resonance value and becomes flat with progressively increasing width. Greater width of the resonance region results in a stronger tunneling and higher front speed. With further increase of B_z, the right edge of the tunneling region moves too far away from the front core into the region where the

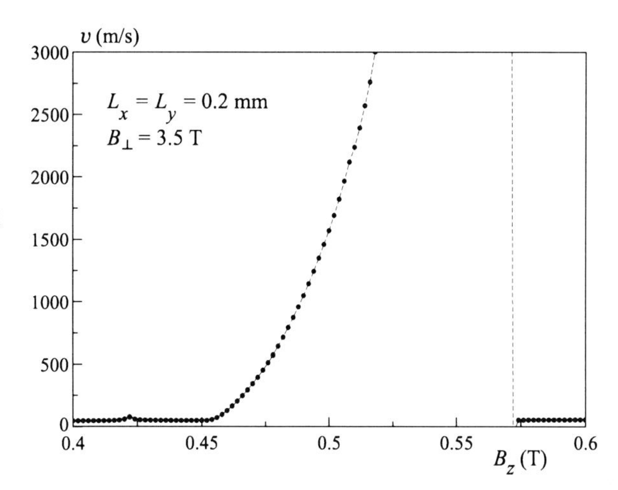

Fig. 6.11 Front speed for a strong transverse field ($B_\perp = 3.5$ T) in the vicinity of the ground-state tunneling resonance at 0.522 T. The small peak on the left is due to the first-excited-state tunneling resonance. *Left* and *right* of the dipolar window the front speed is about 50 m/s

temperature is too low. As the tunneling resonance in question is thermally assisted, it disappears at low temperatures, thus the flat region of $B_{z,\mathrm{tot}}$ cannot spread too far to the right. As a result, the flat configuration of $B_{z,\mathrm{tot}}$ becomes unstable and suddenly $B_{z,\mathrm{tot}}$ changes to the regular shape of the type shown in Fig. 6.6.

If a strong transverse field is applied, the barrier becomes lower and it can completely disappear at a ground-state tunneling resonance. In this case $\Gamma(T, B_k)$ is practically temperature independent and this maximum of the relaxation rate does not disappear at the highest temperatures achieved after burning, T_f. An example is the ground-state tunneling maximum at $B_z = 0.522$T in Fig. 6.5. Although at high temperatures this maximum is hardly visible in the log scale, it is clearly visible in the normal scale in Fig. 5 of Ref. [33]. In such strong transverse fields, the speed of the front becomes very high and spin tunneling plays the dominant role in the front propagation. Figure 6.11 shows a high front speed within a broad dipolar window

$$B_k - B_z^{(k_D)} \le B_z \le B_k + B_z^{(D)} \tag{6.63}$$

having the width of 125.5 mT. The front speed diverges towards the right edge of the dipolar window in accordance with (6.60) and becomes supersonic. A qualitatively similar behavior was observed earlier in calculations for the generic model of $Mn_{12}Ac$, see Fig. 4 of Ref. [32]. In contrast to thermally-assisted tunneling resonances, progressive flattening of $B_{z,\mathrm{tot}}$ at its resonant value is not limited by the temperature before the front since ground-state tunneling occurs already at zero temperature. Thus the front speed diverges at the right edge of the dipolar window, (6.63), where the width of the tunneling region becomes very large.

Comparing the present results with the analytical and numerical results for the cold deflagration, one can see that thermal burning in the central and rear parts of the front are stabilizing the process, so that the laminar solution, (6.60), holds up to the right edge of the dipolar window. There is no breakdown of the laminar regime seen in Fig. 6.8 at $\tilde{W}_{\mathrm{ext}} \simeq 1$.

Another feature of quantum deflagration is complete burning due to the temperature rise, in contrast to the incomplete burning in the cold deflagration, (6.61). Although the speed of the cold deflagration front diverges at $B_z \to B_k + B_z^{(D)}$ (in the laminar regime), the amount of burned metastable population goes to zero, so that the rate of burning remains finite, (6.62). In quantum deflagration burning is complete (up to the equilibrium residual population $n^{(\mathrm{eq})}$ in (6.5)) while the front speed is diverging, so that the rate of burning is diverging, too.

Accordingly, the width of the front becomes very large at $B_z \to B_k + B_z^{(D)}$, in contrast to the width of the cold-deflagration front that remains constant. The structure of the front of the quantum-thermal deflagration near the right border of the dipolar window has a two-tier structure. First goes a fast front of tunneling that reverts a small fraction of the magnetization. The latter leads to heat release that ignites a front of thermal burning that burns all. In the stationary case the speed of the second part of the front is the same but it takes time to develop, thus the width of the whole two-tier front is large. Note that the speed of the quantum deflagration front is not limited by the speed of sound, contrary to the case of detonation [17].

6.3.5 3d Theory of Quantum Deflagration

As mentioned above, the $1d$ theory of fronts of tunneling assumes a flat front that is not well justified because the dipolar field is given by (6.51) only at the symmetry axis. Different values of $B_{z,\mathrm{tot}}$ away from the symmetry axis should self-consistently result in the distribution of the magnetization that depends on all coordinates x, y, z, i.e., in a non-flat front.

On the top of this, there is an instability mechanism for a flat front at a smaller scale due to DDI. In Fig. 6.10 we have seen that, approaching a front of tunneling from before, $B_{z,\mathrm{tot}}$ increases and reaches the resonance value, then it becomes flat. Now, if a small fraction of the surface of a front (going from left to right and changing the magnetization in the positive direction) moves ahead of its neighbors, it produces a *negative* dipolar field on the lagging neighboring parts of the front, as any dipole, see Fig. 6.12. This brings the neighbors further from the resonance, so they tunnel later and their lagging increases. Conversely, lagging portions of the front produce a *positive* dipolar field on the leading part of the front that helps it to propagate faster. (The same mechanism leads to instability of flat domain walls considered in Ref. [29].)

The DDI instability mechanism can potentially destroy any initially flat front of tunneling, making it microscopically rough. The question is whether micro-random dipolar fields produced by a micro-random magnetization in the front are still compatible with resonance tunneling. It is clear that roughness of the front breaks the concept of the adjustment of the system to the resonance, so that the speed of the front should decrease. On the other hand, spins are crossing the resonance, although at random times, so that still there should be a speed-up of the deflagration front near tunneling resonances.

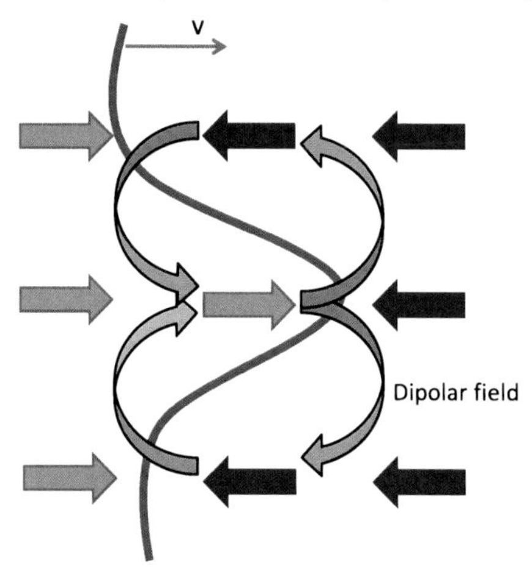

Fig. 6.12 Dipolar instability of a flat front of spin tunneling. A leading part of the front (*in the center*) produces the dipolar fields on its neighbors that slow them down

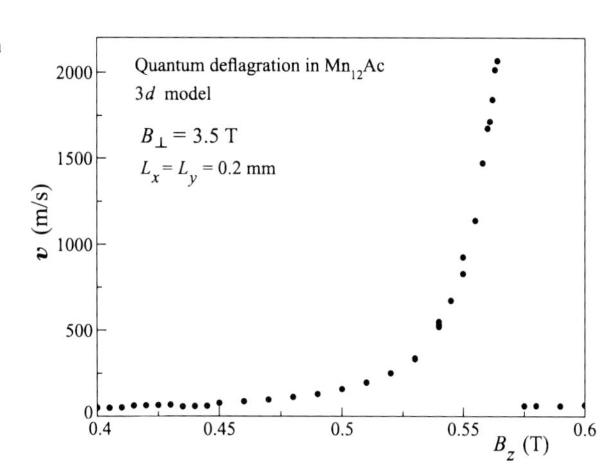

Fig. 6.13 Front speed within the $3d$ model for a strong transverse field ($B_\perp = 3.5$ T) in the vicinity of the ground-state tunneling resonance at $B_z = 0.522$ T

In $3d$ model of quantum deflagration the dipolar field was calculated using (6.55) for crystals of box shape with dimensions $L_x = L_y \ll L_z$ using the relaxation rate Γ for $B_\perp = 3.5$T shown in Fig. 6.5. The crystal was discretized with about 1 million total grid points in all 3 dimensions. The resulting system of first-order nonlinear equations was implemented in Wolfram Mathematica in a vectorized form using a compiled Butcher's 5th-order Runge-Kutta solver with a fixed step.

As expected, roughness of the front due to the dipolar instability has been detected within the dipolar window, (6.63), where the computed front speed is lower than within $1d$ model, Fig. 6.11. Nevertheless, the front speedup due to spin tunneling is still huge, reaching sonic speeds in $Mn_{12}Ac$ on the right of the dipolar window, see Fig. 6.13.

Outside the dipolar window, a regular deflagration with a flat front and front speed $v \simeq 50$ m/s has been found for this value of the transverse field. With en-

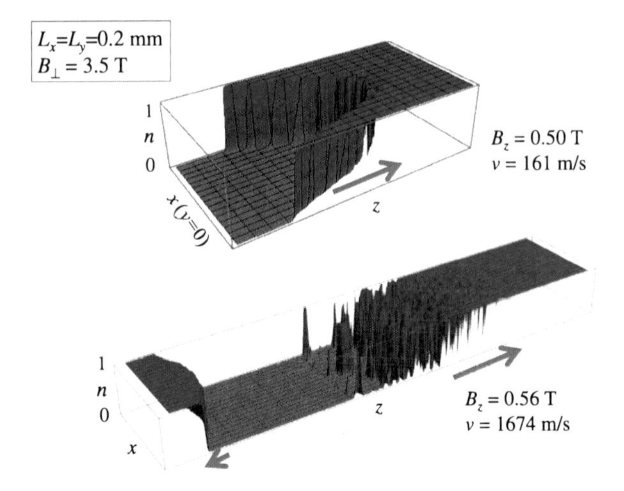

Fig. 6.14 Profile of the metastable population n in the $3d$ model of quantum deflagration for $Mn_{12}Ac$ at $B_\perp = 3.5$ T and $B_z = 0.5$ T (*upper*) and 0.56 T (*lower*)

tering the dipolar window from the left, the front becomes progressively non-flat with its central part leading. Front roughness emerges and increases with the bias. Figure 6.14 shows the profile of the metastable population n for the crystal with $L_x = L_y = 0.2$ mm, as in experiments of Refs. [11–13], for $B_z = 0.5$ T and 0.56 T. The metastable population n is represented as a $3d$ plot as a function of x and z with $y = 0$ at some moment of time. The unburned cold portion of the crystal on the right is shown in blue, while the burned hot part on the left is shown in red. In the upper part of the figure showing the result for $B_z = 0.5$ T the front is essentially non-flat and there is some roughness, especially strong near the symmetry axis. The speed of this front $v = 161$ m/s is already much greater than the speed of the regular deflagration, 50 m/s.

Numerical results for a larger bias $B_z = 0.56$ T and a longer crystal are shown in the lower part of Fig. 6.14. The front has a nearly sonic speed of $v = 1674$ m/s and is very rough, while becoming flat again. The animation of this process looks like precipitation. Ignition of this front occurs at some distance from the left end of the crystal where the resonance condition is fulfilled. From this point, a very fast tunneling front is propagating to the right while a regular slow burning front is propagating to the left.

6.4 Discussion

Regular temperature-driven magnetic deflagration in long crystals of Mn_{12} has been experimentally observed and is relatively well understood. The lack of a quantitative accordance between the theory and experiment can be attributed to still unknown temperature dependence of the thermal diffusivity κ, as well as to the absence of a microscopic theory of relaxation in MM taking into account collective effects such as phonon/photon superradiance and phonon bottleneck.

Effects of spin tunneling on ignition of deflagration and front speed near resonance values of the bias field have been experimentally detected in zero transverse field. However, these effect are due to thermally-assistant tunneling just below the top of the barrier and they are not strong.

In contrast, spin tunneling directly out of the metastable ground state in strong transverse fields can lead to huge effects, such as supersonic quantum deflagration within the dipolar window around tunneling resonances. Unfortunately, creating an initial state for this process is practically difficult. In a strong transverse field also *non-resonant* spin tunneling is rather fast. While the system is being biased to reach the initial state close to the resonance, it is already relaxing and a large portion of the metastable population gets lost before a front of tunneling could start. In addition, non-resonant tunneling in a biased MM leads to heat release that can result in self-ignition if the crystal is thermally insulated.

It would be desirable to employ a fast field sweep to bring the MM into starting position for quantum deflagration in a strong transverse field without deteriorating its state. To observe non-thermal fronts of tunneling, thinner crystals with a good thermal contact to the environment have to be used.

Acknowledgements Part of research on magnetic deflagration presented in this Chapter was conducted jointly with Professor Eugene Chudnovsky. Our students Reem Jaafar and Saaber Shoyeb participated in obtaining some of the results. The author is indebted to Ferran Macia, Pradeep Subedi, and Saül Vélez Centoral for discusions of magnetic deflagration in strong transverse fields. Oliver Rübenkönig and Daniel Lichtblau have provided a great support on vectorization and compilation in Wolfram Mathematica. This work has been supported by the NSF grant No. DMR-1161571 and, in part, under NSF Grants CNS-0958379 and CNS-0855217 and the City University of New York High Performance Computing Center. Eugene Dedits helped me in using the facilities of the CUNY Computing Center.

Note Added in Proof The full 3*d* theory of quantum deflagration in Mn_12Ac has been recently published Ref. [66].

References

1. I. Glassman, *Combustion* (Academic Press, San Diego, 1996)
2. L.D. Landau, E.M. Lifshitz, *Fluid Dynamics* (Pergamon, London, 1987)
3. T. Lis, Acta Crystallogr. B **36**, 2042 (1980)
4. R. Sessoli, D. Gatteschi, A. Caneschi, M.A. Novak, Nature (London) **365**, 141 (1993)
5. C. Paulsen, J.G. Park, in *Quantum Tunneling of Magnetization—QTM'94*, ed. by L. Gunther, B. Barbara (Kluwer, Dordrecht, 1995)
6. F. Fominaya, J. Villain, P. Gaudit, J. Chaussy, A. Caneschi, Phys. Rev. Lett. **79**, 1126 (1997)
7. E. del Barco, J.M. Hernández, M. Sales, J. Tejada, H. Rakoto, J.M. Broto, E.M. Chudnovsky, Phys. Rev. B **60**, 11898 (1999)
8. Y. Suzuki, M.P. Sarachik, E.M. Chudnovsky, S. McHugh, R. Gonzalez-Rubio, N. Avraham, Y. Myasoedov, E. Zeldov, H. Shtrikman, N.E. Chakov, G. Christou, Phys. Rev. Lett. **95**, 147201 (2005)
9. N. Avraham, A. Stern, Y. Suzuki, K.M. Mertes, M.P. Sarachik, E. Zeldov, Y. Myasoedov, H. Shtrikman, E.M. Rumberger, D.N. Hendrickson, N.E. Chakov, G. Christou, Phys. Rev. B **72**, 144428 (2005)

10. A. Hernández-Minguez, J.M. Hernández, F. Macia, A. Garcia-Santiago, J. Tejada, P.V. Santos, Phys. Rev. Lett. **95**, 217205 (2005)
11. S. McHugh, R. Jaafar, M.P. Sarachik, Y. Myasoedov, A. Finkler, H. Shtrikman, E. Zeldov, R. Bagai, G. Christou, Phys. Rev. B **76**(17), 172410 (2007)
12. S. McHugh, R. Jaafar, M.P. Sarachik, Y. Myasoedov, A. Finkler, E. Zeldov, R. Bagai, G. Christou, Phys. Rev. B **80**(2), 024403 (2009)
13. S. McHugh, B. Wen, X. Ma, M.P. Sarachik, Y. Myasoedov, E. Zeldov, R. Bagai, G. Christou, Phys. Rev. B **79**(17), 174413 (2009)
14. S. McHugh, R. Jaafar, M.P. Sarachik, Y. Myasoedov, H. Shtrikman, E. Zeldov, R. Bagai, G. Christou, Phys. Rev. B **79**, 052404 (2009)
15. J. Vanacken, S. Stroobants, M. Malfait, V.V. Moshchalkov, M. Jordi, J. Tejada, R. Amigo, E.M. Chudnovsky, D.A. Garanin, Phys. Rev. B **70**, 220401R (2004)
16. W. Decelle, J. Vanacken, V.V. Moshchalkov, J. Tejada, J.M. Hernández, F. Macià, Phys. Rev. Lett. **102**(2), 027203 (2009)
17. M. Modestov, V. Bychkov, M. Marklund, Phys. Rev. Lett. **107**, 20720 (2011)
18. F. Macià, A. Hernández-Mínguez, G. Abril, J.M. Hernandez, A. García-Santiago, J. Tejada, F. Parisi, P.V. Santos, Phys. Rev. B **76**(17), 174424 (2007)
19. S. Velez, J.M. Hernandez, A. Fernandez, F. Macià, C. Magen, P.A. Algarabel, J. Tejada, E.M. Chudnovsky, Phys. Rev. B **81**, 064437 (2010)
20. S. Vélez, J.M. Hernandez, A. García-Santiago, J. Tejada, V.K. Pecharsky, K.A. Gschneidner, D.L. Schlagel, T.A. Lograsso, P.V. Santos, Phys. Rev. B **85**, 054432 (2012)
21. D.A. Garanin, E.M. Chudnovsky, Phys. Rev. B **76**, 054410 (2007)
22. E.M. Chudnovsky, JETP Lett. **50**, 1035 (1979)
23. M. Enz, R. Schilling, J. Phys. C **19**, L711 (1986)
24. E.M. Chudnovsky, L. Gunther, Phys. Rev. Lett. **60**, 661 (1988)
25. E.M. Chudnovsky, L. Gunther, Phys. Rev. B **37**, 9455 (1988)
26. J.R. Friedman, M.P. Sarachik, J. Tejada, R. Ziolo, Phys. Rev. Lett. **76**, 3830 (1996)
27. J.M. Hernández, X.X. Zhang, F. Luis, J. Bartolomé, J. Tejada, R. Ziolo, Europhys. Lett. **35**, 301 (1996)
28. L. Thomas, F. Lionti, R. Ballou, D. Gatteschi, R. Sessoli, B. Barbara, Nature **383**, 145 (1996)
29. D.A. Garanin, E.M. Chudnovsky, Phys. Rev. B **78**, 174425 (2008)
30. D.A. Garanin, E.M. Chudnovsky, Phys. Rev. Lett. **102**, 097206 (2009)
31. D.A. Garanin, Phys. Rev. B **80**(1), 014406 (2009)
32. D.A. Garanin, R. Jaafar, Phys. Rev. B **81**(18), 180401 (2010)
33. D.A. Garanin, S. Shoyeb, Phys. Rev. B **85**, 094403 (2012)
34. C. Kittel, *Quantum Theory of Solids* (Wiley, New York, 1963)
35. A.M. Gomes, M.A. Novak, R. Sessoli, A. Caneschi, D. Gatteschi, Phys. Rev. B **57**, 5021 (1998)
36. D.A. Garanin, Phys. Rev. B **78**, 020405(R) (2008)
37. N.N. Semenov, Usp. Fiz. Nauk **23**, 251 (1940)
38. M. Modestov, V. Bychkov, M. Marklund, Phys. Rev. B **83**, 214417 (2011)
39. L. Bokacheva, A.D. Kent, M.A. Walters, Phys. Rev. Lett. **85**, 4803 (2000)
40. W. Wernsdorfer, M. Murugesu, G. Christou, Phys. Rev. Lett. **96**(5), 057208 (2006)
41. E.M. Chudnovsky, D.A. Garanin, Phys. Rev. Lett. **79**, 4469 (1997)
42. D.A. Garanin, J. Phys. A **24**, L61 (1991)
43. D.A. Garanin, E.M. Chudnovsky, Phys. Rev. B **56**, 11102 (1997)
44. E.M. Chudnovsky, Phys. Rev. Lett. **92**, 120405 (2004)
45. E.M. Chudnovsky, D.A. Garanin, R. Schilling, Phys. Rev. B **72**, 094426 (2005)
46. D.A. Garanin, in *Advances in Chemical Physics*, vol. 147, ed. by S.A. Rice, A.R. Dinner (Wiley, New York, 2012)
47. F. Luis, J. Bartolomé, J.F. Fernández, Phys. Rev. B **57**, 505 (1998)
48. A. Fort, A. Rettori, J. Villain, D. Gatteschi, R. Sessoli, Phys. Rev. Lett. **80**, 612 (1998)
49. D.A. Garanin, V.V. Ishchenko, L.V. Panina, Teor. Mat. Fiz. **82**, 242 (1990)
50. D.A. Garanin, Phys. Rev. E **54**, 3250 (1996)

51. F. Luis, J. Bartolomé, J.F. Fernández, J. Tejada, J.M. Hernández, X.X. Zhang, R. Ziolo, Phys. Rev. B **55**, 11448 (1997)
52. M.N. Leuenberger, D. Loss, Europhys. Lett. **46**, 692 (1999)
53. R. Dicke, Phys. Rev. **93**, 99 (1954)
54. E.M. Chudnovsky, D.A. Garanin, Phys. Rev. Lett. **89**, 157201 (2002)
55. E.M. Chudnovsky, D.A. Garanin, Phys. Rev. Lett. **93**, 257205 (2004)
56. A. Abragam, A. Bleaney, *Electron Paramagnetic Resonance of Transition Ions* (Clarendon Press, Oxford, 1970)
57. D.A. Garanin, Phys. Rev. B **75**, 094409 (2007)
58. D.A. Garanin, Phys. Rev. B **77**, 024429 (2008)
59. A. Hernández-Minguez, A. Jordi, R. Amigo, A. Garcia-Santiago, J.M. Hernández, J. Tejada, Europhys. Lett. **69**, 270 (2005)
60. O. Shafir, A. Keren, Phys. Rev. B **79**(18), 180404 (2009)
61. L.D. Landau, E.M. Lifshitz, *Electrodynamics of Continuous Media* (Pergamon, London, 1960)
62. D.A. Garanin, R. Schilling, Phys. Rev. B **71**, 184414 (2005)
63. K. Park, M.A. Novotny, N.S. Dalal, S. Hill, P.A. Rikvold, Phys. Rev. B **65**, 014426 (2001)
64. S. Hill, S. Maccagnano, K. Park, R.M. Achey, J.M. North, N.S. Dalal, Phys. Rev. B **65**, 224410 (2002)
65. K. Park, M.A. Novotny, N.S. Dalal, S. Hill, P.A. Rikvold, Phys. Rev. B **66**, 144409 (2002)
66. D.A. Garanin, Phys. Rev. B **88**, 064413 (2013)

Chapter 7
Dipolar Magnetic Order in Crystals of Molecular Nanomagnets

Fernando Luis

Abstract This chapter reviews experimental studies of long-range dipolar magnetic order in crystals of single-molecule magnets. Quantum annealing by a transverse magnetic field enables one to explore the ground state of highly anisotropic SMMs, such as Mn_{12} and Fe_8, both of which order ferromagnetically below $T_c = 0.9$ K and 0.6 K, respectively. In Mn_{12} acetate, molecular tilts caused by the disorder in the orientations of some solvent molecules affect dramatically the character of the field-induced transition, which agrees with the predictions of the random-field Ising model. The existence of a quantum critical point has been shown in crystals of Fe_8 clusters, which are among the best realizations of the archetypical quantum Ising model in a transverse magnetic field.

7.1 Introduction

Dipolar interactions are ubiquitous in Nature. A dipolar magnetic moment $\boldsymbol{\mu}_i$, e.g. a magnetic ion, generates a magnetostatic field that affects other dipoles $\boldsymbol{\mu}_j$ located in its surroundings. The coupling energy between any pair of dipoles separated by a position vector $\mathbf{r}_{ij}$ can be expressed as follows

$$\mathcal{H}_{\mathrm{dip},ij} = -\left[\frac{3(\boldsymbol{\mu}_i \mathbf{r}_{ij})(\boldsymbol{\mu}_j \mathbf{r}_{ij})}{r_{ij}{}^5} - \frac{\boldsymbol{\mu}_i \boldsymbol{\mu}_j}{r_{ij}{}^3}\right] \tag{7.1}$$

Dipolar interactions are known to affect the physical behavior of magnetic materials in a number of ways. They often dominate the line broadening of resonance spectra measured on paramagnets [1]. In magnetically ordered materials, an important manifestation is the formation of magnetic domains pointing along different orientations [2]. However, dipolar interactions often play but a minor role in determining the intrinsic magnetic structure. In order to better understand why this is the case,

F. Luis (✉)
Instituto de Ciencia de Materiales de Aragón and Departamento de Física de la Materia Condensada, CSIC–Universidad de Zaragoza, C/Pedro Cerbuna 12, 50009 Zaragoza, Spain
e-mail: fluis@unizar.es

J. Bartolomé et al. (eds.), *Molecular Magnets*, NanoScience and Technology, DOI 10.1007/978-3-642-40609-6_7,

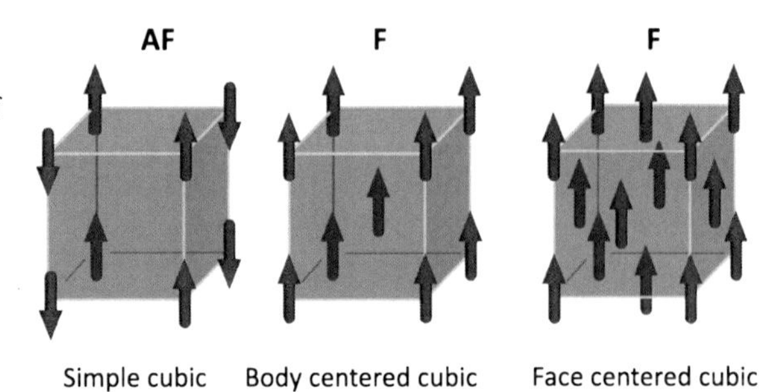

Fig. 7.1 Luttinger and Tisza [4] solutions for the ground-state configurations of interacting magnetic dipoles located on each of the three Bravais cubic lattices

let us consider a specific and simple example: pure metallic iron. Iron is a ferromagnet below a Curie temperature $T_c = 1046$ K. The typical dipolar energy between nearest neighbor Fe atoms amounts to approximately 0.5 K. Clearly, dipolar interactions are much weaker than exchange interactions, of quantum mechanical origin, and the later drive the onset of magnetic order in iron as well and in the vast majority of magnetic materials. The same argument explains why it is so difficult to find systems in Nature showing pure dipolar magnetic order. Even relatively weak exchange couplings, difficult to avoid, might dominate over dipolar interactions.

Brief Survey of Theoretical Studies Not surprisingly, the first steps in the study of dipolar magnetism were almost exclusively of a theoretical nature. Compared with the situation met when exchange interactions are dominant, the problem statement is appealingly simple. Interactions are known, and given by (7.1), and all that needs to be done is to minimize the free energy of a given lattice of dipoles over all possible configurations. However, its numerical solution is complicated by the long-range character of dipolar interactions. Early attempts consisted of numerical calculations (carried out without the aid of a computer!) of the energies of some configurations of classical dipoles located in simple lattices [3]. It was not until 1946 that J.M. Luttinger and L. Tisza found a rigorous method that enables finding the ground state configurations for simple, body-centered, and face-centered cubic lattices [4]. This method was later extended to cover lattices with up to two equivalent dipoles per unit cell [5] and to even more complex lattices in the case of strongly anisotropic (Ising-like) dipoles [6]. The ordering of dipolar Ising crystals was reanalyzed in [7], where a simple expression for the interaction energy between chains of spins pointing along their anisotropy axes was derived. Finally, the ordering temperatures of some materials have been determined by Monte Carlo calculations [7–9].

It follows from these results that the ground state configurations are dictated by lattice symmetry. In cubic systems (see Fig. 7.1), for instance, lattices with low coordination numbers (diamond and simple cubic lattices) order antiferromagnetically, whereas face centered and body centered lattices are ferromagnetic. The same argument applies, within a certain range of parameters (determined by the ratio c/a), for tetragonal and hexagonal lattices [7, 10]. The existence of dipolar ferromagnetism was, however, questioned by Luttinger and Tisza themselves, and thought to depend upon the shape of the specimen [4]. The argument is that the onset of a spontaneous magnetization gives rise, in any finite sample, to an additional increase

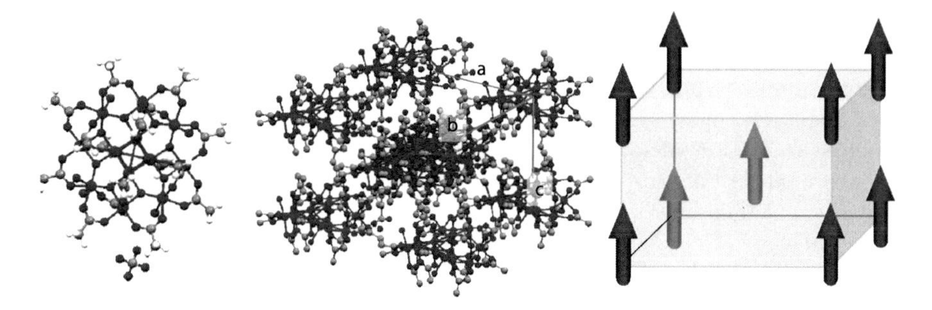

Fig. 7.2 Molecular structure of Mn_{12} acetate (*left*) and molecular packing in the tetragonal unit cell (*center*). This molecular crystal provides a close approximation to a tetragonal lattice of strongly anisotropic (Ising-like) spins, coupled via dipole-dipole interactions (*right*)

in magnetostatic energy that depends on its demagnetizing tensor. The existence of a well-defined ground state for macroscopic lattices at zero magnetic field, independent of the specimen's shape, was demonstrated more than 20 years latter by R.B. Griffiths [11]. This result suggests that dipolar ferromagnets, like any other ferromagnetic material, tend to subdivide into magnetic domains below T_c [12]. Some other important theoretical results worth mentioning here are the existence of important zero-point fluctuations [13] and the prediction that three-dimensional dipolar lattices provide close approximations of mean-field models. In particular, the marginal dimensionality for mean-field behavior is $d^* = 3$ in an Ising dipolar ferromagnet [14].

Experimental Realizations: Single Molecule Magnets For the reasons mentioned above, experimental realizations of dipolar lattices are scarce, even today. Best-known examples are provided by crystals of lanthanide-based compounds [15–22]. Exchange interactions between lanthanide ions are weak, on account of the localization of $4f$ electrons. In lattices with sufficiently separated ions, dipolar interactions might therefore become of comparable and even dominating strength. In these cases, ordering temperatures are often very low, typically below 100 mK. An outstanding exception is represented by $LiHoF_4$, which orders ferromagnetically at $T_c = 1.54$ K [15, 18]. However, in the later case the underlying physics is somewhat complicated by the existence of non-negligible exchange interactions and rather strong hyperfine couplings [23, 24].

Crystals of molecular nanomagnets [25–29] are suitable candidate materials to investigate magnetic order of pure dipolar lattices (see Fig. 7.2). Each of these single-molecule magnets (SMMs) is an electrically neutral entity, in which the magnetic core is surrounded, thus also isolated from its neighbors, by a shell of organic ligands. Many of these molecules have large spins (e.g. $S = 10$ for Mn_{12} and Fe_8 clusters), and therefore large magnetic moments $\mu = g\mu_B S$, where g is the molecular gyromagnetic factor. Dipolar interactions are then relatively strong and often dominate over the very weak, if present at all, exchange interactions. Ordering temperatures are expected to be of the order of 0.5 K or even higher [7], which considerably simplifies the experimental study of the magnetic phase transitions by a variety of techniques, including heat capacity, magnetic susceptibility, and magnetic neutron diffraction.

From a fundamental point of view, these crystals provide close to ideal realizations of physical models, such as the quantum Ising model [30], of broad interest for physics. Dipolar interactions also affect the spin dynamics. This effect is particularly important at very low temperatures, when spin flips occur predominantly via pure tunneling processes. Under these conditions, dipolar bias fields energetically detune states between which tunneling takes place, and magnetic relaxation becomes a collective phenomenon [31]. It follows then that not only the equilibrium state, but also the rate at which this state is attained strongly depend on the onset of magnetic order below T_c [32]. Knowing the equilibrium magnetic state is therefore a necessary pre-requisite to fully understand magnetic relaxation and quantum tunneling phenomena observed at very low temperatures [33–39]. A particularly attractive question is the competition between dipolar interactions, typically weak, and quantum fluctuations, which are strong in molecular nanomagnets and can be made even stronger via the application of an external magnetic field, eventually leading to a quantum phase transition [40, 41].

The information gained via these studies can also be of relevance to other scientific fields and even to applications. In many aspects (e.g., the existence of magnetic memory effects at sufficiently low temperatures, associated with a strong magnetic anisotropy), crystals of SMMs are equivalent to ordered and monodisperse arrays of magnetic nanoparticles. The study of dipolar interactions in the former provides useful information on the nature of the collective magnetic response of coupled nanomagnets [42]. The onset of long-range magnetic order reduces the entropy of the spin lattice, which rapidly vanishes below T_c. This effect ultimately limits the lowest temperature attainable by adiabatic demagnetization methods. Molecular nanomagnets are among the best magnetic coolers at liquid Helium temperatures [43]. The study of dipolar ordering in these materials, and how it depends on crystal symmetry and magnetic anisotropy, is then of practical interest for magnetic refrigeration technologies.

Outline of the Chapter The present chapter is written from an experimental perspective. Its aim is mainly to show, with the help of examples, the existence of dipolar order in some of the most famous single-molecule magnets, in particular Mn_{12} acetate and Fe_8, how these phenomena have been experimentally uncovered, and what physics can be learned from it. Section 7.2 provides a very basic theoretical background on the interactions that play a role in determining the physical behavior of SMMs lattices, and their respective effects. This section introduces also mean-field approximations, which are simple and therefore especially convenient to analyze the results of experiments. Section 7.3 discusses one of the most serious difficulties faced by such experiments and which is related to the slow relaxation of molecular nanomagnets. This section also shows how measurements of the magnetization dynamics and hysteresis can be used to estimate the effective intermolecular interaction fields. Sections 7.4 to 7.6 describe the results of experimental studies of magnetic order performed on several molecular materials. The experiments performed on Mn_{12} (Sect. 7.5) and Fe_8 (Sect. 7.6) illustrate that highly interesting physical phenomena result from the competition between dipolar interactions and transverse magnetic fields. The last Sect. 7.7 summarizes the main conclusions and suggests possible evolutions of this research field.

Most of the results described in this chapter refer to work done, and published [44–49], in the course of the past decade. Yet, it contains a few original aspects too, in particular the determination of the interaction fields in Mn_{12} acetate that is included in Sect. 7.3. Also, the interpretation of some of the experimental results is re-examined on the basis of subsequent theoretical [10] and experimental [50] developments.

7.2 Theoretical Background

7.2.1 Spin Hamiltonian

The spin Hamiltonian of a lattice of SMMs coupled via dipolar interactions can be written as follows

$$\mathcal{H} = \frac{1}{2}\sum_i \sum_{j \neq i} \mathcal{H}_{\mathrm{dip},ij} + \sum_i \mathcal{H}_{0,i} + \sum_i \mathcal{H}_{\mathrm{Z},i} \tag{7.2}$$

where the dipolar interaction Hamiltonian $\mathcal{H}_{\mathrm{dip},ij}$ is given by (7.1), $\mathcal{H}_{0,i}$ gives the magnetic anisotropy of each isolated molecule

$$\mathcal{H}_0 = -DS_z^2 + BS_z^4 + E\left(S_x^2 - S_y^2\right) + \frac{C}{2}\left(S_+^4 + S_-^4\right) + \cdots \tag{7.3}$$

where $D, B, E, C, \ldots$ are anisotropy parameters, and

$$\mathcal{H}_{\mathrm{Z}} = -g\mu_{\mathrm{B}} H_z S_z - g\mu_{\mathrm{B}} H_{\perp}(S_x \cos\phi + S_y \sin\phi) \tag{7.4}$$

describes the Zeeman interaction with an external magnetic field $\mathbf{H}$, having components H_z along the anisotropy axis z and $H_{\perp}$ perpendicular to it, where ϕ is the azimuthal angle of $\mathbf{H}$ in the xy plane.

The zero-field energy level scheme of a generic SMM with Ising-like uniaxial anisotropy (i.e. with $D > 0$ and weak higher-order anisotropies) is schematically shown in Fig. 7.3. "Diagonal" terms (i.e. those commuting with S_z) give rise to a classical energy barrier $U_{\mathrm{cl}} = DS^2 - BS^4$, separating spin-up (i.e. eigenstates of S_z with eigenvalue $m > 0$) from spin-down states (with $m < 0$). Off-diagonal terms (i.e. non-commuting with S_z), induce quantum tunneling between magnetic states $\pm m$, at zero field, and between m and $-m-n$, with n integer, at the "crossing fields"

$$H_{z,n}(m) = n(D/g\mu_{\mathrm{B}})\left\{1 + B\left[m^2 + (m+n)^2\right]/D\right\} \tag{7.5}$$

At these fields, the classical degeneracy between the crossing levels is lifted by a finite quantum tunnel splitting Δ_m.

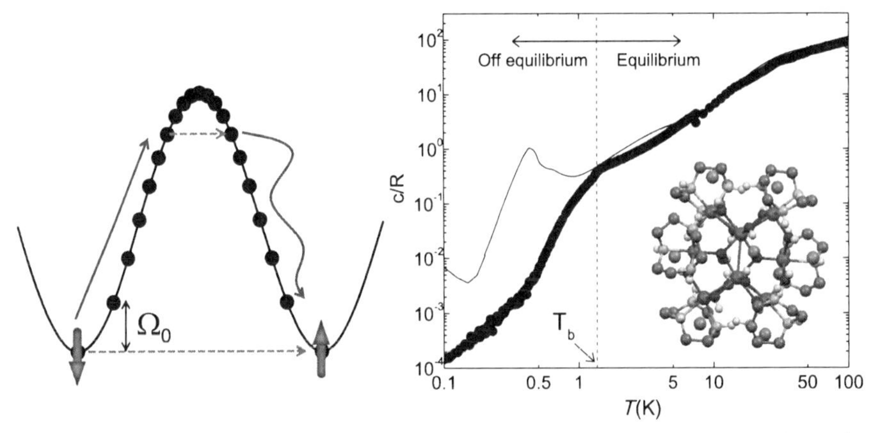

Fig. 7.3 *Left*: Structure of magnetic energy levels of a generic SMM. The spin reversal can take place via a thermally activated mechanism or via pure quantum tunneling processes. *Right*: Zero-field specific heat of Fe_8, whose molecular structure is shown in *the inset*. Above the blocking temperature T_b, the experimental data (*dots*) agree with the equilibrium specific heat, obtained from Monte Carlo calculations (*solid line*). Below T_b, it decreases rapidly, showing no hint of the phase transition to long-range magnetic order

The first term in (7.2) induces, below a critical temperature T_c, a phase transition to a long-range magnetically ordered state, which is mainly determined by crystal symmetry and lattice parameters [7]. Magnetic interactions compete with the polarization induced by an external magnetic field **H**. In addition, and similarly to what happens with exchange-coupled spin systems [51], both the nature of the ensuing magnetic order and T_c are affected by the magnetic anisotropy. A particular case, which is highly relevant to most SMMs, arises when the uniaxial anisotropy is much stronger than dipolar interactions. More specifically, when the zero-field splitting $\Omega_0 \simeq (2S-1)D$ that separates the ground and first excited level doublets of (7.3) is much larger than the characteristic interaction energy, given by $k_B T_c$, the dipolar Hamiltonian (7.1) can be simplified to the following Ising interaction Hamiltonian

$$\mathcal{H}_{\text{dip},ij} \simeq -\left[\frac{3(\mu_{i,z}z_{ij})(\mu_{j,z}z_{ij})}{r_{ij}{}^5} - \frac{\mu_{i,z}\mu_{j,z}}{r_{ij}{}^3}\right] \tag{7.6}$$

involving only S_z.

Zero-field magnetic ground states and ordering temperatures T_c of some specific lattice symmetries, relevant to some particular SMMs systems, have been determined using Monte Carlo calculations based on the Ising Hamiltonian (7.6) [7, 45, 47]. Results of some of these calculations are listed in Table 7.1.

7.2.2 Mean-Field Approximations

An even simpler method to treat the effect of interactions is to make use of a mean-field approximation, which is especially well suited to deal with dipolar magnets

Table 7.1 Magnetic ordering temperatures of some crystals of SMMs. *The fifth and sixth columns* show data calculated, for pure dipolar interactions, using Monte Carlo and mean-field methods, respectively. *The seventh column* provides experimental data. Na stands for data that are "not available"

System	Lattice	Spin	D/k_B (K)	T_c^{MC} (K)	T_c^{MF}	T_c^{exp} (K)
Mn_4Me	Monoclinic	9/2	0.69	0.1	n.a.	0.21(2) [45]
Mn_6	Monoclinic	12	0.013	0.22	n.a.	0.15(1) [44]
Fe_8	Triclinic	10	0.294	0.54 [7]	n.a.	0.60(5) [49]
$Mn_{12}ac$	Tetragonal	10	0.6	0.5 [7]	0.8 [77]	0.9(1) K [46]
Fe_{17}	Trigonal	35/2	0.02	n.a.	n.a.	0.8 [66]

[14]. For simplicity, in the following I consider a lattice of spins ordering ferromagnetically. Within the mean-field approximation, the spin Hamiltonian (7.2) reduces itself to an effective Hamiltonian for a single spin (say spin i)

$$\mathcal{H} = \mathcal{H}_0 - g\mu_B H_z S_z - g\mu_B H_\perp (S_x \cos\phi + S_y \sin\phi) - J_{eff}\langle S_z\rangle_T S_z \tag{7.7}$$

where $\langle S_z\rangle_T$ is the thermal equilibrium average of S_z and

$$J_{eff} = \frac{-(g\mu_B)^2}{2}\sum_{j\neq i}\left(\frac{3z_{ij}^2}{r_{ij}{}^5} - \frac{1}{r_{ij}{}^3}\right) \tag{7.8}$$

is an effective interaction constant. The last term in (7.7) can also be written as a Zeeman interaction $-g\mu_B H_{eff,z} S_z$ with a mean-field magnetic bias

$$H_{eff,z} = \frac{J_{eff}}{g\mu_B}\langle S_z\rangle_T \tag{7.9}$$

The mean-field Hamiltonian (7.7) is appealing for experimentalists, because it allows a relatively easy comparison to different measurable quantities. Above T_c, the intrinsic (i.e. free from demagnetization effects) equilibrium longitudinal magnetic susceptibility $\chi_{i,zz}$ follows Curie-Weiss law

$$\chi_{i,zz} = \frac{C}{T-\theta} \tag{7.10}$$

where C is the Curie constant and $\theta = T_c$ is the Weiss temperature. Notice that (7.10) also applies to dipolar lattices ordering antiferromagnetically. In the latter case, however, $\theta < 0$. Analytical expressions for θ and C can be found for specific limiting situations. For instance, when the magnetic anisotropy is very strong as compared to both dipolar interactions and $k_B T$, each molecular spin behaves effectively as a spin-1/2 system. Under these conditions (i.e. for $D \to \infty$), (7.7) reduces to an effective spin-1/2 Hamiltonian

$$\mathcal{H} \simeq -\mu_B S H_\perp (g_x\sigma_x \cos\phi + g_y\sigma_y \sin\phi) - g_z\mu_B S H_z\sigma_z - J_{eff} S^2\langle\sigma_z\rangle_T\sigma_z \tag{7.11}$$

where the σ's are Pauli spin operators, $g_z \simeq g$, and g_x and g_y depend on the ratio between off diagonal and diagonal anisotropy parameters (i.e. on E/D, C/D, etc). The Curie constant and Weiss temperature then read as follows

$$C = N\frac{(g\mu_B S)^2}{3k_B}$$
$$\theta = \frac{J_{eff}S^2}{k_B} \tag{7.12}$$

where N is the concentration of molecular spins per unit of volume. The susceptibility χ_{powder} of randomly oriented crystals $\chi_{powder} = (1/3)(\chi_{i,xx} + \chi_{i,yy} + \chi_{i,zz})$. Often, especially close to T_c, $\chi_{i,zz} \gg \chi_{i,xx}$, $\chi_{i,yy}$. Therefore, the susceptibility of powdered samples also follows Curie-Weiss law (7.10).

Strictly speaking, (7.10) applies to the case of an infinitely long cylindrical sample, whose long axis coincides with z. For real samples of finite size, demagnetizing effects play a role [52–54]. The susceptibility that is actually measured in a experiment in then approximately given by the following expression

$$\chi_{zz} \simeq \frac{\chi_{i,zz}}{1 + \chi_{i,zz}\widetilde{N}_{zz}} \tag{7.13}$$

where it has been considered, for simplicity, that the z axis corresponds to a principal axis of the demagnetization tensor $\widetilde{N}$. Notice that, at $T = \theta$, χ_{zz} no longer diverges but approaches $\chi_{max} = 1/N_{zz}$.

An additional attractive feature of mean-field models is that they can readily include effects of quantum fluctuations, induced by either the magnetic anisotropy or transverse magnetic fields [46], and of molecular disorder [10, 50], both of which are cumbersome to deal with using Monte Carlo calculations. These effects give rise to interesting physical phenomena and are also particularly relevant to experimental situations met with some molecular crystals, such as those described below in Sects. 7.5 and 7.6.

7.3 Dipolar Order vs. Single-Molecule Magnet Behavior

7.3.1 Magnetic Order and Relaxation Towards Thermal Equilibrium

The above considerations about magnetic ordering apply only provided that spins reach thermal equilibrium, i.e. the state of minimum free energy, below T_c. Relaxation to equilibrium is brought about by the coupling of spins with vibrations of the crystal lattice, which acts as a thermal bath [55–57]. The rate depends on the strength of spin-phonon couplings but also on the structure of magnetic energy levels and the nature of the energy eigenstates. Here, the magnetic anisotropy plays a second, very important role. In many of the best-known single-molecule magnets (SMMs), and as it has been described in detail in previous chapters of this book, relaxation becomes in fact hindered at low temperatures by the presence of high anisotropy energy barriers. This question represents, in fact, one of the most serious difficulties encountered in the search for dipolar magnetic order in crystals of SMMs.

For temperatures not much lower than the zero-field splitting Ω_0, relaxation to thermal equilibrium proceeds via thermally activated processes, whose characteristic relaxation time $\tau \simeq \tau_0 \exp(U/k_B T)$ increases exponentially with decreasing temperature [59]. For any given experimental time τ_e, spins "freeze", i.e. they deviate from thermal equilibrium below a superparamagnetic "blocking" temperature $T_b = U/k_B \ln(\tau_e/\tau_0)$. Here, the pre-factor τ_0 gives the order of magnitude of excited levels lifetimes and U is usually smaller than U_{cl} because spins can flip by tunneling via intermediate states [60–63]. Pure ground state tunneling events might provide also an alternative path for the spin system to approach long-range ordering [32]. However, these processes are usually very slow. For instance, tunneling times measured on Fe_8 are of order 10^4 s [36], while in Mn_{12} acetate they are probably longer than 2 months [64]. Therefore, often $T_b > T_c$ and the underlying magnetic order remains hidden.

The situation can be best illustrated with the help of a specific example. Figure 7.3 shows the specific heat c of Fe_8 measured at zero field [39, 65]. Monte Carlo simulations predict a maximum in c signalling the onset of ferromagnetic order at $T_c \simeq 0.5$ K [7]. However, experimental data deviate from equilibrium already at $T_b = 1.3$ K, decreasing exponentially with T and showing no evidence whatsoever for the existence of a phase transition.

The search for dipolar order must therefore be oriented towards crystals of molecular nanomagnets with sufficiently fast spin-lattice relaxation, i.e. those having $T_b < T_c$. Funnily, the goal is just the opposite to that of finding single-molecules with long-lasting magnetic memory, which has been the main stream of activity in this research field. A remarkable intermediate situation was found in crystals of Fe_{17} SMMs, with a very high spin $S = 35/2$ [66, 67]. These clusters can be packed in two different crystal structures, of cubic and trigonal symmetries, respectively. The critical temperatures associated with dipolar magnetic order in these lattices are different, with $T_c(\text{cubic}) < T_c(\text{trigonal})$. In the cubic case, a situation similar to that described above for Fe_8 arises, thus the system behaves as a SMM with a blocking temperature $T_b \simeq 0.5$ K. In the trigonal case, $T_c \simeq 0.8$ K, thus larger than T_b. As a result, both the equilibrium heat capacity and magnetic susceptibility show clear indications of the onset of long-range dipolar order. Further examples in which equilibrium conditions can be attained down to sufficiently low temperatures are described in detail in Sects. 7.4 to 7.6.

7.3.2 Influence of Dipolar Interactions on Magnetic Relaxation and Spin Tunneling

Dipolar interactions modify also the nature and rates of magnetic relaxation processes. In the paramagnetic state, magnetic fields $\mathbf{H}_d$ vary from one lattice point to another. Near a crossing magnetic field, $H_{z,\mathrm{n}}(m)$, the longitudinal $H_{\mathrm{d},z}$ acting on a given molecule detunes energetically states m and $-m-n$, which would otherwise be in resonance. The effect is to block quantum tunneling processes between these states, the more so the smaller the ratio $\Delta_m/\xi_{\mathrm{d}}(m)$, where the bias $\xi_{\mathrm{d}} = g\mu_{\mathrm{B}}|m-m'|H_{\mathrm{d},z}$. As a result, thermally activated tunneling takes place predominantly via spin levels for which this ratio is not too far from unity [61].

The effect of dipolar interactions becomes even more dramatic at very low temperatures, when only the ground state doublet ($\pm S$ at $H=0$) is populated. Since Δ_S is usually many orders of magnitude smaller than the typical $\xi_{\mathrm{d}}(S)$, only those molecules for which the local bias is either smaller than Δ_S or can be compensated by hyperfine interactions with magnetic nuclei are able to flip their spin [31, 68]. Relaxation becomes then a purely collective process, because the tunneling of each spin changes the local fields acting on other crystal sites. The rate and time evolution depend on the symmetry of the lattice and also on whether the system is evolving towards a paramagnetic or a magnetically ordered state [32, 69, 70].

Transverse dipolar field components $H_{\mathrm{d},x}$ and $H_{\mathrm{d},y}$ affect also spin tunneling and relaxation processes. Off-diagonal anisotropy terms in (7.3) are even. Therefore, they can only connect states m and m' provided that $|m-m'|$ is even too [71]. This condition applies at zero field, but not at some of the crossing fields defined by (7.14). In particular, tunneling would be strictly forbidden at crossing fields with odd "n". Transverse dipolar fields can contribute to break down such "selection rules", as they enable the quantum mixing between any pair of states [61]. This effect explains why magnetization steps (see Fig. 7.4) are observed at all crossing fields even in very precisely aligned crystals that are free from molecular disorder [72].

7.3.3 Experimental Determination of the Average Interaction Fields

The first of the effects described above provides a suitable method to measure the effective interaction field $H_{\mathrm{eff},z}$ (cf. (7.9)). The method makes use of the strong sensitivity of quantum tunneling to the presence of even small bias magnetic fields. Tunneling resonances occur at well defined *local* magnetic fields $H_{z,n}(m)$ given by (7.5). However, the magnetic bias field acting on each molecule in a crystal consists of the applied field H_z plus contributions arising from the interactions with other molecules. The resonant field must therefore fulfill the following approximate condition

$$H_{z,n}(m) = H_z - \widetilde{N}_{zz}M_z + H_{\mathrm{eff},z} \tag{7.14}$$

where $M_z = Ng\mu_{\mathrm{B}}\langle S_z\rangle_T$ is the volumic longitudinal magnetization. It can be expected that $H_{\mathrm{eff},z}$ depends on the spin configuration of a crystal, i.e. on M_z, thus the external field H_z that fulfills condition (7.14) also does.

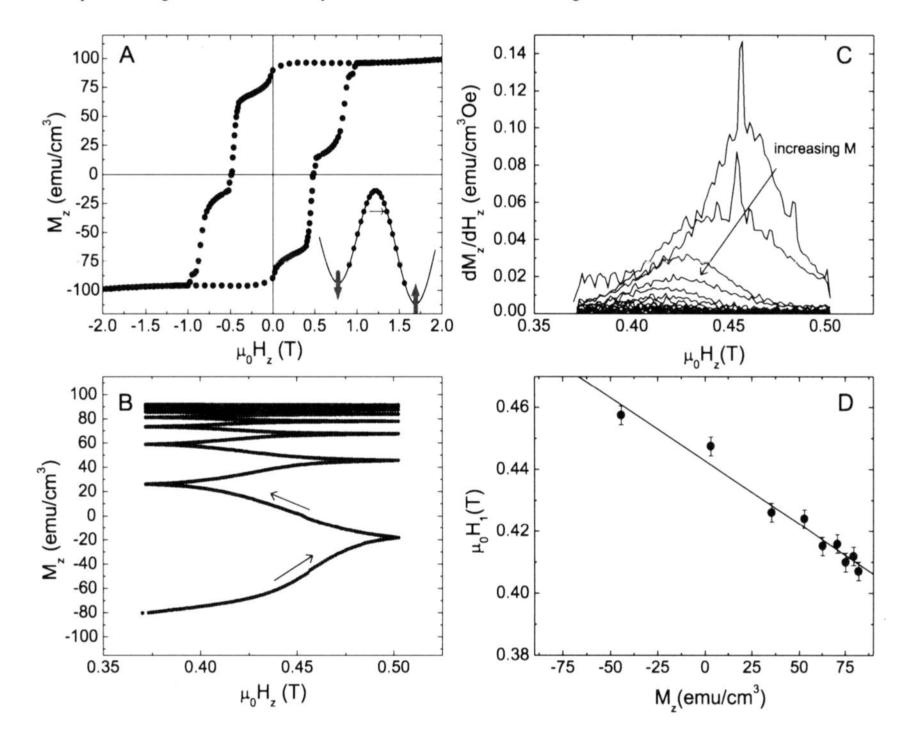

Fig. 7.4 **A**: Magnetization hysteresis loop of Mn_{12} acetate measured at $T = 2.5$ K. *The inset* shows the structure of magnetic energy levels of this molecule at the first crossing field H_1, which corresponds to the magnetization step observed near $\mu_0 H_z = 0.45$ T. **B**: Magnetization of Mn_{12} measured as the magnetic field is swept back and forth across this tunneling resonance. **C**: Magnetization derivative determined from these data. **D**: Position of dM_z/dH_z maxima (resonant fields) as a function of magnetization

This dependence can be explored experimentally by sweeping the magnetic field back and forth across a given crossing field. Figure 7.4 shows the magnetization measured as this procedure is repeated near the first crossing field ($n = 1$) of Mn_{12} acetate, at $T = 2.5$ K. The magnetization step, associated with this first tunneling resonance, shifts towards lower H_z as M_z increases. In fact, the dependence is close to linear, thus showing that the effective $H_{\text{eff},z}$ is nearly proportional to M_z too. Correcting from the demagnetization factor of the crystal, (7.14) gives $H_{\text{eff},z} \simeq \lambda M_z$, with $\lambda \equiv J_{\text{eff}}/N(g\mu_B)^2 \simeq 6$. For a magnetically polarized crystal of Mn_{12} acetate, with $M_z = M_s \simeq 96$ G, the maximum H_{eff} amounts then to approximately 575 Oe. Taking into account the experimental uncertainties involved (mainly associated with the accuracy in the determination of the demagnetization factor) this value agrees well with $H_{\text{eff}} = 515 \pm 85$ Oe, reported in Ref. [73]. In the latter work, the hysteresis loop of fast relaxing Mn_{12} molecules [74] was used to monitor the magnetic field created by the standard, slower relaxing ones.

These results give also the opportunity to estimate the effective, or mean-field, interaction constant $J_{\text{eff}}/k_B \simeq 7.5 \times 10^{-3}$ K. For Mn_{12} clusters, $\Omega_0/k_B \simeq 19$ K, thus it is much larger than T_c. Mean-field equations (7.10) and (7.12) are therefore applicable. The above value of J_{eff} gives then rise to a critical temperature $T_c = 0.75$ K for Mn_{12} acetate, close to the experimental $T_c \sim 0.9$ K (see Sect. 7.5).

7.4 Dipolar Order of Molecular Nanomagnets with Low Magnetic Anisotropy. Ferromagnetism in Mn_6

One of the simplest ways to obtain a dipolar magnet is to look for high-spin molecules having sufficiently weak magnetic anisotropy, thus also low energy barriers opposing the spin reversal. In this section, I briefly describe results of experiments performed on one of such molecules, $Mn_6O_4Br_4(Et_2dbm)_6$, hereafter abbreviated as Mn_6 [44, 47, 75].

The molecular core of Mn_6, shown in the inset of Fig. 7.5, is a highly symmetric octahedron of Mn^{3+} ions (with spin $s = 2$) that are ferromagnetically coupled via strong *intra*-cluster super-exchange interactions. Its ground magnetic state is a $S = 12$ multiplet. The net magnetocrystalline anisotropy of this cluster proves to be very small, with $D \simeq 0.013$ K [44, 47]. The classical energy barrier separating spin-up and spin-down states is then $U_{cl}/k_B \simeq 1.9$ K, much smaller than $U_{cl}/k_B \simeq 70$ K of Mn_{12}. Mn_6 crystallizes in a monoclinic lattice with 4 molecules per unit cell [75] bound together only by Van der Waals forces. *Inter*-cluster super-exchange interactions are therefore expected to be negligible.

As a result of its weak magnetic anisotropy, the equilibrium magnetic susceptibility and specific heat of Mn_6 can be measured down to very low temperatures. Curves measured for $H = 0$ are shown in Fig. 7.5. Contributions associated with lattice vibrations and hyperfine interactions dominate c above 2 K and below 100 mK, respectively. Between these two limits, c is mainly due to the thermal population of molecular spin levels, split by the magnetic anisotropy and dipole-dipole interactions (cf. (7.2)). This magnetic contribution shows a sharp peak at 0.15(2) K. The magnetic entropy change, estimated from data measured between 0.08 K and 4 K, amounts to $3.4k_B$ per molecule, thus very close to the maximum entropy $\Delta S_m = k_B \ln(2S+1) = 3.22k_B$ of a $S = 12$ spin multiplet. It therefore seems appropriate to assign the peak in c to the onset of long-range magnetic order.

It is worth pointing out that the magnetic anisotropy of Mn_6, despite its weakness, leaves its mark on the nature of the long-range order that arises below T_c. The magnetic entropy change measured between 0.08 K and T_c amounts to about $1k_B$ per spin, thus not far above $\Delta S_m = k_B \ln(2) = 0.7k_B$ that is expected for an effective spin-1/2 system. This shows that, because of the low value of T_c, mainly the lowest energy spin states (with $m = \pm 12$) take part in the magnetic ordering.

Information on the character of the magnetic order, i.e. whether it corresponds to a ferro- or antiferromagnetic phase, can be obtained from the ac magnetic susceptibility data shown on the right-hand side of Fig. 7.5. The real susceptibility component χ' shows a sharp maximum at $T_c = 0.161(2)$ K, close to the ordering temperature estimated from heat capacity measurements. These data are compared with the

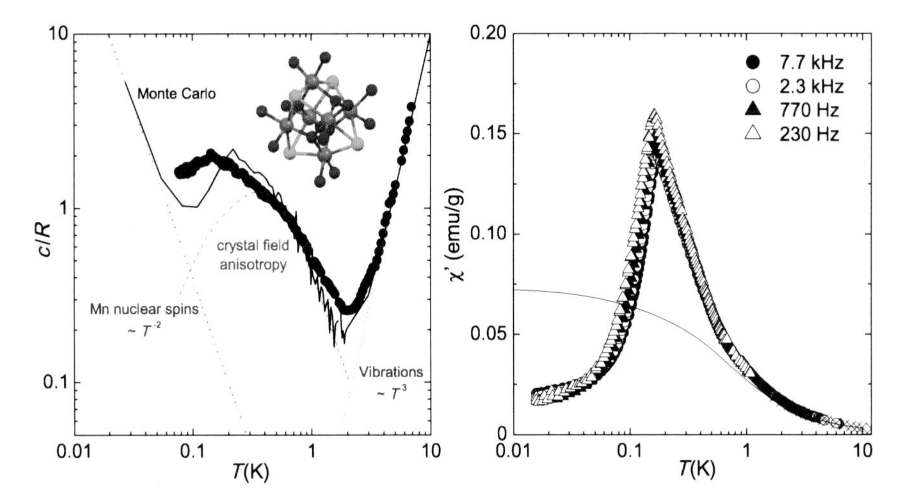

Fig. 7.5 *Left*: *Dots*, zero-field specific heat of Mn_6; *dotted line*, phonon contribution; *dashed line*: Schottky contribution due to crystal field splitting of the $S = 12$ multiplet as calculated with (7.3) for $D/k_B = 0.013$ K; *dotted curve*: nuclear contribution expected from the ^{55}Mn nuclear spins. *Solid line*: equilibrium specific heat derived from Monte Carlo calculations, and including all previous contributions as well as the effects of dipole-dipole interactions. *The inset* shows a sketch of the symmetric octahedral core of each Mn_6 molecule, with total spin $S = 12$. *Right*: Real component of the ac susceptibility of Mn_6 measured at several frequencies. *The solid line* gives the paramagnetic susceptibility of non interacting Mn_6 clusters. These calculations include the effects of the zero-field splitting and of demagnetizing fields

paramagnetic susceptibility of Mn_6, calculated by taking into account the effects of the magnetic anisotropy and of the sample's demagnetization factor. The experimental susceptibility lies clearly above this prediction, thus suggesting that the magnetic order in Mn_6 is ferromagnetic, i.e., that $\theta > 0$ in (7.10). Figure 7.6 shows indeed that, above T_c, the intrinsic magnetic susceptibility χ_i, corrected for demagnetization effects, follows accurately Curie-Weiss law, with $C = 0.034(1)$ emu K/g Oe and $\theta = 0.20(3)$ K. These data agree with the fact that three-dimensional dipolar lattices must be close approximations of mean-field models. The ferromagnetic nature of the ordered phase is also confirmed by the fact that relatively weak magnetic fields completely suppress the heat capacity maximum [44, 47].

The dipolar magnetic order in Mn_6 has been investigated by means of Monte Carlo simulations, which are described in detail in [44, 47]. As it has been have argued in Sect. 7.2, because $U_{cl}/k_B \gg T_c$ only states with $m = \pm 12$ are appreciably populated at and below T_c. This justifies the use of the Ising Hamiltonian (7.6) to describe the magnetic ordering of Mn_6 molecular nanomagnets.

Monte Carlo simulations show that the ground state is ferromagnetically ordered, as observed, and predict a shape for c that is in reasonably good agreement with the experiment. The solid line in Fig. 7.5 shows c calculated assuming that all molecular anisotropy axes (z) point along one of the two nearly equivalent short axes of the actual lattice. Similar results were obtained for other orientations of the anisotropy axes. These simulations give $T_c = 0.22$ K, which is slightly above the experimental

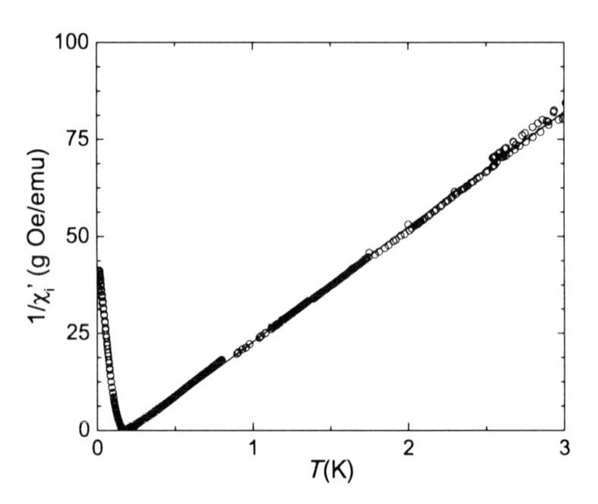

Fig. 7.6 Reciprocal magnetic susceptibility of Mn_6 corrected for demagnetization effects. *The solid line* is a least-squares fit of a Curie-Weiss law to the data measured above 0.3 K

$T_c = 0.161(2)$ K. As it was argued in [47], this difference can be assigned to the finite value of the anisotropy. Model calculations, performed for the same crystal structure but assuming classical Heisenberg spins with varying anisotropy show that different ferromagnetic ground states are possible, depending on the competition between local crystal field effects and long-range dipolar interactions. The strong dependence of variation of T_c on the sign and orientation of the magnetic anisotropy, as well as the form of the calculated and observed specific heat anomalies turn out to be specific for dipolar interactions, and differ widely from the analogues for usual ferromagnets, coupled via super-exchange interactions [51].

Spin Dynamics Close to T_c Ac susceptibility data provide also interesting information on the dynamics of spins close to and below T_c. The maximum value of χ' is seen to weakly vary with the frequency $\omega/2\pi$ of the ac excitation magnetic field. This variation suggests that, for the highest frequencies employed in these experiments, spins begin to deviate from equilibrium already above T_c. A more dramatic effect is observed below the ordering temperature. The real susceptibility component χ' decreases rapidly, thus suggesting that the ferromagnetic response is also being blocked by slow relaxation processes.

These phenomena can be understood, at least qualitatively, if one takes into account once more the finite magnetic anisotropy of Mn_6 clusters. The superparamagnetic blocking of Mn_6 spins is expected to occur at $T_b \simeq DS^2/k_B \ln(1/\omega\tau_0)$. Setting $\tau_0 = 10^{-8}$ s, which is a typical value found for other SMMs, gives $T_b \simeq 0.25$ K for $\omega/2\pi = 7.7$ kHz. In other words, for $T \to T_c$, the approach to equilibrium begins to be hindered by the anisotropy barrier of each molecular spin. These estimates have been confirmed by recent experiments performed on different derivatives of Mn_6, which show slightly lower values of T_c [76]. In these samples, a frequency-dependent super-paramagnetic blocking is observed below 0.2 K.

Below T_c, the slow magnetic relaxation contributes to "pin" magnetic domain walls. This effect accounts for the sharp decrease observed in the linear magnetic susceptibility. The dynamics associated with the displacement of domain walls in Ising-like dipolar ferromagnets had not been simulated until recently [77]. The present experiments suggest that, in the case of Mn_6, the magnetization dynamics close to T_c seems to be dominated by thermal fluctuations. An interesting question that needs to be addressed by future experimental work is whether domain walls move by flipping one molecular spin at a time or via a collective process.

7.5 Dipolar Order in a Transverse Magnetic Field. Ferromagnetism in Mn_{12} Acetate

7.5.1 Magnetic Ordering Via Pure Quantum Tunneling

For many of the best known SMMs, magnetic anisotropy barriers are so high that, close to T_c, thermally activated spin flips take place in time scales that are much longer than the typical experimental time scales. Under these conditions, only pure spin tunneling events contribute to the magnetization dynamics. The precise mechanism by which quantum tunneling enables the spin system to exchange energy with the lattice is not yet fully understood. However, in spite of their intrinsically quantum nature and the fact that they are independent of temperature, experiments show that these quantum fluctuations are nevertheless able to bring the spin system to equilibrium with the thermal bath [45, 65, 78, 79]. These processes enable also the onset of long-range magnetic order in crystals of SMMs. However, as it has been mentioned above, they are also rather slow, with time scales of the order of many hours for Fe_8 or even months, as it is the case for Mn_{12} clusters.

Quantum tunneling can be, to some extent, controlled by chemical design. The symmetry of the cluster magnetic cores determines the structure of the spin Hamiltonian (7.3). Lowering the molecular symmetry allows the presence of lower order off-diagonal terms, which contribute to enhance quantum tunneling probabilities. In clusters with a Mn_4O_3X cubane magnetic core, this effect has been induced via the chemical binding to different ligands X. Then, while highly symmetric Mn_4O_3Cl and Mn_4O_3OAc clusters [80] show the typical SMM behaviour, with blocking temperatures in the vicinity of 1 K, the spins of a strongly distorted Mn_4O_3MeOAc [81] remain in equilibrium down to very low temperatures. In the latter sample, the heat capacity shows the onset of long-range magnetic order at $T_c = 0.2$ K [45]. This value is found to be larger than the maximum critical temperature compatible with dipolar interactions. Therefore, in this case super-exchange interactions probably play a non-negligible role. This example shows that conclusions on the existence of pure dipolar order cannot be drawn from qualitative arguments alone, and that a quantitative comparison with theoretical predictions are always necessary. Ferromagnetic order has also been observed in crystals of low symmetry Ni_4 clusters, which show one of the highest tunneling rates (of order 10^5 s^{-1}) ever measured [48, 76].

7.5.2 Quantum Annealing

An additional trick, based on the above considerations, can be played in crystals of SMMs having their magnetic anisotropy axes aligned along given crystallographic

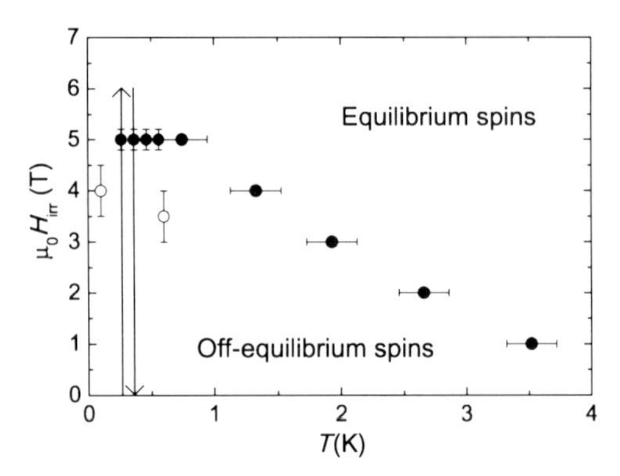

Fig. 7.7 Irreversibility transverse magnetic field separating equilibrium and non-equilibrium conditions of Mn_{12} acetate spins. *Solid* and *open dots* have been determined from specific heat data (experimental time constant ~ 1 s) and magnetic neutron diffraction experiments (experimental time constant $\sim 10^4$ s), respectively. *The arrows* show schematically the quantum annealing protocol employed to explore the existence of long-range magnetic order at very low temperatures

directions. Transverse components of the Zeeman interaction (7.4), i.e. those associated with magnetic field components H_x and H_y, also induce quantum tunneling of the spins. Since off-diagonal terms play, to some extent, a role comparable to that of a kinetic energy in the tunneling of a material particle, the magnetic field enables then to "tune" the effective tunneling mass. This ability has been used to directly detect the existence of a quantum tunnel splitting [38, 39] and to induce quantum interference phenomena between different tunneling trajectories [37]. Naturally, it can also be applied to explore the existence of a magnetically ordered phase.

The basic protocol for this "quantum annealing" (see also [82, 83]) is shown in Fig. 7.7. By increasing the transverse magnetic field $H_\perp$, tunneling probabilities are rapidly enhanced, thus at some point spins are able to reach thermal equilibrium with the lattice. If at this temperature and field the spin system remains ferromagnetically ordered, a net magnetization will be recorded that will "freeze" as the magnetic field is set back to zero through the irreversibility field H_{irr}. The latter field, thus also the result of the quantum annealing process, depend on the experimental probe and its characteristic time scales. This dependence is shown in Fig. 7.7 that compares data derived for Mn_{12} acetate using heat capacity [84] and magnetic neutron diffraction experiments [46].

7.5.3 The Quantum Ising Model

The control of quantum tunneling fluctuations by an external magnetic field offers an additional and very attractive possibility for fundamental physical studies. As it has been discussed in Sect. 7.2, dipole-dipole interactions between highly anisotropic spins (with $D \to \infty$) can be approximated by a spin-1/2 Ising Hamiltonian. In the presence of a transverse magnetic field, a crystal of perfectly oriented

SMMs can therefore provide a material realization of the quantum Ising model [30]. The spin Hamiltonian of this model reads as follows

$$\mathcal{H} = -\frac{S^2}{2}\sum_i \sum_{j\neq i} J_{ij}\sigma_{i,z}\sigma_{j,z} - \Delta \sum_i \sigma_{i,x} \quad (7.15)$$

where J_{ij} are longitudinal couplings (here of dipolar origin) and Δ is the ground-state tunnel splitting which depends on and vanishes with $H_\perp$. Equation (7.15) represents the archetypical (and arguably the simplest) model for a quantum phase transition [40, 41]. The classical long-range order that exists for $H_\perp = \Delta = 0$ (ferromagnetic or anti-ferromagnetic) competes with field-induced quantum fluctuations. The magnetic phase diagram, representing (T_c, Δ_c) [or, equivalently, (T_c, H_c)] points at which magnetic order is suppressed, can be calculated using the mean-field approximation (7.11). The magnetic phase boundary between the ordered and paramagnetic phases is defined by the following equation

$$\frac{k_B T_c(H_\perp = 0)}{\Delta_c} = \coth\left(\frac{\Delta_c}{k_B T_c}\right) \quad (7.16)$$

where $T_c(H_\perp = 0) = \theta$ is given by (7.12). At $T = 0$, magnetic order is completely destroyed at $\Delta_c = k_B T_c(H_\perp = 0)$.

Quantum phase transitions have been extensively studied in recent years. Examples include the superconductor insulator transition in cuprates [85–87], the onset of antiferromagnetism in heavy fermions [88], the pressure driven insulator-metal transition in V_2O_3 [89], and the magnetic transitions driven by field ($LiHoYF_4$ [90]) or concentration (Cr_xV_{1-x} alloys [91]). In addition to their intrinsic interest, a plethora of new properties arise at nonzero temperature.

In spite of this intense activity, pure realizations of the quantum Ising model with magnetic materials are very scarce. As it happens with dipolar magnetism in general, lanthanide-based insulators seem to be a natural choice for these studies [90, 92]. However, the strong hyperfine interactions seriously limit the observation of the intrinsic quantum criticality in these materials [23, 24]. Crystals of single molecule magnets, for which hyperfine interactions are typically much weaker, are then very attractive candidates.

7.5.4 Magnetic Order in Mn_{12} Acetate

Neutron Diffraction Experiments The cluster of Mn_{12} acetate [93], the first and most extensively studied member of the family of single-molecule magnets, is shown in Fig. 7.2. It contains 12 manganese atoms linked via oxygen atoms, with a sharply-defined and *monodisperse* size. At low temperatures, each of them exhibits slow magnetic relaxation and hysteresis, due to the combination of an $S = 10$ magnetic ground state with appreciable uniaxial magnetic anisotropy. Finally, they organize to form tetragonal molecular crystals. Monte Carlo simulations [7], as well as mean field calculations [77], predict that Mn_{12} acetate must order ferromagnetically as a result of dipolar interactions between molecular spins. The critical temperatures derived from these calculations are $T_c = 0.5$ K and 0.8 K, respectively. Therefore, these crystals seem to offer a nearly perfect realization of the quantum Ising model (7.15). To which extent this is indeed the case will be discussed in the following.

Not surprisingly, detecting the presence of long-range magnetic order in Mn_{12} faces some important experimental difficulties. Spin reversal via resonant quantum tunneling [33–35, 64] becomes extremely slow at low temperatures (of order two months at $T = 2$ K). For the time scales $\sim 10^2$–10^4 s of a typical experiment $T_b \sim$ 3 K, thus much higher than the ordering temperature T_c. Equilibrium conditions can be explored via the application of the quantum annealing protocol described above. Magnetic diffraction of thermal neutrons is a suitable tool for these studies because it can probe different components of the magnetization vector, in particular M_z [94], in the presence of a transverse magnetic field. In addition to this, diffraction patterns provide a very accurate determination of the crystal's orientation. And finally, the typical data acquisition times required to obtain reasonably good results are very long, which gives rise to smaller values of H_{irr} (see Fig. 7.7). In the experiments whose results are described below [46], a $\sim 0.5 \times 0.5 \times 1.5$ mm^3 single crystal of deuterated Mn_{12} acetate was attached to the mixing chamber of a ^{3}He-He^4 dilution refrigerator with its **c** axis perpendicular (up to a maximum deviation of about 0.1(1) degrees) to the magnetic field.

Given the strong magnetic anisotropy of Mn_{12} clusters, the magnetization is confined in the plane defined by **c** and **H**, with components M_z and $M_\perp$, respectively. At 4 K, that is, in the paramagnetic state, $M_z = 0$ and $M_\perp$ is proportional to $H_\perp$. For $T \leq 900$ mK, by contrast, a large additional contribution to the magnetic diffraction intensities shows up for $\mu_0 H_\perp < 5$ T, but only provided that $H_\perp$ is first raised above H_{irr} at each temperature. As shown in Fig. 7.8, this contribution reflects the onset of a non-zero spontaneous M_z below $T_c = 0.9(1)$ K. The latter value is close to the Weiss temperature $\theta \simeq 0.8(1)$ K extracted from the extrapolation of $1/\chi'_{zz}$ data measured above T_b (Fig. 7.8). These data suggest that Mn_{12} acetate does indeed order ferromagnetically, as predicted. The experimental T_c is in good agreement with mean-field calculations [77]. However, it is nearly a factor two larger than the critical temperature derived from Monte Carlo calculations for pure dipolar interactions [7]. Therefore, the presence of weak super-exchange interactions contributing to enhance the magnetic ordering temperature of Mn_{12} acetate cannot be completely ruled-out. The same conclusion was derived from the analysis of the susceptibilities and Weiss temperatures of different Mn_{12} derivatives [54].

Another remarkable finding, shown in Fig. 7.9, is the strong dependence of M_z on $H_\perp$. At the minimum temperature $T = 47$ mK, M_z is approximately zero for $\mu_0 H_\perp > 5.5(5)$ T and then it increases when decreasing $\mu_0 H_\perp$, reaching $16\mu_B$ per molecule at zero field. These results show that a transverse magnetic field tends to suppress the ferromagnetic order. The T_c–H_c magnetic phase diagram of Mn_{12} acetate is shown on the right-hand panel of Fig. 7.9. A ferromagnetic phase exists for sufficiently low temperatures and transverse magnetic fields. The qualitative resemblance between the effects caused by temperature and field is typical of systems

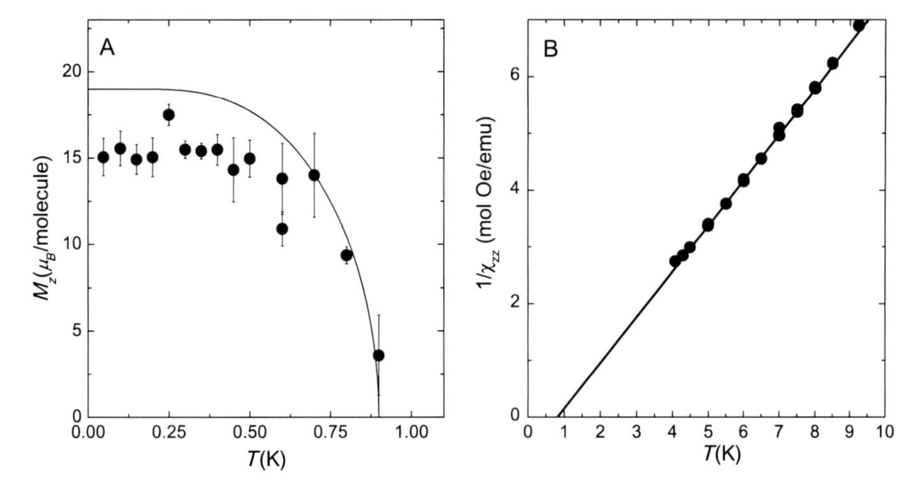

Fig. 7.8 *Left*: Longitudinal magnetization M_z of Mn_{12} acetate obtained from neutron diffraction data measured at $\mu_0 H_\perp = 0$ after decreasing the transverse magnetic field from 6 T at each temperature. *The solid line* is a calculation (for a perfect orientation of the magnetic field perpendicular to the easy axes of all molecules) that includes interactions via the mean-field Hamiltonian (7.17). *Right*: Reciprocal parallel susceptibility measured at $T > 4.5$ K (i.e. above T_b) along the **c** crystallographic axis. *The solid line* is a least-squares linear fit, giving $\theta = 0.8(1)$ K

undergoing a quantum phase transition. However, as it is argued in the following, understanding the true nature of the field-dependent transition can only be achieved by a quantitative comparison to theoretical simulations.

Comparison to the Quantum Ising Model Predictions For a perfectly oriented crystal of Mn_{12} molecules in a transverse magnetic field, the mean-field Hamiltonian (7.7) can be written as

$$\begin{aligned}\mathcal{H} = &- DS_z^2 + BS_z^4 + \frac{C}{2}\left(S_+^4 + S_-^4\right) \\ &- g\mu_B H_\perp (S_x \cos\phi + S_y \sin\phi) - J_{\text{eff}} \langle S_z \rangle S_z \end{aligned} \tag{7.17}$$

Spectroscopic measurements [95–100] give $g = 1.9$, $D/k_B = 0.6$ K, $B/k_B = -10^{-3}$ K, and $C/k_B = -6.1 \times 10^{-5}$ K. Experiments performed on single crystals [99] provide also the orientation of the fourth-order anisotropy axes x and y with respect to the crystallographic axes **a** and **b**. In the neutron diffraction experiments, **H** was approximately parallel to the $1\bar{1}0$ crystallographic direction, which corresponds to $\phi \simeq \pi/4$. The mean-field constant J_{eff} was set to $9 \times 10^{-3} k_B$ that, according to (7.12), fits the experimental $T_c = 0.9$ K. The above J_{eff} value is close to $7.5 \times 10^{-3} k_B$ determined from quantum tunneling experiments described in Sect. 7.3.3.

Predictions for M_z as a function of temperature and magnetic field that follow from (7.17) are shown in Figs. 7.8 and 7.9(A). These calculations account reasonably well for the temperature dependence of M_z measured at $H_\perp = 0$. The fact that

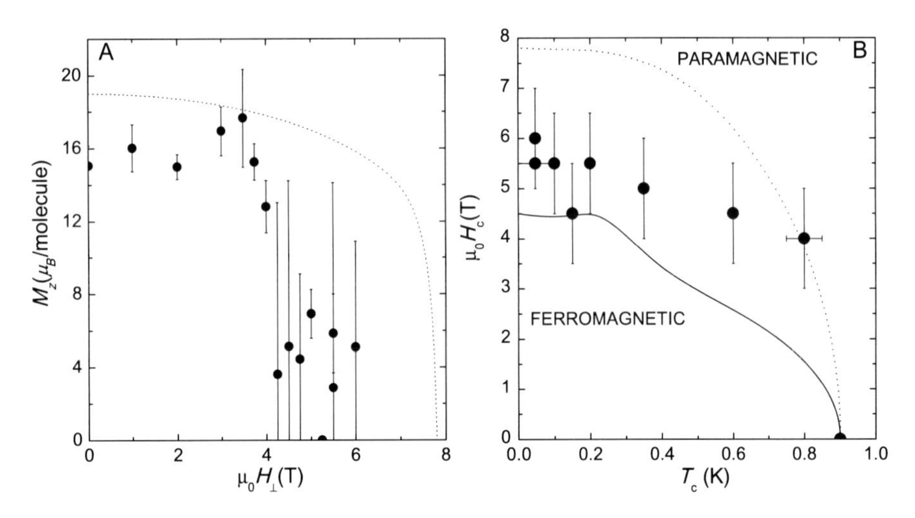

Fig. 7.9 *Left*: Longitudinal magnetization M_z of Mn_{12} acetate measured while decreasing the transverse magnetic field $\mu_0 H_\perp$ from 6 T at $T = 47$ mK. *The dotted line* has been calculated using (7.17) and the parameters given in the text. *Right*: Magnetic phase diagram of Mn_{12} acetate. *The dotted line* was obtained using the mean-field Hamiltonian (7.17) for perfectly aligned anisotropy axes. *The solid line* is the mean field prediction following from the random-field Hamiltonian (7.18), which includes effects of molecular disorder [50]

M_z remains smaller than the saturation magnetization of $19\mu_B$ per molecule even at $T \to 0$ can be ascribed to non equilibrium effects. It probably arises from reversed spins that remain frozen as the magnetic field is reduced below $\mu_0 H_{\rm irr} \simeq 4$ T, because quantum tunneling rates become then extremely slow. However, the same model fails to account for the field-dependent behavior. In particular, the zero-temperature critical field $\mu_0 H_c(T = 0)$, at which quantum fluctuations finally destroy the long-range ferromagnetic order, is close to 8 T, thus considerably higher than the experimental $\mu_0 H_c \simeq 5.5$ T. The discrepancy manifests itself also in the shape of the magnetic phase diagram at low temperatures, shown in Fig. 7.9(B).

Molecular Disorder : Random-Field Magnetism in Mn_{12} Acetate In the original analysis of the neutron diffraction experiments [46], the field-dependent magnetization was fitted by introducing a large and positive fourth-order off-diagonal parameter C, which "helps" the magnetic field in generating sufficiently strong quantum fluctuations. Disorder in the orientation of acetic acid solvent molecules can lower the local symmetry of Mn_{12} molecules and give rise, for some of them, to additional off-diagonal terms, such as $E(S_x^2 - S_y^2)$, not allowed for the ideal molecular symmetry [101]. The presence of such terms has been put into evidence by magnetic relaxation [102, 103] and spectroscopic experiments [99]. However, introducing such terms in (7.17) cannot, by itself, account for either the value of H_c or the magnetic-field diagram that are experimentally observed.

Molecular disorder can, however, affect ferromagnetism in a different, subtle manner. Some of the different isomers, associated with given orientations of the interstitial molecules with respect to Mn_{12} cores, have their easy axes z tilted with respect to the crystallographic **c** axis [101]. The tilting angles δ have been estimated by several experimental methods and turn out to be rather small, of the order of 1 deg., or even less [99, 104]. At zero field, it is therefore expected that their influence on the ferromagnetic order be small. However, their presence makes itself felt when a magnetic field is applied perpendicular to **c**. As it was first pointed out by Millis and co-workers [10], some molecular sites then "see" a nonzero bias field H_z, which for $H_\perp \geq 3$ become already stronger than the maximum H_{eff} associated with intermolecular magnetic interactions. Furthermore, the bias is randomly distributed among the different sites.

In order to describe these effects, the mean-field Hamiltonian for each molecule at site $\mathbf{r_i}$ must include an additional random-field term [10, 50]

$$\mathcal{H}(\mathbf{r_i}) = -DS_z^2 + BS_z^4 + \frac{C}{2}\left(S_+^4 + S_-^4\right) - g\mu_{\mathrm{B}} H_\perp(\mathbf{r_i})(S_x \cos\phi + S_y \sin\phi) - g\mu_{\mathrm{B}} H_z(\mathbf{r_i}) S_z - J_{\mathrm{eff}}\langle S_z \rangle S_z \quad (7.18)$$

As can be expected, the effect of disorder is to suppress magnetic order for applied magnetic field values H that are significantly smaller than the critical field of the pure quantum Ising model (7.17). The solid line in Fig. 7.9(B) shows the magnetic phase diagram derived [50] from (7.18), using the distribution of random easy axes tilts calculated by Park and co-workers [105]. In [50], it was shown that this model gives a fair account of the Weiss temperatures determined from the extrapolation of the reciprocal susceptibility (see Fig. 7.8), although its predictions tend to slightly overestimate H_c at any temperature. It can be seen that it also provides a better description of the low-T/high-$H_\perp$ behavior obtained from magnetic neutron diffraction experiments. However, the discrepancy between experimental and theoretical values of H_c is even larger than that derived from magnetic data [50]. This suggests that the degree of interstitial disorder can be different for different crystals of Mn_{12} acetate and that these differences manifest themselves in the magnetic phase diagram.

These results illustrate the rich physical behavior of Mn_{12} acetate in the presence of a transverse magnetic field. This system provides a unique opportunity to investigate the interplay between dipolar interactions and randomness and represents one of the best material realizations of the random-field Ising model known to date.

7.6 Magnetic Order and Quantum Phase Transition in Fe_8

The previous section illustrates some of the difficulties met in the search of pure quantum phase transitions with SMMs: molecular disorder turns Mn_{12} acetate into a realization of the classical random-field Ising model. In this section, I review experimental work performed on a crystal of Fe_8 SMMs [49]. This molecular material

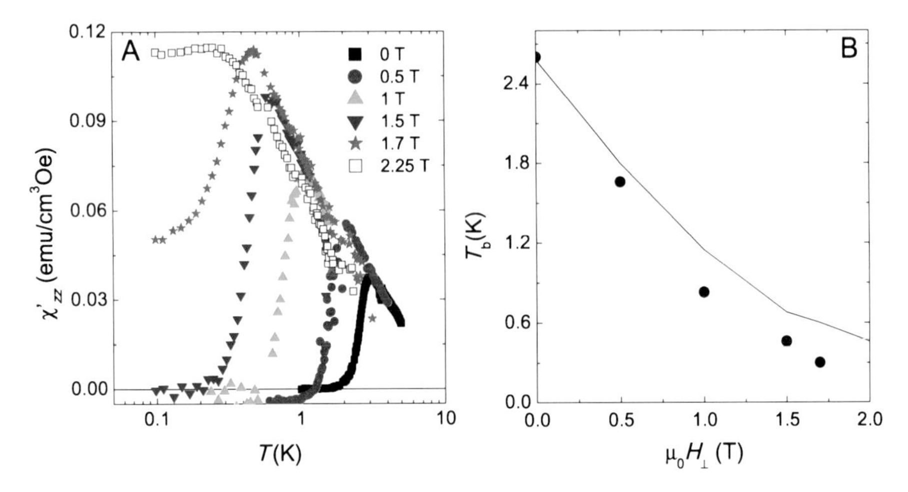

Fig. 7.10 **A**: Longitudinal in-phase ac susceptibility of Fe_8 measured at $\omega/2\pi = 333$ Hz and for different values of the transverse magnetic field $H_\perp$. **B**: Shift of the superparamagnetic blocking temperatures with increasing $H_\perp$. *The solid line* shows theoretical predictions for quantum spin-phonon relaxation that follow from Pauli's master equation as described in [61]

[106] possesses some properties that make it especially well suited for these studies, *viz* (i) classical Monte Carlo (MC) simulations suggest a ferromagnetic ground state with $T_c = 0.54$ K [7] (ii) hyperfine interactions are much smaller than both the magnetic anisotropy and dipolar interactions, thus they cannot perturb quantum dynamics of SMMs and (iii) disorder is weak enough to avoid sizable random fields.

Each Fe_8 molecule (brief for $[(C_6H_{15}N_3)_6Fe_8O_2(OH)_{12}]$) has a spin $S = 10$ and a strong uniaxial magnetic anisotropy [107]. It can be described by Hamiltonian (7.3) with $D/k_B = 0.294$ K, $E/k_B = 0.046$ K, and $g = 2$ [107, 108]. x, y and z correspond to the hard, medium and easy magnetization axes that, in the triclinic crystal structure of Fe_8, are common to all molecules [109].

Ac magnetic susceptibility experiments, reported in [49], were performed down to 90 mK on a 1.6 mg single crystal of approximate dimensions $1 \times 2 \times 1$ mm^3. The magnetic easy axis z was oriented approximately parallel to the ac excitation magnetic field. Therefore, these experiments give access to the longitudinal linear magnetic response that is expected to diverge close to a magnetic phase transition. The dc magnetic field was then carefully aligned with respect to the crystal axes with the help of a 9 T $\times$ 1 T $\times$ 1 T superconducting vector magnet, using the strong dependence of the paramagnetic χ'_{zz} on the magnetic field orientation [49]. It was found that $\vec{H}$ is perpendicular ($\pm 0.05°$) to z and close ($\phi \simeq 68°$) to the medium y axis.

As expected for a high-anisotropy SMM, the ac susceptibility (Fig. 7.10) of Fe_8 deviates from equilibrium for low $H_\perp$ and low T, as shown by the vanishing of χ'_{zz}. The superparamagnetic blocking temperature T_b strongly depends on frequency. However, even for the lowest available frequencies, T_b remains much higher than

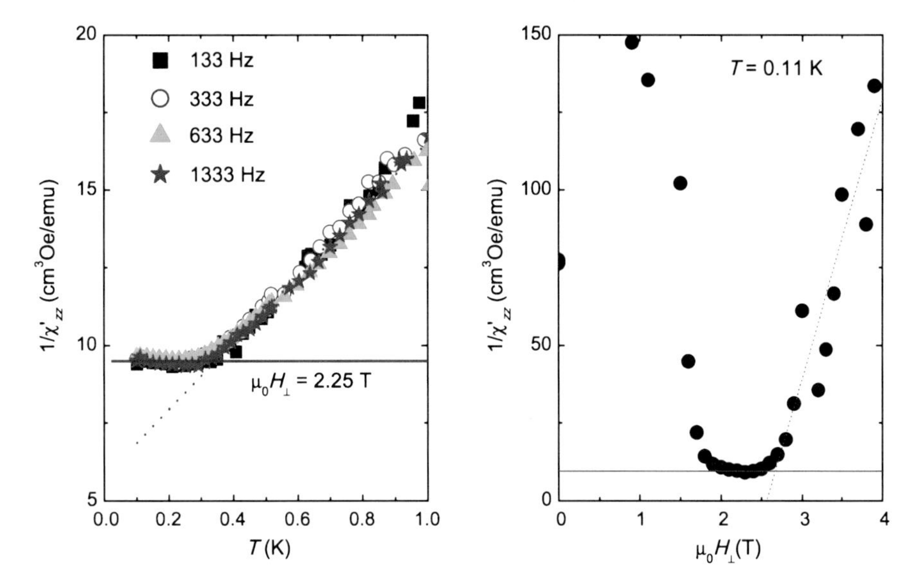

Fig. 7.11 Reciprocal in-phase ac susceptibility measured at $\mu_0 H_\perp = 2.25$ T as a function of T (*left*) and at $T = 0.110$ K as a function of $H_\perp$ (*right*). The crossovers between the "Curie-Weiss" law, observed at either high T or high $H_\perp$ (*dotted blue lines*), and the ferromagnetic limit $1/\chi'_{max} = \widetilde{N}_{zz}$ (*solid red lines*) give T_c (= 0.34(1) K) and $\mu_0 H_c$ (= 2.65(5) T), respectively

1 K, thus also higher than the expected critical temperature. The same applies to heat capacity experiments that have been discussed in Sect. 7.3.1 (cf Fig. 7.3).

As with Mn_{12} acetate, this situation can be reversed by enhancing quantum spin fluctuations via the application of $H_\perp$. As shown in Fig. 7.10, increasing $H_\perp$ reduces T_b, thus showing that spins are able to attain thermal equilibrium at progressively lower temperatures. It is interesting to mention also that, besides enhancing the spin dynamics, the magnetic field also lowers the paramagnetic susceptibility. This decrease can be associated with the reduction of the effective S_z by quantum fluctuations as well as with the decrease in the paramagnetic Weiss temperature (see below). Both effects become more noticeable for $H_\perp \geq 1$ T, as seen in Fig. 7.10.

Experiments show that χ'_{zz} becomes independent of frequency, thus it reaches full equilibrium, for $\mu_0 H_\perp \geq 2$ T. Above this field, $T_b \leq 0.1$ K. Under these conditions, it is possible to explore the existence of a magnetic phase transition in Fe_8 and study its critical behavior. As shown in Fig. 7.11, $1/\chi'_{zz}$ measured at $\mu_0 H_\perp = 2.25$ T follows the Curie-Weiss law at sufficiently high T, becoming independent of T below 0.34 K, which we take as the critical temperature T_c at this field. Furthermore, the saturation value $1/\chi_{max} = 9.5(5)$ cm^{3}Oe/emu agrees well with the demagnetizing factor of our sample $\widetilde{N}_{zz} = 10(1)$ cm^{3}Oe/emu. As discussed in Sect. 7.2.2, this is the behavior expected for an equilibrium ferromagnetic phase transition. Additional evidence supporting the existence of a transition to a ferromagnetic phase is found in the results of neutron diffraction experiments, similar to those described above for the case of Mn_{12} acetate [49].

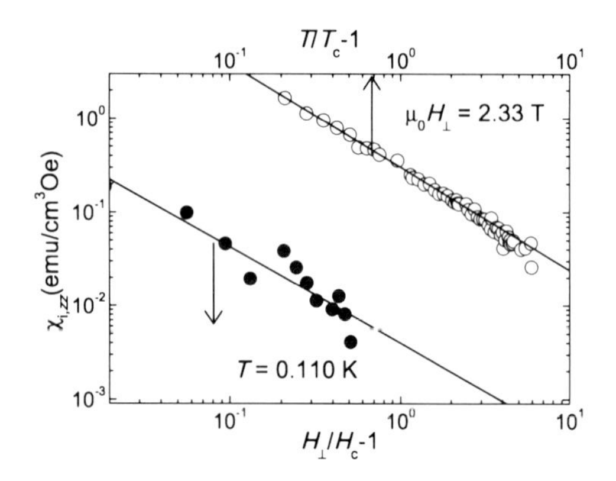

Fig. 7.12 Log-log plot of demagnetization-corrected $\chi'_{i,zz}$ of Fe_8 vs the reduced temperature (for $\mu_0 H_\perp = 2.33$ T with $T_c = 0.31$ K, ○) and field (at $T = 0.110$ K with $\mu_0 H_c = 2.65$ T, •). The linear fits give critical exponents $\gamma_{cl} \simeq 1.1(1)$ and $\gamma_{qu} \simeq 1.0(1)$

The ferromagnetic character of the low temperature phase agrees with theoretical predictions for the magnetic order resulting from dipole-dipole interactions [7, 110]. However, to find out if this transition is dominantly driven by such interactions, one needs to compare also the experimental and theoretical values of T_c. Unfortunately, equilibrium properties cannot be measured in Fe_8 below 1 K for $H_\perp = 0$. Yet, it is still possible to estimate $T_c \simeq \theta$, using θ determined from the Curie-Weiss fit of the reciprocal susceptibility measured above T_b. This method is particularly appropriate here, as the limiting value $1/\chi_{max}$ is known from experiments performed for $\mu_0 H_\perp \geq 2$ T (see, for instance, Fig. 7.11). It gives $T_c(H_\perp = 0) = 0.60(5)$ K, in very good agreement with the theoretical $T_c = 0.54$ K derived from Monte Carlo calculations [7]. It can then be safely concluded that Fe_8 becomes a pure dipolar ferromagnet at very low temperatures.

The reciprocal susceptibility shows a very similar behavior when $H_\perp$ is varied at constant T (Fig. 7.11). Again, $1/\chi'_{zz}$ depends linearly on $H_\perp$ until it saturates (to the same value $\simeq \widetilde{N}_{zz}$) below $\mu_0 H_c = 2.65(5)$ T, which we take as the critical magnetic field at $T = 110$ mK. These experiments evidence that, also in Fe_8, a sufficiently strong transverse magnetic field can destroy ferromagnetic order. However, as we shall see below, the nature of this transition is qualitatively different from that observed in Mn_{12} acetate. Before discussing this question in more detail, it is worth examining the critical behavior of the susceptibility, i.e. its temperature and field dependencies close to the phase transition.

The intrinsic susceptibility $\chi'_{i,zz}$, corrected from demagnetizing effects, is plotted vs the reduced temperature $(T/T_c - 1)$ (at $\mu_0 H_\perp = 2.33$ T) and field $(H_\perp/H_c - 1)$ (at $T = 110$ mK) in Fig. 7.12. Under equilibrium conditions, $\chi'_{i,zz}$ should follow, as it approximately does, the power laws

$$\chi'_{i,zz} = \left(\frac{T - T_c}{T_c}\right)^{-\gamma_{cl}}, \qquad \chi'_{i,zz} = \left(\frac{H_\perp - H_c}{H_c}\right)^{-\gamma_{qu}} \tag{7.19}$$

The slopes give critical exponents $\gamma_{cl} = 1.1(1)$ and $\gamma_{qu} = 1.0(1)$, in good agreement with $\gamma = 1$ of the mean-field universality class. This result agrees with the prediction that the marginal dimensionality for mean-field behavior is $d^* = 3$ in an Ising

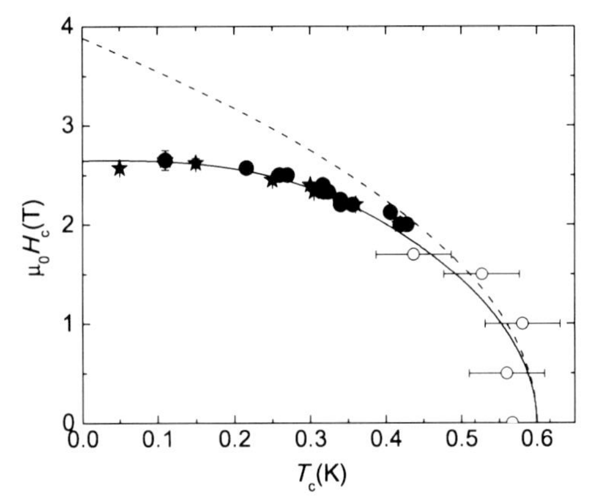

Fig. 7.13 H_c–T_c phase diagram determined from the linear extrapolation of $1/\chi'_{zz}$ to $1/\chi_{max}$. • and ∘ correspond to $T_c > T_b$ and $T_c < T_b$, respectively; ⋆, data determined from susceptibility scaling plots, as those shown in Fig. 7.12. *Solid line*, quantum mean-field calculation of the phase boundary using (7.7) and the parameters given in the text; *dashed line*, classical phase diagram, derived from Monte Carlo simulations

dipolar ferromagnet [14] and with the fact that the critical exponents for the field-induced transition at $T \to 0$ become equivalent to those of the classical transition in $(d+1)$ dimensions [111].

The H_c-T_c magnetic phase diagram of Fe_8 is shown in Fig. 7.13. Each data point in this diagram was obtained by linearly extrapolating $1/\chi'_{zz}$, measured either as a function of temperature at a fixed $H_\perp$ or as a function of magnetic field at constant T. A third method, which provides equivalent results, consists of using the scaling plots of Fig. 7.12 to determine T_c independently. As expected, T_c decreases when quantum fluctuations increase, i.e. with increasing $H_\perp$, thus ferromagnetism survives only for sufficiently low temperatures or magnetic fields.

The experiments can be compared with predictions following from the $S = 10$ quantum Ising model (7.7), using the fact that all anisotropy parameters as well as the magnetic field orientation are accurately known. As Fig. 7.13 shows, a very good fit is obtained for $J_{eff}/k_B = 6 \times 10^{-3}$ K, which, following (7.12), gives T_c equal to the experimental value of 0.6 K at zero field. Classical Monte Carlo simulations of the same model give, by contrast, the classical phase boundary shown by the dashed line in Fig. 7.13. This boundary is well approximated by $H_c(T_c) = H_c(0)[1 - T_c/T_c(H_\perp = 0)]^{1/2}$. In this model $H_c(0)$ equals the anisotropy field $H_K = 2[D - E(\sin^2\phi - \cos^2\phi)]/g\mu_B S \simeq 3.8$ T, which clearly overestimates the experimental critical field due to the absence of quantum fluctuations. The existence of a quantum critical point in Fe_8 can therefore be safely concluded. In summary, these results show that Fe_8 provides a close approximation of the archetypical quantum Ising model in a transverse magnetic field. Recently, the magnetic field dependence of the high-T susceptibility of Mn_{12}-acetate-MeOH has also been found to be in agreement with the quantum Ising model [112]. This high symmetry Mn_{12}-acetate variant has the same spin structure, anisotropy and similar lattice constants to the original Mn_{12}-acetate but has minimal solvent disorder [112, 113]. An important implication of this study is that magnetic order in two chemically very similar SMMs can be described by distinct physical models.

7.7 Conclusions and Outlook

The results reviewed in this chapter show that, despite their name, the physics of SMMs deviates from the image of an isolated molecule, especially at very low temperatures. In a crystal of SMMs, dipolar interactions induce the onset of long-range order. These materials provide therefore examples of pure dipolar magnets, of which so few exist in Nature.

Especially attractive are studies of long-range order in the presence of a transverse magnetic field, as those described in Sects. 7.5 and 7.6. The magnetic ground state results then from a subtle competition between dipolar couplings, quantum fluctuations, and random bias magnetic fields caused by molecular tilts, that is, by local disorder. The underlying physics is very rich, and depends qualitatively on the relative energy scales of these three interactions. In Mn_{12} acetate, random fields generated by molecular tilts dominantly suppresses ferromagnetism. By contrast, disorder-free Fe_8 undergoes a quantum phase transition at $T \to 0$, purely induced by quantum fluctuations generated by the transverse magnetic field. Within this interpretation, M_z vanishes above the critical field because the magnetic ground state becomes a quantum superposition of 'spin-up' and 'spin-down' states, a mesoscopic magnetic "Schrödinger's cat".

Molecular materials offer the possibility to realize in the lab two archetypical models, with broad interest for Magnetism and Solid State Physics: the random-field Ising model and the quantum Ising model. In this respect, molecular systems are appealing because properties such as the spin, magnetic anisotropy, and lattice symmetry can be controlled, to some extent, by chemical design. These possibility might enable experimentalists to explore situations which have not been realized yet, such as low-dimensional (i.e planes or chains) dipolar lattices, for which important deviations from the mean-field behavior can be expected [114], or situations with finite anisotropies that cannot be described by an Ising interaction Hamiltonian.

Quantum entanglement is enhanced near a quantum phase transition [115]. In molecular nanomagnets, the long-range character of the dominant dipolar interactions might lead to new sources of multipartite entanglement, thus change its range with respect to that found in spin systems with dominant nearest neighbor interactions. Entanglement is one of the resources for quantum computation and communication [116]. In my opinion, the measure and characterization of spin entanglement, and the study how quantum information propagates across a crystal of SMMs near the T_c–H_c phase boundary provide fascinating, yet unexplored, topics for research.

The above considerations refer to equilibrium magnetic properties of molecular crystals. However, SMMs are famous for displaying fascinating dynamical phenomena, such as hysteresis, i.e. magnetic memory, and quantum spin tunneling. The experiments described in the present chapter, an a few others, suggest that pure quantum tunneling processes, despite their inherently temperature-independent character, are nevertheless able to bring the spin system into its thermal equilibrium state, be it paramagnetic or magnetically ordered. How this mechanism actually works and, in particular, how energy is exchanged between spins and phonons, is not clear yet and deserves to be investigated further. An interesting, related question is how magnetic correlations grow below T_c, especially when spins are only able to flip by tunneling. This question has been addressed by Monte Carlo simulations [32] performed on the basis of the Prokof'ev and Stamp model for pure quantum tunneling [31], but needs to be tested experimentally. Another relatively unexplored area, especially from the experimental point of view, refers to the structure of domain walls in dipolar ferromagnets and their classical or quantum dynamics.

Close to a phase transition, the system dynamics tends to suffer from a "critical slowing down" [117]. The study of such non-equilibrium critical phenomena came to the fore when it was shown that they give information on the formation and the structure of defects in the early Universe and that some experiments could be carried out on real systems available at the laboratory [117]. An example is the formation of vortices in the vicinity of the (classical) phase transition of superfluid Helium at $T_\lambda = 2.14$ K [118]. Crystals of SMMs offer the possibility to investigate non-equilibrium spin dynamics (e.g. the nucleation of domain walls) across a quantum critical point ($T \simeq 0$, H_c). These studies can reveal the influence of quantum fluctuations [119] and might be relevant to the implementation of adiabatic quantum computation schemes [120] in crystals of SMMs.

References

1. J.H. van Vleck, Phys. Rev. **74**, 1168 (1948)
2. C. Kittel, Rev. Mod. Phys. **21**, 541 (1949)
3. J.A. Sauer, Phys. Rev. **57**, 142 (1940)
4. J.M. Luttinger, L. Tisza, Phys. Rev. **70**, 954 (1946)
5. Th. Niemeijer, H.W.J. Blöte, Physica **67**, 125 (1973)
6. S.K. Misra, Phys. Rev. B **14**, 5065 (1976)
7. J.F. Fernández, J.J. Alonso, Phys. Rev. B **62**, 53 (2000). Ibid., **65**, 189901(E) (2000)
8. H.-J. Xu, B. Bergersen, Z. Rácz, J. Phys. Condens. Matter **4**, 2035 (1992)
9. J.P. Bouchaud, P.G. Zérah, Phys. Rev. B **47**, 9095 (1993)
10. A.J. Millis, A.D. Kent, M.P. Sarachik, Y. Yeshurun, Phys. Rev. B **81**, 024423 (2010)
11. R.B. Griffiths, Phys. Rev. **176**, 655 (1968)
12. C. Kittel, Phys. Rev. **82**, 965 (1951)
13. M.H. Cohen, F. Keffer, Phys. Rev. **99**, 1135 (1955)
14. A. Aharony, Phys. Rev. B **8**, 3363 (1973)
15. A.H. Cooke, D.A. Jones, J.F.A. Silva, M.R. Wells, J. Phys. C, Solid State Phys. **8**, 4083 (1975)
16. L.M. Holmes, J. Als-Nielsen, H.J. Guggenheim, Phys. Rev. B **12**, 180 (1975)
17. J. Als-Nielsen, Phys. Rev. Lett. **37**, 1161 (1976)
18. G. Mennenga, L.J. de Jongh, W.J. Huiskamp, J. Magn. Magn. Mater. **44**, 59 (1984)
19. D.H. Reich, T.F. Rosenbaum, G. Aeppli, H.J. Guggenheim, Phys. Rev. B **34**, 4956 (1986)
20. M.R. Roser, L.R. Corruccini, Phys. Rev. Lett. **65**, 1064 (1990)

21. M.R. Roser, J. Xu, S.J. White, L.R. Corruccini, Phys. Rev. B **45**, 12337 (1992)
22. S.J. White, M.R. Roser, J. Xu, J.T. van der Noorda, L.R. Corruccini, Phys. Rev. Lett. **71**, 3553 (1993)
23. H.M. Ronnow et al., Science **308**, 389 (2005)
24. M. Schechter, P.C.E. Stamp, Phys. Rev. B **78**, 054438 (2008)
25. R. Sessoli, D. Gatteschi, A. Caneschi, M.A. Novak, Nature (London) **365**, 141 (1993)
26. D. Gatteschi, A. Caneschi, L. Pardi, R. Sessoli, Science **265**, 1054 (1994)
27. G. Christou, D. Gatteschi, D.N. Hendrickson, R. Sessoli, Mater. Res. Soc. Bull. **25**, 26 (2000)
28. D. Gatteschi, R. Sessoli, Angew. Chem., Int. Ed. Engl. **42**, 268 (2003)
29. D. Gatteschi, R. Sessoli, J. Villain, *Molecular Nanomagnets*, 1st edn. (Oxford University Press, Oxford, 2006)
30. R.B. Stinchcombe, J. Phys. C, Solid State Phys. **6**, 2459 (1973)
31. N.V. Prokof'ev, P.C.E. Stamp, Phys. Rev. Lett. **80**, 5794 (1998)
32. J.F. Fernández, Phys. Rev. B **66**, 064423 (2002)
33. J.R. Friedman, M.P. Sarachik, J. Tejada, R. Ziolo, Phys. Rev. Lett. **76**, 3830 (1996)
34. J.M. Hernández, X.X. Zhang, F. Luis, J. Bartolomé, J. Tejada, R. Ziolo, Europhys. Lett. **35**, 301 (1996)
35. L. Thomas, F. Lionti, R. Ballou, D. Gatteschi, R. Sessoli, B. Barbara, Nature **383**, 145 (1996)
36. C. Sangregorio, T. Ohm, C. Paulsen, R. Sessoli, D. Gatteschi, Phys. Rev. Lett. **78**, 4645 (1997)
37. W. Wernsdorfer, R. Sessoli, Science **284**, 133 (1999)
38. E. del Barco, N. Vernier, J.M. Hernández, J. Tejada, E.M. Chudnovsky, E. Molins, G. Bellessa, Europhys. Lett. **47**, 722 (1999)
39. F. Luis, F.L. Mettes, J. Tejada, D. Gatteschi, L.J. de Jongh, Phys. Rev. Lett. **85**, 4377 (2000)
40. J.A. Hertz, Phys. Rev. B **14**, 1165 (1976)
41. S. Sachdev, *Quantum Phase Transitions* (Cambridge University Press, Cambridge, 1999)
42. S. Morup, M.F. Hansen, C. Frandsen, Beilstein J. Nanotechnol. **1**, 182 (2010)
43. M. Evangelisti, E.K. Brechin, Dalton Trans. **39**, 4672 (2010)
44. A. Morello, F.L. Mettes, F. Luis, J.F. Fernández, J. Krzystek, G. Aromí, G. Christou, L.J. de Jongh, Phys. Rev. Lett. **90**, 017206 (2003)
45. M. Evangelisti, F. Luis, F.L. Mettes, N. Aliaga, G. Aromí, J.J. Alonso, G. Christou, L.J. de Jongh, Phys. Rev. Lett. **93**, 117202 (2004)
46. F. Luis, J. Campo, J. Gómez, G.J. McIntyre, J. Luzón, D. Ruiz-Molina, Phys. Rev. Lett. **95**, 227202 (2005)
47. A. Morello, F.L. Mettes, O.N. Bakharev, H.B. Brom, L.J. de Jongh, F. Luis, J.F. Fernández, G. Aromí, Phys. Rev. B **73**, 134406 (2006)
48. G. Aromí, E. Bouwman, E. Burzurí, Ch. Carbonera, J. Krzystek, F. Luis, C. Schlegel, J. van Slageren, S. Tanase, S.J. Teat, Chemistry **14**, 11158 (2008)
49. E. Burzurí, F. Luis, B. Barbara, R. Ballou, E. Ressouche, O. Montero, J. Campo, S. Maegawa, Phys. Rev. Lett. **107**, 097203 (2011)
50. B. Wen, P. Subedi, L. Bo, Y. Yeshurun, M.P. Sarachik, A.D. Kent, A.J. Millis, C. Lampropoulos, G. Christou, Phys. Rev. B **82**, 014406 (2010)
51. L.J. de Jongh, A.R. Miedema, Adv. Phys. **23**, 1 (1974)
52. A. Aharoni, *Introduction to the Theory of Ferromagnetism*, 2nd edn. (Oxford University Press, Oxford, 2000)
53. D.A. Garanin, Phys. Rev. B **81**, 220408(R) (2010)
54. S. Li, L. Bo, B. Wen, M.P. Sarachik, P. Subedi, A.D. Kent, Y. Yeshurun, A.J. Millis, C. Lampropoulos, S. Mukherjee, G. Christou, Phys. Rev. B **82**, 174405 (2010)
55. R. de, L. Kronig, Physica **6**, 33 (1939)
56. J.H. Van Vleck, Phys. Rev. **57**, 426 (1940)
57. R. Orbach, Proc. R. Soc. Lond. Ser. A, Math. Phys. Sci. **264**, 456 (1961). Ibid., **264**, 485 (1961)
58. R. Orbach, Proc. R. Soc. Lond. Ser. A, Math. Phys. Sci. **264**, 485 (1961)
59. J. Villain, F. Hartman-Boutron, R. Sessoli, A. Rettori, Europhys. Lett. **27**, 159 (1994)

60. D.A. Garanin, E.M. Chudnovsky, Phys. Rev. B **56**, 11102 (1997)
61. F. Luis, J. Bartolomé, J.F. Fernández, Phys. Rev. B **57**, 505 (1998)
62. A. Fort, A. Rettori, J. Villain, D. Gatteschi, R. Sessoli, Phys. Rev. Lett. **80**, 612 (1998)
63. M.N. Leuenberger, D. Loss, Phys. Rev. B **61**, 1286 (2000)
64. L. Thomas, A. Caneschi, B. Barbara, Phys. Rev. Lett. **83**, 2398 (1999)
65. M. Evangelisti, F. Luis, F.L. Mettes, R. Sessoli, L.J. de Jongh, Phys. Rev. Lett. **95**, 227206 (2005)
66. M. Evangelisti, A. Candini, A. Ghirri, M. Affronte, G.W. Powell, I.A. Gass, P.A. Wood, S. Parsons, E.K. Brechin, D. Collison, S.L. Heath, Phys. Rev. Lett. **97**, 167202 (2006)
67. C. Vecchini, D.H. Ryan, L.M.D. Cranswick, M. Evangelisti, W. Kockelmann, P.G. Radaelli, A. Candini, M. Affronte, I.A. Gass, E.K. Brechin, O. Moze, Phys. Rev. B **77**, 224403 (2008)
68. N.V. Prokof'ev, P.C.E. Stamp, Rep. Prog. Phys. **63**, 669 (2000)
69. J.F. Fernández, J.J. Alonso, Phys. Rev. Lett. **91**, 047202 (2003). See also I.S. Tupitsyn, P.C.E. Stamp, Ibid., **92**, 119701 (2004); J.F. Fernández, J.J. Alonso, Ibid., **92**, 119702 (2004)
70. J.F. Fernández, J.J. Alonso, Phys. Rev. B **69**, 024411 (2004)
71. D.A. Garanin, J. Phys. A, Math. Gen. **24**, L61 (1991)
72. W. Wernsdorfer, M. Murugesu, G. Christou, Phys. Rev. Lett. **96**, 057208 (2006)
73. S. McHugh, R. Jaafar, M.P. Sarachik, Y. Myasoedov, H. Shtrikman, E. Zeldov, R. Bagai, G. Christou, Phys. Rev. B **79**, 052404 (2009)
74. D. Ruiz-Molina, Z.S. Sun, B. Albela, K. Folting, J. Ribas, G. Christou, D.N. Hendrickson, Angew. Chem., Int. Ed. Engl. **37**, 300 (1998)
75. G. Aromí, M.J.Knapp.J.-P. Claude, J.C. Huffman, D.N. Hendrickson, G. Christou, J. Am. Chem. Soc. **121**, 5489 (1999)
76. E. Burzurí, Dissertation, University of Zaragoza (2011)
77. D.A. Garanin, E.M. Chudnovsky, Phys. Rev. B **78**, 174425 (2008)
78. A. Morello, O.N. Bakharev, H.B. Brom, R. Sessoli, L.J. de Jongh, Phys. Rev. Lett. **93**, 197202 (2004)
79. F. Luis, M.J. Martínez-Pérez, O. Montero, E. Coronado, S. Cardona-Serra, C. Martí-Gastaldo, J.M. Clemente-Juan, J. Ses, D. Drung, T. Schurig, Phys. Rev. B **82**, 060403(R) (2010)
80. S.M.J. Aubin, N.R. Dilley, L. Pardi, J. Krzystek, M.W. Wemple, L.C. Brunel, M.B. Maple, G. Christou, D.N. Hendrickson, J. Am. Chem. Soc. **120**, 4991 (1998)
81. N. Aliaga, K. Folting, D.N. Hendrickson, G. Christou, Polyhedron **20**, 1273 (2001)
82. T. Kadowaki, H. Nishimori, Phys. Rev. E **58**, 5355 (1998)
83. J. Brooke et al., Science **284**, 779 (1999)
84. F.L. Mettes, F. Luis, L.J. de Jongh, Phys. Rev. B **64**, 174411 (2001)
85. S. Sachdev, J. Ye, Phys. Rev. Lett. **69**, 2411 (1992)
86. A.V. Chubukov, S. Sachdev, Phys. Rev. Lett. **71**, 169 (1993)
87. A. Sokol, D. Pines, Phys. Rev. Lett. **71**, 2813 (1993)
88. Q. Si, F. Steglich, Science **329**, 1161 (2010)
89. S.A. Carter, T.F. Rosenbaum, J.M. Honig, J. Spalek, Phys. Rev. Lett. **67**, 3440 (1991)
90. D. Bitko, T.F. Rosenbaum, G. Aeppli, Phys. Rev. Lett. **77**, 940 (1996)
91. A. Yeh et al., Nature **419**, 459–462 (2002)
92. P. Stasiak, M.J.P. Gingras, Phys. Rev. B **78**, 224412 (2008)
93. T. Lis, Acta Crystallogr. B **36**, 2042 (1980)
94. R.A. Robinson, P.J. Brown, D.N. Argyriou, D.N. Hendrickson, S.M.J. Aubin, J. Phys. Condens. Mater. **12**, 2805 (2000)
95. A.L. Barra, D. Gatteschi, R. Sessoli, Phys. Rev. B **56**, 8192 (1997)
96. S. Hill et al., Phys. Rev. Lett. **80**, 2453 (1998)
97. I. Mirebeau et al., Phys. Rev. Lett. **83**, 628 (1999)
98. S. Hill, R.S. Edwards, S.I. Jones, N.S. Dalal, J.M. North, Phys. Rev. Lett. **90**, 217204 (2003)
99. S. Takahashi, R.S. Edwards, J.M. North, S. Hill, N.S. Dalal, Phys. Rev. B **70**, 094429 (2004)
100. R. Bircher, G. Chaboussant, A. Sieber, H.U. Güdel, H. Mutka, Phys. Rev. B **70**, 212413 (2004)

101. A. Cornia, R. Sessoli, L. Sorace, D. Gatteschi, A.L. Barra, C. Daiguebonne, Phys. Rev. Lett. **89**, 257201 (2002)
102. E. del Barco, A.D. Kent, E.M. Rumberger, D.N. Hendrickson, G. Christou, Phys. Rev. Lett. **91**, 047203 (2003)
103. E. del Barco, A.D. Kent, S. Hill, J.M. North, N.S. Dalal, E.M. Rumberger, D.N. Hendrickson, N. Chakov, G. Christou, J. Low Temp. Phys. **140**, 119 (2005)
104. E. Burzurí, Ch. Carbonera, F. Luis, D. Ruiz-Molina, C. Lampropoulos, G. Christou, Phys. Rev. B **80**, 224428 (2009)
105. K. Park, T. Baruah, N. Bernstein, M.R. Pederson, Phys. Rev. B **69**, 144426 (2004)
106. K. Wieghardt, K. Pohl, I. Jibril, G. Huttner, Angew. Chem., Int. Ed. Engl. **23**, 77 (1984)
107. A.L. Barra et al., Europhys. Lett. **35**, 133 (1996)
108. R. Caciuffo et al., Phys. Rev. Lett. **81**, 4744 (1998)
109. M. Ueda et al., J. Phys. Soc. Jpn. **70**, 3084 (2001)
110. X. Martínez-Hidalgo, E.M. Chudnovsky, A. Aharony, Europhys. Lett. **55**, 273 (2001)
111. R.J. Elliot, P. Pfeuty, C. Wood, Phys. Rev. Lett. **25**, 443 (1970)
112. P. Subedi, A.D. Kent, B. Wen, M.P. Sarachik, Y. Yeshurun, A.J. Millis, S. Mukherjee, G. Christou, Phys. Rev. B **85**, 134441 (2012)
113. G. Redler, C. Lampropoulos, S. Datta, C. Koo, T.C. Stamatatos, N.E. Chakov, G. Christou, S. Hill, Phys. Rev. B **80**, 094408 (2009)
114. J.F. Fernández, J.J. Alonso, Phys. Rev. B **76**, 014403 (2007)
115. A. Osterloh, L. Amico, G. Falci, R. Fazio, Nature **416**, 608 (2002)
116. M.A. Nielsen, I.L. Chuang, *Quantum Computation and Quantum Information*, 1st edn. (Cambridge University Press, Cambridge, 2000)
117. W.H. Zurek, Nature **317**, 505 (1985)
118. P.C. Hendry, N.S. Lawson, R. Lee, P. Mc Clintok, C. Williams, Nature **368**, 315 (1994)
119. W.H. Zurek, U. Dorner, P. Zoller, Phys. Rev. Lett. **95**, 105701 (2005)
120. E. Farhi, J. Goldstone, S. Gutmann, J. Lapan, A. Lundgren, D. Preda, Science **292**, 472 (2001)

Chapter 8
Single-Chain Magnets

Dante Gatteschi and Alessandro Vindigni

Abstract Single-chain magnets are molecular spin chains displaying slow relaxation of the magnetisation on a macroscopic time scale. To this similarity with single-molecule magnets they own their name. In this chapter the distinctive features of single-chain magnets as opposed to their precursors will be pinpointed. In particular, we will show how their behaviour is dictated by the physics of thermally-excited domain walls. The basic concepts needed to understand and model single-chain magnets will also be reviewed.

8.1 Introduction

The observation of magnetic hysteresis of molecular origin in Single-Molecule Magnets (SMMs) is considered one of the most relevant achievements in nanomagnetism [1, 2]. Fundamental aspects related to quantum tunnelling of the magnetisation have been thoroughly discussed in the previous chapters. On a more practical perspective, that observation rendered the molecular approach one of the possible routes to realizing bistable nano-objects, suitable for magnetic storage or quantum-computing applications. In spite of many efforts, the highest blocking temperature attained by SMMs remains, still nowadays, in the liquid-helium temperature range. The idea that one-dimensional (1D) structures of coupled paramagnetic ions might afford higher blocking temperatures started developing at the end of the nineties and the first examples of slowly relaxing 1D systems were reported at the beginning of

D. Gatteschi (✉)
Department of Chemistry, University of Florence, Via della Lastruccia 3, 50019 Sesto Fiorentino, Italy
e-mail: dante.gatteschi@unifi.it

D. Gatteschi
INSTM, Via G. Giusti 9, 50121 Florence, Italy

A. Vindigni
Laboratory for Solid State Physics, Swiss Federal Institute of Technology, ETH Zurich, Wolfgang-Pauli-Str. 16, 8093 Zurich, Switzerland
e-mail: vindigni@phys.ethz.ch

J. Bartolomé et al. (eds.), *Molecular Magnets*, NanoScience and Technology,
DOI 10.1007/978-3-642-40609-6_8,

the new century [3, 4]. The resulting molecular systems have been dabbed Single-Chain Magnets (SCMs) in order to evidence analogies with their precursors, SMMs, while remarking—at the same time—the 1D character. In some cases, SMMs themselves have been employed as building blocks for such 1D magnetic lattices [5]. With the aim of increasing the blocking temperature as much as possible, different synthesis strategies have been followed to obtain some type of magnetic anisotropy at the level of building blocks or of the coupling among them. In the present chapter we will be dealing with uniaxial anisotropies only, though this requirement is not strict for the observation of SCM behaviour [7–9].

A distinctive feature of 1D magnetic systems is the development of short-range correlations upon cooling. This makes them substantially different from both paramagnets and bulk magnets. Should one establish an analogy between classical magnetic ordering and phases of matter, paramagnets would be identified with perfect gases while bulk magnets with solids. Pushing this naive analogy further, spin chains would be associated with liquids, specifically in the temperature range in which short-range correlations extend over several lattice units. The degree of spatial correlation is quantified by the correlation length. In molecular chains consisting of magnetic building bocks with uniaxial anisotropy, the correlation length typically diverges exponentially with decreasing temperature. From a snapshot taken at finite temperature, any chain would appear as a collection of randomly oriented magnetic domains[1] separated by domain walls (DWs). The average size of those domains is of the order of the correlation length. This pictorial, but essentially correct, scenario is consistent with thermally-driven diffusion of DWs. In this sense, the response of a SCM to a tiny a.c. field is expected to be determined by the time needed to adjust the size of domains to the external stimulus. A robust scaling argument associates the characteristic time of this readjustment with the time elapsed while a DW diffuses over a distance proportional to the correlation length. Within this idealized picture, the relaxation time of the magnetisation is expected to scale with temperature like the square of the correlation length.

The qualitative description given above applies to the ideal case of infinite chains and small applied fields. The first hypothesis is practically never fulfilled in real systems. In fact, the number of magnetic centres interacting consecutively is typically limited to 10^2–10^4 by naturally occurring defects, non-magnetic impurities or lattice dislocations [10–12]. A SCM may thus behave as if it extended indefinitely only when the correlation length is much smaller than the average distance between successive defects. Upon lowering the temperature, a crossover is expected at which the correlation length becomes of the order of the average distance among defects. Below this crossover temperature, spins enclosed between two successive defects are parallel with each other and no DW is present at equilibrium. In this *finite-size regime* relaxation is somewhat equivalent to magnetisation reversal in nanoparticles or nanowires, which may occur via Néel-Brown uniform rotation or by droplet-nucleation mechanism [13].

[1]These soft, fluctuating domains should not be confused with Weiss domains encountered in magnetically ordered phases.

All the mentioned mechanisms represent potential channels for relaxation in SCMs. Which one is favoured depends on the experimental conditions: temperature, applied field and amount of defects in the sample. The random-walk argument which relates the correlation length to the relaxation time holds in the linear-response regime, i.e., when such tiny fields are applied that just slight deviations from thermodynamic equilibrium are induced. On the contrary, relaxation from a saturated configuration typically entails far-from-equilibrium dynamics. In this type of experiments nucleation of soliton-antisoliton pairs or of a single DW adjacent to a defect possibly initiates the relaxation process. Néel-Brown uniform rotation practically represent an alternative channel for relaxation only for very short segments of chain, encountered in samples in which finite-size effects have been enhanced by doping with non-magnetic impurities [10, 11].

It should not be forgotten that molecular spin chains are packed in three-dimensional crystals. Though several synthesis strategies may be followed to minimize interactions among chains, at least the dipolar interaction cannot be suppressed completely. Therefore, below some temperature, a 3D magnetically ordered phase is expected to appear. Whether such a phase is observed or not in a specific compound depends on how long the relaxation time is at the transition temperature [14]. Generally, when the time needed for the system to equilibrate is much longer than experimental time scales, the distinctive features of the underlying equilibrium phase, possibly ordered, cannot be evidenced. For weakly interacting spin chains, the expected transition temperature is much higher than interchain interaction in Kelvin units ($kT_C \gg J'$ with the forthcoming notation). In fact, the 3D-ordering process is "assisted" by the development of strong short-range correlations inside each chain [15–17]. However, in realistic samples, defects prevent the intrachain correlation length from diverging indefinitely, which eventually lowers the transition temperature to the ordered phase. In several SCMs slow dynamics was observed down to few Kelvins, before 3D ordering took place, right because of the presence of defects and non-magnetic impurities.

Both SMMs and SCMs are characterized by slow dynamics of molecular origin, acting at macroscopic time scales and in the absence of 3D magnetic ordering. Even if impurities play a crucial role in SCMs, they usually do not bring enough disorder to give rise to spin-glass behaviour. Consistently, slow dynamics is typically characterized by a single time scale which does not display a super-Arrhenius behaviour at any temperature [18–20]. Besides preventing the onset of 3D magnetic ordering, the increase of relaxation time with cooling usually leads to complete blocking before genuine quantum effects become evident [21–23].

From what written till now, it should be clear that many effects may interplay in determining the magnetic behaviour of spin chains. We will focus on those systems in which slow dynamics can be ascribed to each single chain and does not originate from cooperative 3D interactions.

The goal of this chapter is that of highlighting the properties of SCMs with a critical view to what has been done and what still deserves further investigation. We will not try to cover in detail all the representative literature, for which the reader is addressed to excellent reviews [6–9]. Although SCMs have been widely

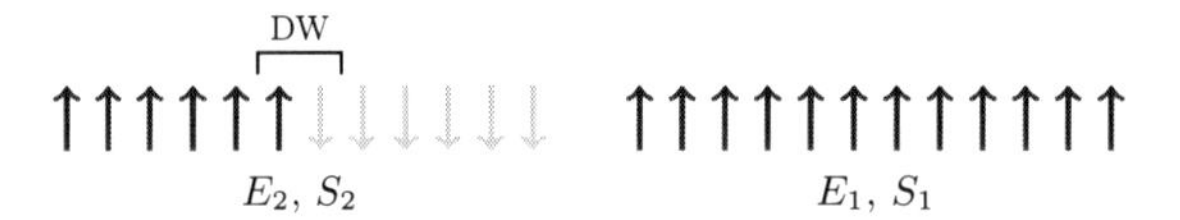

Fig. 8.1 Sketch representing the configurations whose free-energy difference is evaluated in the text: a ferromagnetic ground state with all the spins parallel to each other (*right*) and a configuration consisting of two domains with opposite spin alignment (*left*)

investigated, the interest of the physics community has not been comparable to that shown for SMMs. This is partially due to the fact that the novelty of SCMs compared to traditional 1D spin systems hardly emerged. Here we attempt to provide and efficient overview of the essential, novel physics of SCMs and hope that the final comment will be more benevolent. The chapter is organised as follows: Section 8.2 will cover the basic aspects of classical spin chains; the chemical frame will be discussed within a bottom-up or building-block approach in Sect. 8.3; in Sect. 8.4 the spin Hamiltonians typically used to rationalise the physical properties of SCMs will be introduced; relevant extensions of the Glauber model developed in the SCM context without and with defects will be treated in Sects. 8.5 and 8.6, respectively; phenomenological arguments not contained in the Glauber model but relevant for understanding SCMs will be discussed in Sect. 8.7; a short section on perspectives will conclude the chapter.

8.2 Thermal Equilibrium and Slow Dynamics in Ideal SCMs

In this section the peculiarities of classical spin chains with uniaxial anisotropy that directly affect the physics of SCMs will be recalled. Indeed, the distinctive feature of SCMs is that of approaching thermodynamic equilibrium slowly. By *slowly* we mean that relaxation time becomes longer than milliseconds at temperatures of the order of 10 K or lower. The reference equilibrium state to be reached is also relevant. As already mentioned, as long as 3D interactions are negligible, no magnetisation is expected in zero applied field at thermodynamic equilibrium. Long-range magnetic order may be destroyed by thermally-excited spin waves or DWs either. The first ones are effective in the absence of anisotropy, according to the Mermin–Wagner theorem [24, 25]. The fact that disordering is, instead, driven by DWs in the presence of anisotropy can be easily understood recalling an argument presented in the Landau–Lifshitz series [26]. Let us consider a group of N spins that preferentially point along the same direction, say up or down. For the moment we assume the axes of easy anisotropy to be collinear, as represented schematically in Fig. 8.1. We evaluate the variation of the free energy associated with the creation of a DW starting from a configuration with all the spins parallel to each other. Creating a DW increases the energy by a factor $E_2 - E_1 = \mathcal{E}_{\mathrm{dw}}$. On the other hand, such a DW may occupy N different positions in the spin chain, so that the relative entropy increase scales as $S_2 - S_1 = k\ln(N)$. The free-energy difference between the two configurations sketched in Fig. 8.1 is roughly $\Delta F \simeq \mathcal{E}_{\mathrm{dw}} - kT\ln(N)$. When the thermodynamic limit $N \to \infty$ is taken, one immediately realizes that it is always convenient to split the system into groups of parallel spins. As a consequence, long-range magnetic order is destroyed at any finite temperature. In principle, in an infinite chain, the same mechanism may allow creating an indefinite number of DWs. However, the average distance among them does depend on temperature and it is inversely proportional to the correlation length [27]. It is worth remarking that in the text-book argument given above the following assumptions have been implicitly made:

1. DWs extended just only over one lattice unit
2. spin-wave excitations were not considered
3. the thermodynamic limit was taken.

Whether the first hypothesis is fulfilled or not depends on the relative strength of exchange interaction and magnetic-anisotropy energy. This can be discussed more concretely by considering the classical Heisenberg model with uniaxial anisotropy:

$$\mathcal{H}_{\mathrm{H}} = -\sum_{i=1}^{N}\left[J\boldsymbol{S}_i \cdot \boldsymbol{S}_{i+1} + D\left(S_i^z\right)^2\right], \tag{8.1}$$

where $\boldsymbol{S}_i$ are classical spins, J and D the exchange and the magnetic-anisotropy energy, respectively; $|\boldsymbol{S}_i| = 1$ will be assumed henceforth. Though it does not entail the complexity of many real systems, Hamiltonian (8.1) is a useful reference to discuss the physics of SCMs. To the aim of distinguishing between two simple types of DWs, we fix $D > 0$ and $J > 0$. With Hamiltonian (8.1), DWs can be larger than one lattice spacing. In fact, the actual DW profile results from the *competition* between the exchange energy (which is minimized by broadening the wall) and the anisotropy energy (which favours a sharp wall). Domain walls whose structure develop over more lattice units will be referred as *broad*; these are opposed to *sharp* DWs in which the local magnetisation changes abruptly its sign, within one lattice distance. The energy associated with a broad DW is $\mathcal{E}_{\mathrm{dw}} = 2\sqrt{2DJ}$ [28], namely the energy needed to create one soliton "particle" in the spin chain [29]. For sharp DWs, one obtains $\mathcal{E}_{\mathrm{dw}} = 2J$, as per the Ising model. The crossover from sharp- to broad-wall occurs at $D/J = 2/3$ [30–32]. The analytic formula for broad-DW energy, $\mathcal{E}_{\mathrm{dw}} = 2\sqrt{2DJ}$, was obtained in the continuum formalism and gets less and less accurate as the threshold ratio is approached from below, $D/J \to (2/3)^{-}$.

If the Landau's argument is rephrased for DW excitations of finite thickness $w = \sqrt{J/2D}$, the counting of equivalent configurations with the same energy needs to be modified and—in turn—the entropy contribution $S_2 - S_1 = k\ln(N/w)$. In this case, splitting the uniform configuration into domains becomes convenient as soon as the number of spins exceeds the product $w e^{\mathcal{E}_{\mathrm{dw}}/kT}$. The latter threshold gives an estimate of the average number of consecutive spins that can be found aligned at a given temperature. To the leading order, the correlation length scales in the same way at low temperature: $\xi \sim w e^{\mathcal{E}_{\mathrm{dw}}/kT}$. The energy $\mathcal{E}_{\mathrm{dw}}$ represents the natural "unit" which controls the divergence of the correlation length. Thus, in classical

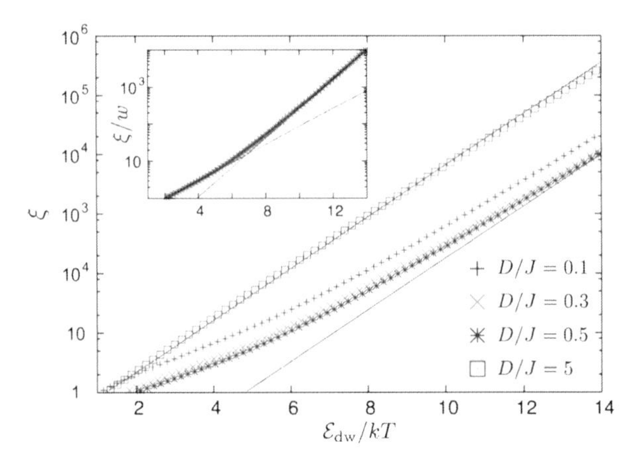

Fig. 8.2 Log-linear plot of ξ in lattice units computed with the transfer-matrix technique as function of the ratio $\mathcal{E}_{dw}/kT$ for different values of D/J. For $D/J = 0.1$ (*red crosses*), 0.3 (*green crosses*), 0.5 (*blue stars*) DWs are *broad* and $\mathcal{E}_{dw}$ has been computed numerically on a discrete lattice. For $D/J = 5$ (*open squares*) DWs are *sharp* and $\mathcal{E}_{dw} = 2J$ has been used. *The two solid lines* give the "reference" behaviour $\xi \sim e^{\Delta_\xi/kT}$ which is indeed followed when DWs are *sharp* ($D/J = 5$) but not when DWs broaden. *Inset*: $\Lambda = \xi/w$ is plotted as a function of $\mathcal{E}_{dw}/kT$ for the values of D/J consistent with *broad* DWs. The universality of Λ is highlighted by the data collapsing. *Solid lines* evidence the decrease of Δ_ξ with increasing temperature [34]

spin chains with uniaxial anisotropy the characteristic exponential divergence of ξ is closely related to the fact that ferromagnetism is destroyed by thermally excited DWs.

In contrast to the Ising model [33], the classical spin Hamiltonian (8.1) can also host spin-wave excitations, besides DWs. Due to the interaction between spin waves and broad DWs an additional temperature-dependent factor appears in front of the exponential in the low-temperature expansion of the correlation length [35]. Moreover, spin waves renormalise the DW energy at intermediate temperatures. The net result of the complicated interplay between thermalised spin waves and DWs is that the energy barrier controlling the divergence of ξ (usually called Δ_ξ in SCM literature [6–9]) is generally smaller than $\mathcal{E}_{dw}$ and takes different values depending on the temperature range in which it is measured [34]. A similar effect was reported for the activation energy of 2π sine-Gordon solitons in Mn^{2+}-radical spin chains [36]. Figure 8.2 highlights how Δ_ξ is constant and equal to $\mathcal{E}_{dw}$ for sharp DWs, while it varies significantly for broad DWs. However, the correlation length in units of w keeps depending only the ratio $\mathcal{E}_{dw}/kT$, i.e., $\xi/w = \Lambda(\mathcal{E}_{dw}/kT)$. The inset shows that the curves corresponding to broad DWs indeed collapse onto each other when the ratio ξ/w is plotted as a function of $\mathcal{E}_{dw}/kT$.

As mentioned in the introduction, in realistic spin chains the divergence of the correlation length is always hindered by the presence of defects and non-magnetic impurities. This implies that results derived taking the thermodynamic limit, $N \to \infty$, do not hold down to indefinitely low temperatures. If we assume—for the time being—an idealized scenario in which such defects do not occur, a certain number of DWs shall be present at any finite temperature. A simple random-walk argument then relates the relaxation time to the correlation length: within a time τ a DW performs a random walk over a distance proportional to ξ [37]. In other words, the relation

$$\xi^2 \simeq 2D_s\tau \tag{8.2}$$

can be assumed, with D_s being the diffusion coefficient. The latter generally increases with increasing temperature. Moreover, it is expected to depend on temperature differently for sharp or broad diffusing DWs. When presenting the Glauber model we will see that D_s can also be interpreted as the attempt frequency to flip a spin adjacent to a sharp DW.

Summarizing, the presence of uniaxial anisotropy produces an exponential divergence of the correlation length with decreasing temperature. As the relaxation time is related to ξ by a random-walk argument, it is also expected to diverge likewise, so that

$$\xi \sim e^{\Delta_\xi/kT} \qquad \tau \sim e^{\Delta_\tau/kT}. \tag{8.3}$$

In ideal 1D magnetic systems [29, 38] the correlation length is proportional to the product of temperature by static susceptibility (measured in zero field):

$$\chi_{eq}T \sim \xi. \tag{8.4}$$

The relaxation time can, instead, be obtained from dynamic susceptibility measurements as follows

$$\chi(\omega, T) = \frac{\chi_{eq}}{1 - i\omega\tau}, \tag{8.5}$$

where ω is the frequency of the oscillating applied field and χ_{eq} is the static susceptibility.[2] Both real and imaginary parts of $\chi(\omega, T)$ display a maximum for $\omega\tau = 1$. The basic experimental characterization of SCMs essentially reduces to determining the temperature dependence of ξ and τ, which is—in principle—possible thanks to (8.4) and (8.5).

Even within the idealized scenario presented in this section, the way in which Δ_ξ and Δ_τ defined in (8.3) relate to the Hamiltonian parameters J and D depends on the DW thickness, w, and on the temperature range in which such energy barriers are measured. Besides this, model Hamiltonians of real SCMs may differ significantly from (8.1). In the next section we will recall some synthesis strategies that have been followed to produce different SCMs. The features of the employed building blocks and the type of coupling among them eventually decide which model is more appropriate to describe a specific SCM.

[2]The more general Cole-Cole equation is needed when relaxation is not characterised by a single τ or to account for adiabatic contribution to χ [39].

8.3 Tailoring SCMs by Building-Block Approach

The initial interest in Molecular Magnets stemmed from the attempt to design molecular systems displaying long-range magnetic order. However, after more than 30 years of attempts there are only two room-temperature molecular magnets and matters are no better for liquid-nitrogen temperatures [40, 41]. To have long-range order it is necessary to build 2D or 3D structures of centres magnetically coupled. This is difficult with molecular bricks since the number of coordination sites which are available to propagate the exchange coupling in different directions is small due to the presence of capping ligands. Such bricks are then more suitable to produce low-dimensional systems, like clusters of metal ions (zero dimensional) [2] or spin chains. These systems do not display long-range order but still show a variety of interesting phenomena, including SCM behaviour. It is pedagogically useful to imagine that synthesizing a SCM is like assembling bricks with a magnetic functionality and a structural functionality. Usually, the latter is provided by organic molecules and the former by metal ions. Building blocks need to be chosen and arranged in a structure which maximizes the intrachain and minimizes the interchain interactions. Bricks with magnetic functionality must be coupled ferro- or ferri-magnetically and control of the magnetic anisotropy must be achieved. Chemists are not yet able to have that detailed control but progress is fast and serendipity always helps.

Some centres of the building blocks shall be magnetically active, which implies the presence of unpaired electrons that are formally assigned to magnetic orbitals, either p, d, or f. In organic radicals the unpaired electrons normally belong to p orbitals: these are external orbitals which strongly interact with the environment. For this reason such electrons hardly remain unpaired but rather tend to couple with electrons of other molecules in covalent bonds, which eventually explains why few stable organic radicals exist. In the following, we will mostly refer to nitronyl nitroxide radicals (NITR), whose structure is shown in Fig. 8.3(a). The unpaired electron is delocalised on the group O-N-C-N-O and, from the magnetic point of view, basically behaves as a free electron. Its magnetic moment is essentially spin determined, with little orbital contribution due to small spin-orbit coupling. This implies low magnetic anisotropy which is the final blow for purely organic SCMs.

NITR radicals have the right geometry for bridging two metal ions through their equivalent oxygen atoms (extremes of the O-N-C-N-O fragment in Fig. 8.3(a)). The above considerations suggest that NITR radicals are not appropriate for being used alone, but they become excellent bricks for SCMs when coordinated to metal ions [6–9, 42]. In fact, the interaction of the p orbitals with the d (or f) orbitals can be strong, of direct type, both ferro and antiferromagnetic in nature.

Transition-metal ions provide good magnetic bricks. As anticipated in the introduction, we will limit ourselves to consider SCMs possessing uniaxial anisotropy at the brick level. In molecular systems, magnetic anisotropy is closely related to the fact that the surrounding of metal ions is not spherically symmetric. Figure 8.4(a) shows a generic metal atom (M) in an octahedral environment of ligands. Oxygen atoms occupy the vertices of the octahedron. In the group $M(hfac)_2$, for instance, two oxygens of each hexafluoroacetylacetonate (hfac) ligand coordinate to M, thus

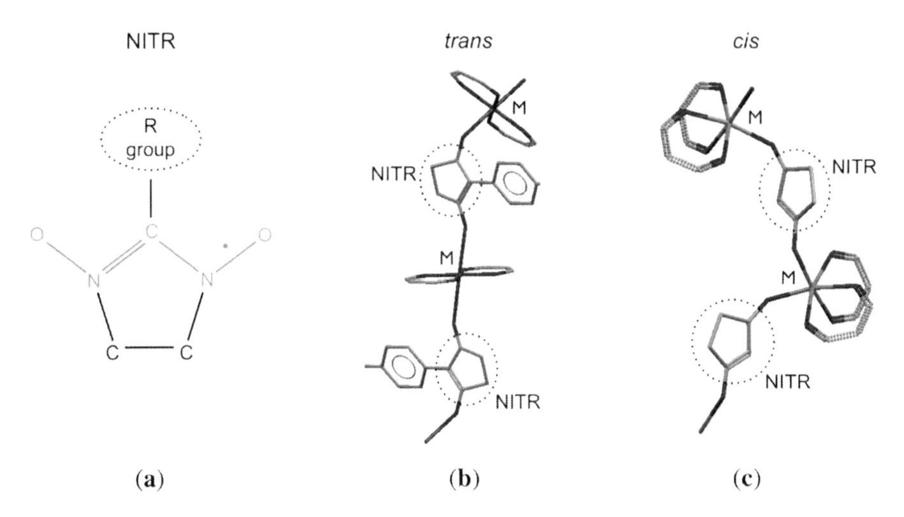

Fig. 8.3 (**a**) schematic structure of the NITR radical: the unpaired electron is delocalised over the O-N-C-N-O fragment (coloured), which is magnetically active. (**b**) and (**c**) show two possible realisations of direct exchange coupling between the electron of each NITR radical and a metal ion (intrachain interaction): each M can be bond to two NITR groups through oxygens occupying either *trans* (**b**) or *cis* (**c**) positions in the coordination octahedron (see Fig. 8.4)

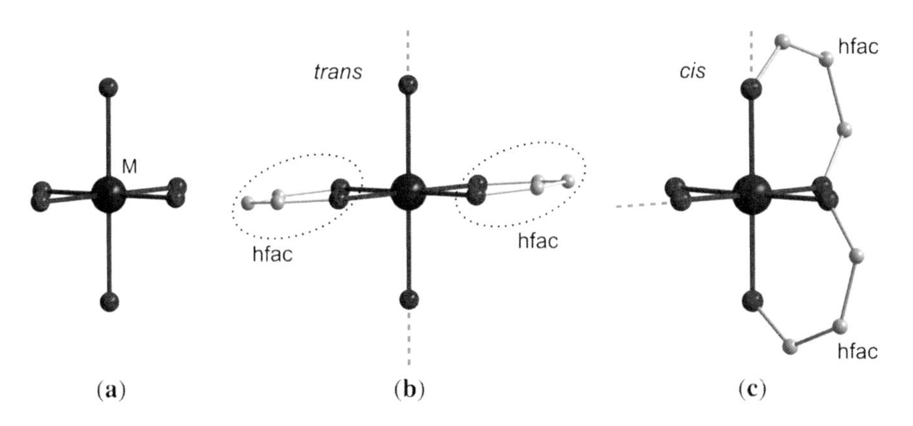

Fig. 8.4 Sketch of a metal ion (*purple spheres*) in an octahedral environment of oxygen-donating ligands (*red spheres* representing oxygens). (**a**) metal-oxide coordination in an extended solid. (**b**) and (**c**) $M(hfac)_2$ moiety with two empty coordination sites in *trans* position (**b**) and in *cis* position (**c**); the CF3 groups of hfac ligands are not shown for clarity sake. *Green dashed lines* indicate the directions along which the intrachain exchange coupling mediated by a different ligand (e.g., NITR radical) may propagate

occupying two neighbouring vertices of the octahedron per hfac molecule. The two remaining, empty coordination sites can be in either *trans* or *cis* position (Fig. 8.4(b) and (c), respectively) and may host oxygens of other ligands that can be used to connect different $M(hfac)_2$ moieties. The choice of NITR to bridge those moieties creates a strong, direct exchange coupling between M and the electron delocalised on each O-N-C-N-O group (intrachain interaction, J). Consistently with the two possible coordination configurations of $M(hfac)_2$ sketched in Fig. 8.4, the segments connecting the metal ions with NITR oxygens may form an angle of 180° (*trans*) or 90° (*cis*). Such segments specify the direction along which the intrachain interaction propagates. The bulky hfac groups prevent efficient interchain exchange coupling. The residual interchain interaction J' has mainly dipolar origin and it is, typically, from 3 to 6 orders of magnitude smaller than the intrachain interaction. For this reason, $M(hfac)_2$ moieties are perfectly suited for realizing isolated spin chains ($|J'/J| < 10^{-3}$). The interaction between successive magnetic bricks can either be ferro- or ferrimagnetic and give rise to straight or zig-zag structures. Besides, magnetic bricks are often characterized by low symmetry with the metal ions occupying general positions in the unit cells, which does not impose limitations to the orientation of anisotropy axes. Therefore, in practice, full collinearity among anisotropy axes is more an exception rather than the rule [6–9].

Since SCM behaviour requires some magnetic anisotropy, the orbital momentum must not be completely quenched. The surviving component may be associated with single-ion anisotropy or with pair-spin interaction. In the former case, the residual orbital contribution can show up in a g tensor different from the free-electron one and/or in the zero-field splitting. With a large periodic table it is amazing that only cobalt and manganese, with some iron and nickel have been used. Mn^{3+} is an example of anisotropy determined by zero-field splitting; while in Co^{2+} the anisotropy is associated with the g tensor [43]. The crystal-field theory is the simplest way to describe the ground and low-lying levels of a transition-metal ion. The Hamiltonian can be expressed as a sum of terms:

$$\mathcal{H} = \mathcal{H}_0 + \mathcal{H}_{\mathrm{ee}} + \mathcal{H}_{\mathrm{CF}} + \mathcal{H}_{\mathrm{LS}}, \tag{8.6}$$

where $\mathcal{H}_0$ is the origin of the electron configuration $(3\mathrm{d})^n$, $\mathcal{H}_{\mathrm{ee}}$ is the electron-electron repulsion, $\mathcal{H}_{\mathrm{CF}}$ is the crystal-field term and $\mathcal{H}_{\mathrm{LS}}$ is the spin-orbit coupling. For 3d ions $\mathcal{H}_{\mathrm{ee}}$ and $\mathcal{H}_{\mathrm{CF}}$ are comparable and larger than spin-orbit coupling. It is customary to neglect in first approximation the spin-orbit coupling which is introduced later as a perturbation. Mn^{3+} has a $(3\mathrm{d})^4$ valence-electron configuration which in octahedral symmetry yields a $^5\mathrm{E_g}$ ground state. This is unstable to phonon coupling (Jahn-Teller theorem), which lowers the symmetry to D_{4h}, namely to a tetragonally elongated coordination. The ground state $^5\mathrm{A_{1g}}$, in zero-order approximation, is five-fold degenerate (no orbital degeneracy and $S = 2$ spin multiplet). The spin-orbit *perturbation* yields no contribution in the first order, but to the second order it admixes excited states with the ground state. This removes the degeneracy of the spin multiplet and produces anisotropic components in the g tensor. Its effect is usually summarized introducing an effective single-ion spin Hamiltonian of the form:

$$\mathcal{H} = -D\hat{S}_z^2 - \mu_{\mathrm{B}}\boldsymbol{B}\mathsf{g}\hat{\boldsymbol{S}}, \tag{8.7}$$

where g is a symmetric tensor. The first term is responsible for the zero-field splitting of the $2S + 1$ levels. It is often referred to as crystal-field term, even though this is misleading because it is not the crystal field which splits the levels but rather the spin-orbit coupling. The spin-Hamiltonian parameters are determined by the spin-orbit coupling constant λ and by the degree of mixing between the $^5A_{1g}$ ground state and the excited states induced by $\mathcal{H}_{LS}$ [44]. The lowering of the symmetry produces axially symmetric g and D tensors[3] whose components are related, to the leading order, through the following formula:

$$D = D_z - D_{x,y} = \lambda(g_{x,y} - g_z) = \lambda \Delta g. \tag{8.8}$$

8.4 Realistic Spin Hamiltonians for Single-Chain Magnets

So far we have neglected the coupling among spin pairs, which can be written as

$$\mathcal{H}_{\text{exch}} = -\hat{\boldsymbol{S}}_p \mathsf{J} \hat{\boldsymbol{S}}_k, \tag{8.9}$$

where $\hat{\boldsymbol{S}}_p$ and $\hat{\boldsymbol{S}}_k$ are effective spin operators of any two interacting magnetic bricks and J is a generic 3-by-3 matrix. Limiting ourself to intrachain spin-spin coupling, we can neglect the contribution due to dipolar interaction which is typically much smaller than the exchange one. When pair-spin interaction involves transition metals whose ground state is not orbitally degenerate, the isotropic contribution to the J tensor dominates. As mentioned before, second-order perturbation theory prescribes that the ground-state wave functions be modified because of the admixing with excited states mediated by spin-orbit coupling. When the corrected wave functions of the bricks p and k are employed to compute the exchange integral, the anisotropic and antisymmetric contributions emerge. The former is proportional to $(\Delta g/g)^2$, while the latter is proportional to $\Delta g/g$. This ratio is usually much smaller than one, thus the antisymmetric term—if allowed by symmetry—is expected to dominate with respect to the anisotropic exchange. For our purposes, it will be enough to know that anisotropic contributions to J can be neglected when the g tensor is fairly isotropic, as for Mn^{2+}, Mn^{3+}, high-spin Fe^{3+}, etc.

When magnetic bricks comprise transition metals with orbitally degenerate ground state, predicting the properties of the g, D and J tensors on simple footing becomes extremely complicated [45]. One possibility is that of considering just a symmetric exchange tensor, which is then diagonal on a proper basis with principal values J_x, J_y and J_z. If compatible with symmetry, an antisymmetric, Dzyaloshinskii-Moriya term may be added independently.

In 1D magnetic systems realized by coupling radicals with neighbouring transition-metal ions, anisotropic terms in J may originate only from the metal atoms. The first successful examples consisted in ferrimagnetic chains of general

[3] Without loss of generality g and D can be assumed symmetric. Consequently, they are diagonal on a proper reference frame with eigenvalues g_x, g_y, g_z and D_x, D_y, D_z. The same notation will be used for the J tensor.

formula $Mn(hfac)_2NITR$ [17]. The radical is isotropic and so is Mn^{2+}, therefore J is expected to be proportional to the identity. Indeed, these systems represented text-book examples of 1D Heisenberg ferrimagnets described by the Hamiltonian

$$\mathcal{H}_{\text{Mn-rad}} = -J \sum_{p=1}^{N/2} \hat{S}_{2p} \cdot (\hat{s}_{2p+1} + \hat{s}_{2p-1}), \tag{8.10}$$

where $\hat{S}_{2p}$ stand for Mn^{2+} spin operators (lying at even sites $2p$ with $S_{2p} = 5/2$), while $\hat{s}_{2p+1}$ are the radical spin-one-half operators. J is negative and tends to orient the nearest-neighbouring spins antiparallel to each other. The temperature dependence of the static susceptibility was fitted using the Seiden model [46] with $|J|$ in the range 300–475 K depending on the substituent R on the radical[4] [47]. In the Seiden model the Mn spins are replaced by classical vectors, which—in the absence of field and single-ion anisotropy—makes the model analytically solvable. Due to the large value of the coupling between Mn^{2+} and NITR radicals, strong pair-spin correlations develop, which is highlighted by a divergence of the correlation length at low temperature (proportional to $|J|/T$ and not exponential like in spin chains with uniaxial anisotropy [47]). In the presence of such strong intrachain correlations even a tiny interchain interaction J' may induce 3D ordering [15]. In $Mn(hfac)_2(NITiPr)$ this happens at $T_C = 7.6$ K [17]. ESR and NMR studies provided evidence of spin-diffusion effect allowing for an estimate of the ratio between inter- and intrachain exchange interaction of the order $|J'/J| = 2 \times 10^{-6}$ [16].

This example confirms that combining transition metals with organic radicals is a powerful strategy for designing ideal 1D systems. An additional ingredient is needed to realize a SCM: magnetic anisotropy. This may easily be introduced by replacing Mn^{2+} with Mn^{3+}. Recently, the observation of slow relaxation consistent with SCM features was reported for ferrimagnetic spin chains consisting of Mn^{3+} and TCNQ or TCNE[5] organic radicals [9, 48–50]. The relatively large multiplicity of Mn^{3+} spins, $S = 2$, allows justifying their replacement by classical vectors. Thus, the Seiden model is still a good starting point for describing the magnetic properties of these systems, provided that single-ion-anisotropy terms are added. Even if the modelling aspects are well-defined, the rationalization of Mn^{3+}-radical SCMs is complicated by the fact that $J \gg D$, meaning that the relevant excitations are broad DWs.

The extreme anisotropic g tensor obtained for Co^{2+} in a tetragonally compressed symmetry suggests that its coupling with NITR be, to leading order, of the Ising type. This idea led to the synthesis of the first compound showing SCM behaviour: $Co(hfac)_2(NITPhOMe)$ [3]. Experimental results pertaining slow dynamics have

---

[4] Henceforth, energies will be expressed in Kelvin units to make it easier to compare them with thermal energy. The conversion factor to SI coincides with the Boltzmann constant k : 1 K = $1.3806503 \times 10^{-23}$ J.

[5] Acronyms stand for tetracyanoquinodimethane (TCNQ) and tetracyanoethylene (TCNE).

shown a substantial agreement with the kinetic version of the Ising model developed by Glauber [51]. Unfortunately, up to date, the static properties have not been successfully modelled yet. The first reason is that above 40 K treating Co^{2+} as an effective $S = 1/2$ is not legitimate (the energy separation between the ground-state Kramers doublet and the excited multiplets is about 100 K). A second reason relates to the helical structure of this compound, because of which the elementary magnetic cell contains 3 Co^{2+} and 3 radical spins. Apart from the question of reproducing its static properties, it is instructive to give a closer look at the Hamiltonian of this system to show how non-collinearity can be modelled in general. For temperatures lower than 40 K, a reasonable Hamiltonian for the $Co(hfac)_2(NITPhOMe)$ chain is given by

$$\mathcal{H}_{\text{Co-rad}} = -\sum_{p=1}^{N/(2N_r)} \sum_{r=1}^{N_r} \big[\hat{\boldsymbol{S}}_{p,2r} \mathsf{J}_{2r} (\hat{\boldsymbol{s}}_{p,2r+1} + \hat{\boldsymbol{s}}_{p,2r-1}) + \mu_{\text{B}}(\boldsymbol{B} \mathsf{g}_{2r} \hat{\boldsymbol{S}}_{p,2r} + g\boldsymbol{B} \cdot \hat{\boldsymbol{s}}_{p,2r+1})\big], \tag{8.11}$$

where both $\hat{\boldsymbol{S}}_{p,2r}$ and $\hat{\boldsymbol{s}}_{p,2r+1}$ are spin one-half operators associated with Co^{2+} ions and radicals, respectively. p represents the magnetic cell index while r spans the inequivalent Co^{2+} atoms inside each cell (with $\hat{\boldsymbol{s}}_{p,2N_r+1} = \hat{\boldsymbol{s}}_{p+1,1}$). For the specific case, r takes $N_r = 3$ different values which correspond to different orientations of the principal axes along which the J and g tensors are diagonal. If spin projections are expressed in the crystal frame, the tensors appearing in Hamiltonian (8.11) are built applying a standard $O(3)$ rotation to the diagonal tensors [43, 55, 56]. Formally, r in J_{2r} and g_{2r} labels different sets of rotation angles. The Landé factor of the radical is isotropic and thus independent of r.

When spins $S > 1/2$ are considered, a magnetic brick may possess some single-ion anisotropy, which implies that also the D tensor needs to be rotated in non-collinear systems.

The thermodynamic properties of classical spin chains with nearest-neighbour interactions can be efficiently computed by means of the transfer-matrix method. Letting the general Hamiltonian be $\mathcal{H} = -kT \sum_p V(\boldsymbol{S}_p, \boldsymbol{S}_{p+1})$, the partition function $\mathcal{Z}$ is obtained integrating over all the possible directions along which each unitary vector $\boldsymbol{S}_p$ may point:

$$\mathcal{Z} = \int d\Omega_1 \int d\Omega_2 \cdots \int \mathrm{e}^{V(\boldsymbol{S}_1,\boldsymbol{S}_2)} \mathrm{e}^{V(\boldsymbol{S}_2,\boldsymbol{S}_3)} \cdots \mathrm{e}^{V(\boldsymbol{S}_N,\boldsymbol{S}_1)} d\Omega_N. \tag{8.12}$$

Defining the transfer kernel as $\mathcal{K}(\boldsymbol{S}_p, \boldsymbol{S}_{p+1}) = \mathrm{e}^{V(\boldsymbol{S}_p,\boldsymbol{S}_{p+1})}$ and assuming periodic boundary conditions, the partition function $\mathcal{Z}$ can be recasted into the trace of the N-th power of $\mathcal{K}(\boldsymbol{S}_p, \boldsymbol{S}_{p+1})$:

$$\mathcal{Z} = \int d\Omega_1 \int d\Omega_2 \cdots \int \mathcal{K}(\boldsymbol{S}_1, \boldsymbol{S}_2)\mathcal{K}(\boldsymbol{S}_2, \boldsymbol{S}_3) \cdots \mathcal{K}(\boldsymbol{S}_N, \boldsymbol{S}_1) d\Omega_N = \mathrm{Tr}\{\mathcal{K}^{\mathrm{N}}\}. \tag{8.13}$$

When the transfer kernel is expressed on a basis of eigenfunctions, the partition function reduces to a sum of eigenvalues $\mathcal{Z} = \sum_m \lambda_m^N$, where $\psi_m(\boldsymbol{S}_p)$ and λ_m are solutions of the following eigenvalue problem:

$$\int \mathcal{K}(\boldsymbol{S}_p, \boldsymbol{S}_{p+1})\psi_m(\boldsymbol{S}_{p+1})d\Omega_{p+1} = \lambda_m \psi_m(\boldsymbol{S}_p). \tag{8.14}$$

For kernels that can be written in a symmetric form with respect to the exchange $\boldsymbol{S}_p \leftrightarrow \boldsymbol{S}_{p+1}$ the spectral theorem warrants that eigenvalues are real.[6] They are also positive, because the transfer kernel is a positive function of $\boldsymbol{S}_p$ and $\boldsymbol{S}_{p+1}$, and upper bounded so that they can be ordered from the largest to the smallest one: $\lambda_0 > \lambda_1 > \lambda_2 > \cdots$. In the thermodynamic limit the asymptotic behaviour of the partition function (8.13) is dominated by the largest eigenvalue λ_0, $\mathcal{Z} \simeq \lambda_0^N$, meaning that the free energy per spin is given by $f = -kT \log \lambda_0$. Macroscopic experimental observables are obtained as derivatives of f, but this method allows computing microscopic averages as well. Apart from some fortunate cases [25], (8.14) needs to be solved numerically by sampling the unitary sphere with a finite number of special points. This number can be increased dynamically untill the desired precision is reached. Even though it may not be transparent from our description, a new eigenvalue problem ought to be solved for any computed temperature or applied field. Referring the reader to the existing literature for implementation details [52–55], we remark that the transfer-matrix method allows computing the magnetic properties of classical spin chains more efficiently than, e.g., standard Metropolis Monte Carlo. This makes it possible to fit spin Hamiltonian parameters directly to experimental data sets. The major drawback is that the number of spin variables that appear in the kernel scales like the range of interaction (2 for nearest-neighbour, 4 for second nearest-neighbour interaction, etc.), which finally affects the complexity of the eigenvalue problem in (8.14).

The transfer-matrix method can easily be extended to models in which classical and quantum spins alternate, like in the Seiden model [46]. Noting that the quantum-spin operators are not directly coupled with each other, one can integrate out their degrees of freedom independently. In fact, a generic quantum spin located at site $2p+1$ experiences an effective "field" $kT\boldsymbol{h}_{2p,2p+2} = J(\boldsymbol{S}_{2p} + \boldsymbol{S}_{2p+2}) + \mu_B g \boldsymbol{B}$. The corresponding energy levels are $\pm kT|\boldsymbol{h}_{2p,2p+2}|$, which depend *parametrically* on the orientation of the two classical spins, $\boldsymbol{S}_{2p}$ and $\boldsymbol{S}_{2p+2}$. After tracing over the quantum degrees of freedom, one is left with the kernel

$$\mathcal{K}(\boldsymbol{S}_{2p}, \boldsymbol{S}_{2p+2}) = 2\cosh(|\boldsymbol{h}_{2p,2p+2}|)\exp\left(\frac{\mu_B \boldsymbol{B} g \boldsymbol{S}_{2p}}{kT}\right)\exp\left(\frac{D(S_{2p}^z)^2}{kT}\right) \tag{8.15}$$

where the single-ion anisotropy and Zeeman term acting on the classical spins have been added. The kernel (8.15) may be used to compute, e.g., the equilibrium suceptibility of Mn^{3+}-radical chains [48–50]. To the aim of sketching how to proceed for

[6]In the general, non-symmetric case left and right eigenfunctions of $\mathcal{K}(\boldsymbol{S}_p, \boldsymbol{S}_{p+1})$ have to be considered, but the basic ideas of the transfer-matrix method remain the same.

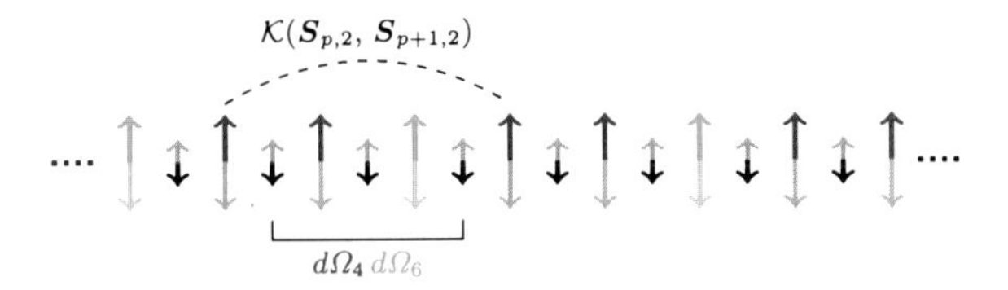

Fig. 8.5 Sketch of the periodicity associated with Hamiltonian (8.11): *small, black arrows* represent radical spins $\hat{s}_{p,2r+1}$, while the *large coloured arrows* represent the metal-ion ones $\hat{S}_{p,2r}$. The kernel (8.16) depends only on $S_{p,2}$ and $S_{p+1,2}$ (*blue arrows*) because an integration over the internal degrees of freedom $S_{p,4}$ (*red arrows*) and $S_{p,6}$ (*green arrows*) has been performed

modelling non-collinearity, let us substitute the spins $\hat{S}_{p,2r}$ in Hamiltonian (8.11) by classical vectors.[7] Even after integrating out the radical degrees of freedom, the are still 3 *non-equivalent* classical spins in each magnetic unit cell, resulting in 3 different kernels if $\boldsymbol{B}$ is applied along a generic direction: $\mathcal{K}(S_{p,2}, S_{p,4})$, $\mathcal{K}(S_{p,4}, S_{p,6})$ and $\mathcal{K}(S_{p,6}, S_{p+1,2})$. The role of the kernel (8.15) is played by

$$\mathcal{K}(S_{p,2}, S_{p+1,2}) = \int d\Omega_4 \int \mathcal{K}(S_{p,2}, S_{p,4})\mathcal{K}(S_{p,4}, S_{p,6})\mathcal{K}(S_{p,6}, S_{p+1,2})d\Omega_6, \tag{8.16}$$

obtained by tracing over the degrees of freedom internal to the considered cell, $d\Omega_4$ and $d\Omega_6$.[8] The way in which the kernel is built is sketched pictorially in Fig. 8.5. In the thermodynamics limit the partition function is given by $\mathcal{Z} \simeq \lambda_0^{N/(2N_r)}$, where the number of spins have been replaced by the number of unit cells $N/(2N_r)$.

Due to non-collinearity, a strong anisotropy in the D, g or J tensors may not necessarily be evident at the macroscopic level [56]. More concretely, having similar saturation values for the magnetisation along different crystallographic directions may still be compatible with a strong uniaxial character at the level of individual bricks. Non-collinearity is also consistent with an inversion of the directions of easy and hard magnetisation by increasing temperature [55] or with the vanishing of the correlation length for some specific applied fields [57].

In passing, we note that finite-size effects can be taken into account in the general transfer-matrix framework [52] as well as interchain interactions if treated at the mean-field level [15, 58, 59].

In the cases in which one of the principal values of the D or J tensors is much larger than the other two (say $J_z \gg J_x, J_y$), spin operators can be substituted by two-valued classical variables $\sigma_p = \pm 1$. In this way, the problem reduces to the

---

[7] Even though this is not justified for the specific case of Co^{2+}, the classical approximation allows us to discuss the general formalism.

[8] Actually, the choice of the unit cell is not unique: one might integrate over any pair of internal degrees of freedom $d\Omega_{2r}$. This turns necessary in order to compute microscopic averages of individual spin components.

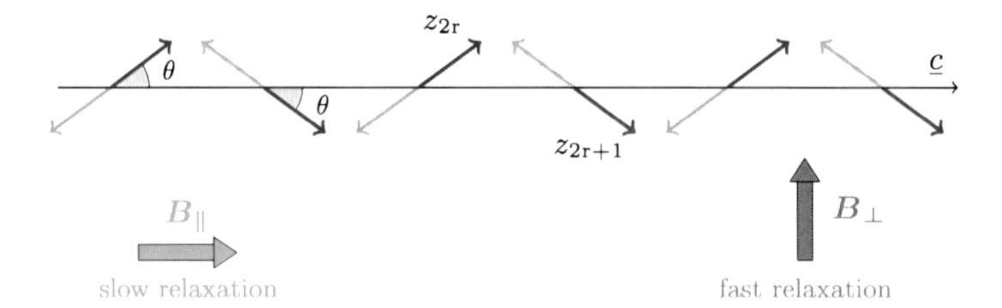

Fig. 8.6 Sketch of a two-fold non-collinear Ising chain. The local anisotropy axes have been chosen coplanar for simplicity and form an angle θ with the chain axis $\underline{c}$. For $\theta < \pi/4$ and $J > 0$, slow relaxation is expected only when $\boldsymbol{B}$ is applied parallel to the chain axis

Ising Hamiltonian

$$\mathcal{H} = -J \sum_{p=1}^{N} \sigma_p \sigma_{p+1} - \mu_B B \sum_{p=1}^{N} g_p \sigma_p, \tag{8.17}$$

in which J and g_p may contain information about non-collinearity. In Fig. 8.6 a sketch of a two-fold, non-collinear Ising chain is shown. Assuming that both the g and J tensors have only one non-zero component along their principal axes, the corresponding parameters in Hamiltonian (8.17) are given by $J = \cos(2\theta) J_z$ and $g_p = \cos(\theta) g_z$ if $\boldsymbol{B}$ is parallel to the chain axis ($\underline{c}$ axis in Fig. 8.6) while $g_p = (-1)^p \sin(\theta) g_z$ if $\boldsymbol{B}$ is perpendicular to the chain.[9] The framework in which static properties of SCMs can be modelled seems to be well-defined. However, it should not be forgotten that the genuine 1D static behaviour can be accessed only above a certain temperature T_b, dependent on the specific experiment, below which slow dynamics starts playing a major role. Moreover, finite-size effects or 3D inter-chain interactions may come into play at higher temperature than T_b [14]. On the high-temperature side, distinctive 1D features (short-range order) smear out in the isotropic paramagnetic phase. All these phenomena set limitations to the applicability of any equilibrium 1D model.

8.5 Glauber Model and Single-Chain Magnets

In this section we will assume the anisotropy at the brick level to be large enough that Hamiltonian (8.17) suffices to discuss the important features of slow dynamics. A kinetic version of the Ising model was proposed by J.R. Glauber in 1963 [51]. As based on stochastic dynamics, this model relates to coarse-grained dynamics, typically some orders of magnitude longer than a Larmor period.

Following Glauber, let $P(\underline{\sigma}, t)$ be the probability of occurrence of some configuration $\underline{\sigma} = \sigma_1, \ldots, \sigma_p, \ldots, \sigma_N$ at time t and $w_{\sigma_p \to -\sigma_p}$ the probability of reversing

[9] The reader is addressed to Ref. [43, 45] for a more rigorous treatment.

the p-th spin per unit time. The master equation of the problem, thus, reads

$$\frac{\mathrm{d}}{\mathrm{d}t}P(\underline{\sigma},t) = -\sum_{p=1}^{N} w_{\sigma_p\to-\sigma_p} P(\sigma_1,\ldots,\sigma_p,\ldots,\sigma_N,t) + \sum_{p=1}^{N} w_{-\sigma_p\to\sigma_p} P(\sigma_1,\ldots,-\sigma_p,\ldots,\sigma_N,t). \tag{8.18}$$

To model the magnetisation dynamics and a.c. susceptibility it is not necessary to solve (8.18): one can limit oneself to single-spin averages $s_p(t)$

$$s_p(t) = \sum_{\{\underline{\sigma}\}} \sigma_p P(\underline{\sigma},t). \tag{8.19}$$

It can be shown that spin averages fulfill the differential equation

$$\frac{\mathrm{d}s_p}{\mathrm{d}t} = -2\langle \sigma_p w_{\sigma_p\to-\sigma_p}\rangle, \tag{8.20}$$

where $\langle\ldots\rangle$ denotes, again, the time-dependent average performed by means of $P(\underline{\sigma},t)$. In order that dynamics drives the system towards Boltzmann equilibrium, the detailed-balance condition shall hold

$$\frac{P_{\mathrm{eq}}(\sigma_1,\ldots,-\sigma_p,\ldots,\sigma_N)}{P_{\mathrm{eq}}(\sigma_1,\ldots,\sigma_p,\ldots,\sigma_N)} = \frac{w_{\sigma_p\to-\sigma_p}}{w_{-\sigma_p\to\sigma_p}}. \tag{8.21}$$

The equilibrium probabilities on the left-hand side of (8.21) are obtained reversing the p-th spin while leaving the other $(N-1)$ unchanged. Their ratio can be written as follows

$$\frac{P_{\mathrm{eq}}(\sigma_1,\ldots,-\sigma_p,\ldots,\sigma_N)}{P_{\mathrm{eq}}(\sigma_1,\ldots,\sigma_p,\ldots,\sigma_N)} = \frac{\exp[-\kappa\sigma_p(\sigma_{p-1}+\sigma_{p+1})]\exp(-h_p\sigma_p)}{\exp[\kappa\sigma_p(\sigma_{p-1}+\sigma_{p+1})]\exp(h_p\sigma_p)}$$
$$= \frac{[1-\frac{1}{2}\sigma_p(\sigma_{p-1}+\sigma_{p+1})\tanh(2\kappa)][1-\sigma_p\tanh(h_p)]}{[1+\frac{1}{2}\sigma_p(\sigma_{p-1}+\sigma_{p+1})\tanh(2\kappa)][1+\sigma_p\tanh(h_p)]} \tag{8.22}$$

with $\kappa = J/kT$ and $h_p = \mu_{\mathrm{B}}Bg_p/kT$. The above relation suggests the following form for the transition probability:

$$w_{\sigma_p\to-\sigma_p} = \frac{1}{2}\alpha\left[1-\frac{1}{2}\gamma\sigma_p(\sigma_{p-1}+\sigma_{p+1})\right]\left[1-\sigma_p\tanh(h_p)\right], \tag{8.23}$$

with $\gamma = \tanh(2\kappa)$, so that detailed balance (8.21) is automatically fulfilled. Equation (8.23) corresponds to Glauber's original choice; other transition probabilities fulfilling (8.21) could be chosen [60, 61], but—to our knowledge—they have not been considered in the context of SCMs. Note that the parameter α entering (8.23) sets the natural time unit of the model. It can be interpreted as the attempt frequency of an isolated spin, i.e., the transition probability for vanishing exchange coupling, $J = 0$. Already Suzuki and Kubo commented that, in general, α should depend on temperature [62]. We will come back to this important point further on. Combining (8.20) and (8.23), a set of differential equations for spin averages is obtained

$$\frac{1}{\alpha}\frac{\mathrm{d}s_p}{\mathrm{d}t} = -\left[s_p - \frac{\gamma}{2}(s_{p-1} + s_{p+1})\right] + \left[1 - \frac{\gamma}{2}(r_{p-1,p} + r_{p,p+1})\right]\tanh(h_p), \quad (8.24)$$

where $r_{p,l} = \langle\sigma_p\sigma_l\rangle$. This means that the knowledge of pair-spin correlations is needed to solve (8.24). In turn, the knowledge of three-spin correlations is needed to obtain $r_{p,l}$ and so on. In other words, (8.24) is the first one of an infinite hierarchy of kinetic equations [51, 63]. A judicious truncation of this series is, therefore, required in order to get analytic results which could easily be compared with experiments. In the following, we will analyse different decoupling schemes, related to different physically relevant situations. Another crucial point concerns the choice of boundary conditions for the system (8.24). A realistic SCM consists of a collection of open arrays of spins coupled via the exchange interaction. The length distribution of these arrays is determined by the spatial distribution of defects in a sample. In this sense, open boundary conditions give a more accurate description of SCM dynamics than periodic boundary conditions. However, we start considering periodic boundary conditions because calculations are less involved but still provide insight into the essential features which are not affected by the presence of defects.

When no external field is applied, $h_p = 0$, the dependence on $r_{p,l}$ disappears from (8.24) that then reduces to a linear system of first-order differential equations. The corresponding eigenvalue problem involves a circulant matrix and is diagonalized by a discrete Fourier transform. A general solution takes the form $s_p = \sum_q \tilde{s}_q \mathrm{e}^{iqp}\mathrm{e}^{-\lambda_q t}$, with

$$\lambda_q = \alpha(1 - \gamma\cos q) \quad (8.25)$$

and $q = 0, 2\pi/N, \ldots, 2\pi(N-1)/N$ set by periodic boundary conditions. The initial configuration determines, instead, the Fourier amplitudes. If the system is prepared into a ferromagnetic saturated state with $s_p = 1$ for every p, the only nonzero Fourier component corresponds to $q = 0$, that is $\tilde{s}_0 = 1/N$. Accordingly, the magnetisation is expected to follow a mono-exponential relaxation with a characteristic time scale $\tau = 1/[\alpha(1-\gamma)]$. For ferromagnetic coupling, $J > 0$, τ diverges exponentially at low temperature like e^{4K}. Because of this divergence, some ferromagnetic ordering may persist over macroscopic time scales in the absence of applied field. In this sense, the work of Glauber has foresaw what would be observed in SCMs about forty years later. The realization of these systems gave the opportunity to generalize the original Glauber model to include features of realistic SCMs and specific experiments.

The response to a tiny a.c. field $B = B'\mathrm{e}^{-i\omega t}$ is modelled by linearising the hyperbolic tangent in (8.24). But this does not eliminate the dependence on pair-spin correlations. Already Glauber circumvented this problem replacing $r_{p-1,p}$ and $r_{p,p+1}$

by their equilibrium average, equal to $\tanh(\kappa)$ [51]. Limiting himself to equivalent spins ($g_p = g$ independent of the site), he predicted an a.c. susceptibility of the form (8.5), provided that χ_{eq} was taken as the static susceptibility of the Ising model and $\tau = 1/[\alpha(1-\gamma)]$. As mentioned in the previous section, non-collinearity among local anisotropy axes is more the rule rather than an exception. This affects Hamiltonian (8.17) through the site-dependent Landé factor. The spatial periodicity of g_p defines the magnetic unit cell. It is worth remarking that the periodicity of g_p generally depends on the direction along which the magnetic field is applied. The simplest case of a two-fold non-collinear Ising chain is sketched in Fig. 8.6. With relatively small effort, an analytic formula for the a.c. susceptibility can be derived, which accounts for non-collinearity or non-equivalence of magnetic centres [64]. For $\omega \ll \alpha$, a resonant behaviour, i.e. a frequency-dependent peak in $\chi(\omega)$, is expected only when the field is applied along specific crystallographic directions (e.g., the $\underline{c}$ axis in Fig. 8.6). In particular, those directions are the ones along which the ground-state magnetisation is uncompensated. This prediction for the dynamic response of non-collinear Ising chains was indeed supported by experiment [56, 64, 65].

For several years the truncation schemes summarized above had represented the starting point for generalizations of the Glauber model which aimed at giving a better account for the characteristics of real SCMs. Then, the restriction to zero-field a.c. susceptibility prevented from modelling the dependence of relaxation time on static applied field. A breakthrough was represented by the work of Coulon and co-workers [66] who actualized the local-equilibrium approximation for pair-spin correlations proposed by Huang in the seventies [63]. Let us start from refreshing the main ideas of local-equilibrium approximation for the case of periodic boundary conditions, as treated by Huang. Equivalent magnetic moments, $g_p = g$, coupled ferromagnetically will be assumed. With these hypotheses, the single-spin averages are independent of the site at thermodynamic equilibrium and read

$$m = \langle \sigma_p \rangle_{eq} = \frac{\sinh(h)}{\Delta^{1/2}} \quad \text{with } \Delta = e^{-4\kappa} + \sinh^2(h), \tag{8.26}$$

where $\langle \ldots \rangle_{eq}$ stands for equilibrium average, given by Boltzmann statistics. Translation invariance holds also for nearest-neighbour pair-spin correlations, of our interest, which are given by

$$\Gamma = \langle \sigma_{p+1} \sigma_p \rangle_{eq} = \frac{\sinh^2(h)}{\Delta} + \frac{e^{-4\kappa}[\cosh(h) - \Delta^{1/2}]}{\Delta[\cosh(h) + \Delta^{1/2}]}. \tag{8.27}$$

By means of (8.26), $\sinh(h)$ and $\cosh(h)$ appearing in the nearest-neighbour correlation can be expressed in terms of m and $e^{-4\kappa}$, which yields

$$\Gamma = 1 - \frac{2(1-m^2)}{1 + \sqrt{m^2 + (1-m^2)e^{4\kappa}}}. \tag{8.28}$$

In two physically relevant situations translational invariance may be assumed for time-dependent spin averages, s_p, as well. The first one corresponds to having equal initial conditions for all spins: $s_p(0) = \mu$, with $-1 \leq \mu \leq 1$. Since $g_p = g$ has been assumed, this initial condition is simply realized when a magnetic field (possibly zero) has been switched on far in the past ($t \to -\infty$) and changed to some different value at time $t = 0$. The second situation is a typical a.c. susceptibility experiment, for which only the stationary response to a tiny external drift is relevant. In these two cases, time-dependent spin averages become independent of the site and the label p can be dropped from the variables s_p in (8.24). The local-equilibrium approximation consists in assuming that (8.28), which establishes a closed relation between equilibrium spin averages and nearest-neighbour correlations, holds true for time-dependent averages as well, namely out of equilibrium. Equation (8.24), thus, simplifies as

$$\frac{1}{\alpha}\frac{\mathrm{d}s}{\mathrm{d}t} = -(1-\gamma)s + (1-\gamma\Gamma)\tanh(h), \tag{8.29}$$

where Γ is given by (8.28) with m is replaced by s (time-dependent average). Within the Glauber model, the local-equilibrium approximation is nothing but a trick to truncate the hierarchy of kinetic equations. The resulting equations of motion are generally non-linear, the non-linearity arising from $\Gamma[s]$. Fortunately enough, equation (8.29) can be solved analytically [63]. More importantly, for $t \to \infty$ the exact steady-state solution is recovered. For instance, a mean-field truncation scheme might alternatively be assumed, setting $\Gamma = s^2$, but this would not reproduce the exact steady-state solution. Note that local-equilibrium approximation does not require small applied fields. For what concerns SCMs, much interest relates to the study of linear departures from equilibrium. Following Pini and co-workers [67], let us split the field into a static contribution of any intensity ($h_0 = \mu_B B_0 g/kT$) plus an oscillating field of much smaller intensity B' and with frequency ω: $h = h_0 + h'\mathrm{e}^{-i\omega t}$. As a consequence, s is expected to deviate slightly from its equilibrium value, $m(T, B_0)$, and (8.29) can be linearised as follows:

$$\frac{1}{\alpha}\frac{\mathrm{d}\delta s}{\mathrm{d}t} = -\big(1-\gamma+2\gamma\tanh^2(h_0)\big)\delta s + \big(1-\tanh^2(h_0)\big)h'\mathrm{e}^{-i\omega t}, \tag{8.30}$$

where $\delta s = s(t) - m$ and the fact that $\Gamma[s] \approx \Gamma[m] + (d\Gamma/dm)_{\mathrm{eq}}\delta s$ with $(d\Gamma/dm)_{\mathrm{eq}} = 2\tanh(h_0)$ has been used. The stationary behaviour is obtained inserting the trial solution $\delta s = \widetilde{\delta s}\mathrm{e}^{-i\omega t}$ in (8.30), which yields the a.c. susceptibility. The resulting formula is equivalent to (8.5) and χ_{eq} is the susceptibility that would be obtained by differentiating m in (8.26) with respect to B. This matching is a direct consequence of the fact that local-equilibrium approximation provides the exact steady-state solution for s. The relaxation time is, instead, given by the inverse of the prefactor of δs in (8.30):

$$\tau = \frac{1}{\alpha(1-\gamma+2\gamma\tanh^2(h_0))}. \tag{8.31}$$

The Glauber relaxation time is recovered in the limit $h_0 \to 0$ and, as already pointed out, diverges exponentially upon lowering temperature. Note that the net effect of a static field is that of removing such a divergence, though the dependence of the relaxation time on B_0 is much less dramatic than on temperature.

The Glauber model was extended to weakly interacting spin chains by Žumer [68]. Similarly to Scalapino [15], he treated the interchain interaction as a mean field, limiting his analysis to the critical region around the transition to a 3D ordered phase. Equation (8.31) may allow generalizing Žumer's results to lower temperatures, away from the critical region. A joint theoretical and experimental investigation of this phenomenon would provide important information on the critical behaviour of SCMs [14]. A realistic model should, however, take into account finite-size effects induced by the presence of defects.

8.6 Glauber Model for Finite Chains

Though it may sound somewhat technical, the study of finite-size effects have played a central role in theoretical and experimental characterization of SCMs. As a first step, open boundary conditions need to be considered, which makes the transition probability of extremal spins take the form

$$\begin{aligned} w_{\sigma_1 \to -\sigma_1} &= \frac{1}{2}\alpha[1 - \eta\sigma_1\sigma_2]\big[1 - \sigma_1 \tanh(h_1)\big] \\ w_{\sigma_N \to -\sigma_N} &= \frac{1}{2}\alpha[1 - \eta\sigma_N\sigma_{N-1}]\big[1 - \sigma_N \tanh(h_N)\big], \end{aligned} \tag{8.32}$$

with $\eta = \tanh(\kappa)$ (and $\kappa = J/kT$), obtained again from the detailed-balance condition. The kinetic equations for spin located at boundaries are modified accordingly:

$$\begin{aligned} \frac{1}{\alpha}\frac{\mathrm{d}s_1}{\mathrm{d}t} &= -(s_1 - \eta s_2) + (1 - \eta r_{1,2}) \tanh(h_1) \\ \frac{1}{\alpha}\frac{\mathrm{d}s_N}{\mathrm{d}t} &= -(s_N - \eta s_{N-1}) + (1 - \eta r_{N-1,N}) \tanh(h_N). \end{aligned} \tag{8.33}$$

In the absence of external field, the characteristic time scales can be deduced by inserting the trial solution $s_p = (\mathcal{A}_p \mathrm{e}^{ipq} + \mathcal{A}_r \mathrm{e}^{-ipq})\mathrm{e}^{-\lambda_q t}$ into system (8.24) that still holds for bulk spins, with labels $2 \leq p \leq N-1$. The relation between λ_q and q remains the same as in (8.25) but the values taken by q are different from the case of periodic boundary conditions. Due to the loss of translation invariance, both amplitudes $\mathcal{A}_p$ and $\mathcal{A}_r$ must be considered. A pair of equations for these amplitudes are obtained inserting the trial solution into (8.33), with λ_q given by (8.25). For $B = 0$, this is a homogeneous system that only admits the trivial solution $\mathcal{A}_p = \mathcal{A}_r = 0$ unless the determinant of the coefficients of $\mathcal{A}_p$ and $\mathcal{A}_r$ is zero. By requiring this, the following implicit equation for the values of q is obtained [69, 70]:

$$\tan\big[(N-1)q\big] = -\frac{2\hat{\xi}\tan q}{1 - \hat{\xi}^2 \tan^2 q} \tag{8.34}$$

with $\hat{\xi} = \eta/(\gamma - \eta)$. The $q = 0$ solution has to be rejected because it is independent of N for every temperature, which is not physical. The remaining solutions will be labelled with ν, i.e., λ_ν. For ferromagnetic coupling, $J > 0$, the eigenfrequency corresponding to the slowest time scale can be expanded for low temperatures to get [71]

$$\lambda_1 = \frac{2\alpha}{N-1} e^{-2\kappa} + \mathcal{O}\left(e^{-4\kappa}\right). \tag{8.35}$$

The previous expansion contributed significantly to understanding SCMs. From (8.35) one expects the slowest degrees of freedom of the system to equilibrate with a relaxation time $\tau_N \sim N e^{2\kappa}$. The fact that the energy barrier at the exponent is halved with respect to Glauber's result suggests that, at low temperature, relaxation is driven by nucleation of a DW from a boundary. At higher temperatures, the Glauber behaviour is recovered. This happens when the correlation length becomes significantly smaller than N and physics becomes independent of boundary conditions. Thus, in real systems, the relaxation time is expected to diverge like $e^{4\kappa}$ at high temperatures—when ξ is much smaller than the average distance among defects—and like $e^{2\kappa}$ at low temperatures. The experimental observation of such a crossover represented an important step in establishing that SCM behaviour could, indeed, be described properly in the framework of Glauber dynamics [4, 6]. When $\xi \gg N$, the first step of relaxation is analogous to the nucleation of a critical droplet to reverse the magnetisation in metallic nanowires or elongated nanoparticles [13]. Depending on geometrical characteristics of the sample, non-uniform magnetisation reversal may be favoured with respect to the standard Néel-Brown mechanism (uniform rotation). The latter is known to follow an Arrhenius law, $\tau \sim e^{\Delta_\tau / kT}$, with an energy barrier proportional to the sample volume. To the leading order, the temperature dependence is of the Arrhenius type also in the case of non-uniform magnetisation reversal, but Δ_τ typically does not depend on the sample size. This fact directly originates from the local character of DW excitations that serve as nuclei to initiate magnetisation reversal (relaxation), both in metallic nanowires and in SCMs at low temperature.

After being nucleated at one boundary, a DW may reach the other end of the chain with probability $\sim 1/N$ by performing an unbiased random walk [31]. This is at the origin of the dependence on N appearing in (8.35) and, consequently, in τ_N. When this is the main channel for relaxation, in real SCMs one would expect to observe a decrease of the pre-exponential factor of the relaxation time by increasing the number of defects (see (8.3)); the energy barrier of the Arrhenius law should, instead, remain constant: $\Delta_{\tau_N} = 2J$. This trend was qualitatively confirmed in $Co(hfac)_2(NITPhOMe)$ compounds in which part of the Co^{2+} ions were substituted, in different amounts, by non-magnetic Zn^{2+} atoms [10, 11]. The fact that the pre-exponential factor increases with the system size is typical of a sizeable time elapsed during DW propagation in the relaxation process. In passing, we note that the opposite trend, i.e., a decrease of the pre-exponential factor of relaxation time with increasing the system size, was predicted for nanowires in which magnetisation reversal is forced to initiate from the bulk (e.g., in toroidal samples or with enhanced anisotropy at the ends) [13]. In that case, the probability to nucleate a soliton-antisoliton pair increases with N and the reversal rate consequently.

The local-equilibrium approximation may also be used to decouple the hierarchy of Glauber equations when a finite field is applied to an open chain. It is convenient to linearise directly (8.24) with respect to $\delta s_p = s_p - m_p$. Note that the equilibrium values m_p are now site-dependent due to the lack of translation invariance. The kinetic equations for δs_p contain the variation of nearest-neighbour correlation functions $\delta r_{p-1,p}$ and $\delta r_{p,p+1}$. For a chain of N equivalent spins, with $g_p = g$, Matsubara and co-workers provided a set of analytic relations to express equilibrium correlations $\langle \sigma_p \sigma_{p+1} \rangle_{\text{eq}}$ as functions of single-spin averages of open chains of different length [72]. If one assumes that such relations still hold true out of equilibrium, pair-spin variations can be written in terms single-spin averages: $\delta r_{p,p+1} = A_{N,p} \delta s_p + B_{N,p} \delta s_{p+1}$ with $A_{N,p}$ and $B_{N,p}$ depending only on equilibrium quantities (the reader is addressed to Ref. [66] for details). With the same convention introduced in (8.30) the response to an a.c. field $B' \mathrm{e}^{-i\omega t}$ superimposed to a static field B_0 is described by a system of linear equations of the form:

$$\frac{\mathrm{d}\boldsymbol{\Sigma}}{\mathrm{d}t} = -M\,\boldsymbol{\Sigma} + \alpha\big(1 - \tanh^2(h_0)\big) h' \mathrm{e}^{-i\omega t}\, \boldsymbol{\Psi}, \tag{8.36}$$

where $\boldsymbol{\Sigma} = (\delta s_1, \ldots, \delta s_N)^{\mathrm{T}}$; the matrix M and the vector $\boldsymbol{\Psi}$ are only functions of equilibrium averages, model parameters, temperature and static field (explicit expressions can be found in Ref. [67]). Let $\boldsymbol{\phi}_\nu$ and λ_ν be the eigenvectors and eigenvalues of M, namely $M\boldsymbol{\phi}_\nu = \lambda_\nu \boldsymbol{\phi}_\nu$. The stationary solution of (8.36) then reads

$$\boldsymbol{\Sigma} = \alpha\big(1 - \tanh^2(h_0)\big) h' \mathrm{e}^{-i\omega t} \sum_\nu \frac{\boldsymbol{\Psi} \cdot \boldsymbol{\phi}_\nu}{\lambda_\nu} \frac{\boldsymbol{\phi}_\nu}{1 - i\omega\tau_\nu} \tag{8.37}$$

with $\tau_\nu = 1/\lambda_\nu$. The dynamic susceptibility is given by $\chi(\omega) = g\mu_{\mathrm{B}} \mathrm{e}^{i\omega t} \sum_p \delta s_p / B'$, where δs_p are the components of the $\boldsymbol{\Sigma}$ vector in (8.37). With respect to the case with periodic boundary conditions, the choice of a site-independent Landé factor does not yield an a.c. response dependent on a single relaxation time. The relative weight of different contributions labelled by ν shall depend on temperature and on the static field B_0. In practice, the matrix M can be diagonalized numerically for any values of B_0 and T. The size of this matrix, N by N, is set by the number of spins in the chain. Realistic values of N fall in the range 10–10^4, meaning that $\chi(\omega)$ can easily be computed with standard diagonalisation routines. Among other things, this allows checking whether a unique relaxation time is dominating the summation (8.37) and thus $\chi(\omega)$. When the distribution of defects in a SCM compound is not peaked, an average over all the possible lengths may be required to compare the theoretical susceptibility with experiments [10, 11, 67]. Analogously, for a comparison with experiments on powder samples an average over all the possible orientations of the applied field with respect to the easy axis might be needed [66]. Due to space limitations, we prefer not to enter the details of those averaging procedures but rather address to the existing literature.

The divergence of relaxation time upon lowering temperature can be interpreted as critical slowing down. The 1D Ising model displays a magnetic phase transition at zero temperature, meaning that the critical point is located at the origin of the

(T, B_0) plane, that is $T = 0$ and $B_0 = 0$. Since the divergence of the correlation length is hampered by defects, it is more appropriate to investigate the critical behaviour of SCMs with finite-size scaling. For $B_0 = 0$, Luscombe et al. noted that the ratio between the relaxation time of a finite chain, τ_N, and that of the infinite chain, τ originally obtained by Glauber, is a *universal* function of $x = N/\xi$, when both N, $\xi \gg 1$:

$$\frac{\tau_N}{\tau} = f(x) = \frac{1}{1 + (\frac{\omega(x)}{x})^2}, \tag{8.38}$$

where $\omega(x)$ is the smallest root of the transcendental equation $\omega \tan(\omega/2) = x$ [70]. By definition, $f(x)$ tends to one for $x \gg 1$; while for $x \ll 1$ it is $f(x) \simeq x/2$. The reader may easily verify this limit by using formulae $\tau \simeq e^{4\kappa}/2\alpha$, $\tau_N \simeq N e^{2\kappa}/2\alpha$ and $\xi \simeq e^{2\kappa}/2$, which hold for N, $\xi \gg 1$ (see (8.35) and (8.25)). More recently, Glauber dynamics of the open chain in presence of realistic fields was studied by Coulon and co-workers [66] who found

$$\frac{\tau_N(B_0 = 0)}{\tau_N(B_0 \neq 0)} = 1 + a^2 h_0^2; \tag{8.39}$$

remarkably, the constant on the right-hand side is given by $a = 2\xi f(x\sqrt{2/3})$, $f(x)$ being the scaling function defined in (8.38). For $x \gg 1$, the limit $a = 2\xi$ is recovered by expanding the hyperbolic tangent in (8.31) (remember that $f(x) \to 1$ in this limit). In the opposite limit, one has $a = \sqrt{2/3}N$ consistently with the work of Schwarz developed in the context of helix-coil transition of polypeptides [73]. The quadratic dependence on B_0 of the ratio of relaxation times in the vicinity of the critical point stated by (8.39) was confirmed by experiment: first, in SCMs made up of repeating trinuclear units, Mn^{3+}-Fe^{3+}-Mn^{3+} and Mn^{3+}-Ni^{2+}-Mn^{3+} [66], later in $Co(hfac)_2$(NITPhOMe) compounds [67]. As pointed out in Ref. [66], the quadratic dependence on B_0 is also expected for SMMs. In fact, when repeating units in a spin chain consist of SMM-like centres an additional dependence on temperature and on B_0 enters the Glauber model through the attempt frequency α. Thus, information about the 1D universality class is somehow contained in the scaling function $f(x)$ rather than in the quadratic take-off of $\tau_N(B_0 \neq 0)$ as a function of the applied field.

In summary, the Glauber model prescribes that precise relations among characteristic energy scales shall hold for a text-book SCM. Recalling (8.3) and (8.4), the barrier controlling the divergence of the correlation length can be directly deduced from static susceptibility measurements at high enough temperature. The last condition is required in order for ξ to be smaller than the distance among defects. When this ceases to hold true, a saturation of the product χT is observed, at low T. According to the Ising model $\Delta_\xi = 2J$, which implies that the energy barrier of the relaxation time is expected to be $\Delta_\tau = 2\Delta_\xi = 4J$ at high temperature and $\Delta_{\tau_N} = \Delta_\xi = 2J$ at low temperature. Moreover, the crossover between the thermodynamic limit ($\xi \ll N$) and the finite-size regime ($\xi \gg N$) should be described by (8.38) and (8.39) in the presence of a static applied field. SCMs represent a class of model systems in which most of these predictions were confirmed. Often, finding a quantitative agreement required *ad-hoc* generalizations of Glauber's idealized picture, without renouncing its basic concepts. Some of those generalizations will be discussed in the next section.

8.7 Beyond the Glauber Model

An important generalisation of the Glauber model relates the temperature dependence of the parameter α [74]. Introduced in (8.23) for the single-spin transition probability, this parameter turns out to be the proportionality coefficient between the low-temperature expansions of the correlation length and the relaxation time: $\tau = 2\xi^2/\alpha$. In other words, for the time scales of interest, one has that $\alpha = 4D_s$, with D_s being the diffusion coefficient for thermally-driven DW motion (see (8.2)). Given this equivalence, we will focus on D_s henceforth. For explaining the experimental results of a SCM made of Mn^{3+}-Ni^{2+}-Mn^{3+} repeating units it was proposed that $D_s \sim e^{-\Delta_A/kT}$, where Δ_A was the global effective anisotropy energy of each unit [4]. The relationship between the energy barriers of the correlation length and the relaxation time was adapted accordingly: $\Delta_\tau = 2\Delta_\xi + \Delta_A$. The last formula has been validated by experiments on a variety of SCMs with sharp DWs. In those cases, it was also found that the energy barrier of τ has to be modified consistently at low temperature, namely $\Delta_{\tau_N} = \Delta_\xi + \Delta_A$.

One minor remark is that the Δ_A contribution to the energy barrier of the relaxation time is justified only when some single-ion anisotropy is present. For instance, we have seen that Co^{2+} in distorted octahedral environment is usually assumed to behave as an effective spin one-half at low temperature. This assumption is not consistent with a finite Δ_A for SCMs based on Co^{2+}. More importantly, the picture appears more blurred for broad DWs. Let us refer again to Hamiltonian (8.1). As mentioned in Sect. 8.2, for $J > D$ the correlation length in units of DW width is a universal function of the temperature expressed in units of DW energy: $\xi/w = \Lambda(\mathcal{E}_{dw}/kT)$. It has been known since the eighties that a spin wave can propagate across a broad DW acquiring a phase shift[10] [75]. In order to conserve the total magnetisation at short time scales, the DW is left displaced after this scattering event [76]. When many of such events occur incoherently and involve thermalised spin waves, the resulting DW motion may be assimilated with that of a Brownian particle. Indeed, still in the eighties, it was shown that D_s scales like the square of the ratio $kT/\mathcal{E}_{dw}$ in the absence of damping [77, 78] and linearly when some damping term is included [79, 80]. Dimensional analysis suggests to complete the latter result as

$$D_s \propto \frac{w^2}{\tau_d}\frac{kT}{\mathcal{E}_{dw}}, \tag{8.40}$$

with τ_d being a characteristic time scale of the problem, associated with short-time dynamics. Equation (8.40) is consistent with recent numerical results reported in

[10] Recently, magnonic applications of DWs which exploit such a phase shift has been proposed in the context of metallic nanowires [81–83].

Ref. [34]. In the same paper, the activated behaviour of D_{s} expected for sharp DWs was recovered as well. A qualitative argument for the different temperature dependence of D_{s} expected for sharp and broad DWs can be given starting from zero-temperature dynamics. In the continuum formalism one finds that a field of any intensity applied along the easy axis is able to move a broad DW [28, 75, 84–87]. In the opposite limit, it was shown that a finite threshold field is needed to let a sharp DW propagate [30]. In this case, translating a DW requires local modifications of the spin profile, which create an effective Peierls potential. This potential is periodic with respect to the position of the DW centre and the difference between its minima and maxima decreases exponentially with increasing the DW width [32, 88, 89], till it vanishes in the continuum limit. It seems, therefore, plausible to expect a thermally-activated diffusion coefficient *only* for sharp DWs.

While for sharp DWs the relaxation time depends on J, D and T independently, our present understanding of SCMs suggests that τ should depend only on the ratio $\mathcal{E}_{\mathrm{dw}}/kT$ for broad DWs. This can be readily deduced by relating τ to the correlation length $\xi = w\Lambda(\mathcal{E}_{\mathrm{dw}}/kT)$ by means of the random-walk argument and (8.40) (remember that this argument holds only for $\xi \ll N$) [34].

The standard theoretical framework to deal with magnetisation dynamics is the Landau-Lifshitz-Gilbert (LLG) equation. In that context, one expects τ_{d} introduced in (8.40) to be of the order of the dumping time: $\tau_{\mathrm{d}} \simeq (1+\alpha_{\mathrm{G}}^2)/(\alpha_{\mathrm{G}}\gamma_0 H_{\mathrm{A}}) \simeq \hbar/(2D\alpha_{\mathrm{G}})$, where $\alpha_{\mathrm{G}} \ll 1$ is the Gilbert damping [90], γ_0 is the gyromagnetic factor and H_{A} the anisotropy field. For values of D that are realistic for SCMs, $\hbar/D$ falls in the picosecond range while the damping constant is typically $\alpha_{\mathrm{G}} = 10^{-1}$–$10^{-4}$. As slow dynamics is usually probed in SCMs at time scales longer than milliseconds, it clearly pertains to long-time behaviour in the language of LLG equation. Moreover, since physics of SCMs is dictated by thermal fluctuations, a stochastic noise should be included in numerical simulations [91–95]. In spite of the enormous improvements experienced in computational capabilities [96–98], performing a stochastic-dynamic simulation which covers a time window of several orders of magnitudes still remains prohibitive. In this sense, the brute-force approach to SCMs dynamics does not seem promising for the next future.

With respect to the sharp-wall case, there is no analogous of the Glauber's formalism for SCMs with broad DWs. Experimental realizations basically consist of ferrimagnetic chains alternating Mn^{3+} with an organic radical [48–50]. A reasonable model is the one which produces the kernel (8.15), where Mn spins are treated as classical vectors. In the experimentally accessible region Δ_ξ can be much smaller than $\mathcal{E}_{\mathrm{dw}}$—up to about half of it—due to spin-waves renormalisation [9]; while for sharp DWs one has $\Delta_\xi = \mathcal{E}_{\mathrm{dw}}$ at any temperature $kT < J$ [34]. This fact needs to be taken into account in the experimental characterization of SCMs with broad DWs (see Fig. 8.2). For what concerns the barrier of the relaxation time, the available experimental results yield Δ_τ about 10–20% times larger than $\mathcal{E}_{\mathrm{dw}}$[11] [9, 48–50]: much smaller than twice the DW energy at $T = 0$ as predicted by Glauber for the Ising

[11]For these mixed chains $\mathcal{E}_{\mathrm{dw}} = 2\sqrt{JD}$, with a factor 2 of difference with respect to Hamiltonian (8.1) [9].

model. Making a definitive statement about the origin of energy scales involved in dynamics is not possible yet. Defects probably affect the nucleation and diffusion of broad DWs differently with respect to the Ising limit. In metallic nanowires, for instance, defects act as pinning centres for DWs or vortices. In SCMs a similar phenomenon may induce a reduction of DW mobility, namely D_S. Another possibility is that DWs may preferentially be nucleated at defects because it is energetically favourable.[12] Only a thorough characterisation of SCMs with broad DWs in which the concentration of defects may be controlled could allow answering those questions. At the same time, such a study would provide important information about the joint effect of defects and thermal fluctuations. This would also be relevant for DW dynamics in metallic nanowires that are typically described by the very same classical Heisenberg Hamiltonian (8.1) [13, 32, 85–87, 99–106].

8.8 Conclusion and Perspectives

The title of the review we wrote about five years ago was "Single-chain magnets: where to from here?" [42]. The idea was that of reviewing critically what had been done in the synthetic, experimental and theoretical fields. The analysis indicated that the hunt for high-temperature blocking magnets was going to continue. This has been confirmed but with the explosion of the interest for Lanthanides with the challenging difficulties associated with the large unquenched orbital moment [107]. Much more work shall be done especially in theory. Another field which is developing fast is that of ab initio, DFT calculations which are rapidly complementing/substituting Ligand-field approaches [55, 108, 109]. Far-from-equilibrium dynamics and aging [18–20, 110, 111] as well as the interplay between SCM behaviour and quantum effects [21–23, 112] call for a more systematic investigation. The comparison of the properties of molecular nanomagnets with elongated magnetic nanoparticles and magnetic nanowires has been stated a few times throughout the chapter [13, 34]. We feel that SCMs can provide good insight into the finite-temperature behaviour of such nanosystems. Finally, as a matter of facts, molecular systems have already entered the domains of spintronics [113] and quantum computing [114, 115]. In future, besides their traditional role as model systems, SCMs can possibly find their place in those applicative research contexts.

Acknowledgements This chapter is the result of several collaborations and fruitful, sometimes animated discussions with many colleagues. For this valuable contribution we are sincerely grateful to R. Sessoli, M.G. Pini, A. Rettori, L. Bogani, R. Clérac, C. Coulon, M. Verdaguer, J. Villain, V. Pianet, T.T. Michaels, B. Sangiorgio, G. Venturi, H. Miyasaka, W. Wernsdorfer, and O.V. Billoni. We would also like to thank L. Sorace, M.G. Pini, F. Totti and L.G. De Pietro for the precious help provided in the editing phase and their patient and careful reading of the manuscript.

[12] Accommodating the DW centre onto a defect reduces the anisotropy energy and $\mathcal{E}_{dw}$ consequently.

References

1. D. Gatteschi, R. Sessoli, Angew. Chem., Int. Ed. Engl. **42**, 268 (2003)
2. D. Gatteschi, R. Sessoli, J. Villain, *Molecular Nanomagnets* (Oxford University Press, Oxford, 2006)
3. A. Caneschi et al., Angew. Chem., Int. Ed. Engl. **40**, 1760 (2001)
4. C. Coulon et al., Phys. Rev. B **69**, 132408 (2004)
5. L. Lecren et al., J. Am. Chem. Soc. **129**(16), 5045 (2007)
6. C. Coulon, H. Miyasaka, R. Clérac, Struct. Bond. **122**, 163 (2006)
7. R. Lescouëzec et al., Coord. Chem. Rev. **249**, 2691 (2005)
8. H.-L. Sun, Z.-M. Wang, S. Gao, Coord. Chem. Rev. **254**, 1081 (2010)
9. W. Zhang et al., RSC Adv. (2013). doi:10.1039/C2RA22675H
10. L. Bogani et al., Phys. Rev. Lett. **92**, 207204 (2004)
11. A. Vindigni et al., Appl. Phys. Lett. **87**, 073102 (2005)
12. P. Gambardella et al., Nature **416**, 301 (2002)
13. H.B. Braun, Adv. Phys. **61**, 1–116 (2012)
14. C. Coulon et al., Phys. Rev. Lett. **102**, 167204 (2009)
15. D.J. Scalapino, Y. Imry, P. Pincus, Phys. Rev. B **11**, 2042 (1975)
16. F. Ferrero et al., J. Am. Chem. Soc. **113**, 8410 (1991)
17. A. Caneschi et al., Inorg. Chem. **28**, 1976 (1989)
18. F. Stickel, E.W. Fischer, R. Richert, J. Chem. Phys. **102**, 6251 (1995)
19. E. Vincent et al., in *Complex Behaviour of Glassy Systems*, ed. by M. Rubi, C. Perez-Vicente. Lecture Notes in Physics, vol. 492 (Springer, Berlin, 1997), p. 184
20. A. Cavagna, Phys. Rep. **476**, 51 (2009)
21. S. Sachdev, Science **288**, 475 (2000)
22. R. Coldea et al., Science **327**, 177 (2010)
23. J. Simon et al., Nature **472**, 307 (2011)
24. N.D. Mermin, H. Wagner, Phys. Rev. Lett. **17**, 1133 (1966)
25. M.E. Fisher, Am. J. Phys. **32**, 343 (1964)
26. L.D. Landau, E.M. Lifshitz, *Statistical Physics* (Pergamon, Oxford, 1986)
27. J.A. Krumhansl, J.R. Schriffer, Phys. Rev. B **11**, 3535 (1975)
28. U. Enz, Helv. Phys. Acta **37**, 245 (1964)
29. H.J. Mikeska, M. Steiner, Adv. Phys. **40**, 191 (1991)
30. B. Barbara, J. Phys. (Paris) **34**, 139 (1973)
31. A. Vindigni, Inorg. Chim. Acta **361**, 3731 (2008)
32. P. Yan, G.E.W. Bauer, Phys. Rev. Lett. **109**, 087202 (2012)
33. E. Ising, Z. Phys. **31**, 253 (1925)
34. O.V. Billoni et al., Phys. Rev. B **84**, 064415 (2011)
35. H.C. Fogedby, P. Hedegard, A. Svane, J. Phys. C **17**, 3475 (1984)
36. F. Ferrero et al., Mol. Phys. **85**, 1073 (1995)
37. R. Cordery, S. Sarker, J. Tobochnik, Phys. Rev. B **24**, 5402(R) (1981)
38. M. Steiner, J. Villain, C. Windsor, Adv. Phys. **25**, 87 (1976)
39. K.S. Cole, R.H. Cole, J. Chem. Phys. **9**, 341 (1941)
40. J.S. Miller, D. Gatteschi, Chem. Soc. Rev. **40**, 3065 (2011)
41. J.S. Miller, Chem. Soc. Rev. **40**, 3266 (2011)
42. L. Bogani et al., J. Mater. Chem. **18**, 4750 (2008)
43. A.V. Palii et al., J. Am. Chem. Soc. **130**, 14729 (2008)
44. A.L. Barra et al., Angew. Chem., Int. Ed. Engl. **36**, 2329 (1997)
45. A. Palii et al., Chem. Soc. Rev. **40**, 3130–3156 (2011)
46. J. Seiden, J. Phys. Lett. **44**, 947 (1983)
47. A. Caneschi et al., Inorg. Chem. **27**, 1756 (1988)
48. H. Miyasaka et al., Chem. Eur. J. **27**, 7028 (2006)
49. M. Balanda et al., Phys. Rev. B **74**, 224421 (2006)
50. R. Ishikawa et al., Inorg. Chem. **51**, 9123 (2012)

51. R.J. Glauber, J. Math. Phys. **4**, 294 (1963)
52. A. Vindigni et al., Appl. Phys. A **82**, 385 (2006)
53. M. Blume, P. Heller, N.A. Lurie, Phys. Rev. B **11**, 4483 (1975)
54. R. Pandit, C. Tannous, Phys. Rev. B **28**, 281 (1982)
55. K. Bernot et al., Phys. Rev. B **79**, 134419 (2009)
56. A. Caneschi et al., Europhys. Lett. **58**, 771 (2002)
57. A. Vindigni, N. Regnault, Th. Jolicoeur, Phys. Rev. B **70**, 134423 (2004)
58. M. Kardar, Phys. Rev. B **28**, 244 (1983)
59. D. Mukamel, S. Ruffo, N. Schreiber, Phys. Rev. Lett. **95**, 240604 (2005)
60. G.O. Berim, E. Ruckenstein, J. Chem. Phys. **119**, 9640 (2003)
61. M. Einax, M. Schulz, J. Chem. Phys. **115**, 2282 (2001)
62. M. Suzuki, R. Kubo, J. Phys. Soc. Jpn. **24**, 51 (1968)
63. H.W. Huang, Phys. Rev. A **8**, 2553 (1973)
64. A. Vindigni, M.G. Pini, J. Phys. Condens. Matter **21**, 236007 (2009)
65. K. Bernot et al., J. Am. Chem. Soc. **130**(5), 1619 (2008)
66. C. Coulon et al., Phys. Rev. B **76**, 214422 (2007)
67. M.G. Pini et al., Phys. Rev. B **84**, 094444 (2011)
68. S. Zŭmer, Phys. Rev. B **21**, 1298 (1980)
69. D. Dhar, M. Barma, J. Stat. Phys. **22**, 259 (1980)
70. J.H. Luscombe, M. Luban, J.P. Reynolds, Phys. Rev. E **53**, 5852 (1996)
71. J.K.L. da Silva et al., Phys. Rev. E **52**, 4527 (1995)
72. F. Matsubara, K. Yoshimura, S. Katsura, Can. J. Phys. **51**, 1053 (1973)
73. G. Schwarz, Biopolymers **6**, 873 (1968)
74. J. Shen et al., Phys. Rev. B **56**, 2340 (1997)
75. H.C. Fogedby, *Theoretical Aspects of Mainly Low Dimensional Magnetic Systems* (Springer, Berlin, 1980)
76. P. Yan, X.S. Wang, X.R. Wang, Phys. Rev. Lett. **107**, 177207 (2011)
77. N. Theodorakopoulos, E.W. Weller, Phys. Rev. B **38**, 2749 (1988)
78. K. Fesser, Z. Phys. B **39**, 47 (1980)
79. M. Salerno, E. Joergensen, M.R. Samuelsen, Phys. Rev. B **30**, 2635 (1984)
80. D.J. Kaup, E. Osman, Phys. Rev. B **33**, 1762 (1986)
81. R. Hertel, W. Wulfhekel, J. Kirschner, Phys. Rev. Lett. **93**, 257202 (2004)
82. C. Bayer et al., IEEE Trans. Magn. **41**, 3094 (2005)
83. S. Macke, D. Goll, J. Phys. Conf. Ser. **200**, 042015 (2010)
84. N.L. Schryer, L.R. Walker, J. Appl. Phys. **45**, 5406 (1974)
85. O. Boulle, G. Malinowski, M. Kläui, Mater. Sci. Eng., R Rep. **72**, 159 (2011)
86. A. Thiaville et al., J. Appl. Phys. **95**, 7049 (2004)
87. J. Yang et al., Phys. Rev. B **77**, 014413 (2008)
88. H.R. Hilzinger, H. Kronmüller, Phys. Status Solidi **54**, 593 (1972)
89. K.S. Novoselov et al., Nature (London) **426**, 812 (2003)
90. T. Gilbert, IEEE Trans. Magn. **40**, 3443 (2004)
91. D. Hinzke, U. Nowak, Phys. Rev. B **61**, 6734 (2000)
92. U. Nowak, R.W. Chantrell, E.C. Kennedy, Phys. Rev. Lett. **84**, 163 (2000)
93. O. Chubykalo et al., Phys. Rev. B **67**, 064422 (2003)
94. W.T. Cheng et al., Phys. Rev. Lett. **96**, 067208 (2006)
95. D.A. Stariolo, O.V. Billoni, J. Magn. Magn. Mater. **316**, 49 (2007)
96. R. Chang et al., J. Appl. Phys. **109**, 07D358 (2011)
97. D. Pinna et al., Appl. Phys. Lett. **101**, 262401 (2012)
98. D. Pinna, A.D. Kent, D.L. Stein, J. Appl. Phys. **114**, 033901 (2013). arXiv:1210.7675
99. U. Atxitia et al., Phys. Rev. B **82**, 134440 (2010)
100. M. Kläui et al., Phys. Rev. Lett. **95**, 026601 (2005)
101. A. Vanhaverbeke, A. Bischof, R. Allenspach, Phys. Rev. Lett. **101**, 107202 (2008)
102. E. Saitoh et al., Nature **432**, 203 (2004)
103. G. Tatara, H. Kohno, Phys. Rev. Lett. **92**, 086601 (2004)

104. S.S. Parkin, M. Hayashi, L. Thomas, Science **320**, 190 (2008)
105. M. Hayashi et al., Science **320**, 209 (2008)
106. X. Jiang et al., Nano Lett. **11**, 96 (2011)
107. L. Sorace, C. Benelli, D. Gatteschi, Chem. Soc. Rev. **40**, 3092 (2011)
108. A.V. Postnikov, J. Kortus, M.R. Pederson, Phys. Status Solidi B **243**, 2533 (2006)
109. E. Heintze et al., Nat. Mater. (2012). doi:10.1038/nmat3498
110. J.J. Brey, A. Prados, Phys. Rev. E **53**, 458 (1996)
111. A. Prados, J.J. Brey, B. Sánchez-Rey, Europhys. Lett. **40**, 13 (1997)
112. W. Wernsdorfer et al., Phys. Rev. Lett. **95**, 237203 (2005)
113. L. Bogani, W. Wernsdorfer, Nat. Mater. **7**, 179 (2008)
114. M.N. Leuenberger, D. Loss, Nature **410**, 789 (2001)
115. S. Loth et al., Nat. Phys. **6**, 340 (2010)

Chapter 9
Magnetism of Metal Phthalocyanines

Juan Bartolomé, Carlos Monton, and Ivan K. Schuller

Abstract Metal-phthalocyanine (MPc) are uniquely suited for the exploration of the intrinsic mechanisms which gives rise to molecular magnetism. In this chapter, we review the structural and magnetic properties of bulk crystal, thin film and single MPcs molecules adsorbed on different substrates. Traditional magnetic measurements and new techniques like x-ray magnetic circular dichroism show that the magnetic behavior of MPc molecules is strongly related with the electronic ground state of the central metal atom hybridized with the ligand states (intra-molecular interaction). In bulk and thin films, with stacked molecules, intermolecular exchange interactions between magnetic M atoms regulates their magnetic properties. Moreover experimental results show that the magnetic properties of single molecules are strongly affected by the electronic coupling to the supporting substrate.

9.1 Introduction

Since their discovery and later systematic studies of their molecular structure [1, 2], Phthalocyanines [3–5] have been subject of research because of their multiple applications such as dyes, catalysts and coatings. At present they are one of the most studied organic materials for possible applications in nanodevices and spintronics [6, 7]. This chapter describes the magnetic properties of Metal Phthalocyanines in bulk solid state phases, thin films and isolated molecules on various substrates. It also summarizes their recently expanding applications in molecular magnetism with a future perspective given at the end.

J. Bartolomé (✉)
Instituto de Ciencia de Materiales de Aragón and Departamento de Física de la Materia Condensada, CSIC–Universidad de Zaragoza, C/Pedro Cerbuna 12, 50009 Zaragoza, Spain
e-mail: barto@unizar.es

C. Monton · I.K. Schuller
Center for Advanced Nanoscience, Department of Physics, University of California San Diego, 9500 Gilman Drive, La Jolla, CA 92093, USA

J. Bartolomé et al. (eds.), *Molecular Magnets*, NanoScience and Technology,
DOI 10.1007/978-3-642-40609-6_9,

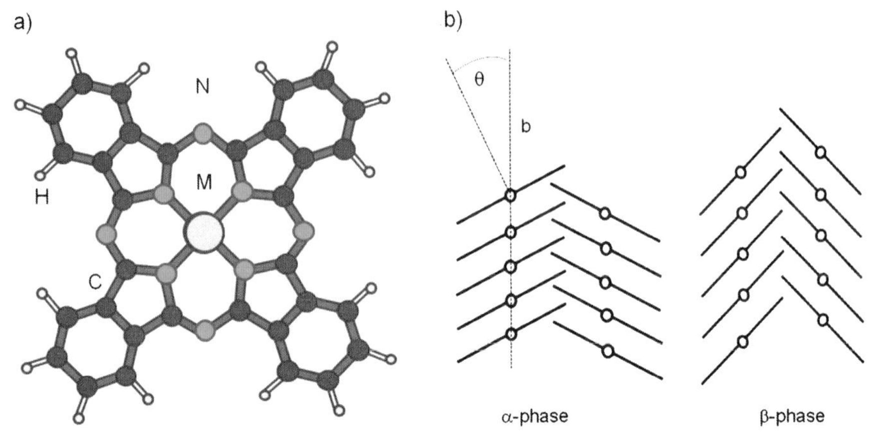

Fig. 9.1 (**a**) M-Phthalocyanine molecule. (**b**) Schematic stacking of the herring-bone α- and β-phases. θ, angle between the z axis of the molecule and the b axis of the crystal structure

9.2 Solid State MPcs

The MPc is a macrocyclic planar aromatic molecule (Pc) in which a central metal atom (M) is bound to the organic structure through four inwardly projecting nitrogen centers (Fig. 9.1(a)). Interestingly, in these molecules a large number of M substitutions are possible, giving rise to special physical and chemical properties [8]. In the present chapter only the transition metal substitutions M = Mn, Fe, Co, Ni and Cu compounds will be reviewed since they are the most relevant MPcs in the field of molecular magnetism.

Bulk MPcs crystals grow in high aspect ratio needle-shapes, a consequence of their strong anisotropic molecular structures. This anisotropy arises from the van-der-Waals molecule-molecule interactions which are greatest when the plate-like molecules are face-to-face rather than side-to-side. Such an affinity gives rise to molecule stacking in which the central metal atoms form one-dimensional (1D) chains. The stacking axis is defined by the direction of these chains, which define the b-axis, as shown in Fig. 9.1(b). In some MPcs strong intrachain and weak interchain coupling between metal atoms exist resulting in new anisotropic, optical, electrical and magnetic properties [7].

The angle between the b-axis and the normal to the plane of the molecule θ, together with the intermolecular distance, gives rise to different polymorphs, the most abundant being the α and β-phases (see Fig. 9.1(b)). The β-phase is mostly found in bulk crystals is stable and characterized by $\theta \sim 45°$. In contrast, the α-phase is metastable with $\theta \sim 25°$ angle found in bulk and often in thin films at room temperature. Table 9.1, summarizes the available structural data for solid (bulk) phase and thin film MPc molecules.

The MPc´s magnetic properties depend basically on the electronic ground state of the M substitution, which, in turn, are determined by the nearest neighbor coordination and MPc polymorphism. The molecular symmetry of MPc is very close to D_{4h} point group (see Fig. 9.1(a)), which in a simple approximation implies that the

Table 9.1 Phases and structural parameters of the MPc molecules in bulk and film phases. (θ: angle between b-axis and normal to the molecule; β: angle between ab and ac planes)

MPc	S. Group	a (Å)	b (Å)	c (Å)	θ (deg)	β (deg)	References
α'-MnPc (film)	Rhombohedral	17.60	17.60	17.6	25.0	82.0	[9]
β-MnPc	$P2_1a$	19.40	4.76	14.61	47.9	120.7	[10]
α-FePc	Monoclinic	25.50	3.77	25.2	26.5	90.0	[11]
β-FePc	$P2_1a$	19.39	4.78	14.60	47.3	120.8	[10]
α'-CoPc (film)	C2/c	25.88	3.75	24.08	26.5	90.4	[12]
β-CoPc	$P2_1c$	14.54	4.77	19.35	47.3	120.8	[13]
α'-NiPc (film)	C2/c	26.15	3.79	24.26	~26.5	94.8	[12]
β-NiPc	$P2_1a$	19.90	4.71	14.90	~46.5	121.9	[14]
α'-CuPc (film)	C2/c	25.92	3.79	23.92	26.5	90.4	[12, 15]
β-CuPc	$P2_1a$	19.60	4.79	14.60	46.5	120.6	[14, 15]

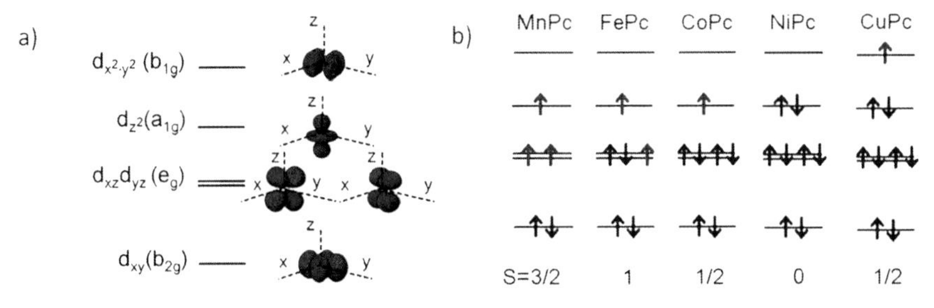

Fig. 9.2 (**a**) Metal-d orbitals with respect to the Pc molecule axes (notation as irreducible representations in D_{4h} symmetry). Spatial electron probability density of the orbitals. Blue and red identify the different complex wave function phases. (**b**) Electron filling scheme for MPc's (*black*) paired electrons and (*red*) un-paired electrons. Below, total spin due to the unpaired electrons

M d-orbitals can be classified according to a square-planar ligand-field. The mono-electronic d-states are denoted by the irreducible representation under which each transforms in the D_{4h} symmetry (and in the Cartesian coordinates notation). With this the five metal d-orbitals transform as: $a_{1g}(d_{z^2})$, $b_{1g}(d_{x^2-y^2})$, $e_g(d_{zx}, d_{yz})$ and $b_{2g}(d_{xy})$ (see Fig. 9.2(a)). The different M substitutions supply the electrons that fill consecutively these states (see Fig. 9.2(b)).

The M atom electronic states are hybridized with the phthalocyanine molecular orbitals (MO), generating the total MO, with dominant 3d electron character arising from M. The gas phase MOs for M = Mn, Fe, Co, Ni and Cu have been calculated in a series of works [16, 17] using density functional theory (DFT). These calculations incorporate hybrid states with π and σ character. The most relevant MO states to the magnetic properties are the ^3e_g, a_{1u}, a_{2u} and the $^4e_g * \pi$ antibonding states, which may hybridize with d-orbitals as seen in Fig. 9.3. The e_g MO states results from the interaction between the Fe (d_{xz}, d_{yz}) and the nitrogen (N)-p_z states of the delocalized π system. The a_{1g} MO state results from the interaction of Fe d_{z^2} states and the N-s and N-p_{xy} states, pointing toward the metal center (i.e. parallel

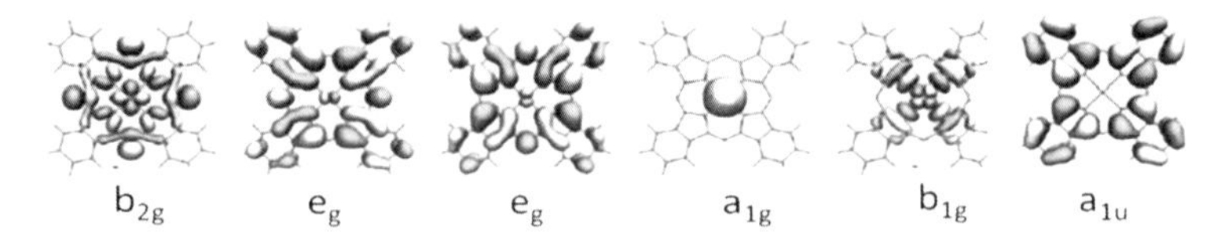

Fig. 9.3 Electron probability density of the molecular orbital states formed by the coupling of M and ligand p-states identified by the irreducible representation with respect to the D_{4h} point group [19, 20] (Reprinted Figure with permission from Betti et al. [20]. Copyright 2012 by the American Chemical Society)

to the substrate). The b_{2g} MO results from the interaction of Fe d_{xy} states and the N-p_{xy} states, orthogonal to the metal center. In the same way the b_{1g} MO state is the combination of the Fe $d_{x^2-y^2}$ states with the N-p_{xy} states pointing toward the metal center. The overlap between the metal and organic states determines the relative energy ordering of these MO states. Besides, for example, in CuPc the a_{1u} MO state is the highest occupied molecular orbital (HOMO) and the $^2e_g(\pi)$ orbital is the lowest unoccupied molecular orbital (LUMO) [18]. These MO states have wavefunctions fully localized at the macrocycles. The energy level configuration of the MO states for the transition metal substitutions was calculated for M = Fe, Co, Ni and Zn, where the LUMO and HOMO levels have been predicted showing that 3d electrons are more localized in MPc with closed shells.

Tables 9.2 and 9.3 show the experimental and theoretical information available regarding the d-electron ground states for MPcs in α and β-phases. The ground state wavefunction, made of coupled monoelectronic d-states, is described in parenthesis by the irreducible representation (irrep) under which it transforms in D_{4h} symmetry, and by the upper right index by its electron occupancy N, (irrep)^N (see Tables 9.2 and 9.3, 3rd column) on one hand, and as the irreducible representation of the total spin configuration in terms of the corresponding irreducible representation and its spin degeneracy n, ^{n}Xirrep (see Tables 9.2 and 9.3, 4th column), on the other hand. Some results in Table 9.3 include the ligand state occupation, as given by Ref. [18]. In many cases there are still disagreements in the reported information (i.e. β-MnPc and β-FePc ground states occupancy). Moreover, with the advent of new experimental techniques like x-ray magnetic circular dichroism (XMCD) some of these experimental conclusions are under revision.

The description of the magnetic properties of the MPc molecules is best done in terms of localized moments at the M atom. In the solid phases, there are two different crystallographic sites for the M atom. Therefore, in crystalline bulk phases, the spin Hamiltonian which describes the MPc molecules is represented by:

$$H = -2\sum_{ij} J\hat{S}_i \cdot \hat{S}_j - 2\sum_{ij} J'\hat{S}_i \cdot \hat{S}_j + \sum_j D\left(\hat{S}_j^x\right)^2 + \sum_k D\left(\hat{S}_k^y\right)^2 \qquad (9.1)$$

This equation includes both, intrachain (J) and interchain (J') exchange interactions, and single-ion crystal field (D) anisotropy terms at the two different crystallographic sites (third and fourth terms).

Table 9.2 Ground state orbital occupancy determined by experimental techniques. MP refers to magnetic properties, MS magnetic susceptibility, MCD magnetic circular dichroism, XRD x-ray diffraction, XANES x-ray absorption near edge structure, XAS x-ray absorption spectroscopy, XMDC x-ray magnetic circular dichroism, PE photoemission and NMR nuclear magnetic resonance. J/k_B is the intrachain exchange. D/k_B is the crystal field parameter

MPc	Exp. technique	Ground state occupancy	Ground state	Spin	Tc (K)	J/k_B	D/k_B	Refs.
β-MnPc	MP	$(b_{2g})^2(e_g)^2(a_{1g})^1$		3/2	8.3(c)	11	28	[21]
β-MnPc	MS	$(b_{2g})^2(e_g)^2(a_{1g})^1$	$^4A_{2g}$-^4E_g	3/2	(c)			[22]
β-MnPc	MP			3/2	10(c)			[23]
α-MnPc	MP			3/2	(e)			[23]
MnPc(a)	MCD	$(e_g)^3(b_{2g})^1(a_{1g})^1$	^4E_g	3/2				[24]
MnPc(b)	XAS, PE	$(e_g)^3(b_{2g})^1(a_{1g})^1$ $(e_g)^2(b_{2g})^2(a_{1g})^1$		3/2				[25]
β-FePc	XRD	$(b_{2g})^2(e_g)^3(a_{1g})^1$	E_{gA}	1	(d)			[26]
β-FePc	MS	$(b_{2g})^2(e_g)^3(a_{1g})^1$	^3E_A	1	(d)			[27]
β-FePc	XANES	$(a_{1g})^2(e_g)^3\ (b_{2g})^1$	^3E	1	(d)			[28]
β-FePc	MS		^3B_g	1	(d)			[29]
α-FePc	MP	$(b_{2g})^2(e_g)^3(a_{1g})^1$	E_{gA}	1	5(c)	25.7 76(h)	53.2	[30]
FePc(b)	XAS, PE	$(b_{2g})^2(e_g)^3(a_{1g})^1$ $(b_{2g})^2(e_g)^2(a_{1g})^2$		1				[30]
β-CoPc	MS	$(b_{2g})^2(e_g)^4(a_{1g})^1$		1/2	(d)			[31–33]
α-CoPc	XAS, XMCD	$(e_g)^{3.8}(b_{2g})^2(a_{1g})^{1.2}$	$^2A_{1g}$-^2E_g	1/2				[34]
CoPc(b)	XAS, PE	$(e_g)^4(b_{2g})^2(a_{1g})^1$		1/2	(d)			[25]
β-CuPc	MP			1/2	(d)	0		[23]
β-CuPc	MS	$(b_{2g})^2(e_g)^4(a_{1g})^2(b_{1g})^1$		1/2	(d)			[31, 33]
β-CuPc	NMR				(g)	0.286		[35]
α-CuPc(b)	MP			1/2	(e)	~ 1.5		[23]
CuPc(b)	XAS, PE	$(b_{2g})^2(e_g)^4(a_{1g})^2(b_{1g})^1$		1/2	(d)			[25]
β-NiPc	MS	$(b_{2g})^2(e_g)^4(a_{1g})^2$		0	(f)			[31, 36]

(a) In Ar matrix, (b) thin film, (c) ferromagnetic, (d) paramagnetic, (e) antiferromagnetic, (f) diamagnetic, (g) one dimensional chain, (h) intrachain interaction (J/k_B) obtained from the soliton-kink model for $s = 1/2$

In general, most of the bulk solid state MPcs remain paramagnetic down to the lowest achievable temperatures because the molecule-molecule interaction is too weak to sustain long range order. However NiPc, β-MnPc and α-FePc are exceptions. Both polymorphs of NiPc are diamagnetic on account of their completely filled orbitals. On the contrary, β-MnPc and α-FePc polymorphs, develop long range ferromagnetism below an ordering temperature, T_c, caused by weak interchain interactions (see Tables 9.2 and 9.3). As a summary of the current understanding of the MPc's magnetic properties, we describe briefly the properties of each MPc compound.

Table 9.3 Ground state orbital occupancy obtained from theory. DFT refers to density functional theory

MPc	Theoretical method	Ground state occupancy	Ground state	Spin	References
β-MnPc	AB initio	$(b_{2g})^1(e_g)^3(a_{1g})^1$	^4E_g	3/2	[37]
$MnPc^{(im)}$	DFT		^4E_g	3/2	[38]
β-FePc	DFT		$^3B_{2g}/^3A_g$		[17]
β-FePc	DFT	$(b_{2g})^2(a_{1u})^2(a_{1g})^2(1e_g)^2$	$^3A_{2g}$	1	[18]
β-FePc	AB initio	$(b_{2g})^1(e_g)^4(a_{1g})^1/(b_{2g})^2(e_g)^3(a_{1g})^1$	$^3E_g/^3B_{2g}$	1	[37]
β-CoPc	AB initio	$(b_{2g})^2(e_g)^4(a_{1g})^1$	$^2A_{1g}$	1/2	[37]
β-CoPc	DFT	$(a_{1g})^2(a_{1u})^2(1e_g)^3$	^1E_g	1/2	[18]
$CoPc^{(im)}$	DFT		$^2A_{1g}$	1/2	[38]
β-CuPc	DFT		$^2B_{1g}$	1/2	[18]
β-NiPc	DFT		$^1A_{1g}$	0	[18]

$^{(im)}$ Isolated molecule

Mn-Phthalocyanine This compound presents both ferromagnetism (FM) and antiferromagnetism (AFM) in the β or α-phase polymorphs, respectively. In MnPc the intrachain superexchange interactions via the organic ring (Pc)π MO compete in sign. For example, the Mn-Mn FM coupling is promoted by d-electrons a_{1g}–a_{1g} interactions via the e_g filled π MO, while the AFM coupling is due to d-electrons e_g–e_g interaction via the e_g MO. In the early 70's Barraclough [22] found that in β-MnPc the intrachain FM interaction prevails with $J/k_B = 11$K for Mn(II) $S = 3/2$. Recently, Kataoka et al. [39] using XMCD confirmed the proposed FM coupling and determined that the ground state of β-MnPc is ^4E_g, where the a_{1g} is the HOMO level.

Crystal field splitting also plays an important role in the MPc's magnetic properties. In β-MnPc for example, the crystal field splits the Mn(II) $S = 3/2$ quadruplet into two doublets. This can be deduced from (9.1). Since the crystal field splitting is positive [22], $D = 20$ cm^{-1}, and the exchange field is $B_{ex} = 48$ T, below T_c the competing exchange interaction and the positive crystal electric field pull down the energy of the electronic state $S_z = -3/2$ below the $S_z = -1/2$ one, turning it into the ground state. Alternatively, the $S_z = -3/2$ ground state may be explained in terms of a negative effective crystal field, $D' < 0$, as proposed by Miyoshi et al. [33] In β-MnPc the interchain interactions are strong enough to give rise to FM ordering at $T_c = 8.3$ K (see Table 9.2). The magnetic structure of single crystals was deduced from neutron diffraction experiments by Mitra et al. [21]. They proposed that MnPc molecules order in a canted structure with an easy magnetization axis on the ac plane.

Fe-Phthalocyanine Direct M-M ferromagnetic interaction may occur when there is overlap between two orbital states in each metal atom, a half-filled and an empty orbital, or a half-filled and a full orbital [40]. In α-FePc direct exchange is dominant since the latter case is applicable, and the mechanism may be effective to yield ferromagnetic interaction. Indeed, the orbital configuration of each Fe is E_{gA} $(b_{2g})^2(e_g)^3(a_{1g})^1$ (Table 9.2). Therefore, the $(e_g)^3$ orbital doublet of one Fe is triply occupied, thus it has a half filled orbital and a full orbital. It may overlap with the homologous in the nearest neighboring Fe since the interatomic distance is small. Since the Fe-Fe distance in α-FePc (0.38 nm) is smaller than in β-FePc (0.47 nm), direct FM exchange is most likely responsible for the FM correlations within the Fe chain with $J/k_B = 25.7$ K, for Fe(II) $S = 1$, while weak interchain interactions give rise to long range order transition at $T_C = 5$ K. In contrast, β-FePc remains paramagnetic above 2 K. The reason is that the crystal field parameter above 70 K is positive, $D/k_B = 53.2$ K, with $S = 1$, in other words the ground state corresponds to $S_z = 0$ while the excited state is $S_z = \pm 1$. Therefore, the ground state is non-magnetic, and the observed paramagnetism corresponds to the thermal population of the excited doublet [27]. In the α-phase, on the other hand, the intrachain exchange interaction is $J/k_B = 25.7$ K and it splits the upper doublet $S_z = \pm 1$ lowering the $S_z = -1$ level 44 K down in energy, close enough to match the $S_z = 0$ level. As a result, at low temperatures this system behaves as a $S = 1/2$ effective spin [30].

Since the metastable α-phase FePc is difficult to obtain in single crystal form, there are few studies of its magnetic properties. Evangelisti et al. [30], using magnetic measurement and Mössbauer spectroscopy, found FM behavior below $T = 10$ K in α-FePc. In contrast to the behavior of β-MnPc, the α-phase of FePc shows an unusual slow relaxation which resembles the one-dimensional slow relaxation process attributed to domain wall excitations along weakly coupled FM chains. These domain wall excitations (solitons) arise in an Ising system, i.e., when the single ion anisotropy is high compared to the intrachain interaction. Under these conditions the domain wall, labeled as a "kink", is just the separation by one lattice constant of spin-up and spin-down domains.

Two types of excitations are possible in Ising chains, kink-pair excitation (Fig. 9.4(a)), and single kink excitation at the end of the chain or at defects (Fig. 9.4(b)). Each excitation type presents the following energies $E_{a2} = 2J$ and $E_{a1} = J$ respectively. Filoti et al. [41] successfully explained the observed slow magnetic relaxation dynamics using the soliton-kink model. They took into account that α-FePc satisfies Ising chain conditions and that its magnetic ground state below T_c can be described with an effective $S = 1/2$ and an intrachain exchange interaction $J/k_B = 76 \pm 2$ K.

The temperature dependence of the relaxation time constant can be described by an Arrhenius law, $\tau = \tau_0 \exp(-E_a/k_B T)$, where E_a is the soliton activation energy. The temperature dependence of the frequency dependent ($10 < f < 5000$ Hz) susceptibility shows two peaks arising from a single or double kink excitations with $E_{a1} = 72$ K and $E_{a2} = 116$ K respectively, and the same $\tau_0 = 2 \times 10^{-11}$ seconds pre-factor time constant. (See Fig. 9.4(c)). The temperature dependent Mössbauer spectra show an excess electron spin flip linewidth broadening due to the propagation of the kink (or double kinks) along the chain. The flip rate Γ_ω is proportional to the product of the wall density and the average wall velocity ($n_S \times v_S$). As a consequence, the electronic spins fluctuate and the Mössbauer spectra are broadened via the hyperfine interaction with the Fe nuclei as Γ_ω approaches the Larmor

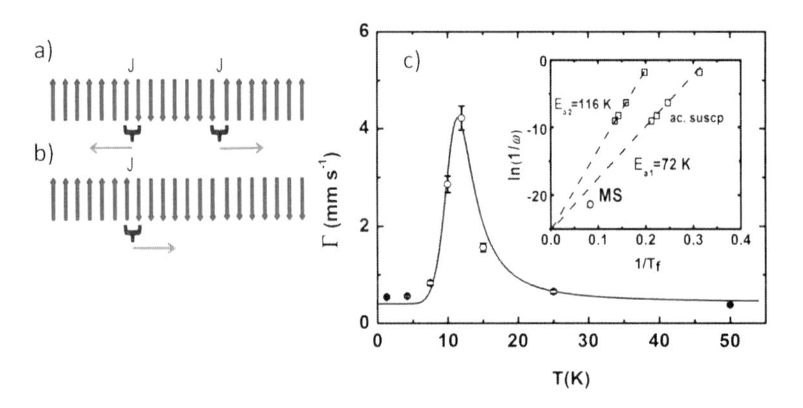

Fig. 9.4 (**a**) Double kink soliton, (**b**) single kink soliton. Keys indicate the domain wall (DW), arrows indicate the DW motion direction. (**c**) Experimental Mössbauer spectra linewidth as a function of temperature. *Dashed line* is a fit to the temperature dependence of the linewidth due to the soliton single-kink. *Inset*: Inverse of the relaxation rate, with single- and double kink soliton activation energies, as determined from a.c. susceptibility measurements (*squares*) performed at B = 800 G, and from the Mössbauer spectra [41] (*circle*)

frequency ω_L. Its temperature dependence is given by the same Arrhenius law as the ac susceptibility. The excess Mössbauer line width broadening due to solitons is given by $\Delta\Gamma \propto \Gamma_\omega/(\omega_L^2 + \Gamma_\omega^2)$ (inset of Fig. 9.4(c)). The double kink process on the other hand cannot be observed in Mössbauer spectroscopy, probably because its excitation energy falls beyond its frequency window. The spin fluctuations above T_c can also be deduced from the Single Chain Magnet model proposed in this book (Chap. 8). The relaxation processes are then described in terms of Glauber´s model for the relaxation of 1D classical chains [42]. Within this model the spin transition probability depends on the local field experienced by the spin and an Arrhenius law is predicted with an activation energy $E_a = \mathrm{J}$ for Ising S = 1/2 spins, as applicable to α-FePc. Thus soliton excitation scheme explains satisfactorily the peculiar slow relaxation found below T_c in α-FePc.

Co-Phthalocyanine Both α- and β-CoPc are paramagnetic down to the lowest measured temperature. The g tensor components of both phases were determined using electron spin resonance (ESR) at 77 K of α-ZnPc and β-ZnPc diamagnetic matrices, with some Zn atoms substituted with Co(II). In both phases $g_\perp > g_{//}$, i.e. Co presents planar anisotropy. This anisotropy is more accentuated in the β-phase where the ratio $g_\perp/g_{//}$ is larger [43]. The effect of the different N adjacent positions with respect to the Co atom in the α- and β-CoPc is also detected with emission Mössbauer spectroscopy since the chemical shift and the quadrupole splitting are larger for the β-phase [44]. The anisotropic character of Co in the Pc environment is also evidenced by magnetic susceptibility. Powder β-CoPc shows a rounded maximum, characteristic of an antiferromagnetic coupled chain with S = 1/2. These results imply that although the intrachain interaction is rather strong ($\mathrm{J/k_B} = -2.3$ K), low dimensionality inhibits the establishment of long range order down to 1.8 K [33].

Cu-Phthalocyanine β-CuPc is paramagnetic down to $T = 1.7$ K, the lowest temperatures investigated so far. This paramagnetism is due to thermal fluctuation of 1D character, with a non-negligible intrachain interaction. Using NMR proton spectroscopy [35], it was determined that its spin is $S = 1/2$, and that the electron-nuclear dipolar interaction couples the proton nuclear spins of the molecule directly to the well localized electronic spins at the Cu sites. The analysis of the spin-spin relaxation time is also explained in terms of an $S = 1/2$ isotropic Heisenberg interaction with an intrachain exchange constant of $|J/k_B| = 0.286K$. An upper limit of the interchain interaction found yields a ratio $|J/J'| \geq 6.4 \times 10^3$. This limit is obtained from the lack of 1-D to 3-D crossover of the spin-lattice relaxation $T_1(\omega)$ down to the lowest measured temperature. For completeness sake, let us mention that XAS and XMCD measurements in bulk powder β-CuPc at the Cu $L_{2,3}$ edge are available [45]. The XAS features just a simple peak as expected for Cu which has an almost filled 3d orbital.

9.3 MPc Thin Films

Many different techniques, like Langmuir Blodget synthesis, spin coating [46] and organic molecular beam epitaxy (OMBE) [47] have been used to grow MPc thin films. Among these, OMBE produces MPc films with different crystalline orientation and order which depends on the substrate type. In general Cu-, Fe-, Mn-, Co- and Ni-Pc molecular planes stack parallel ("lying") to the substrate surface when deposited on Au, [48] Ag, [49–51] Cu [52]or Pd (this work), and tilted so that the MPc molecule plane is nearly perpendicular to the substrate, as has been found for FePc deposited on sapphire, [47], or on polycrystalline substrates [53] and low work function metals like Al and V [54].

Figure 9.5 shows molecular orientations in the standing and lying configurations. In spite of the molecular orientation, MPc films always grow in the lying configuration in most of metallic substrates when the temperature is kept below 200 °C during deposition. The angle θ corresponds to the stacking found in the bulk α-phase [9], however this phase is not identical since the herringbone structure is not observed. Different orientations of the adjacent MPc molecules have been reported to occur, [55, 56] therefore, we denote this structure as α'-phase. MPc films allow exploration of the anisotropic properties related to this phase.

The x ray absortion (XANES) and photoemission (PE) spectra of NiPc, CoPc and FePc thin films have been measured and a qualitative assignment of the peaks done, which show that most of the HOMO and LUMO electronic states are mainly of 3d character [58]. Angular studies using X-ray linear polarized absorption (XLPA) and X-ray magnetic circular dichroism (XMCD), at the M ion $L_{2,3}$ edges (2p $\rightarrow$ 3d electron excitations) helps relate magnetic properties to the electronic structure.

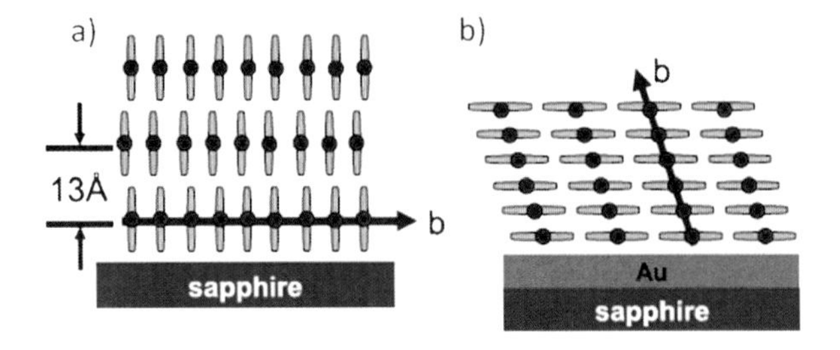

Fig. 9.5 MPc thin film stacking on a supporting surface. (**a**) Standing configuration, (**b**) lying configuration [57]

A simple analysis based on the angle dependent selection rules for the dipolar 2p $\rightarrow$ 3d excitations, has been proposed [59] to explain the XLPA and XMCD spectra. This method is based on the so-called "lighthouse-effect", namely that the absorption intensity is proportional to the number of empty valence states in the direction of the electric field of the incoming linearly polarized x-ray photons. The inset in Fig. 9.6(a) shows an incoming linearly polarized x-ray beam with the electric field contained in the incidence plane at grazing incidence. In this geometry, for MPc in the lying configuration, the electric field vector is perpendicular to the molecule plane (z axis), while for normal incidence the electric field is parallel to the molecule's xy-plane. Since the a_{1g} (d_{z^2}) orbital along the z axis has a larger density of hole states than in the xy-plane, a strong intensity variation is expected as a function of incident angle. This variation reflects the number of empty states in this specific orbital. Similar considerations can be used for the other three ligand field split 3d-states. In addition, increasing the x-ray photon energy (i.e. varying the energy of the incident x-ray beam) allows determination of the empty states above the Fermi level (see Fig. 9.6(a)). The indexing is aided by the calculated spin-split molecular orbital energy level scheme (Fig. 9.6(b)).

This analysis, first applied to a FePc α'-phase thin layer [60], implied that above the Fermi level, the 2p $\rightarrow$ 3d electron transitions to the 3d empty orbital may be indexed as follows: for increasing photon energy, to minority spin empty e_g state, with some mixing of the a_{1g} state, and to the a_{1g} state, the next excitations correspond to transitions to the e_g state in the minority antibonding states, and finally to the b_{1g} majority and minority antibonding states (see Figs. 9.6(a) and 9.6(b)).

In the following subsections the magnetic properties of different MPc compounds films are discussed.

Mn-Phthalocyanine Epitaxial thin films of MnPc were grown on a hydrogen-terminated Si [H-Si(111)] substrate. In these thin films, the molecules stack in columns up to 40 molecular layers, with the stacking axis forming an angle of $\theta = 26.5°$ with respect to the substrate plane normal, and with an intermolecular distance close to d = 3.3 Å. The angle θ corresponds to the α'-phase and becomes less definite for thicker films [9]. Such a highly textured film growth produces striking magnetic anisotropy, with the easy axis perpendicular to the substrate and antiferromagnetic intrachain interactions, in radical contrast to the ferromagnetism of bulk β-MnPc phase. This different magnetic behavior can be explained within the same model of d-electron orbitals overlap as for β-MnPc. However, the relative Mn and N

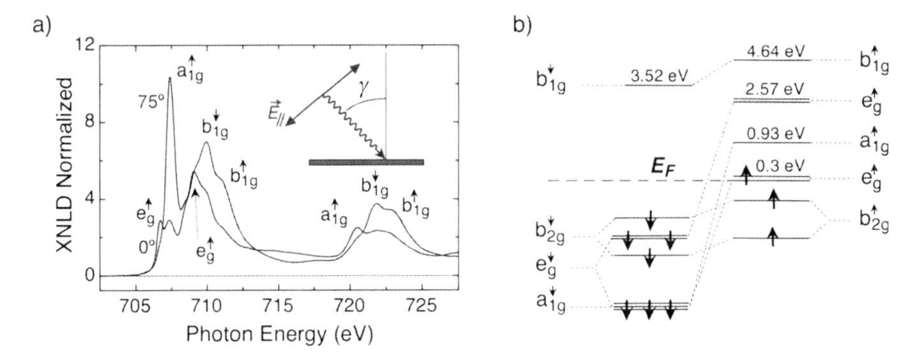

Fig. 9.6 (**a**) X-ray linear polarized absorption at the Fe L2,3 edges of FePc at different incident angles; $\gamma = 0$ (*black line*) and 75° (*red line*). The peaks are classified according to selection rules for X-ray dipolar (2p → 3d) transitions. Inset shows the incidence angle γ and the electric field vector $E_{//}$. (**b**) Spin split molecular orbital energy level scheme of FePc, where only the states with a relevant component of 3d weight ($w_{3d} > 0.05$) have been included. E_F is the Fermi energy. *Arrows* indicate the electron occupation and spin direction at each energy level [60]

positions differ between the thin film and bulk phases; the ferromagnetic exchange path via the a_g–$e_g(\pi)$ becomes weaker, while the antiferromagnetic exchange path e_g–$e_g(\pi)$, via the $e_g\pi$ MO prevails.

Fe-Phthalocyanine These films have been extensively studied since their structure and texture strongly affect their magnetic properties. The "standing" and "lying" stacking of the molecules (Fig. 9.5) can be controlled by the type of substrate as described at the beginning of this section.

In the standing case, depending on the substrate temperature during deposition, AFM and x-ray diffraction [47, 61] show that asymmetrical, elongated grains are formed. Between room temperature and 200 °C the grains consist of α'-phase FePc chains with characteristic lengths ranging from 100 to 3000 molecules, and a typical width of 25 to 80 chains [57]. When the substrate temperature is above 200 °C, the chains organize within the grain in the β-phase [62].

Below 4.5 K, the α'-phase grains order magnetically and give rise to "wasp-like" hysteresis loops. This peculiar magnetic behavior has been explained using the Preisach model with a bi-modal coercive field distribution [63] as found by optical magnetic circular dichroism (MCD) at 2 K. However, part of the hysteresis loop opening may come from slow magnetic relaxation. In contrast, the β-phase grains do not present long range order, as expected from the paramagnetism down to the lowest temperature of the bulk β-phase.

FePc film grows in the "lying" configuration on sapphire substrates covered with a 40 nm nominal thickness Au buffer layer. Annealing for one hour at 300 °C improves surface quality. For optimal crystalline growth and grain size, the substrate is kept at 150 °C during deposition [47]. The lying configuration was determined using grazing incidence XLPA at the N-K edge. The spectra corresponding to the horizontal ($\vec{E}$ field near the substrate normal) and vertical ($\vec{E}$ parallel to the substrate)

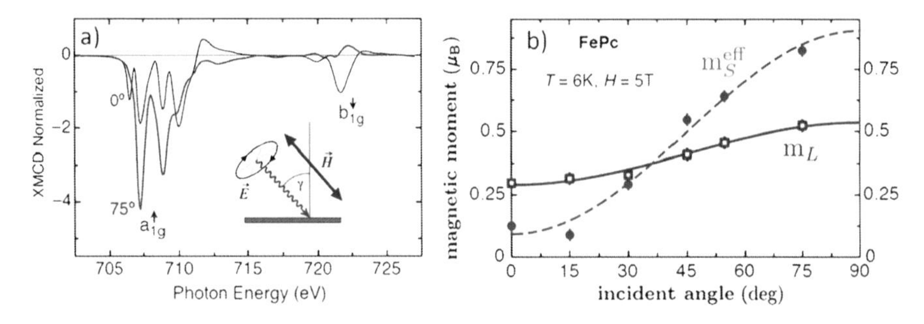

Fig. 9.7 (**a**) XMCD at the Fe $L_{2,3}$ edges of FePc, at different incident angles; $\gamma = 0$ (*black line*) and 75° (*red line*). The applied field (B = 5 T) and the helicity of the beam are parallel for every γ. (**b**) Fe spin (*blue line*) and orbital moments (*red line*) as a function of incident angle [60]

polarizations are completely different. The near disappearance of the π^* resonances for $\vec{E}$ in the molecular plane demonstrates that the four N atoms lay parallel to the substrate [60]. The molecules stack parallel to the substrate, forming chains along an axis which forms an angle of 26.5° with respect to the normal to the substrate.

The magnetization of FePc film, with the molecules parallel to the substrate, is highly anisotropic; at $T = 1.8$ K, with the magnetic field parallel to the substrate (parallel to the plane of the molecule) a magnetic hysteresis loop is observed. This loop nearly collapses when the applied field is perpendicular to the substrate plane.

The "lying" FePc film configuration has also been studied using XMCD at the Fe $L_{2,3}$ edges as a function of the incident angle. At $T = 6$ K (slightly above T_c in the paramagnetic region), a magnetic anisotropy is observed when the magnetic moments are polarized by the applied field in the parallel and perpendicular directions with respect to the substrate [60] (see Fig. 9.7(a)).

Information on the orbital and spin components of the magnetic moment of the absorbing atom projected along the field direction for a given incidence angle γ is obtained from the sum rules at the $L_{2,3}$ edges [64, 65],

$$\frac{m_L(\gamma)}{\mu_B} = -\frac{2n_h}{r} \int (\Delta\mu_{L3} + \Delta\mu_{L2}) dE \tag{9.2}$$

$$\frac{m_S^{\text{eff}}(\gamma)}{\mu_B} = -\frac{3n_h}{r} \int (\Delta\mu_{L3} - 2\Delta\mu_{L2}) dE \tag{9.3}$$

where $\Delta\mu_{L2,3} = \mu_{L2,3}^{-} - \mu_{L2,3}^{+}$, and $\mu_{L2,3}^{-}$ and $\mu_{L2,3}^{+}$ are the absorption measured with left ($-$) and right ($+$) circularly polarized light. The number of d holes above the Fermi energy is given by n_h.

After applying the sum rules analysis, the results are fitted to the following equations:

$$\begin{aligned} m_L &= m_L^z \cos^2\gamma + m_L^{\text{xy}} \sin^2\gamma, \\ m_S^{\text{eff}} &= m_S - 7\left(m_T^z \cos^2\gamma + m_T^{xy} \sin^2\gamma\right), \end{aligned} \tag{9.4}$$

where γ is the incident angle (Fig. 9.7(b)). The magnetic moment parameters are collected in Table 9.4. The most striking result is the easy-plane anisotropy of the orbital component of the magnetic moment, m_L^{xy} (parallel to the molecule plane), and the very large ratio with respect to the spin component, m_L/m_S. This implies that in FePc there is a highly unquenched orbital moment, with a larger component in the molecule plane. The intra-atomic dipolar term m_T is not negligible in this anisotropic FePc layered film and its contribution to the XMCD signal is associated to the planar symmetry of the ligand field.

The origin of this anisotropy is related to the orbital degeneracy of the HOMO, d-electron e_g level. In the α'-phase FePc thin layer the electronic structure can be determined from XLPA measured at the Fe $L_{2,3}$ edges as a function of incident angle, at $T = 6$ K [60]. Using the selection rules for these transitions, the spectral peaks could be indexed in terms of the spin-split molecular orbital energy level scheme [66]. Ab initio density functional calculations (DFT), including ligand field interactions, intra atomic exchange and the hybridization with the four N atoms, provide the energy levels (Fig. 9.6(b)) which explain quantitatively the measured XANES and XMCD. Comparison of these spectra to the results of the calculation suggests that an e_g orbital doublet with three electrons lies at the Fermi energy level (E_F) with the LUMO at an energy 0.3 eV above E_F and an empty state a_{1g} at 0.93 eV. The hole-hole interaction between the e_g and a_{1g} levels and the spin-orbit coupling, split the degenerate states into three doublets with a ground state with $\langle L_z \rangle = \pm\hbar$ [41]. A second order perturbation, for example, any interaction that lowers the symmetry from D_{4h} to D_{2h}, splits this doublet yielding an occupied and an unoccupied e_g level with a_{1g} mixing. The orbital moment of Fe(II) is largely unquenched, thus explaining qualitatively the large orbital moment found. The presence of the all-important e_g partially unoccupied level has been observed in electron energy-loss spectroscopy (EELS) measurements on a free-standing FePc film, in the form of a low energy excitation from the $a_{1u}(\pi)$ ligand states into the Fe (e_g) d-states [67].

The FePc electronic state has been obtained from DFT calculations of the isolated and interacting molecule cases. In the former case the ground state is found to be $^3A_{2g}$. In the latter case, the columnar stacking of the FePc molecules in α'-FePc gives the ^3E_g as the ground state because of the hybridization between MO orbitals along the b-axis. From these results one can conclude that exchange interaction has a strong effect on the electronic ground state [68].

Co-Phthalocyanine CoPc films deposited on Au(111) have been studied by XAS [69], XMCD [34] and XLPA [19]. The first two imply that the electronic ground state is a mix of $^2A_{1g}$ and ^2E_g states. However, the XLPA data analysis of the L_3 spectra indicates that the electronic ground state of CoPc is similar to that of FePc, except for the absence of the low energy e_g excitation. Since Co has one more electron than Fe, it occupies the lowest available hole located at the e_g level filling it completely. As a consequence, only one electron remains uncoupled giving rise to a total spin $S = 1/2$. For this reason, the orbital degeneracy is lost and the orbital moment may be quenched.

The XMCD at the $L_{2,3}$ edges with a $5T$ applied field as a function of incident angle, was used to determine the magnetic moment components [19] (see Table 9.4)

Table 9.4 Atomic orbital and spin component moments of M in MPcs, as determined from XMCD experiments. EDM refers to easy direction of magnetization

MPc	EDM	m_L^z	m_L^{xy}	m_L	m_S	m_T^z	m_L/m_S	m_T	Ref.
β-MnPc				0.21	1.45			1.66	[39]
α'-FePc	x, y	0.29(5)	0.53(4)	0.45	0.64(5)	0.074(5)	0.70(4)	1.19	[60]
α'-CoPc	x, y	0.01(3)	0.07(3)	0.05(3)	0.10(4)	0.02(3)	0.5	0.15	[19]
α'-CuPc	z	0.05(2)	0.01(1)	0.02	0.21(2)	−0.08(5)	0.1	0.23	[19]
CuPc 1 ML	z	0.20	0.045	0.096		−0.268	0.096	1.096	[45]

of a CoPc film. A strong reduction in the magnetic moment is evident, coming from a strong orbital quenching, as expected for a A_{1g} ground state. However, orbital intermixing due to spin-orbit coupling may give rise to a small contribution to the orbital moment. The easy axis magnetic anisotropy is found to be in-plane, as for FePc.

Another mechanism giving rise to small magnetic moments is the AFM intrachain coupling, already active at short range, though attenuated by thermal fluctuations. Even though this is an expected feature, as discussed in the previous section, it is noteworthy that an AFM intrachain coupling with $J/k_B = -208$ K (18 meV) has been reported using inelastic electron tunneling microscopy (IETS) [70]. We believe this is erroneous, since it differs from the reported interactions for other MPcs by 3 orders of magnitude (see Table 9.2).

Cu-Phthalocyanine In thin CuPc films there is only one strong XLPA peak at the L_3 edge that arises from the p → d electron transition to the only hole state available at the b_{1g} high energy level. XMCD shows one peak that corresponds to the excitation of a 2p electron to that hole. The analysis, in terms of the sum rules, indicates an out-of-plane easy magnetic anisotropy axis. In addition, the orbital component of the magnetic moment is very small and the spin component corresponds to $S = 1/2$. Since the orbital moment is nearly quenched for the b_{1g} state, the strong uniaxial anisotropy originates in the dipolar intra-atomic term m_T of the effective spin moment and it reflects the planar b_{1g} orbital which governs the magnetic properties of this compound [19].

9.4 MPc Molecules Adsorbed on Substrates

The very robust adsorption of MPc molecules on a wide variety of substrates, has allowed to perform non-destructive x-ray absorption experiments. Low temperature Scanning Probe Microscopy allows inducing modifications at the single molecule level. The possibility of manipulating, modifying, relocating, and constructing structures at the atomic level has been essential to develop understanding of the interaction mechanisms at the molecule-substrate interface.

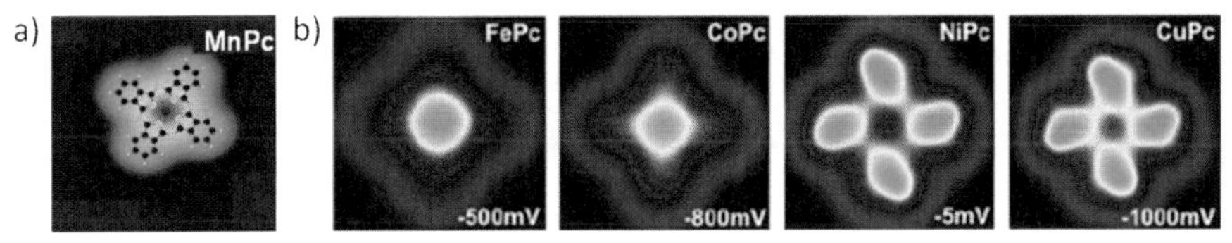

Fig. 9.8 STM topographic images of MPcs. (**a**) Image of MnPc deposited on Pb [71] (Reprinted Figure (**a**) with permission from Fu et al. [71]. Copyright 2007 by the American Physical Society). The MnPc molecular structure is superimposed. (**b**) Bias voltage dependent images of the M = Fe, Co, Ni and Cu Pcs deposited on Au(100). Note the chiral contrast for the Ni and Cu cases [51] (Reprinted Figure (**b**) with permission from Mugurza et al. [51]. Copyright 2012 by the American Physical Society)

MPcs have attracted much attention because of their self-assembling capacity on substrates [72]. For a metal substrate, hybridization of the $3d_{z^2}$ states of the metal (M) center in the MPc and the electronic states of the substrate forces the position of M on the substrate, while the orientation of the molecule is mostly influenced by the interaction of the N atoms and the surface. The relative orientation of the molecules is driven by intermolecular forces. It is now clear that the substrate on which the molecule is adsorbed plays a fundamental role in defining its magnetic state. The final state of the molecule depends on whether the substrate is a ferromagnet, a metal or a semiconductor.

Typically, a single MPc molecule adsorbed on a metallic surface appears in STM as a four-leaf clover shape with a protuberant (Fe, Co) or depressed (Cu, Ni) spot at its center (see Fig. 9.8). The MOs of the π conjugated macrocycles interact with the conducting substrate and modify the total spin of the molecule. The mechanism underlying these phenomena is known as the Kondo effect [73]. This effect is produced by resonant scattering coming from the hybridization of the magnetic impurity, in this particular case the MPc molecule, with a continuum of electronic states, in the conducting substrate. As a result, the net effect is the generation of a collective non-magnetic singlet state. The Kondo effect occurs below a characteristic temperature T_K, which defines the boundary between resonant and non-resonant behavior. When molecules assemble in clusters, the intermolecular interactions may compete with the Kondo effect and the molecules may become magnetic. The first report on Kondo screening of MPcs was done by Gao et al. [74] on FePc. In a short period after his report many researchers have dealt with this subject. Below we review some of the most prominent results on the magnetism of adsorbed MPcs on different substrates.

Mn-Phthalocyanine Single MnPc molecules were deposited by sublimation on top of Pb(111) nanoislands [71]. The molecule image consists of a four lobe cross with a protrusion at its center. The thickness of the Pb nanoisland affects strongly the Kondo resonance, with T_K oscillating as a function of the number of Pb monolayers. This feature originates in the strong confinement of the Pb electrons in films thickness ranging from 2 ML to 22 ML. By comparing STM spectra to simulations, it was concluded that the magnetic moment on Mn decreases from $3\mu_B$, for an isolated molecule, to $0.99\mu_B$ when adsorbed on Pb. The contribution to the magnetic

Table 9.5 Kondo temperature , T_K, and calculated magnetic moment, m, located at the metal atom in M-Phthalocyanine molecules adsorbed on a single-crystal substrate

Sample	Substrate	T_K (K)	$m(\mu_B)$	Reference
MnPc	Pb(111)	23–419	0.99	[71]
FePc	Au(111)	2.6 ± 1.4		[75]
Dehydrogenated CoPc	Au(111)	208	1.03	[76]
CuPc	Ag(100)	27 ± 2		[51]

moment arising from the d_{xy} orbital survives since it is less hybridized with the substrate energy bands.

Fe-Phthalocyanine Single FePc molecules which lay parallel on Au(111) single crystal surface show up as a four-leaf clover, with the organic lobes as the leafs. The clover is oriented in two configurations, at 15° with respect to each other. A bright spot protrusion, at the molecule center, is related to the d-orbital character near the Fermi surface [74]. This feature is related to the strong coupling of the a_{1g} and e_g (perpendicular) orbitals near the Fermi level with the tip states [77]. The STM spectra at the single molecule center position shows a Fano-type resonance characteristic of the Kondo effect. From its temperature dependence, the Kondo temperature T_K of the molecule and substrate collective singlet state can be determined [75] (see Table 9.5). As the density of molecules increases on the substrate, molecules self-assemble in planar clusters, forming a 2D Kondo superlattice on the metal surface. The STS spectra at the borders of these superlattices has a Fano resonance similar to that of a single molecule, however, below T_K the Fano resonance intensity at the center of the cluster is reduced and is split into two peaks. This feature is caused by the oscillatory Ruddermann-Kittel-Kasuya-Yosida (RKKY) interaction, via the conduction electrons, which generates an antiferromagnetic coupling and consequently, AFM correlations between the spins [75].

In contrast, no Kondo effect is detected when a single FePc molecule is deposited on Ag(100). This is caused by a stronger interaction of the a_{1g} and e_g perpendicular states with the substrate [51]. Basically due to the interaction between the FePc molecules and Ag, one electron is transferred from the substrate. The MPc states ^2e_g and e_g are mixed due to hybridization with the substrate while a_{1g} is the MO with the highest degree of hybridization. This modifies the charge distribution in the molecule, which consequently reduces the Fe spin value.

These results are in agreement with XMCD performed on a 1ML FePc on Au(111) [19, 34], which shows a reduced (but not completely quenched) Fe magnetic moment (Fig. 9.9). This conclusion arises from the charge transfer between the d^6 Fe configuration and the metal substrate through the d_{z^2} orbital, assuming a weak mixing with the d^7 configuration. Under this assumption, the net result is that the spin of the bound electron from the substrate couples AFM to the two lowest states of Fe, and as a consequence yields a total spin $S = 1/2$.

The conductivity of the substrate also plays an important role in the magnetic state of the adsorbed molecule. In a single FePc molecule deposited on a clean,

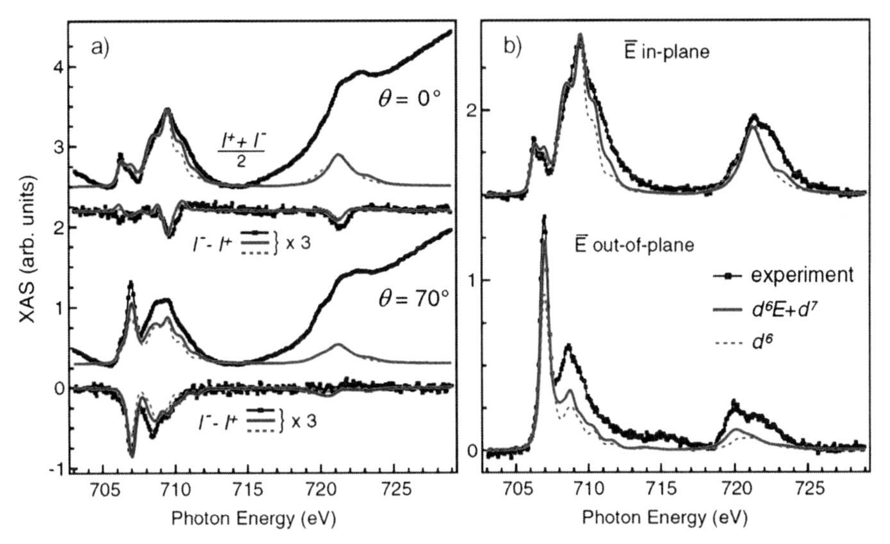

Fig. 9.9 (**a**) Circular polarized XAS (*upper*) and XMCD (*lower*) at the Fe $L_{2,3}$ edges on a 1ML FePc on Au(111) at two different incident angles θ (*black lines*). (**b**) X-ray linear polarized absorption in two polarizations. The x-ray electric field vector is either fully in-plane or out-of-plane at an angle of 20° with respect to the surface normal (*red lines*) simulations [34] (Reprinted Figure with permission from Stepanow et al. [34]. Copyright 2011 by the American Physical Society)

metallic Cu(110) substrate, the Fe magnetic state changes from the bulk $S = 1$ to $S = 0$. In contrast, when deposited on semiconducting, oxidized, Cu surface (Cu(110)(2 × 1)-O), the $S = 1$ state is retained, although with a different ligand field splitting than in the bulk [78].

A recent interesting development is the deposition of FePc on graphene, supported by Ir(111). FePc sub-monolayers lay flat on graphene with a $8 \pm 10°$ angular tilt with respect to the graphene surface. Thicker deposition leads to less ordered island formation [79]. XMCD measurements, at the Fe $L_{2,3}$ edge, show clear anisotropic dichroism, indicating that magnetic properties of FePc are affected by the interaction with graphene. An increase in the planar anisotropy (more intensity due to the in-plane orbitals) with respect to α'-FePc in the thin film case (Sect. 9.3. and Table 9.4) is observed. On the other hand, the total moment of the Fe atom decreases. Both the moment and anisotropy decrease with increasing thickness [80].

Co-Phthalocyanine As for FePc deposited on Ag, STM of CoPc shows a protusion at the CoPc molecule center [51]. However, in CoPc deposited on Au(111) or Ag(100) there is no Kondo effect [51, 76]. In fact, XMCD of 1 ML CoPc shows that valence fluctuations quench the Co moment [34]. Charge transfer between the molecule and the substrate accounts for the absence of magnetic moment. This occurs because a Co excited level is occupied by an extra substrate electron which intermixes with the ground state. As a result there is a reduced a_{1g} level occupation, and the total coupled state is a non-magnetic singlet state [51]; i.e. with $S = 0$. In CoPc deposited on Cu(111), the N 1s XAS spectra imply that an electronic charge redistribution, compatible with charge transfer from the substrate to the molecule, takes place upon adsorption [52].

Single CoPc molecules couple FM to Co nanoislands as shown by spin polarized STM. The magnetization density observed in this case reaches a maximum close to the Co atoms, although there is some oppositely oriented magnetic moment at the N and C atoms. The compensation of these moments leads to the quenching of the total moment [81]. Moreover, when 1ML CoPc is deposited on a metallic Fe FM film, a small but distinguishable XMCD component is detected, which indicates the presence of a non-zero moment parallel to the Fe substrate magnetization [52]. In fact, no XMCD signal is observed on a 1ML of CoPc deposited on Au(110) [19]. When the thickness increases to 6 ML this XMCD component completely disappears, as in the thick film case. In fact, no XMCD signal is observed on a 1ML of CoPc deposited on Au(110) [19]. This result is compatible with DFT calculations of the electronic structure and van der Waals (vdW) forces to determine the total molecule spin state. Although the electron supplied by the substrate fills the d_{z^2} MO state, the distortions produced by the vdW forces give rise to a spin redistribution so that spin splitting is recovered due to the molecule-surface bonding at the ligand portions of the molecule [52, 82].

A very interesting development in the field of MPcs adsorbed on metallic substrates is the possibility of manipulating the Pc ligands in order to modify the magnetic state of the molecule/substrate magnetic state. Pioneering work has shown that an STM tip may induce dehydrogenation of a single CoPc adsorbed on Au(111) [76]. The as-deposited CoPc molecule shows no Kondo effect, while in the dehydrogenated molecule there is an onset of Kondo effect, with $T_K \approx 208$ K. The STM tip removes the external H atoms of the CoPc molecule and as a consequence, favors chemical binding with the substrate via the ligands. The molecule deforms from a flat shape, prior to dehydrogenation, into a four-legged concave cap-like table with the concavity towards the surface, after H removal (Fig. 9.10(a)). The magnetic state of CoPc transforms from initially non-magnetic, to magnetic in the H-trimmed bound molecule, with a moment of $1.09\mu_B$. Curiously, such a moment is larger than that of a single Co atom directly bound to a Au atom on the Au(111) substrate. The coupling of the Co atom in the molecule with the substrate, via the H-trimmed Phthalocyanine molecule ligands, is stronger than in the direct Co-Au coupling case [76, 83].

Cu-Phthalocyanine Adsorption of sub-monolayers CuPc on Au(111), Ag(111) and Cu(111) studied with low-energy electron diffraction [84], show various degrees of binding. The binding is weak on Au (physisorption), where no charge transfer is observed. On Ag it is more intense (weak chemisorption), with charge transfer that gives rise to intermolecular repulsion. When deposited on Cu the interaction with the substrate is the strongest (strong chemisorption), and the intermolecular interaction is attractive. The symmetry reduction from fourfold to twofold caused by the different filling of the LUMO state induces an electrostatic quadrupole moment that, in turn, generates an attractive intermolecular force. These attractive forces prevail over the repulsive ones and there is a net effective attractive interaction.

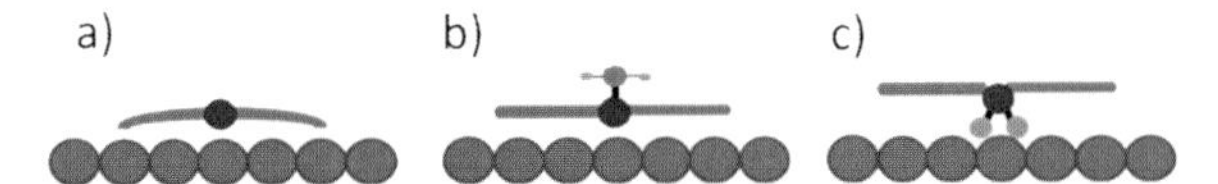

Fig. 9.10 Schematic diagram of MPc molecule deposited on a substrate (*blue circles* represent the substrate's atoms): (**a**) dehydrogenated CoPc on Au [76]. (**b**) NH_3/FePc/Au. Note the NH_3 is located in the external side of the FePc molecule [85]. (**c**) FePc/(η^2-O_2)/Ag. Note the O_2 group sandwiched between the FePc molecule and the substrate [86]

STM of CuPc adsorbed on Ag shows a depression at its center. The spin state of Cu is $S = 1/2$ and an elastic Kondo resonance is also observed. This Kondo resonance arises from the presence of an unpaired spin located at the macrocycle $^2e_g(\pi)$ orbital. The $S = 1/2$ spin couples with the Cu $S = 1/2$ to generate a $S = 1$ triplet ground state and an excited $S = 0$ singlet. This is confirmed by the presence of a strong Kondo peak at zero bias with $T_K = 27$ K.

These results may be compared to XMCD performed at the Cu $L_{2,3}$ edge on a 1ML thin film on Au, recalling that this technique is only sensitive to the Cu empty d-states. The XMCD measurements on a CuPc 1ML [19, 45] show that about 10 % of the magnetic moment is of orbital origin, and the anisotropy is perpendicular to the molecule plane. Moreover, the value of the magnetic moments is strongly enhanced with respect to that of the Cu in a thin film [19] (see Table 9.4). The anisotropic dipolar term, represented by m_T (Table 9.4) is negative in the CuPc, reflecting the uniaxial anisotropy instead of the planar one observed in FePc and CoPc.

9.5 Perspectives of MPcs

In this Section we mention briefly the research lines on MPcs that have been opened and are gaining impetus in the advancing field towards single molecule magnetic switching. This relates to the possibility of controlling the magnetic state of a molecule by external means and on single molecule spintronics.

Since the seminal work on chemical switching of magnetic properties of molecules deposited on a substrate [67], a renewed activity in the molecule state control has erupted. A planar Fe-TPA_4 (TPA = terephthalate) molecule deposited on Cu can switch the Fe in-plane magnetic anisotropy to out-of-plane when the molecule adsorbs selectively two oxygen atoms (O_2-FeTPA4) [87]. FePc deposited on a metallic substrate, and subject to adsorption of different molecules, has been used in this context also. After deposition on Au, as described in the previous section, FePc is allowed to react with different ligands. In NH_3/FePc/Au there is a weak chemisorption bonding of NH_3 with the Fe atom on the external side of the deposited FePc molecule on Au (Fig. 9.10(b)). This modifies the Fe coordination with a consequent reorganization of the Fe charge, which leads to pairing of electrons and quenching of the spin ($S = 0$). In addition, there is a weakening in the FePc-Au bonding. These conclusions are obtained from the interpretation of XPS measurements at the $2p_{3/2}$ line combined with DFT calculations [85]. For Pyridine/FePc/Au the effect on the molecule-surface bonding and spin quenching is similar. However, in the case of NO/FePc/Au the spin quenching seems to be partial ($S = 1/2$). The mechanism causing the weakening of the FePc- substrate interaction is the capture of the gas molecule on the external side of the adsorbed molecule [88] that causes a redistribution of charges and, in turn, the dehybridization of the Fe d_{z^2} orbitals and the substrate Au(111) states.

The self-assembly of functionalized MPc's on a oxygen-reconstructed Co substrate has been used to create a checkerboard lattice of Mn(III)Pc and fluorinated Fe(II)F_{16}Pc [89]. XMCD, shows that both Fe and Mn are AF coupled to the Co substrate magnetization. In a subsequent step, dosing with NH_3 modifies the electronic state of the 3d metals; it quenches the magnetic moment of Fe ($S = 0$), while it maintains the Mn moment AF coupled to the substrate. The original state is recovered by annealing at 300 K; the Mn moment is ON (AF coupled to Co, while the Fe moment switches from ON ($S = 1/2$, AF coupled to Co) to OFF ($S = 0$). The process can be reversible, thus it allows cyclic switching. This recent achievement corroborates the very active research on the magnetic switching by external chemical activation.

The exposure to oxygen of 1 ML FePc on Ag(110) produces different binding. The most stable configuration, compatible with experimental STM and spectroscopic evidence (XAS, XPS), is the FePc/(η^2-O_2)/Ag one, where η^2-O_2 describes that each of the O atoms is bound to Fe by an Fe-O bond. In this configuration the oxygen is chemisorbed in the interfacial structure between the organic molecule and the Ag support, with the Fe of the Pc molecule placed on top of an Ag atom and the two oxygen Fe-O bonds directed towards two lateral substrate Ag atoms; i.e. the oxygen is encaged between the FePc and the Ag substrate (Fig. 9.10(c)). The effect of this configuration on the magnetic properties remains to be studied. Interestingly, the FePc/Ag system acts as a cyclic catalyst in the oxygen reduction reaction, a property that is of great interest as a substitute for Pt catalyst [86]. This implies an expanding basic and applied surface chemistry and physics activity related to molecular switching mechanisms produced by adsorption of atoms and ligands.

The possibility of creating single molecule spintronic devices, such as spin valves or spin-filters, in electrode/MPc/electrode sandwiches has been theoretically and experimentally explored recently. Particularly intriguing is the predictions that a MnPc molecule sandwiched between two semi-infinite armchair single walled carbon nanotubes may be a robust, 100 % efficient spin-filter with very high transmission around the Fermi energy [90]. The delocalization of the π type HOMO MnPc molecule states and their "pinning" with the conducting states of the carbon electrodes lead to the formation of efficient conducting channels. The same efficient spin-filter configuration is predicted to occur with the FePc molecule [91].

In addition, giant magnetoresistance (GMR) has been reported to occur through a single H_2Pc molecule. This was claimed for a molecule on a Co metal island deposited on Cu(111) surface FM electrode, with a Co coated tip as the second FM electrode. GMR as high as 60 % was measured. This was proposed to be due to the LUMO molecular states coupling to highly polarized spin minority states of Co

[92]. GMR was also found for a H_2Pc molecule sandwiched between AFM Mn as one electrode and an Fe coated ferromagnetic tip [6]. Thus, the fabrication of a single molecule phthalocyanine spin-valve may not be too far.

In the introduction of this chapter we set out to cover the magnetic properties of transition metal MPcs. However, M substitution is much richer and goes beyond the simple MPcs described above. Of particular interest to this book is the recent work on double decker phthalocyanines ($LnPc_2$), where Ln is a rare earth. In this type of molecule, the Ln(III) atom is sandwiched between two Pc molecules, rotated by 45° with respect to each other. The Ln = Tb and Dy compounds hysteresis cycle exhibits slow relaxation, similar to that of Single Molecule Magnets (SMM) formed by clusters. Interestingly, in this case however a single magnetic atom is responsible for the phenomenon, which has been therefore denoted as a "single ion magnet" (SIM) [93, 94]. The low temperature hysteresis curves show steps at certain fixed fields, similar to those in SMMs discussed in other Chapters, which have been explained as due to the existence of Magnetic Quantum Tunneling (MQT). In the very low temperature hysteresis loop in a single crystal, these occur in $TbPc_2$, less clearly in $DyPc_2$ [95], and later in $HoPc_2$ [96]. To explain the regularly spaced steps in applied magnetic field in the Ho case, and irregularly in the Tb case, a different mechanism is necessary. The steps due to magnetic tunneling resonance in SMM clusters (see Chaps. 1 and 2), take place when the field split electronic spin levels cross. In the $LnPc_2$'s this type of crossing yields to very high crossing fields compared to those observed. In fact, the non-zero nuclear spin of the Ln lends a new magnetic degree of freedom. The interactions acting on the 4f electronic ground state are the ligand field which splits the Ln free electronic states. This gives rise to a highly uniaxial anisotropic electronic spin ground state, the hyperfine interaction coupling it with the nuclear spin I and the nuclear quadrupole interaction term. Then, the steps in the $LnPc_2$ low temperature hysteresis curves occur at those applied fields for which the total energy levels (i.e. the entangled electronic J and nuclear spin states I), $\phi = |J_z\rangle|I_z\rangle$, become degenerate. Moreover, the hyperfine and the quadrupolar interactions provide the off-diagonal terms in the Hamiltonian to allow resonant tunneling at the so called, "avoided crossing" of levels, which produce the fast relaxation channels at the crossing field. The nuclear spin states degree of freedom play a crucial role for the Quantum Tunneling of $LnPc_2$ but not for the transition metal SMMs [95, 96].

Because of the chemical stability of the MPc adsorbed on different substrates and of the magnetic bi-stability, $TbPc_2$ is a natural candidate as a possible single molecule memory or quantum computing element. $TbPc_2$ has been deposited on polycrystalline Au [97] and Cu(100) substrate covered by ferromagnetic Ni thin capping layer [98]. Molecules are in the "lying" configuration i.e. with the easy magnetization axis perpendicular to the substrate. The ferromagnetic Ni layer couples antiferromagnetically with the Tb magnetic moment and stabilizes its polarization up to room temperature. In contrast to the ferromagnetic coupling present in the transition metal ferromagnetic substrate, discussed in the previous section, in $TbPc_2$ there is an intermediate Pc molecule between the Tb ion and the substrate. As a consequence, direct interaction between Tb and Ni substrate weakens and allows for the AFM superexchange coupling to prevail.

The element selectivity of XMCD has been exploited very nicely in proving the AFM character of this interaction [98]. As a function of applied field, the XMCD at the Ni L_3 and the Tb M_5 edges were measured at fixed photon energy at the XMCD peak. The field applied perpendicular to the substrate overcomes the competing AFM Tb-Ni exchange field until the Tb moment rotates and becomes aligned with the field. This first exciting result opens the possibility of using these molecules as spintronic elements at room temperature, and that new molecules and configurations can be expected to appear in the near future.

Acknowledgements J. Bartolomé acknowledges financial support from the projects MINECO (MAT2011/23791) and DGA IMANA E34. C. Monton and I.K. Schuller acknowledge the financial support of AFOSR FA9550-10-1-0409. The magnetism aspects of this work were supported by the Office of Basic Energy Science, US Department of Energy, under Grant No. DEFG03-87ER-45332. Illuminating discussions with F. Bartolomé and G. Filoti are acknowledged.

References

1. R.P. Linstead, J. Chem. Soc. **1016** (1934). doi:10.1039/jr9340001016
2. R.P. Linstead, A.R. Lowe, J. Chem. Soc. **1022** (1934). doi:10.1039/jr934000122
3. A. Braun, J. Tcherniac, Ber. Dtsch. Chem. Ges. **40**, 2709 (1907)
4. H. de Diesbach, E. von der Weid, Helv. Chim. Acta **10**, 886 (1927)
5. A.G. Dandridge, H.A. Drescher, J. Thomas, Br. Pat. Abstr. **322**, 169 (1929)
6. W. Wulfhekel, T. Miyamachi, S. Schmaus, T.K. Yamada, A.F. Takacs, A. Bagrets, F. Evers, T. Balashov, M. Gruber, V. Davesne, M. Bowen, E. Beaurepaire, in *Spintronics with Single Molecules: 12th IEEE International Conference on Nanotechnology (IEEE-NANO)*, Birmingham, UK (2012)
7. M. Evangelisti, in *Encyclopedia of Supramolecular Chemistry*, ed. by J.L. Atwood, J. Steed (Dekker, New York, 2004), p. 1069
8. M.K. Engel, *The Porphyrin Handbook* (Academic Press, New York, 2003), p. 1
9. H. Yamada, T. Shimada, A. Koma, J. Chem. Phys. **108**, 10256 (1998)
10. J.F. Kirner, W. Dow, W.R. Scheidt, Inorg. Chem. **15**, 1685 (1976)
11. C. Ercolani, C. Neri, J. Chem. Soc. A **1715** (1967). doi:10.1039/j19670001715
12. M. Ashida, N. Uyeda, E. Suito, Bull. Chem. Soc. Jpn. **39**, 2616 (1966)
13. R. Mason, G.A. Williams, P.E. Fielding, J. Chem. Soc., Dalton Trans. **676** (1979). doi:10.1039/dt9790000676
14. J.M. Robertson, J. Chem. Soc. **615** (1935). doi:10.1039/jr9350000615
15. B. Honigman, H.U. Lenne, R. Schrodel, Z. Kristallogr. **122**, 185 (1965)
16. N. Marom, L. Kronik, Appl. Phys. A, Mater. Sci. Process. **95**, 159 (2009)
17. N. Marom, L. Kronik, Appl. Phys. A, Mater. Sci. Process. **95**, 165 (2009)
18. M.S. Liao, S. Scheiner, J. Chem. Phys. **114**, 9780 (2001)
19. P. Gargiani, G. Rossi, R. Biagi, V. Corradini, M. Pedio, S. Fortuna, A. Calzolari, S. Fabris, J.C. Cezar, N.B. Brookes, M.G. Betti, Phys. Rev. B **87**, 165407 (2013)
20. M.G. Betti, P. Gargiani, C. Mariani, S. Turchini, N. Zema, S. Fortuna, A. Calzolari, S. Fabris, J. Phys. Chem. C **116**, 8657 (2012)
21. S. Mitra, A.K. Gregson, W.E. Hatfield, R.R. Weller, Inorg. Chem. **22**, 1729 (1983)
22. C.G. Barraclough, R.L. Martin, S. Mitra, R.C. Sherwood, J. Chem. Phys. **53**, 1638 (1970)
23. S. Heutz, C. Mitra, W. Wu, A.J. Fisher, A. Kerridge, M. Stoneham, T.H. Harker, J. Gardener, H.H. Tseng, T.S. Jones, C. Renner, G. Aeppli, Adv. Mater. **19**, 3618 (2007)
24. B.E. Williamson, T.C. Vancott, M.E. Boyle, G.C. Misener, M.J. Stillman, P.N. Schatz, J. Am. Chem. Soc. **114**, 2412 (1992)

25. T. Kroll, R. Kraus, R. Schonfelder, V.Y. Aristov, O.V. Molodtsova, P. Hoffmann, M. Knupfer, J. Chem. Phys. **137**, 054306 (2012)
26. P. Coppens, L. Li, N.J. Zhu, J. Am. Chem. Soc. **105**, 6173 (1983)
27. B.W. Dale, R.J.P. Williams, C.E. Johnson, T.L. Thorp, J. Chem. Phys. **49**, 3441 (1968)
28. P.S. Miedema, S. Stepanow, P. Gambardella, F.M.F. de Groot, J. Phys. Conf. Ser. **190**, 012143 (2009)
29. C.G. Barraclough, R.L. Martin, S. Mitra, R.C. Sherwood, J. Chem. Phys. **53**, 1643 (1970)
30. M. Evangelisti, J. Bartolome, L.J. de Jongh, G. Filoti, Phys. Rev. B **66**, 144410 (2002)
31. N. Ishikawa, Struct. Bond. **135**, 211 (2010)
32. A.B.P. Lever, J. Chem. Soc. **1821** (1965). doi:10.1039/jr9650001821
33. H. Miyoshi, Bull. Chem. Soc. Jpn. **47**, 561 (1974)
34. S. Stepanow, P.S. Miedema, A. Mugarza, G. Ceballos, P. Moras, J.C. Cezar, C. Carbone, F.M.F. de Groot, P. Gambardella, Phys. Rev. B **83**, 220401 (2011)
35. S. Lee, M. Yudkowsky, W.P. Halperin, M.Y. Ogawa, B.M. Hoffman, Phys. Rev. B **35**, 5003 (1987)
36. H. Senff, W. Klemm, J. Prakt. Chem. **154**, 73 (1939)
37. P.A. Reynolds, B.N. Figgis, Inorg. Chem. **30**, 2294 (1991)
38. Y. Kitaoka, T. Sakai, K. Nakamura, T. Akiyama, T. Ito, J. Appl. Phys. **113**, 17E130 (2013)
39. T. Kataoka, Y. Sakamoto, Y. Yamazaki, V.R. Singh, A. Fujimori, Y. Takeda, T. Ohkochi, S.I. Fujimori, T. Okane, Y. Saitoh, H. Yamagami, A. Tanaka, Solid State Commun. **152**, 806 (2012)
40. J.B. Goodenough, Magnetism and the chemical bond, in *Interscience Monograph on Chemistry*, ed. by F.A. Cotton (Willey, New York, 1966)
41. G. Filoti, M.D. Kuz'min, J. Bartolome, Phys. Rev. B **74**, 134420 (2006)
42. R.J. Glauber, J. Math. Phys. **4**, 294 (1963)
43. J.M. Assour, W.K. Kahn, J. Am. Chem. Soc. **87**, 207 (1965)
44. T.S. Srivasta, J.L. Przybyli, A. Nath, Inorg. Chem. **13**, 1562 (1974)
45. S. Stepanow, A. Mugarza, G. Ceballos, P. Moras, J.C. Cezar, C. Carbone, P. Gambardella, Phys. Rev. B **82**, 014405 (2010)
46. M.J. Cook, Pure Appl. Chem. **71**, 2145 (1999)
47. C.W. Miller, A. Sharoni, G. Liu, C.N. Colesniuc, B. Fruhberger, I.K. Schuller, Phys. Rev. B **72**, 104113 (2005)
48. T. Gredig, K.P. Gentry, C.N. Colesniuc, I.K. Schuller, J. Mater. Sci. **45**, 5032 (2010)
49. T. Takami, C. Carrizales, K.W. Hipps, Surf. Sci. **603**, 3201 (2009)
50. K. Manandhar, K.T. Park, S. Ma, J. Hrbek, Surf. Sci. **603**, 636 (2009)
51. A. Mugarza, R. Robles, C. Krull, R. Korytar, N. Lorente, P. Gambardella, Phys. Rev. B **85**, 155437 (2012)
52. E. Annese, J. Fujii, I. Vobornik, G. Panaccione, G. Rossi, Phys. Rev. B **84**, 174443 (2011)
53. H. Peisert, T. Schwieger, J.M. Auerhammer, M. Knupfer, M.S. Golden, J. Fink, P.R. Bressler, M. Mast, J. Appl. Phys. **90**, 466 (2001)
54. C. Monton, I. Valmianski, I.K. Schuller, Appl. Phys. Lett. **101**, 133304 (2012)
55. X. Chen, Y.S. Fu, S.H. Ji, T. Zhang, P. Cheng, X.C. Ma, X.L. Zou, W.H. Duan, J.F. Jia, Q.K. Xue, Phys. Rev. Lett. **101**, 197208 (2008)
56. M. Casarin, M. Di Marino, D. Forrer, M. Sambi, F. Sedona, E. Tondello, A. Vittadini, V. Barone, M. Pavone, J. Phys. Chem. C **114**, 2144 (2010)
57. T. Gredig, M. Werber, J.L. Guerra, E.A. Silverstein, M.P. Byrne, B.G. Cacha, J. Supercond. Nov. Magn. **25**, 2199 (2012)
58. K.A. Simonov, A.S. Vinogradov, M.M. Brzhezinskaya, A.B. Preobrajenski, A.V. Generalov, A.Y. Klyushin, Appl. Surf. Sci. **267**, 132 (2013)
59. J. Stohr, H. Konig, Phys. Rev. Lett. **75**, 3748 (1995)
60. J. Bartolome, F. Bartolome, L.M. Garcia, G. Filoti, T. Gredig, C.N. Colesniuc, I.K. Schuller, J.C. Cezar, Phys. Rev. B **81**, 195405 (2010)
61. G. Liu, T. Gredig, I.K. Schuller, Europhys. Lett. **83**, 56001 (2008)
62. K.P. Gentry, T. Gredig, I.K. Schuller, Phys. Rev. B **80**, 174118 (2009)

63. T. Gredig, C.N. Colesniuc, S.A. Crooker, I.K. Schuller, Phys. Rev. B **86**, 014409 (2012)
64. B.T. Thole, P. Carra, F. Sette, G. Vanderlaan, Phys. Rev. Lett. **68**, 1943 (1992)
65. P. Carra, B.T. Thole, M. Altarelli, X.D. Wang, Phys. Rev. Lett. **70**, 694 (1993)
66. M.D. Kuz'min, R. Hayn, V. Oison, Phys. Rev. B **79**, 024413 (2009)
67. A. König, F. Roth, R. Kraus, M. Knupfer, J. Chem. Phys. **130**, 214503 (2009)
68. K. Nakamura, Y. Kitaoka, T. Akiyama, T. Ito, M. Weinert, A.J. Freeman, Phys. Rev. B **85**, 235129 (2012)
69. T. Kroll, V.Y. Aristov, O.V. Molodtsova, Y.A. Ossipyan, D.V. Vyalikh, B. Buchner, M. Knupfer, J. Phys. Chem. A **113**, 8917 (2009)
70. X. Chen, Y.-S. Fu, S. H. Ji, T. Zhang, P. Cheng, X. C. Ma, X.-L. Zou, W.-H. Duan, J.-F. Jia, Q.-K. Xue, Phys. Rev. Lett. **101**, 197208 (2008)
71. Y.S. Fu, S.H. Ji, X. Chen, X.C. Ma, R. Wu, C.C. Wang, W.H. Duan, X.H. Qiu, B. Sun, P. Zhang, J.F. Jia, Q.K. Xue, Phys. Rev. Lett. **99**, 256601 (2007)
72. R. Otero, J.M. Gallego, A.L.V. de Parga, N. Martin, R. Miranda, Adv. Mater. **23**, 5148 (2011)
73. J. Kondo, Prog. Theor. Phys. **32**, 37 (1964)
74. L. Gao, W. Ji, Y.B. Hu, Z.H. Cheng, Z.T. Deng, Q. Liu, N. Jiang, X. Lin, W. Guo, S.X. Du, W.A. Hofer, X.C. Xie, H.J. Gao, Phys. Rev. Lett. **99**, 106402 (2007)
75. N. Tsukahara, S. Shiraki, S. Itou, N. Ohta, N. Takagi, M. Kawai, Phys. Rev. Lett. **106**, 187201 (2011)
76. A.D. Zhao, Q.X. Li, L. Chen, H.J. Xiang, W.H. Wang, S. Pan, B. Wang, X.D. Xiao, J.L. Yang, J.G. Hou, Q.S. Zhu, Science **309**, 1542 (2005)
77. X. Lu, K.W. Hipps, X.D. Wang, U. Mazur, J. Am. Chem. Soc. **118**, 7197 (1996)
78. N. Tsukahara, K.I. Noto, M. Ohara, S. Shiraki, N. Takagi, Y. Takata, J. Miyawaki, M. Taguchi, A. Chainani, S. Shin, M. Kawai, Phys. Rev. Lett. **102**, 167203 (2009)
79. M. Scardamaglia, G. Forte, S. Lizzit, A. Baraldi, P. Lacovig, R. Larciprete, C. Mariani, M.G. Betti, J. Nanopart. Res. **13**, 6013 (2011)
80. M. Scardamaglia, S. Lisi, S. Lizzit, A. Baraldi, R. Larciprete, C. Mariani, M.G. Betti, J. Phys. Chem. C **117**, 3019 (2012)
81. C. Iacovita, M.V. Rastei, B.W. Heinrich, T. Brumme, J. Kortus, L. Limot, J.P. Bucher, Phys. Rev. Lett. **101**, 116602 (2008)
82. J. Brede, N. Atodiresei, S. Kuck, P. Lazic, V. Caciuc, Y. Morikawa, G. Hoffmann, S. Blugel, R. Wiesendanger, Phys. Rev. Lett. **105**, 047204 (2010)
83. M.F. Crommie, Science **309**, 1501 (2005)
84. B. Stadtmuller, I. Kroger, F. Reinert, C. Kumpf, Phys. Rev. B **83**, 085416 (2011)
85. C. Isvoranu, B. Wang, E. Ataman, K. Schulte, J. Knudsen, J.N. Andersen, M.L. Bocquet, J. Schnadt, J. Chem. Phys. **134**, 114710 (2011)
86. F. Sedona, M. Di Marino, D. Forrer, A. Vittadini, M. Casarin, A. Cossaro, L. Floreano, A. Verdini, M. Sambi, Nat. Mater. **11**, 970 (2012)
87. P. Gambardella, S. Stepanow, A. Dmitriev, J. Honolka, F.M.F. de Groot, M. Lingenfelder, S. Sen Gupta, D.D. Sarma, P. Bencok, S. Stanescu, S. Clair, S. Pons, N. Lin, A.P. Seitsonen, H. Brune, J.V. Barth, K. Kern, Nat. Mater. **8**, 189 (2009)
88. C. Isvoranu, B. Wang, K. Schulte, E. Ataman, J. Knudsen, J.N. Andersen, M.L. Bocquet, J. Schnadt, J. Phys. Condens. Matter **22**, 472002 (2010)
89. C. Wäckerlin, J. Nowakowski, S.-X. Liu, M. Jaggi, D. Siewert, J. Girovsky, A. Shchyrba, T. Hählen, A. Kleibert, P.M. Oppeneer, F. Nolting, S. Decurtins, T.A. Jung, N. Ballav, Adv. Mater. **25**, 2404 (2013)
90. X. Shen, L.L. Sun, E. Benassi, Z.Y. Shen, X.Y. Zhao, S. Sanvito, S.M. Hou, J. Chem. Phys. **132**, 054703 (2010)
91. X. Shen, L.L. Sun, Z.L. Yi, E. Benassi, R.X. Zhang, Z.Y. Shen, S. Sanvito, S.M. Hou, Phys. Chem. Chem. Phys. **12**, 10805 (2010)
92. S. Schmaus, A. Bagrets, Y. Nahas, T.K. Yamada, A. Bork, M. Bowen, E. Beaurepaire, F. Evers, W. Wulfhekel, Nat. Nanotechnol. **6**, 185 (2011)
93. N. Ishikawa, M. Sugita, T. Ishikawa, S.Y. Koshihara, Y. Kaizu, J. Am. Chem. Soc. **125**, 8694 (2003)

94. N. Ishikawa, M. Sugita, N. Tanaka, T. Ishikawa, S.Y. Koshihara, Y. Kaizu, Inorg. Chem. **43**, 5498 (2004)
95. N. Ishikawa, M. Sugita, W. Wernsdorfer, Angew. Chem., Int. Ed. Engl. **44**, 2931 (2005)
96. N. Ishikawa, M. Sugita, W. Wernsdorfer, J. Am. Chem. Soc. **127**, 3650 (2005)
97. L. Margheriti, D. Chiappe, M. Mannini, P.E. Car, P. Sainctavit, M.A. Arrio, F.B. de Mongeot, J.C. Cezar, F.M. Piras, A. Magnani, E. Otero, A. Caneschi, R. Sessoli, Adv. Mater. **22**, 5488 (2010)
98. A. Lodi Rizzini, C. Krull, T. Balashov, J.J. Kavich, A. Mugarza, P.S. Miedema, P.K. Thakur, V. Sessi, S. Klyatskaya, M. Ruben, S. Stepanow, P. Gambardella, Phys. Rev. Lett. **107**, 177205 (2011)

Part III
Applications

Chapter 10
Potentialities of Molecular Nanomagnets for Information Technologies

Marco Affronte and Filippo Troiani

Abstract The possibility of tailoring their functionalities at the molecular scale makes molecular nanomagnets interesting for applications in information technologies where the race for extreme miniaturization will soon lead at requiring components of few nanometers in size. Properties like the magnetic bistability or the switchability by external stimuli actually allow one to mimic, at the molecular scale, basic operations commonly used in computers while embedding magnetic molecules in suitable electronic circuits allows the fabrication of novel spintronic devices. Even more challenging is the control and the exploitation of quantum properties in molecular spin clusters that may allow the encoding of quantum information with molecules. These concepts are substantiated by many achievements obtained in the recent years and presented in this chapter along with some perspectives and next challenges for the future.

10.1 Introduction

In this chapter we consider possibilities for exploiting molecular nanomagnets (MNM) for Information and Communication Technologies (ICT). The basic ideas are essentially grounded on the parallelism between the way molecular nanomagnets and computer components work. For instance, a molecule with a double-well potential due to magnetic anisotropy mimics the behaviour of a magnetic register at the nanoscale; a switchable molecule can be seen as an element in a logic gate or in a spintronic device, while the dynamics of a molecular spin perfectly maps operations required by quantum gates.

Before discussing in detail all these mechanisms and related potentialities for applications, we notice that the search of these parallelisms follows similar activities

M. Affronte (✉)
Universitá di Modena e Reggio Emilia, via G. Campi 213/A, 41125 Modena, Italy
e-mail: marco.affronte@unimore.it

F. Troiani
Institute NanoSciences, CNR, via G. Campi 213/A, 41125 Modena, Italy
e-mail: filippo.troiani@unimore.it

J. Bartolomé et al. (eds.), *Molecular Magnets*, NanoScience and Technology,
DOI 10.1007/978-3-642-40609-6_10, © Springer-Verlag Berlin Heidelberg 2014

done in the past in different fields of research. It is interesting to notice, for instance, that the pioneers of Information Science had first to figure out fundamental relationships between some algebras and the functioning of electronic switches and relays in order to built a machine capable to solve problems. Claude Shannon in the 1930's first established the relation between Boolean logic and the functioning of electronic devices that is still at the basis of a modern computers in spite of the revolution due to the introduction of semiconductors in ICT in the late Fifties. Similarly, Richard Feynman noticed, in 1982, that the evolution of a quantum system cannot be efficiently simulated by classical ones, and thus by conventional computers. Since the initial Feynman's proposal, several quantum systems, such as photons, quantum dots, semiconducting circuits, trapped ions or impurities in solids have been studied with the aim to fabricate quantum computers. Yet the most suitable quantum hardware has not been definitively found, and molecular spin clusters may well contribute to this search.

In a different field, the similarities between the charge transport mechanisms through organometallic groups with the functioning of specific electronic devices have been noticed, thus providing sufficient ground to propose the use of these molecules as electronic devices. That was the start of molecular electronics about 20 years ago. In chemistry, the Nobel laureate J.M. Lehn noticed that the growth of (supra-) molecular assemblies follows specific laws, and that these can be used to solve complex problems.

From these examples we learn that, in order to find applications of MNM in ICT, we have to understand and control how molecular spin clusters work and establish a parallelism between these and the operational mode of devices in ICT.

The race for faster, smaller and more versatile devices for ICT is well known and it is a great stimulus for the search of new and smarter materials, with novel functionalities or better performances with respect to CMOS (Complementary Metal Oxides Semiconductors) or to the magnetic metals used in hard disks. Molecules constitute pre-assembled functional units, that can efficiently work at the nanoscale. Coordination and supramolecular chemistry allow one to tailor the magnetic states and properties of MNM, that are real systems on which it is possible to observe and control quantum phenomena at single molecule level. These arguments provide further motivation to study molecular nanomagnets with the aim of finding efficient functional materials working at the nanoscale. The challenges that we have to face to provide efficient solutions to ICT problems are, however, very tough. In this chapter we discuss some recent activities ongoing in this interdisciplinary field, report achievements obtained in the last few years, and discuss targets for the future. The ambitious goal of this chapter is to touch many aspects of the problem in order to have a broad view of this emerging field. Many topics mentioned in this chapter certainly deserve a much deeper discussion. Some of them are central in other chapters of this book and they will be treated more extensively there. For others, we shall just mention the essential literature and refer the reader to dedicated textbooks.

The chapter is organized as follows: in Sect. 10.2 we introduce some basic concepts in classical and quantum computation; in Sect. 10.3 we review some of the main issues and trends in ICT; In Sect. 10.4 we discuss the implementation of quantum information processing with MNM. While the former two Sections will have

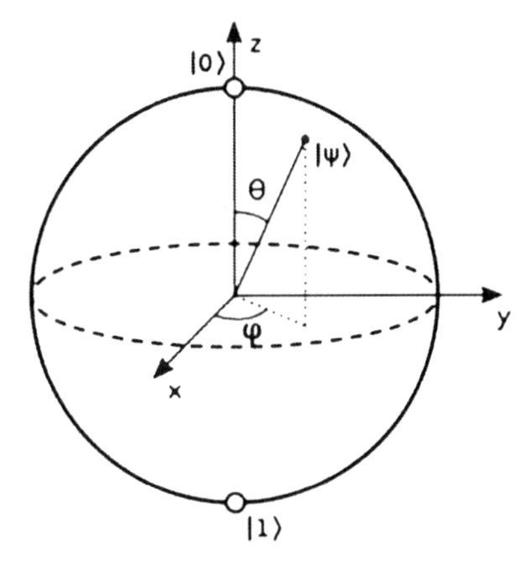

Fig. 10.1 Bloch sphere representing a qubit state $|\psi\rangle$. The north and south poles correspond to the states $|0\rangle$ and $|1\rangle$ states, respectively. Points on the sphere ($r = 1$) represent pure states $|\psi\rangle = \cos(\theta/2)|0\rangle + e^{i\varphi}\sin(\theta/2)|1\rangle$. Points in the sphere ($r < 1$) correspond to mixed states

a very general character, in the latter one we focus on some more specific aspects, more related to our own research activity.

10.2 Classical and Quantum Bits

C. Shannon, one of the pioneers of information theory, first introduced the word *bit* to name binary digit in 1948. Modern computers still work with elementary *bits* of information, represented by two-valued variables and these can be implemented by bistable systems. The ON/OFF states of transistors, the opposite polarizations of light, or the UP/DOWN magnetization states of small magnets may well represent a *bit*. The hysteresis loop of a magnet with two states of remanent magnetization (UP and DOWN) was actually one of the first methods for data recording: in 1898 a Danish engineer V. Poulsen used a magnetic wire to store information. Since then, bistability in many other systems, such as ferroelectrics, non linear optical media, capacitors or transistors were used to store and process bits.

There are many forms of bistability in molecular systems, such as spin crossover [1], valence tautomers [2], single molecule magnets with anisotropy barrier [3], etc: that's why it is straightforward to imagine them as elementary units for data encoding. There are also systems that present multi-states. It is possible to encode multiple bits with the same unit, although this is a less common practice due to a reduced reliability in writing, reading and storing information: each of these processes needs to be robust enough to limit errors.

In quantum mechanics, we learn to deal with quantum states of physical systems. Quantum states are represented by wavefunctions $|\psi\rangle$. The simplest case is a quantum state represented in a basis of two eigenvalues $|0\rangle$ and $|1\rangle$, which allows to define the elementary quantum bit or *qubit*. The difference between a classical bit and a quantum bit is therefore clear: the quantum state may exist in any superposition of the two states of the basis set $|\psi\rangle = \alpha|0\rangle + \beta|1\rangle$, while the classical bit is either in state 0 or 1. Quantum bits can be implemented by many physical systems,

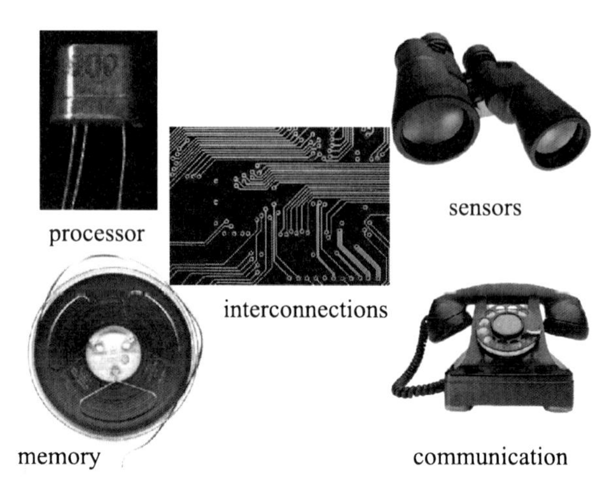

Fig. 10.2 Schematics of basic elements in a computing machine

such as one electron in a quantum dot, or a photon with two polarization states, or a nuclear or electron spin $s = 1/2$. The latter is well described by *spinors* which can be represented using a Bloch sphere (see Fig. 10.1).

Computing Machines The variety (number and type) of tasks performed by computers changes continuously. The same is—even more—true for the their components, yet we can identify some common items. Interestingly, some basic operations and units in modern laptops are essentially the same that one could find in the early machines (Fig. 10.2). In 1936 Alan Turing, a British mathematician, proposed a prototype of a computing machine that is still used to fix basic operations and elements [4]. The Turing machine consists of a memory unit, such as an infinite tape, on which symbols can be written and read in unit cells by a head. A finite sequence of instructions is performed by a processor on these symbols. An alternative description considers electronic circuits that comprise active and passive elements. Among these, interconnections now deserve considerable attention as critical elements in terms of efficiency (e.g., power dissipation or speed), and for the need to build complex architectures. Besides these basic elements, interfaces with external world (e.g. communication, sensors, displays) also play an important role in modern computers. Most of these elements (memories, processors, interconnections, communication, sensors, etc.) are also found in quantum technologies. Therefore in a quantum computer we need registers to store and others to efficiently process quantum information or to communicate these at different length scales. With this general scheme (Fig. 10.2) in mind, we can envisage possible applications of molecular nanomagnets in a computing machine.

Memory Cells Bistable molecules can act as registers where information can be recorded for short time or "permanently". In the latter case, the general requirement for commercial memories is that information must be retained for at least 10 years at ambient conditions. The benchmark for nanomagnets are FePt nanoparticles, that have the highest magnetocrystalline anisotropy (10^8 erg/cm^3) know so far, besides showing high chemical stability at high temperatures. Three nanometer FePt particles actually have a stability ratio $\Delta E / k_B T > 50$ at room temperature, thus satisfying the requirement for commercial memories. So, the first issue for molecular nanomagnets is to stabilize their magnetization at room temperature.

In spin crossover systems, atoms of first-row transition metals with electronic configurations d^4-d^7 in an octahedral symmetry can adopt two different electronic ground states, according to the occupation of the d orbitals split into the e_g and t_{2g} subsets. The mechanism of bistability originates from the interplay between orbital levels and lattice vibrations and, as such, it may actually work at room temperature [1]. Recently, great progresses have been achieved in nano-structuring Prussian Blue Analogues [5], and intense research is also devoted to demonstrate spin transition in Fe compounds at single molecule level and at room temperature [6].

Isomers of organic compounds that can readily interconvert [7] or, more specifically, valence tautomers comprising transition metals [8, 9] present chemical (and magnetic) bistability which—in principle—can also be used to store information. Typically, these are systems based on radicals using polyoxolene molecules which can be in the paramagnetic $S = 1/2$ semiquinone forms or in the diamagnetic cathecolate or quinone form. Their interactions with metal ion species which undergo internal charge transfers may be obtained, thus providing magnetic bistability at finite temperature.

The mechanisms controlling relaxation of the magnetization in single molecule magnets have been largely studied in the last two decades and are discussed in detail in Ref. [3] (see also Fig. 10.3). Typically, the uniaxial magnetic anisotropy gives rise to a double well of energy levels in the ground state multiplet, that include states with opposite magnetization Fig. 10.3(a). The highest anisotropy barriers have been recently found in rare earths embedded in organic shells (single ion magnets) [10]. Much effort is currently devoted to control the retention of magnetization through chemical and physical interactions of a single molecule with an active surface, which also looks a viable route for room temperature operation [11].

A general issue in view of using magnetic memories is how to "write" and "read" information at the nm scale. In a transistor we can "write" 0 and 1 by changing the gate voltage and we can "read" such information by measuring the current passing into the channel. The conventional way to write information in a magnetic memory is to use magnetic fields. Yet, as magnets get smaller and smaller, it becomes more difficult to address them individually by localized magnetic fields (this is more difficult than with electric fields). For instance, it is not trivial to design heads or small antennas that may generate magnetic fields confined within 10 nm [12]. Moreover, magnetic fields generated by electrical currents waste a lot of energy. These two are additional problems that need to be solved in order to make nanomagnets competitive in the race for ultra high density memories.

A conventional way to read a magnetic bit is to use highly sensitive magnetic flux detectors in proximity of the nanomagnets. In commercial hard disks, spin valves, based on celebrated enhanced (giant/colossal) magnetoresistance effects, are normally used [13]. Whether this scheme of detection can be extended to the ultimate

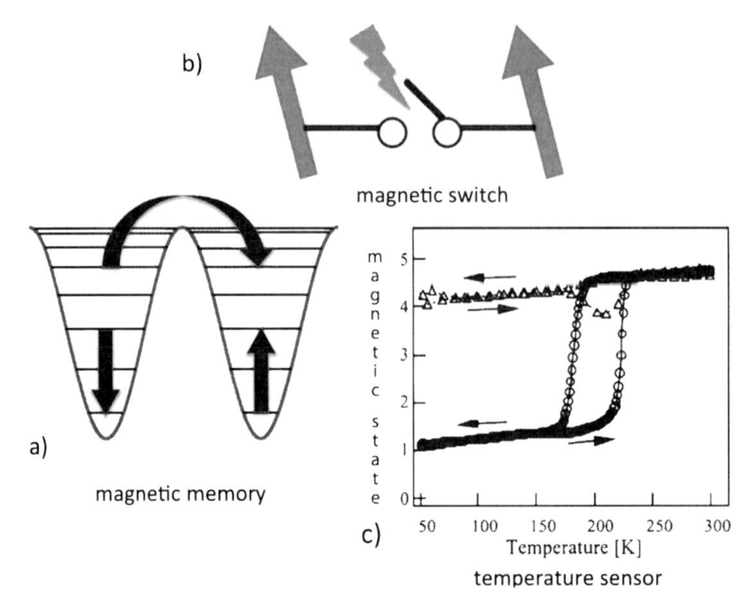

Fig. 10.3 Switching mechanisms and possible use of molecular switches: (**a**) magnetic bistability induced by anisotropy in a single molecule (ion) magnet: in this case an external magnetic field can be used to switch the magnetization from UP to DOWN and the other way around; (**b**) molecular switch between two magnetic registers: an optical or electrical stimulus can be used to switch ON and OFF the coupling between two magnetic registers; (**c**) temperature-induced magnetic switch in FeCo Prussian blue [9]: when the sample is cooled, the high temperature phase with the Fe(III)-CN-Co(II)$^{-HS}$ structure is trapped without relaxing back to the low-temperature phase with the Fe(II)-CN-Co(III)$^{-LS}$ structure (from Ref. [9]). The magnetic state of these molecular systems can be used to probe temperature

limit of single molecule level is an open issue, that was recently addressed by the realization of tiny devices based on carbon nanotubes [14] or graphene [15]. Alternatively, single molecule transistors have been demonstrated to detect the magnetic state of a single molecule inserted between two electrodes [14, 16]. In both cases, the molecular spin state is converted into electronic information.

Writing and reading quantum bits are certainly more difficult tasks, so that we have to define efficient methods and protocols to do these jobs. Besides what has been mentioned above for classical bits, we have to control the phase of the quantum state, that is the α and β coefficients in the bra-ket notation used above. To appreciate the difficulty of this task, we may consider the fascinating experiments of spin-polarized STM recently presented [17–19], where it was possible to write and read the UP and DOWN state of each single magnetic atom but the quantum information (phase) appears washed out by the method used in current experiments. At first glance, writing and reading magnetic states in a molecule seems easier than in atoms, due to the larger size of the molecules, but this is not straightforward: besides technical problems to deposit molecules on clean surfaces, recent experiments and theoretical works showed that the magnetic polarization of the organic ligands may well be different from that of the metal ions, due to the interaction with the

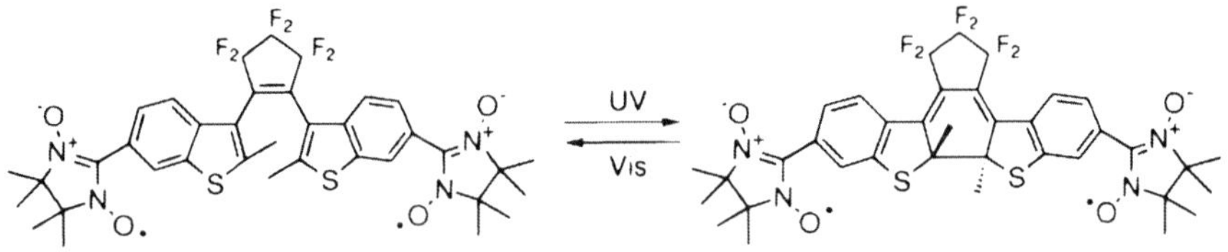

Fig. 10.4 Diarylethene ligand as reversible photo-cromatic switch between two radicals (from Ref. [22])

metal substrate [20]. This further puzzles the protocol of reading the magnetic state of an isolated molecule by STM tips.

Molecular Switches Switching molecules and interconversion between molecular states are fascinating topics that keep attracting the interest of chemists and physicists after several years of intense research [21]. Within the context of this chapter, we are interested in molecules switching their *magnetic* states/properties, although many molecules are known as electronic switches [21]. Incidently, the switch of magnetic states/properties is often related to the electronic ones. In principle, switchable magnetic molecules can be exploited in different ways for ICT (e.g., information storage, processing, sensing etc.) as depicted in Fig. 10.3. For instance, bistable molecules capable of retaining their magnetization for long time can be used as memory cells (Fig. 10.3(a)), as previously discussed. Alternatively, switchable organometallic groups can be inserted between two molecular spins to set ON or OFF magnetic interactions (Fig. 10.3(b)).

In the previous paragraph we have mentioned the conventional switch of a magnetic molecule under the application of a magnetic field (writing process). Much interest has been attracted by the optical switch of molecular states due to energy or electron transfer [21]. The reversible processes can be photo-isomerization or photo-cyclization but there are also redox-based molecular switches, rotaxanes or catenates, chiro-optical switches or molecular systems that function by virtue of photo-chemical reaction [21]. A prototypical case is realized by a photocromatic diarylethene ligand that magnetically links two spins (for instance two nitronyl nitroxides) at the edges (Fig. 10.4). The magnetic coupling can be switched on and off reversibly by using UV or visible radiation [22–24]. In this way, the magnetic state of the two radicals is controlled by an external stimulus via the switchable ligand. There are also linkers with switchable metal centers that can do the same job. This is the case of luminescent Re compounds, dimers (such as Cu_2 or Ni_2 acetates) with spectroscopically accessible separation between active and passive magnetic states or, alternatively, Ru_2 dimers with redox properties and switchability [21]. Much interest has been recently devoted to organic free radicals which are stable on surfaces and present interesting switchable electronic, optical and magnetic properties [25].

Besides the application of magnetic field or light, the spin state of the metal center can be also changed by different external stimuli such as temperature or pressure. In the large family of spin crossover compounds (more than 200) there are molecular systems for which the spin transition is induced by temperature [1, 6] (see

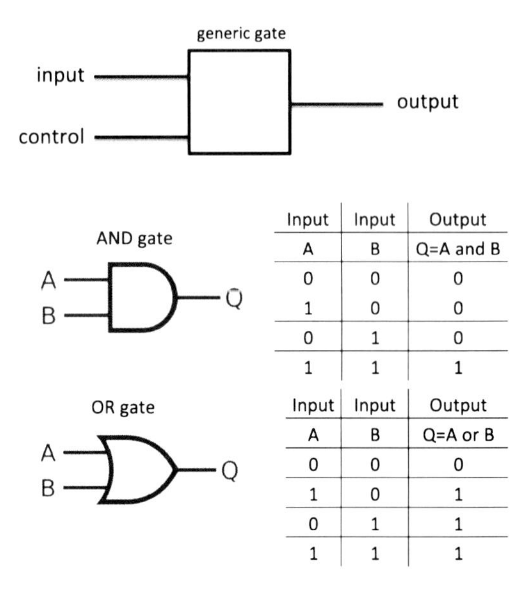

Fig. 10.5 *Top*: Generic logic gate comprises one (or more) *input(s)*, a *control input* and an *ouput*. Note that information may only flow from the input to the output. *Bottom*: two-bits gates AND and OR and related truth tables explaining their functioning

Fig. 10.3(c)). Since the switching parameter is an external stimulus, these systems have been proposed for the realization of molecular sensors [26].

Interconnections These are further essential elements for the realization of a computing machine. More specifically, we are interested in molecular linkers—not necessarily switchable—that can transmit *magnetic* information at nanometric scale. Organic materials, made of light elements, have long spin coherence time due to the weak spin-orbit interaction. Consider, for instance, the prototypical case of spin information propagation through aromatic linkers. These have been intensively studied for the realization of permanent bulk molecular magnets. The strength of the interactions between two spins linked by an aromatic bridge, has been found to obey some general empirical rules [27]: (1) the larger the number of bonds which compose the interaction path, the smaller is the interaction; (2) charge and spin polarization induced by bonding to a metal site proceed in an alternating fashion in aromatic cycles; (3) since a magnetic interaction can be sustained by different bond paths in the linker, the strength of the interaction should depend on whether constructive or destructive quantum interference between paths with different lengths arise [28]. Our recent studies have shown how aromatic groups may also link large molecular spin clusters [29, 30]. Moreover, the magnetic coupling can be mechanically switched on and off by twisting the angle between two aromatic groups [30], similarly to what happens for the electronic communication channels in the same aromatic systems [31].

Logic Gates Conventional gates perform Boolean logic operations such as AND, NOT, OR, whose functioning is represented by truth tables reported in Fig. 10.5. Typically, we can recognize basic elements in a gate: *input(s)*, *control* and *ouput* in such a way that the output is determined by the inputs upon the condition(s) given by the control (Fig. 10.5). Due to their electronic and optical capabilities, several molecular groups have been considered as true elementary logic gates [21]. More complex operations can be performed by assembling elementary Boolean gates and it can be demonstrated that a minimum set of elementary gates can constitute a universal computing machine, able to solve a large number of problems. An alternative approach is to realize more sophisticated gates that perform only specialized operations but with higher efficiency.

A similar situation is found for quantum gates: there are basic quantum gates that constitute a universal set and perform a wide class of quantum operations. An exhaustive description of these is given in textbooks of quantum computation like that of Nielsen and Chuang [4]. Single-electron spins represent a prototypical example of a two-level system. The qubit encoding in composite and complex spin systems—such as molecular nanomagnets—is less straightforward. In order to make MNMs competitive candidates for quantum computation we have to compare their performances with those of other, well established, candidate quantum systems. Some peculiarities of *molecular* spins exist and we shall discuss them in Sect. 10.4.

For the sake of completeness, we note that quantum computation is not the only way in which a molecular system can efficiently process information. Alternative computational methods have been proposed and still attract much interest, especially to solve complex problems. J.M. Lehn first noticed how some chemical reactions perform specific codes and sequences to give rise to supramolecular structures [32]. Similarly, the way biological systems (DNA) perform complicated operations, such as recognition, is highly efficient [33]. Another novel way of computing is that performed by neural networks [34]. Arrays of molecular spins interacting by exchange coupling work as quantum cellular automata that may efficiently solve specific problems [35]. The contribution of molecular nanomagnets to these and other alternative ways of computing has not been explored yet.

10.3 Issues, Trends and Benchmarks of Information Technologies

An alternative way to figure out how magnetic molecules and related synthetic and characterization methods can be used in ICT is to identify main problems and trends in current technologies. This approach to the problem may better help to fix goals and benchmarks. To have an idea on the design and realization of today devices we refer to the roadmap of international organizations on nanotechnologies [36] or to review articles on nanoelectronics [37] or spintronics [13].

Size Reduction The race towards smaller and smaller devices is well know in ICT. For semiconductors, the 25 nm-technology is currently in production while, for commercially available hard disks with storage density of 50 Gbit/in^2, the size of an elementary magnetic register is $\sim$1000 nm^2. At present, industries are developing hard disks with 1 Tbit/in^2 storage density and this implies magnetic cells of $\sim$100 nm^2 in size. We'll soon approach the *nm* scale, i.e. the *molecular* scale!

The trend towards extreme miniaturization has several consequences and the threshold of 10 nm seems to be critical for several reasons. The first one is related to the fabrication processes. Although radiation with shorter and shorter wavelengths may well be used to define devices with resolution down to nm scale, lithographic methods are becoming more and more sophisticated and expensive so bottom-up fabrication methods can represent an interesting alternative [38]. Therefore, besides the individuation of specific molecules, any achievement on the definition of products, methods and protocols for deposition of functional molecules on surfaces is of potential interest for ICT. This search also stimulates the application and the development of methods for the chemical and magnetic characterization of arrays of molecules on surface [39]. Self assembly of functional units is certainly interesting for the production of scalable devices. While the evaporation and organization of simple molecules, like metal phthalocianines, on surfaces is well established, self assembling of large and complex magnetic molecules on surfaces has been demonstrated only recently [40], along with the chemical stability and preservation of their functionalities [41]. Moreover, the realization of nanoarchitectures and devices requires precise positioning over large area and selective deposition of molecules. Several tools and methods for nano-structuring and decorating surface with molecules (nano-particles) are currently under study [39]. Among these, dip-pen lithography allows us to deposit few molecules with nanometric precision . A further requirement for realization of devices is the control of orientation, more specifically the control magnetic anisotropy with respect to the substrate. On this line, important results have been obtained for Fe_4 high spin molecules [42] and for Cr_7Ni low spin rings [43].

Size Effects As just mentioned, extreme miniaturization implies several technological challenges. Yet we have to realize that there are also fundamental limitations that occur at the *nm* scale and that cannot be simply overcome by technical means. We deal here with size effects that, in first instance, induce a discrete pattern of energy levels and states. In semiconductors, energy bands are replaced by discrete levels in quantum dots so single electron devices can be considered as the natural evolution of transistors in nano-electronics. From this viewpoint, molecules with well defined patters of energy levels and states represent a unique playground to test and develop new ideas in magnetism. Besides, molecules are better than nanoparticles, that typically have irregular shape and size dispersion. However, organo-metallic molecules are fragile so a crucial step is to demonstrate that individual molecules preserve their characteristics in different environments, like for instance on surfaces or in ambient conditions. These are big challenges. Yet, similarly to previous experiences in other fields, the study of size effects in molecules present unique opportunities from which we can learn a lot and hope that this knowledge can be transferred to suitable systems in the future.

A further consequence of the reduction in size is the appearance of quantum phenomena. Discrete levels and states for molecules or quantum dots are intrinsically quantum in nature and should be treated as such. In the presence of quantum coherence, properties like interference and entanglement can be exploited for new generations of quantum devices. This evolution implies new concepts and extra value can

Table 10.1 Comparison of typical switching times between electronic and magnetic devices and different molecular processes

System	Switching process	Switching time (ns)
Electronic device	Electrical	$\sim 10^{-1}$
Magnetic bit in HD [44]	Magnetic	~ 1
SMM [3]	Magnetic	10
Ru(II)-complexes [45]	Redox	$>10^3$
Molecular motors [46]	Mechanical	10^6–10^9
Spin crossover [1, 26]	Optical	10^2
Valence tautomers [9]	Optical	10^{-3}–10^2
Photo-chromic switches [21]	Optical	10^{-4}–1

be actually added to conventional computation. On the other hand, quantum states are fragile and we have to learn to manipulate them in coherent way, as discussed in the following section.

Switching Rates Performances in ICT are measured in time scale so, bit writing or communication rates are benchmarks to be considered. To fix some numbers, modern electronic processors write bits with rates of few tenths of *ns*, i.e. few GHz. Without going into many details on the functioning of each device, Table 10.1 directly compares the order of magnitude of switching times of electronic devices and magnetic memories that are currently in the market with typical rates of molecular switching processes.

Power Dissipation One of the most critical factors that currently limits performance of electronic devices is the power consumption and the consequent dissipation of heat. There are impressive numbers here. For instance, about 5–10 % of whole national electricity budget in the U.S. is dissipated by computers and communications today and this fraction will double by the year 2020. The power density dissipated by an electronic processor attains 100 W/cm^2, that is more than what is dissipated by a hot plate to cook eggs! Trends show that the situation is getting even worse for electronic nanodevices due to the increase of active devices density and the intrinsically reduced thermal conductivity [47]. In electronics, we can distinguish several sources of power consumption: (i) dynamic switching processes; (ii) leakage power (related to maintain device in standby mode); (iii) Joule effect in the electrical connections. Power consumption is given by the energy required to activate a process (e.g. a switch) *times* the rate at which this process occurs. It has been shown that the minimum energy required to switch a binary device is $3k_BT$ [47]. In electronics, information is related to the intensity of charge current so a switch is very consuming: at present, fast electronic devices require $\sim 10^3 k_BT$. To flip a spin S bound by an exchange coupling J requires an energy of order of JS^2, while a thermally activated process of magnetization reversal of the same spin in a anisotropy double well requires DS^2. In order to guarantee that information is sufficiently stable against thermal fluctuations, it is generally required that the magnetic barrier is $\ln(\Delta/k_BT) > 50$ so, in principle, magnetic switch requires less energy than an electronic device to operate at room temperature. Magnetic information is also non volatile, so they do not require energy for standby mode (that is the case of all magnetic memories, for instance). It is also worth noting that magnetic bits, being vectors, are more efficient than charge current (scalar), but interconversions between spin and charge information are scarcely efficient processes and it should be avoided. This consideration leads to the idea of realizing "all spin logic" computing machine [48], that represents a *monolithic* spintronic machine (see Fig. 10.6).

Besides these general arguments, it is interesting to consider how specific spintronic devices work. A prototypical spin-FET is not consuming less than an equivalent MOSFET [49]. MRAM, currently in the market, make use of a magnetic field generated by an electrical current to switch the magnetic memory and thus they require more power than the equivalent electronic RAM. The use of spin torque may well reduce the energy required to switch a magnetic bit, and such type of solution make spintronics promising for the future [50]. Whether these concepts can be extended down to the molecular level is an open question.

Nano-architectures So far we have considered planar arrangements of devices, memories and sensors as the natural choice. While this may hold for the present, recent trends suggest that 3D architectures can become attractive in the near future. A first reason for this is related to efficient packing: the density of devices achievable in 3D is by far denser than any arrangement in 2D. Secondly, realization of interconnects is recognized as one of the most difficult tasks in electronics. Thus, exploiting the vertical dimension to connect devices and memories represents an advantage with respect to any planar arrangement. The idea of employing 3D architecture is not new and several proposals have been put forward in the past but, obviously, these have to face difficulties in the realization using current technology. Yet some solutions have been recently demonstrated to be feasible and efficient and can be kept in mind when we deal with molecular nanomagnets. For instance, the introduction of holographic methods to optically address 3D arrays of memories is a technology available in the market since 2006 [51]. In principle, this method can be applied to optically switchable molecular crystals or metallorganic frameworks (MOF). Another way in which molecular science may contribute is to develop methods for hierarchical 3D growth of functional molecular layers. Along this line, the realization of hybrid architectures that alternate, for instance, carbon-based conductors with functional molecules looks also particularly attractive.

Spintronics and Quantum Computation We have already mentioned the fields of *Spintronics* and of *Quantum Computation* (QC). When talking about spins, quite often we mix the two since the common effort is actually devoted to exploit the spin as information carrier. There are, however, some differences between these fields, starting from the fact that, historically, the former was developed by the community studying electron transport in magnetic metals while the latter was born—and

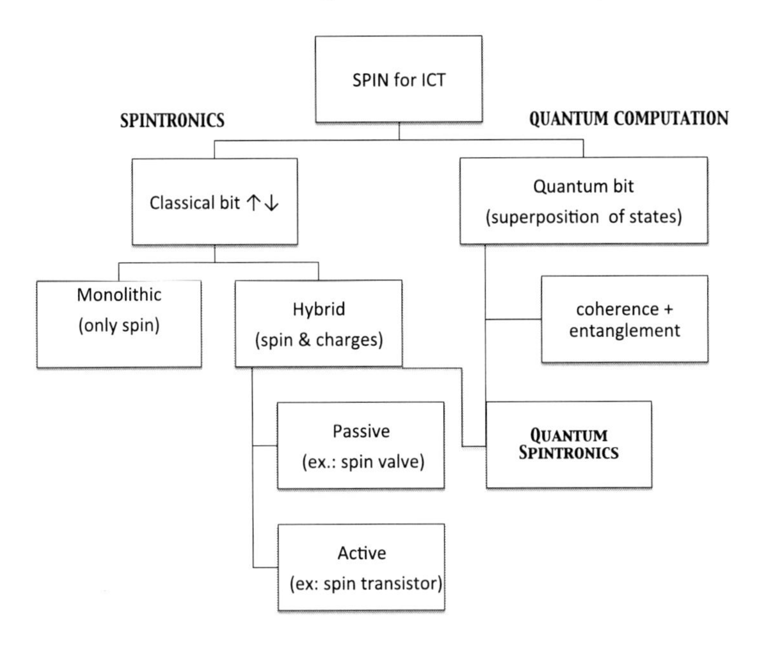

Fig. 10.6 Spintronics, quantum computation and quantum spintronics

it is still—strongly related to diverse branches of quantum physics (quantum optics, low-dimensional semiconductors, superconducting circuits, atomic physics, etc.). Following the scheme in Fig. 10.6, as suggested by S. Bandyopadhyay and M. Cahay [13], we can notice that:

- for QC, the spin is a quantum observable which can exist as superposition of states, while it is simply treated as tiny magnet with only the North and South poles in Spintronics;
- the control of coherent spin dynamics is essential for QC: the phase of the wavevector contains an essential part of the quantum information. This is not required in conventional Spintronics;
- QC exploits resources, like entanglement, which are typical features of quantum systems, while this is not the case for Spintronics.

Therefore, QC is much more powerful but also much more demanding as compared to Spintronics. Conventional Spintronics makes use of both the charge and the spin of electrons. Ideally, a device which does not need to interconvert spin information in charge current, like the "All Spin Logic" proposal described in Ref. [48], would be a "Monolithic" spin technology. We can also distinguish between passive devices, like spin valves, for which information is essentially carried by charge current and the magnetism of the polarizers is used only to control it, from active devices, like spin-FET, where information is carried also by the spin current (see Fig. 10.7). A new branch has also emerged in the last few years and it aims at exploiting the interconversion of information contained in a (single) spin in a charge signal while

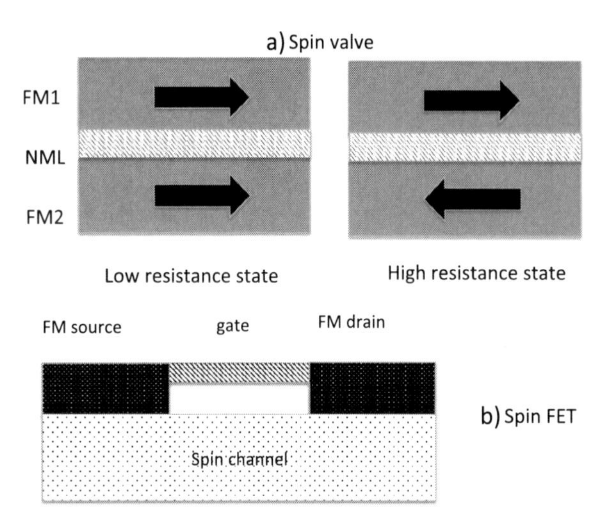

Fig. 10.7 Schematic view of prototypical spintronic devices: (**a**) spin valve (*upper pannel*) and (**b**) spin FET (Field Effect Transistor, *lower panel*). The spin valve is essentially made by two ferromagnets sandwiching a non-magnetic layer (NML). The first ferromagnet (FM1) injects spin polarized current through the NML, which is then collected by the second ferromagnet (FM2). The resistance of the valve depends on the relative orientation of magnetization of the two FM polarizers [13]. The magnetization controls the current, that is the true carrier of information. This is thus a passive spin device (see Fig. 10.6). In the spin Field Effect Transitor, a spin polarized current is injected to a spin channel by a ferromagnetic electrode (source). The polarization of the spin current is controlled by the gate, which can be controlled either by an electric or by a magnetic field. The ferromagnetic drain detects the polarization of the spin current at the end of the channel. The information variable is the spin current, and the control is made directly on this. Therefore, the device is an active one [13] (see Fig. 10.6)

preserving the coherence of the wavefunction during the spin manipulation. This field, known as Quantum Spintronics, started with fascinating experiments on quantum dots and on spin impurities in Si and it is now becoming a realistic avenue with pioneering experiments performed on single magnetic molecules, as discussed in another chapter of this book.

These distinctions, although not universally accepted/used, should help the reader to appreciate the different requirements needed to realize different types of technology.

10.4 Quantum Computation

Spin Cluster Qubits As widely engineerable and coherent systems, molecular nanomagnets have been initially proposed by Leuenberg and Loss in 2001 for the encoding and manipulation of quantum information [52]. Scalable proposals for general-purpose quantum computation are based on the use of each MNM as a spin-cluster qubit [53, 54]. The molecules suited for the qubit encoding typically consist

Table 10.2 Ground state spin S of various molecular spin clusters, their measured decoherence times (if references are not specified, information are taken from private communications)

Molecule	Ground state	Decoherence time (μs)
Cr_7Ni [56]	$S = 1/2$	15 μs at 2K [57]
V_{15} [58]	$S = 3/2$	0.2 μs at 4K
Cu_3 [59]	$S = 1/2$	1 μs at 1.5K
Nitronil nitroxide radicals	$S = 1/2$	3 μs at 70K
Malonyl radicals [60]	$S = 1/2$	1 μs
Polyoxometallates [61, 62]	$S > 1/2$	1 μs at 1.5K
Fe_4 [63]	$S = 5$	0.64 μs at 2K
Fe_8 [64]	$S = 10$	0.7 μs at 1K
Mn_{12} [65]	$S = 10$	
Er ions [66]	$J = 15/2$	1 μs at 2K
Tb_2 [67–70]	$J = 6$	0.1 μs at 4K

of spin clusters with antiferromagnetic exchange interaction, and with an uncompensated spin that results in an $S = 1/2$ ground doublet. Radicals with $S = 1/2$, well known for their sharp line-shape in EPR spectra, have also shown great potentialities as molecular building block for the implementation of quantum gates [55]. High spin molecular spin clusters or molecular derivatives of single lanthanide ions working as effective two-level systems might in principle be considered, as long as two basic requirements are fulfilled. Firstly, it must be possible to efficiently rotate the two lowest states $|0\rangle$ and $|1\rangle$ into one another by external fields (as is the case for the $M = \pm 1/2$ states, that are coupled to each other by magnetic dipole transitions). Secondly, occupation of the additional levels must be suppressed in the initialization process, and kept negligible throughout the qubit manipulation. This implies that the applied fields should not induce undesired transitions between the $|0\rangle$ and $|1\rangle$ states and higher-lying levels.

Besides the level structure, molecular nanomagnets can present significant differences in terms of decoherence times. In fact, the coupling to the nuclear spin bath, as well as to phonon modes, can present relevant system-specific features. In addition, the possibility of growing ordered arrays by depositing the molecules on surfaces, or of inducing intermolecular exchange by means of supramolecular bridges can vary significantly from one system to the other. These aspects, that can play an important role in the implementation of quantum information processing, can in fact depend not only on the magnetic core of the molecules, but also on the surrounding shell of organic ligands. In Table 10.2 we report the ground state spin of a number of molecular nanomagnets, as well as their decoherence times, measured at low temperatures. The form in which the molecules can be aggregated within the sample are also reported.

The main advantage of using a spin cluster, rather than a single electron spin, for encoding the qubit was initially identified with the larger spatial extension of the former with respect to the latter one [53]. This would in principle facilitate the individual addressing of the qubits, by reducing the required spatial resolution of the applied fields. With the development of more system-specific proposals, other potential advantages have emerged, resulting from the possible exploitation of the internal degrees of freedom belonging to the spin clusters. These can in fact be exploited to relax some of the most demanding requirements related to the MNM-based implementation of quantum-information processing (switchable interaction between the molecules, individual addressing of the nanomagnets) [71], or to enable the qubit manipulation by means of electric—rather than magnetic—fields [72]. Some of these issues are discussed in more detail in the following.

Coherent Switching of the Qubit-Qubit Coupling The implementation of single- and two-qubit gates [4] requires the individual addressing of the molecules and the coherent switching of intermolecular coupling. In order to be coherent, the switching should take place in timescales shorter than the decoherence time (typically microseconds, see Table 10.2), and it shouldn't involve degrees of freedom other than spin (such as in phonon emission, or other forms of relaxation). The requirements are thus more stringent than in the case of classical switches, mentioned in the previous paragraph. However, different approaches to the qubit manipulation can be envisaged, that allow to partially relax the above requirements. We first consider the ones based on short-range, exchange coupling between MNMs. One possible way to switch the effective coupling between two qubits, in spite of the permanent character of the underlying exchange interaction, is to exploit the different spin textures of the ground- and excited-multiplet states of MNM. For example, the intermolecular bridge can symmetrically couple a nanomagnet A with two spins of a second MNM B that—in the subspace $\{|0\rangle, |1\rangle\}$—have opposite expectation values [71]. This condition is approached by neighboring spins in rings with antiferromagnetic exchange, and can be fully met provided that the two spins in question are equivalent [73], resulting in a vanishing intermolecular exchange. The coupling can be controllably turned on by exciting qubit B to an auxiliary excited state where such cancellation condition doesn't apply. The same principle can be exploited in MNM acting as auxiliary units [74], that mediate the magnetic coupling between spin qubits. In the simplest case, such unit consists of a dimer of antiferromagnetically coupled spins, with a finite amount of S-mixing, such that singlet-triplet transitions are dipole-allowed. This dimer can act as a switch, that prevents or allows the magnetic coupling between the qubits depending on whether the two spins have zero or finite expectation value, as in the $S = 0$ and $S = 1$ states, respectively. We note that within this approach, the internal degrees of freedom that one exploits are the ones of the auxiliary unit, while the qubits themselves need no longer be spin clusters, and can be encoded in single spins. In both the above approaches, the effective switchability of the qubit-qubit coupling requires a fine (though time-independent) tuning of the intra- and inter-molecular exchange interactions. Alternative approaches, based on the dynamic control of the intermolecular exchange through, e.g., electric fields would disclose alternative opportunities for quantum-information processing with MNMs.

Individual Addressing of the Qubits Another crucial aspect is the individual addressing of the molecules, required for their selective manipulation and for the qubit-specific readout of the final state. These operations imply the use of external fields with a high spatial and/or spectral resolution. Two approaches have been proposed for relaxing this requirement and performing the qubit manipulation only by means of global (i.e. spatially homogeneous) fields. One consists in performing universal general-purpose quantum computation with global control schemes [75–77]. Here, the individual addressing of the molecules is achieved by using part of the qubits as auxiliary units. The state of each auxiliary unit only determines whether or not the neighboring qubits—that actually encode information—are to be manipulated by each applied field. Typically, the auxiliary units are all in their $|0\rangle$ state, apart from one (the so-called *control unit*), which is moved next to the qubit to be addressed at each step. The lack of individual control in the global-field approach implies an extra cost in terms of complexity of the hardware, of the qubit encoding and of the manipulation scheme. The presence of auxiliary levels within molecular spin clusters, along with the above mentioned switching mechanism, allows to minimize both these costs [71].

Quantum Simulation A second strategy to cope with the absence of local control is that of reducing the generality of the computational task. This corresponds to implementing a dedicated device, rather than a universal quantum computer, as in the case of quantum simulators [75, 78, 79]. Quantum simulators are devices—initially envisioned by Richard Feynman [80]—that can simulate the dynamics of other quantum systems. A digital simulator is in fact nothing less than a (small) quantum computer, where the time-evolution operator $e^{-iHt/\hbar}$ is decomposed into a discrete sequence of single- and two-qubit quantum gates. In analog quantum simulators, the system interactions are tuned in such a way that its Hamiltonian coincides with (or is equivalent to) that of the target system. Hybrid approaches can also be conceived, where the manipulation consists of a sequence of quantum gates, but the resemblance between simulated and simulating systems (in terms of, e.g., geometry and dimensionality) allows a significant reduction of the physical requirements. In particular, translationally invariant Hamiltonians can be efficiently simulated in arrays of molecular nanomagnets, manipulated by global fields [74]. Here, one of the key resources is represented by the demonstrated capability of synthesizing MNM chains, and of linking neighboring MNMs through intermolecular exchange bridges with a high degree of control. We note that, even if the simulator shares physical features with the simulated system, the local interactions can be completely different in the two cases. In fact, the nanomagnet-based simulator can efficiently mimic the dynamics of prototypical fermion systems, or of chains of spins $s > 1/2$. In all these cases, the translational invariance of the simulated systems allows to implement the simulation by a number of pulses that is independent of the system size.

Cavity-Mediated Coupling While the coupling between MNMs considered above relies on short-range exchange interactions, an alternative approach can be developed, based on long-range couplings between the qubits, mediated by microwave photons of planar cavities. In order for this to be possible, the timescale

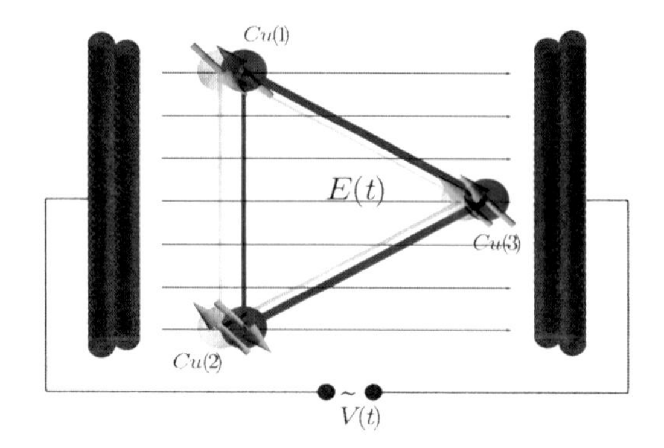

Fig. 10.8 Schematics of a time-dependent electric field applied to the Cu_3 spin triangle. The in-plane introduces a different renormalization of the three exchange couplings, and couples states with equal values of S and of the spin projection S_z (*shaded spin triangle*)

of the cavity-qubit coupling needs to be shorter than those of the dissipative and dephasing processes in the system (including both the cavity and the molecular nanomagnets). This defines the so-called *strong-coupling regime*. With the available stripline cavities, the maximum field intensity and the photon-loss rate don't allow to achieve the strong coupling regime between a single cavity photon and a single spin. The limitations arising from the small value of the spin-photon coupling constant g can however be overcome by replacing single-spins (or a single MNM), with spin ensembles. If an ensemble of identical systems are coupled to the field, the transition amplitude between its two lowest (Dicke) states is $\sqrt{N}$ times larger than that between the two lowest states of each molecule. The strong coupling has been achieved in this way with color centers (more specifically, nitrogen-vacancy defects) in diamond [81, 82]. In this scenario, each qubit would thus correspond to an ensemble of molecular nanomagnets, and the excitation of the $|1\rangle$ state is no longer that of a single molecule, but is rather delocalized in the whole ensemble. The main advantage with respect to the exchange-based approach is that here qubits can be separated by macroscopic distances, making their individual addressing much easier. Besides, the cavity represents an ideal means for coupling heterogeneous systems and degrees of freedom, such as spin and superconducting qubits. Within such hybrid device, different systems can be exploited for different functions, according to their specific features [83]. For example, degrees of freedom that are more protected with respect to the environment can be used for storing information, whereas those that are more efficiently accessible to external means can be used for the manipulation and read-out of quantum information.

Spin-Electric Coupling Spin-based proposals for implementing quantum-information processing are typically based on the use of the spin projection as the computational degree of freedom: $|0\rangle$ and $|1\rangle$ are identified with physical states of equal S (ideally $S = 1/2$) and different S_z. This choice implies the use of pulsed magnetic fields for the qubit manipulation. Molecular nanomagnets offer however additional spin degrees of freedom, such as spin chirality, that can be possibly manipulated by means of pulsed electric fields [72]. Triangles of antiferromagnetically coupled spins, with Dzyaloshinskii-Moriya interaction represent the ideal systems for the exploitation of spin-electric coupling [84] (Fig. 10.8). In fact, one can define there the chirality operator $C_z = (4/\sqrt{3})\mathbf{s}_1 \cdot \mathbf{s}_2 \times \mathbf{s}_3$, that commutes with both $\mathbf{S}^2$ and with the three components of the total spin. An applied electric field introduces an inhomogeneous modulation of the exchange couplings in the triangle, resulting in a spin-electric Hamiltonian that couples states of opposite chirality ($C_z = \pm 1$). The basic and general requirement for the linear spin-electric coupling to be present is the lack of inversion symmetry within each bond of the triangle. In other words, the nanomagnet must exhibit permanent electric dipoles on each of the bonds, while the overall dipole of the molecule doesn't need to be finite. Spin-Hamiltonian calculations show that electric-field induced transitions between states of opposite chirality can also take place in triangles of $s > 1/2$ spins [84]. Rings consisting of an odd number of spins with $N > 3$ can present eigenstates of opposite chirality within the ground $S = 1/2$ quadruplet, but these are not directly coupled to each other by electric-dipole transition. The actual value of the spin-electric coupling in existing molecules is still unknown. Theoretical estimates, based on ab initio density-functional theory calculations, are compatible with manipulation times in the order of 1 ns [85]. Further investigation is indeed required in order to measure such value, and to engineer molecule with large spin-electric couplings [86].

Decoherence-Free Qubit The electron-spin coherence in molecular nanomagnets can be limited by nuclei, phonons, or dipolar interactions [3]. While the former two contributions can significantly depend on the arrangement of the molecules within the sample, hyperfine interactions represent an intrinsic source of decoherence, since they are still present when the molecules are isolated from one another. The chirality qubit presents an important peculiarity that has potential relevance to nanomagnet-based quantum-information processing, namely it can be relatively immune to nuclear-induced decoherence [87]. In fact, the nuclear bath couples efficiently to the total-spin projection of the molecule, and can thus discriminate between $|0\rangle$ and $|1\rangle$, if these are encoded in two molecular states that present different total-spin projections. Degrees of freedom such as spin chirality allow instead to encode $|0\rangle$ and $|1\rangle$ in two physical states with identical spin projections. As a consequence, the terms in hyperfine interactions Hamiltonian that dominate the decoherence of the S_z qubit vanish in the case of the chirality qubit. This results in an enhancement of at least two orders of magnitude of the time scales related to nuclear-related decoherence [87]. Spin chirality is not the only operator whose eigenstates are coupled by the electric field. Within the spin triangle, one could in fact define the partial spin sum S_{12} as an alternative encoding within the ground $S = 1/2$ quadruplet. Here, $|0\rangle$ and $|1\rangle$ coincide with two states with $S_{12} = 0, 1$, and correspond to the eigenstates of an isosceles triangle, where the exchange coupling between s_1 and s_2 differs from that of the remaining pairs. This alternative qubit encoding presents features that are intermediate between those of S_z and C_z. In fact, like the chirality qubit, the total-spin expectation value is the same for the two logical states. However, unlike C_z, the expectation values of the single spins strongly depends on the qubit state (see Fig. 10.9). The simulations performed for the S_{12} qubit provide decoherence times much closer to those of the spin-projection qubit,

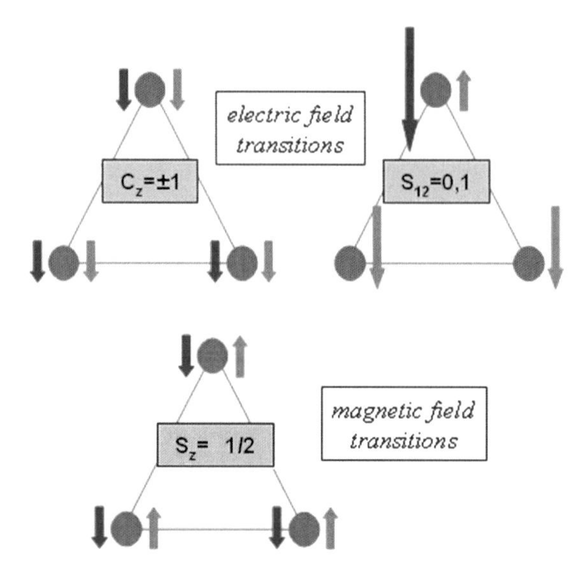

Fig. 10.9 Expectation values of the single-spin operators $s_{i,z}$ corresponding to the two logical states $|0\rangle$ and $|1\rangle$ as a function of the qubit encoding in the $S = 1/2$ subspace of the spin triangle. In the case of chirality (*upper left*), the two states have identical expectation values. In the case of the partial spin sum (*upper right*), the logical states have identical values of the total spin projection, $\langle S_z \rangle = -1/2$, but different expectation values of the individual spins. The more conventional encoding, based on the value of S_z, makes the $|0\rangle$ and $|1\rangle$ states distinguishable both in terms of the total and of the local spin projections

than to those of the chirality qubit. This shows that, in order for the qubit to be substantially immune to decoherence, the two logical states need to be indistinguishable also in terms of their single-spin expectation values.

Entanglement Within a Nanomagnet Quantum entanglement is one of the fundamental resources exploited in quantum-information processing. The capability of generating and detecting entanglement in nanomagnets is therefore one of the preliminary requirements for their use in this context. To some extent, entanglement comes for free in molecular spin clusters. These are in fact prototypical examples of strongly correlated quantum systems, with highly entangled ground states. Quantum correlations are thus expected to persist in the equilibrium state of the molecules up to temperatures of the order of the energy gap between ground and first excited state. In order to detect and quantify such correlations, suitable tools have to be developed. Most of the many result obtained in the field of entanglement quantification and detection in fact apply to qubit systems, i.e. $s = 1/2$ spins, whereas the spins that form nanomagnets are typically $s > 1/2$. One of the most practical means for entanglement detection is represented by entanglement witnesses. These are observables whose expectation value can exceed given thresholds only in the presence of specific forms of entanglement. In spin systems with dominant exchange interaction, energy can be regarded as an entanglement witness. Let's consider for simplicity a chain of N spins s, with nearest-neighbor coupling J. In the absence of quantum correlations between two spins, the lowest possible value of their scalar product is the classical minimum $-s^2$. As a consequence, the minimum of $\langle H \rangle$ is $-J(N-1)s^2$. Quite remarkably, one can derive additional thresholds corresponding to multi-spin correlations. In fact, it can be shown that the ground state of the Heisenberg chain exhibits N−spin correlations (i.e., it's not factorizable, not even partially), and that the absence of k-spin entanglement results in lower bounds E_k of the exchange energy, where E_k are decreasing functions of k [88, 89]. Expectation values of $\langle H \rangle$ lower than E_k thus imply the presence of k-spin entanglement in the equilibrium-state density matrix.

In an homogeneous system, such as a homometallic ring, all pairs of neighboring spins are equivalent, and pairwise entanglement is homogeneously distributed amongst all spin pairs. The controlled introduction of magnetic defects, can spatially modulate quantum correlations, and potentially represents a useful means for engineering entanglement in molecular systems. The underlying physical mechanism is in fact rather intuitive. In order to minimize its exchange energy, each spin pair in the ring tends to a singlet state, which also happens to be a maximally entangled state. The spin s_k, however, cannot simultaneously form a singlet state with both its neighbors, s_{k-1} and s_{k+1}: the two conditions are mutually exclusive. (This can also be regarded as a manifestation of the so-called *entanglement monogamy*.) The ring ground state therefore results from the compromise between the tendency to minimize non-commuting exchange operators (such as $\mathbf{s}_{k-1} \cdot \mathbf{s}_k$ and $\mathbf{s}_k \cdot \mathbf{s}_{k+1}$) in the Hamiltonian. The substitution in the ring of an ion, with spin s, with one belonging to a different chemical element, carrying a spin $s' \neq s$, breaks such balance. It thus introduces a spatial modulation, and more specifically an oscillating dependence of spin-pair entanglement on the index k that identifies the spin pair. In Cr-based antiferromagnetic rings, a whole set of chemical substitutions have been demonstrated in the last years. By comparing these cases, one can show that the amplitude of the above mentioned oscillations increases monotonically with the difference between the impurity spin s_M and that of the Cr ion ($s = 3/2$). Such dependence persists at finite temperature and can be detected by local observables, corresponding to (combinations of) two-spin exchange operators [90].

Entanglement Between Nanomagnets Besides being composite quantum systems, molecular nanomagnets can be regarded as building blocks of supramolecular structures. In this perspective, it is thus useful to investigate entanglement between the collective spin degrees of freedom, such as the total spin projection of each molecule [92]. In order to induce equilibrium-state entanglement between two (or more) MNM, an antiferromagnetic intermolecular exchange interaction is required, whose magnitude is smaller than the working temperature. Once such system has been synthesized, a direct experimental evidence of entanglement can be achieved by using magnetic susceptibility as an entanglement witness [89]. In fact, just like exchange energy, also the variance of the total spin projection—averaged over three orthogonal directions - presents a lower bound that can only be exceeded in the presence of entanglement. Such direct experimental evidence of quantum correlation between molecular spins has been recently provided within dimers of antiferromagnetically-coupled Cr_7Ni rings [91, 92]. A detailed understanding of intermolecular exchange [30] can enhance the range of physical parameters (temperature, magnetic field) where equilibrium-state entanglement is observable. Interestingly, such entanglement between collective degrees of freedom is compatible with that between individual spins within each ring, as can also deduced from general symmetry arguments [90].

10.5 Conclusions and Future Directions

From the very beginning, the possibility to control magnetic bistability in high spin molecules has suggested the use of MNM as elementary cells to store information. More recently, the control of coherent spin dynamics, on one side, and the control of spin entanglement, on the other, have disclosed the possibility of using molecular spin clusters for quantum information processes. In spite of the fact that experimental conditions to manipulate molecular spins remain difficult (but not impossible!) some advantages in using molecular—rather than isolated—spins for quantum information processing have emerged in the last few years. One of this is certainly the flexibility of engineering magnetic links or auxiliary (excited) states at (supra-)molecular level that may allow to design specific quantum devices. Relaxing the requirements for specific quantum operations (e.g. coherence and entanglement) there is still much room to figure out spintronic devices at single molecule level. Here we may exploit either the switchability or the retention of magnetic properties of single molecule as well as the capability to selectively position molecules on substrates. Challenge for the next future is the control of these features at single molecule level which implies much synthetic, experimental and theoretical work and … imagination. Further potentialities of molecular nanomagnets for what concerns communication, sensing or power dissipation at nm scale have not been explored yet but they may deserve attention in future.

Acknowledgements We wish to thank Dr. Alberto Ghirri, Dr. Andrea Candini, Dr. Giulia Lorusso, Dr. Valdis Corradini, Dr. Valerio Bellini, Dr. Ilaria Siloi, Prof. U. del Pennino (CNR and University of Modena, Italy), Prof. S. Carretta, Prof. P. Santini and Prof. G. Amoretti (University of Parma, Italy), Prof. A. Cuccoli and Dr. P. Verrucchi (University of Firenze, I) for stimulating discussions. We also wish to thank Dr. Grigore Timco and Prof. Richard Winpenny (University of Manchester, UK) for all their hints. This work was partially supported by PRIN project of the Italian Ministry of Research and by the US AFOSR/AOARD program, contract FA2386-13-1-4029.

References

1. P. Gütlich, H.A. Goodwin, *Spin Crossover in Transition Metal Compounds* (Springer, Berlin, 2004)
2. P. Gütlich, Y. Garcia, T. Woike, Coord. Chem. Rev. **219–221**, 839 (2001)
3. D. Gatteschi, R. Sessoli, J. Villain, *Molecular Nanomagnets* (Oxford University Press, London, 2006)

4. M.A. Nielsen, I. Chuang, *Quantum Computation and Quantum Information* (Cambridge University Press, Cambridge, 2000)
5. E. Coronado, C. Martí-Gastaldo, E. Navarro-Moratalla, A. Ribera, Inorg. Chem. **49**, 1313 (2010)
6. I. Salitros, N.T. Madhu, R. Boća, J. Pavlik, M. Ruben, Monatsh. Chem. **140**, 695 (2009)
7. A.R. Katritzky, J. Elguero et al., *The Tautomerism of Heterocycles* (Academic Press, New York, 1976)
8. A. Dei, D. Gatteschi, C. Sangregorio, L. Sorace, Acc. Chem. Res. **37**, 827 (2004)
9. O. Sato, J. Tao, Y.-Z. Zhang, Angew. Chem., Int. Ed. Engl. **46**, 2152 (2007)
10. N. Ishikawa, M. Sugita, T. Ishikawa, S. Koshihara, Y. Kaizu, J. Phys. Chem. B **108**, 11265 (2004)
11. H. Wende et al., Nat. Mater. **6**, 516 (2007)
12. D.A. Lidar, J.H. Thywissen, J. Appl. Phys. **96**, 754 (2004)
13. S. Bandyopadhyay, M. Cahay, *Introduction to Spintronics* (CRC Press, Boca Raton, 2008)
14. M. Urdampilleta, S. Klyatskaya, J.P. Cleuziou, M. Ruben, W. Wernsdorfer, Nat. Mater. **10**, 502 (2011)
15. A. Candini, S. Klyatskaya, M. Ruben, W. Wernsdorfer, M. Affronte, Nano Lett. **11**, 2634 (2011)
16. A.S. Zyazin, J.W.G. van den Berg, E.A. Osorio, H.S.J. van der Zant, N.P. Konstantinidis, M. Leijnse, M.R. Wegewijs, F. May, W. Hofstetter, C. Danieli, A. Cornia, Nano Lett. **10**, 3307 (2010)
17. A.A. Khajetoorians, J. Wiebe, B. Chilian, R. Wiesendanger, Science **332**, 1062 (2011)
18. C.F. Hirjibehedin, C.P. Lutz, A.J. Heinrich, Science **312**, 1021 (2006)
19. S. Loth, S. Baumann, C.P. Lutz, D.M. Eigler, A.J. Heinrich, Science **335**, 196 (2012)
20. N. Atodiresei, J. Brede, P. Lazic, V. Caciuc, G. Hoffmann, R. Wiesendanger, S. Blugel, Phys. Rev. Lett. **105**, 066601 (2010)
21. B.L. Feringa, *Molecular Switches* (Wiley, New York, 2001)
22. K. Matsuda, K. Takayama, M. Irie, Inorg. Chem. **43**, 482 (2004)
23. K. Matsuda, M. Irie, J. Am. Chem. Soc. **122**, 7195 (2000)
24. K. Matsuda, M. Matsuo, S. Mizoguti, K. Higashiguchi, M. Irie, J. Phys. Chem. B **106**, 11218 (2002)
25. M. Mas-Torrent, N. Crivillers, C. Rovira, J. Veciana, Chem. Rev. **112**, 2506 (2012)
26. J.F. Létard, P. Guionneau, L. Goux-Capes Top, Curr. Chem. **235**, 221 (2004)
27. D.E. Richardson, H. Taube, J. Am. Chem. Soc. **105**, 40 (1983)
28. V. Marvaud, J.-P. Launay, C. Joachim, Chem. Phys. **177**, 23 (1993)
29. T.B. Faust, V. Bellini, A. Candini, S. Carretta, L. Carthy, B.J. Coe, D. Collison, R.J. Docherty, J. Kenyon, G. Lorusso, J. Machin, E.J.L. McInnes, C.A. Muryn, R.G. Pritchard, S.J. Teat, G.A. Timco, F. Tuna, G.F. Whitehead, W. Wernsdorfer, M. Affronte, R.E.P. Winpenny, Chemistry **17**, 14020 (2011)
30. V. Bellini, G. Lorusso, A. Candini, T.B. Faust, G.A. Timco, W. Wernsdorfer, R.E.P. Winpenny, M. Affronte, Phys. Rev. Lett. **106**, 227205 (2011)
31. L. Venkataraman, J.E. Klare, C. Nuckolls, M.S. Hybertsen, M.L. Steigerwald, Nature **442**, 904 (2006)
32. J.M. Lehn, Eur. J. Chem. **6**(12), 2097 (2000)
33. L.M. Adleman, Science **266**, 1021 (1994)
34. M. van Rossum, *Neural Computation* (2011). Lecture notes
35. F. Troiani, M. Affronte, Chem. Soc. Rev. **40**, 3119 (2011)
36. Strategic Plan of ENIAC at: http://www.eniac.eu/web/documents/general.php
37. K.L. Wang, J. Nanosci. Nanotechnol. **2**, 235 (2002)
38. W. Lu, C.M. Lieber, Nat. Mater. **6**, 841 (2007)
39. N. Domingo, E. Bellido, D. Ruiz-Molina, Chem. Soc. Rev. **41**, 258 (2012)
40. M. Rancan, F. Sedona, M. Di Marino, L. Armelao, M. Sandi, Chem. Commun. **47**, 5744 (2011)

41. A. Ghirri, V. Corradini, V. Bellini, R. Biagi, U. del Pennino, V. de Renzi, J.C. Cezar, C.A. Muryn, G.A. Timco, R.E.P. Winpenny, M. Affronte, ACS Nano **5**, 7090 (2011)
42. M. Mannini, F. Pineider, P. Sainctavit, C. Danieli, E. Otero, C. Sciancalepore, A.M. Talarico, M.A. Arrio, A. Cornia, D. Gatteschi, R. Sessoli, Nat. Mater. **8**, 194 (2009)
43. V. Corradini, A. Ghirri, E. Garlatti, R. Biagi, V. De Renzi, U. del Pennino, V. Bellini, S. Carretta, P. Santini, G. Timco, R.E.P. Winpenny, M. Affronte, Adv. Funct. Mater. **22**, 3706 (2012)
44. M.L. Plumer, J. van Ek, D. Waller, *The Physics of Ultra-high Density Magnetic Recording* (Springer, Berlin, 2001)
45. L.K. Keniley, N. Dupont, L. Ray, J. Ding, K. Kovnir, J.M. Hoyt, A. Hauser, M. Shatruk, Inorg. Chem. **52**, 8040 (2013). doi:10.1021/ic4006949
46. E.R. Kay, D.A. Leigh, F. Zerbetto, Angew. Chem., Int. Ed. Engl. **46**, 72 (2007)
47. E. Pop, Nano Res. **3**, 147 (2010)
48. B. Behin-Aein, D. Datta, S. Salahuddin, S. Datta, Nat. Nanotechnol. **5**, 266 (2010)
49. S. Bandyopadhyay, M. Cahay, Nanotechnology **20**, 412001 (2009)
50. C. Chappert, A. Fert, F. Nguyen Van Dau, Nat. Mater. **6**, 813 (2007)
51. J. Ashley, M.-P. Bernal, G.W. Burr, H. Coufal, H. Guenther, J.A. Hoffnagle, C.M. Jefferson, B. Marcus, R.M. Macfarlane, R.M. Shelby, G.T. Sincerbox, IBM J. Res. Dev. **44**, 341 (2000)
52. M.N. Leuenberg, D. Loss, Nature **410**, 789 (2001)
53. F. Meier, J. Levy, D. Loss, Phys. Rev. B **68**, 134417 (2003)
54. F. Troiani, A. Ghirri, M. Affronte, S. Carretta, P. Santini, G. Amoretti, S. Piligkos, G. Timco, R.E.P. Winpenny, Phys. Rev. Lett. **94**, 207208 (2005)
55. S. Nakazawa, S. Nishida, T. Ise, T. Yoshino, N. Mori, R.D. Rahimi, K. Sato, Y. Morita, K. Toyota, D. Shiomi, M. Kitagawa, H. Hara, P. Carl, P. Hofer, T. Takui, Angew. Chem., Int. Ed. Engl. **51**, 9860 (2012)
56. F. Troiani, A. Ghirri, M. Affronte, S. Carretta, P. Santini, G. Amoretti, S. Piligkos, G. Timco, R.E.P. Winpenny, Phys. Rev. Lett. **94**, 207208 (2005)
57. C.J. Wedge, G.A. Timco, E.T. Spielberg, R.E. George, F. Tuna, S. Rigby, E.J.L. McInnes, R.E.P. Winpenny, S.J. Blundell, A. Ardavan, Phys. Rev. Lett. **108**, 107204 (2012)
58. S. Bertaina, S. Gambarelli, T. Mitra, B. Tsukerblat, A. Müller, B. Barbara, Nature **453**, 203 (2008)
59. K.-Y. Choi, Z. Wang, H. Nojiri, H. van Tol, P. Kumar, P. Lemmens, B.S. Bassil, U. Kortz, N.S. Dalal, Phys. Rev. Lett. **108**, 067206 (2011)
60. K. Sato, S. Nakazawa, R. Rahimi, T. Ise, S. Nishida, T. Yoshino, N. Mori, K. Toyota, D. Shiomi, Y. Yakiyama, Y. Morita, M. Kitagawa, K. Nakasuji, M. Nakahara, H. Hara, P. Carl, P. Hofer, T. Takui, J. Mater. Chem. **19**, 3739 (2009)
61. J. Lehmann, A. Gaita-Arino, E. Coronado, D. Loss, Nat. Nanotechnol. **2**, 312 (2007)
62. M.J. Martínez-Pérez, S. Cardona-Serra, C. Schlegel, F. Moro, P.J. Alonso, H. Prima-García, J.M. Clemente-Juan, M. Evangelisti, A. Gaita-Ariño, J. Sesé, J. van Slageren, E. Coronado, F. Luis, Phys. Rev. Lett. **108**, 247213 (2012)
63. C. Schlegel, J. van Slageren, M. Manoli, E.K. Brechin, M. Dressel, Phys. Rev. Lett. **101**, 147203 (2008)
64. S. Takahashi, I.S. Tupitsyn, J. van Tol, C.C. Beedle, D.N. Hendrickson, P.C.E. Stamp, Nature **476**, 76 (2011)
65. M.N. Leuenberger, D. Loss, Nature **410**, 789 (2001)
66. S. Bertaina, S. Gambardelli, A. Tkachuk, A.N. Kurkin, B. Malkin, A. Stepanov, B. Barbara, Nat. Nanotechnol. **2**, 39 (2007)
67. F. Luis, A. Repollés, M.J. Martonez-Pérez, D. Aguilá, O. Roubeau, D. Zueco, P.J. Alonso, M. Evangelisti, A. Camón, J. Sesé, L.A. Barrios, G. Aromí, Phys. Rev. Lett. **107**, 117203 (2011)
68. E.C. Sañudo, T. Cauchy, E. Ruiz, R.H. Laye, O. Roubeau, S.J. Teat, G. Aromí, Inorg. Chem. **46**, 9045 (2007)
69. L. Barrios, D. Aguilá, O. Roubeau, P. Gamez, J. Ribas-Ariño, S.J. Teat, G. Aromí, Eur. J. Chem. **15**, 11235 (2009)
70. G. Aromí, D. Aguilá, P. Gamez, F. Luis, O. Roubeau, Chem. Soc. Rev. **41**, 537 (2012)

71. F. Troiani, M. Affronte, S. Carretta, P. Santini, G. Amoretti, Phys. Rev. Lett. **94**, 190501 (2005)
72. M. Trif, F. Troiani, D. Stepanenko, D. Loss, Phys. Rev. Lett. **101**, 217201 (2008)
73. S. Carretta, P. Santini, G. Amoretti, F. Troiani, M. Affronte, Phys. Rev. B **76**, 024408 (2008)
74. P. Santini, S. Carretta, F. Troiani, G. Amoretti, Phys. Rev. Lett. **107**, 230502 (2011)
75. S. Lloyd, Science **261**, 1569 (1993)
76. S.C. Benjamin, Phys. Rev. A **61**, 20301(R) (2000)
77. S.C. Benjamin, Phys. Rev. Lett. **88**, 17904 (2001)
78. S. Lloyd, Science **273**, 1073 (1996)
79. I. Buluta, F. Nori, Science **236**, 108 (2009)
80. R. Feynman, Int. J. Theor. Phys. **21**, 467 (1982)
81. Y. Kubo, F.R. Ong, P. Bertet, D. Vion, V. Jacques, D. Zheng, A. Dréau, J.-F. Roch, A. Auffeves, F. Jelezko, J. Wrachtrup, M.F. Barthe, P. Bergonzo, D. Esteve, Phys. Rev. Lett. **105**, 140502 (2010)
82. Y. Kubo, C. Grezes, A. Dewes, T. Umeda, J. Isoya, H. Sumiya, N. Morishita, H. Abe, S. Onoda, T. Ohshima, V. Jacques, A. Dréau, J.-F. Roch, I. Diniz, A. Auffeves, D. Vion, D. Esteve, P. Bertet, Phys. Rev. Lett. **107**, 220501 (2011)
83. A. Imamoglu, Phys. Rev. Lett. **102**, 083602 (2009)
84. M. Trif, F. Troiani, D. Stepanenko, D. Loss, Phys. Rev. B **82**, 045429 (2010)
85. M.F. Islam, J.F. Nossa, C.M. Canali, Phys. Rev. B **82**, 155446 (2010)
86. N. Baadji, M. Piacenza, T. Tugsuz, F.D. Sala, G. Maruccio, S. Sanvito, Nat. Mater. **8**, 813 (2009)
87. F. Troiani, D. Stepanenko, D. Loss, Phys. Rev. B **86**, 161409 (2012)
88. F. Troiani, I. Siloi, Phys. Rev. A **86**, 032330 (2012)
89. O. Gühne, G. Toth, Phys. Rep. **474**, 1 (2009)
90. I. Siloi, F. Troiani, Phys. Rev. B **86**, 224404 (2012)
91. A. Candini, G. Lorusso, F. Troiani, A. Ghirri, S. Carretta, P. Santini, G. Amoretti, C. Muryn, F. Tuna, G. Timco, E.J.L. McInnes, R.E.P. Winpenny, W. Wernsdorfer, M. Affronte, Phys. Rev. Lett. **104**, 034203 (2010)
92. F. Troiani, V. Bellini, A. Candini, G. Lorusso, M. Affronte, Nanotechnology **21**, 274009 (2010)

Chapter 11
Molecular Magnets for Quantum Information Processing

Kevin van Hoogdalem, Dimitrije Stepanenko, and Daniel Loss

Abstract In this chapter we will examine the possibility of utilizing molecular magnets for quantum information processing purposes. We start by giving a brief introduction into quantum computing, and highlight the fundamental differences between classical- and quantum computing. We will introduce the five DiVincenzo criteria for successful physical implementation of a quantum computer, and will use these criteria as a guideline for the remainder of the chapter. We will discuss how one can utilize the spin degrees of freedom in molecular magnets for quantum computation, and introduce the associated ways of controlling the state of the qubit. In this part we will focus mainly on the spin-electric effect, which makes it possible to control the quantum states of spin in molecular magnets by electric means. We will discuss ways to couple the quantum state of two molecular magnets. Next, we will identify and discuss the different decoherence mechanisms that play a role in molecular magnets. We will show that one of the advantages of using molecular magnets as qubits is that it is possible to use degrees of freedom that are more robust against decoherence than those in more traditional qubits. We briefly discuss preparation and read-out of qubit states. Finally, we discuss a proposal to implement Grover's algorithm using molecular magnets.

11.1 Introduction

Conceptually, a computer is a device that takes an input and manipulates it using a predetermined set of deterministic rules to compute a certain output. Both input and output are defined in terms of bits, classical physical systems which can be in one of two different states. These states are typically denoted 0 and 1. The set of rules that a computer uses for a computation, also named the algorithm, can be described by a set of gates. A simple example of a gate is the one-bit NOT-gate, which gives a 1 as output when the input is 0, and vice versa. An example of a two-bit

K. van Hoogdalem · D. Stepanenko · D. Loss (✉)
Department of Physics, University of Basel, Klingelbergstrasse 82, 4056 Basel, Switzerland
e-mail: daniel.loss@unibas.ch

K. van Hoogdalem
e-mail: kevin.vanhoogdalem@unibas.ch

J. Bartolomé et al. (eds.), *Molecular Magnets*, NanoScience and Technology,
DOI 10.1007/978-3-642-40609-6_11, © Springer-Verlag Berlin Heidelberg 2014

gate is the NAND-gate, which gives a 0 as output only if both the input bits are 1, and yields a 1 otherwise. Interestingly, it can be shown that any classical algorithm can be implemented using a combination of NAND-gates only. However, this completeness theorem does not state anything about the time in which a certain problem can be solved. Instead, such questions belong to the field of computational complexity theory [1]. A large class of problems, called NP, contains all the problems for which a candidate solution can be checked in polynomial time. In contrast, the class of problems that can be solved in polynomial time is called P. Whether P is a strict subset of NP is one of the great open problems in mathematics. It is widely believed that there are problems in the difference between P and NP. Some of the candidates were shown to be solvable using a quantum computer, but an efficient solution on a classical computer is unknown. This inability of a classical computer to solve certain problems efficiently is one of the main driving forces behind the study of quantum computation. Heuristically one might argue that, since classical computers are governed by Newtonian mechanics—which is only valid in certain limits of the underlying quantum theory—a quantum computer must have computational power which is at least the same as, and hopefully greater than, that of a classical computer [2]. Different algorithms exist that support the claim that a quantum computer is inherently more powerful than a classical computer. Among these are Deutsch-Jozsa's [3, 4], Grover's [5], and Shor's algorithm [6].

Besides being interesting from this pragmatic point of view, quantum computing is also of fundamental importance in the fields of information theory and computer science. The fact that quantum mechanics plays a role in information theory becomes clear when one realizes that abstract information is always embedded in a physical system, and is therefore governed by physical laws. This was made explicit by Deutsch [7], when he proposed a stricter version of the Church-Turing hypothesis, emphasizing its 'underlying physical assertion'. The original Church-Turing hypothesis loosely states that every function which would naturally be regarded as computable can be computed by the universal Turing machine [8, 9], and this statement can be seen as the basis underlying computer science. In a sense, a universal Turing machine is a theoretical formalization of a computer (with an infinite memory) as we described it previously. Deutsch replaces this hypothesis by his more physical Church-Turing principle: 'Every finitely realizable physical system can be perfectly simulated by a universal model computing machine operating by finite means'. He then went on to show that the universal Turing machine does not fulfill the requirements for a universal model computing machine, while the universal quantum computer, proposed in the same work, is compatible with the principle. In this way, the universal quantum computer takes the role of the universal Turing machine.

The basic unit of information in a quantum computer is a qubit [10]. Like a classical bit, a qubit is a physical two-level system, with basis states denoted by $|0\rangle$ and $|1\rangle$. Unlike a classical bit, however, a qubit is a quantum system. This makes the information stored in a qubit ultimately analog, since a qubit can be in any state $|\psi\rangle = \alpha|0\rangle + \beta|1\rangle$, with α and β complex numbers such that $|\alpha|^2 + |\beta|^2 = 1$. In a quantum computer, a gate will act linearly on a state $|\psi\rangle$, and hence in a sense on

$|0\rangle$ and $|1\rangle$ simultaneously. This quantum parallelism is one of the advantages of a quantum computer. Of course, one must keep in mind that reading out the qubit (measuring the state) collapses the quantum state into one of the basis states $|0\rangle$ or $|1\rangle$, so this parallelism cannot be used trivially. The other key advantage of using quantum computing is the fact that two qubits can be entangled, i.e. there can exist non-classical correlations between two qubits. The final important property of qubits is captured by the no-cloning theorem [11], which states that it is impossible to copy an unknown quantum state. This theorem invalidates the use of classical error-correction methods -which are typically based on redundancy, and therefore require copying of bits- for quantum computation. Instead, one has to resort to quantum error-correction codes that rely upon entanglement and measurement, but do not require an ability to copy an unknown quantum state.

Quantum mechanics dictates that the time evolution of an isolated quantum state is described by a unitary operator. This means that the action of any valid quantum gate must also be described by a unitary operator. In fact, it turns out that this is the only requirement on a valid quantum gate. Consequently, there exists a rich variety of quantum gates: Where the only non-trivial classical one-bit gate is the NOT-gate, any rotation in the one-qubit Hilbert space is a quantum gate. As an important example of a one-qubit gate that has no classical analog we mention the Hademard-gate, which transforms $|0\rangle$ into $(|0\rangle + |1\rangle)/\sqrt{2}$ and $|1\rangle$ into $(|0\rangle - |1\rangle)/\sqrt{2}$. An example of a two-qubit gate is the CNOT-gate, which acts as a NOT-gate on the second qubit when the first qubit is in the state $|1\rangle$, and does nothing otherwise. It can be shown that arbitrary single qubit rotations together with the CNOT-gate are sufficient to implement any two-qubit unitary evolution exactly [12].

After all these theoretical considerations, one might wonder what is actually required to build a physical quantum computer. The requirements have been succinctly summarized by DiVincenzo, in terms of his five DiVincenzo criteria for successful implementation of a quantum computer [2]. In order to have a functional quantum computer we need

- a collection of well-defined physical quantum two-level systems (qubits), which should be well-isolated and scalable, i.e. it should be possible to add qubits at will.
- a procedure to initialize the system in an initial state, for instance $|00\ldots0\rangle$.
- the ability to perform logic operations on the qubits, i.e. one- and two-qubit gates.
- long enough decoherence times compared to the 'clock time' of the quantum computer for quantum error correction to be efficient.
- the ability to read out the final state of the qubit.

Satisfying these criteria in a single system simultaneously has turned out to be quite a tour de force. Although tremendous progress -both theoretical and experimental- towards completion of this goal has been made in a wide variety of different areas of solid state physics, it is at this point not clear which system will turn out to be most suitable. Of all the systems that have been proposed as a basis for qubit, we mention here quantum dots [13, 14], cold trapped ions [15], cavity quantum electrodynamics [14, 16], bulk nuclear magnetic resonance [17], low-capacitance

Josephson junctions [18], donor atoms [19, 20], linear optics [21], color centers in diamond [22–24], carbon nanotubes [25], nanowires [26], and lastly the topic of this chapter: Molecular magnets [27–32].

11.2 Encoding of Qubits in Molecular Magnets

We have seen that information in a quantum computer must be encoded in qubits, i.e. well-defined physical quantum two-level systems. Probably the first candidate for a qubit that comes to mind is a single spin in for example an atom. However, experimentally it would be very challenging to control this single spin, since the length scale on which this control would have to take place is prohibitively small. On the other side of the spectrum, solid state implementations of qubits such as Ref. [13] require fields on the scale of several tens to hundreds of nanometers only, making control of the state easier (though still very hard). However, with the increased size we pay the price of additional sources of decoherence, and a huge effort has been made in recent years to combat these sources. For molecular magnets, the requirements on the spatial scale on which control has to be possible are loosened with respect to those for a single spin, because the typical size of such systems is relatively large. However, molecular magnets are still small as compared to other solid states implementations of qubits. This fact, as well as the possibility of chemically engineering molecular magnets with a wide variety of properties, may make one hopeful that sources of decoherence in molecular magnets can be suppressed. Indeed, we will show later that such suppression is possible by choosing the degree of freedom that encodes the qubit wisely.

On the other hand, since molecular magnets have a complex chemical structure containing many interacting magnetic atoms, it is not a priori clear that it will be possible to identify a well-separated, stable, and easily controllable two-level subspace in the spectrum. As we will show next, the fact that this does in fact turn out to be possible is due to the high symmetry of the molecule and the existence of well-separated energy scales. We have seen in previous chapters that molecular magnets can—to a very good approximation—be described by a collection of coupled spins. The low-energy multiplet of the system is then described by a spin-multiplet with fixed total spin, separated from excited states on an energy scale set by the exchange interaction. This low-energy multiplet has either maximal total spin for ferromagnetically coupled individual spins, or minimal total spin for antiferromagnetically coupled spins. In the latter case, the details of the ground state are then determined by the symmetry of the molecule, and frustration can play an important role.

The first requirement which has to be fulfilled by any qubit-candidate is that the physical system has to show genuine quantum behavior. Quantum behavior of the spin state in molecular magnets has been shown in experiments on quantum tunneling of magnetization [33–40], and shows up in hysteresis curves of ferromagnetic (although similar effects are predicted to occur in anitferromagnetic systems [41, 42]) molecular magnets with large spin and high anisotropy barrier [36, 37, 43–45]. In the absence of external fields, the barrier due to the anisotropy lifts the degeneracy between states with different magnetization, and leads to the existence of long-lived spin states. Transitions between different spin states can be driven in a coherent manner, and manifest themselves as stepwise changes in the magnetization. The fact that the transitions show interference between transition paths and Berry phase effects are a signature of their coherent nature [46–52].

Quantum computing in antiferromagnetically coupled spin clusters was studied in Ref. [29]. In the simplest cases of a spin chain or a bipartite lattice with an odd number of spins the degenerate ground state is a spin doublet with effective total spin 1/2. The total spin can be controlled by an applied magnetic field just as a single spin can, and exchange interaction between two clusters can be introduced by coupling single spins in the two different clusters. A downside of using a collection of spins is that generally decoherence increases with number of spins, unless one manages to encode the qubit in a state which is protected due to symmetry, something we will come back to later. In Ref. [30], Cr-based AFM molecular rings, and specifically Cr_7Ni, were proposed as suitable qubit candidates.

An interesting way of encoding a qubit is offered by geometrically frustrated molecules [32, 53]. Exemplary molecules that display geometric frustration are antiferromagnetic spin rings with an odd number of spins. The simplest example of such a system is given by an equilateral triangular molecule with a spin-1/2 particle at each vertex, such as is for instance realized to a good approximation in Cu_3 (we will use Cu_3 as an abbreviation for the molecule $Na_9[Cu_3Na_3(H_2O)_9(\alpha\text{-}AsW_9O_{33})_2] \cdot 26H_2O$) (see Ref. [54]). Spin rings (of which the spin triangle is the simplest non-trivial example) in general are described by the Heisenberg Hamiltonian with Dzyaloshinskii-Moriya interaction

$$H_0 = \sum_{i=1}^{N} J_{i,i+1}\mathbf{S}_i \cdot \mathbf{S}_{i+1} + \mathbf{D}_{i,i+1} \cdot (\mathbf{S}_i \times \mathbf{S}_{i+1}). \tag{11.1}$$

Here, N is the number of spins in the ring, and $\mathbf{S}_{N+1} = \mathbf{S}_1$. For the triangular molecular magnet $N = 3$. Furthermore, the fact that the point group symmetry of the triangular molecule is D_{3h} imposes the constraints $J_{i,i+1} = J$ and $\mathbf{D}_{i,i+1} = D\hat{\mathbf{z}}$ on the parameters of the Hamiltonian of an planar molecule. Since we are considering antiferromagnetic systems, J is positive. In a Cu_3 molecule, $|J|/k_B \sim 5$ K and $|D|/k_B \sim 0.5$ K. Due to this separation of energy scales, and in the absence of strong magnetic- or electric fields, the Hilbert space containing the 8 eigenstates of the triangular molecule can be split up in a high-energy quadruplet with total spin $\mathbf{S} = 3/2$ and a low-energy quadruplet with total spin $\mathbf{S} = 1/2$. The splitting between the two subspaces is $3J/2$.

In the absence of Dzyaloshinskii-Moriya interaction the low-energy subspace is fourfold degenerate. The eigenstates are given by

$$|1/2, \pm 1\rangle = \frac{1}{\sqrt{3}} \sum_{j=0}^{2} e^{\pm i2\pi j/3} C_3^j |\uparrow\downarrow\downarrow\rangle, \tag{11.2}$$

and $|-1/2, \pm 1\rangle$. The latter states are also given by (11.2) but with all the spins flipped. These states are thusly labeled as $|m_S, m_C\rangle$, with m_S the quantum number belonging to the z projection of the total spin of the triangle, and m_C the z projection of the chirality of the molecular magnet. The chirality operator $\mathbf{C}$ has components

$$\begin{aligned} C_x &= -\frac{2}{3}[\mathbf{S}_1 \cdot \mathbf{S}_2 - 2\mathbf{S}_2 \cdot \mathbf{S}_3 + \mathbf{S}_3 \cdot \mathbf{S}_1], \\ C_y &= \frac{2}{\sqrt{3}}[\mathbf{S}_1 \cdot \mathbf{S}_2 - \mathbf{S}_3 \cdot \mathbf{S}_1], \\ C_z &= \frac{4}{\sqrt{3}}\mathbf{S}_1 \cdot [\mathbf{S}_2 \times \mathbf{S}_3]. \end{aligned} \tag{11.3}$$

The chirality contains information about the relative orientation of the spins that make up the molecule. Like the components of the total spin operator, the components of the chirality operator obey angular momentum commutation relations. It is straightforward to show that the total spin and chirality commute. We will show later that states with opposite chirality are split by an energy gap which is determined by the magnitude of the Dzyaloshinskii-Moriya interaction. Furthermore, we can separate states with opposite total spin by applying a magnetic field. This allows us to choose which doublet makes up the ground state, chirality or total spin. In this way it is possible to either encode the qubit in the total spin of the molecule or in the chirality. Furthermore, even thought the commutation relations of the chirality components are the same as those of the spin components, the transformation properties of spin and chirality under rotations, reflections, and time-reversal do differ. Therefore, interactions of chirality with external fields can not be inferred from the analogy with spins. We will discuss later how using the chirality offers certain benefits with regards to the possibility to control the qubit and with regards to increasing the decoherence time of the qubit.

11.3 Single-Qubit Rotations and the Spin-Electric Effect

If one chooses to encode a qubit in a spin state -be it the spin of an electron in a quantum dot, or the total spin of a molecular magnet- the most intuitive way to implement a one-qubit gate is by utilizing the Zeeman coupling $\mu_B \mathbf{B} \cdot \bar{\bar{g}} \cdot \mathbf{S}$, where $\bar{\bar{g}}$ is the g-tensor. This coupling in principle allows one to perform rotations around an arbitrary axis by applying ESR (electron spin resonance) pulses. Indeed, it has been shown to be possible to implement single spin rotations on a sub-microsecond time scale using ESR techniques in quantum dots [55]. Furthermore, Rabi-oscillations of the magnetic cluster V_{15} have been shown to be possible, also on a sub-microsecond time scale [56]. At the moment, however, it appears experimentally very challenging to increase the temporal- and spatial resolution with which one can control magnetic fields to the point that is required for quantum computation in molecular magnets (i.e. nanosecond time scale and nanometer length scale).

For this reason, a large effort has been made to find alternative ways to control the spin state of molecular magnets. One natural candidate to replace magnetic manipulation is electric control. Strong, local electric fields can be created near a STM tip, and these fields can be rapidly turned on and off by applying an electric voltage to electrodes that are placed close to the molecules that are to be controlled.

Electric manipulation requires a mechanism that gives a sizable spin-electric coupling. In quantum dots, the mechanism behind this coupling is the relativistic spin-orbit interaction (SOI), and experiments that show that it is possible to perform single spin rotations by means of electric dipole spin resonance (EDSR) have been proposed [57] and performed [58]. Unfortunately, the fact that this effect scale with the system size L as L^3 makes them unsuitable for molecular magnets, which are much smaller.

Instead, in Ref. [32], Trif et al. proposed a mechanism that leads to spin-electric coupling in triangular magnetic molecules with spin-orbit interaction and broken inversion symmetry. The mechanism relies on the fact that in such systems an electric field can alter the exchange interaction between a pair of spins within a molecule due to the field's coupling to the dipole moment of the connecting bond.

The lowest order coupling between electric field and the spin state of the triangular molecule is given by the electric-dipole coupling, through the Hamiltonian $H_{\text{e-d}} = -e\sum_i \mathbf{E}\cdot\mathbf{r}_i \equiv -e\mathbf{E}\cdot\mathbf{R}$. Here, e is the electron charge and $\mathbf{r}_i$ is the position of the i-th electron. The total dipole moment of the molecule is given by $-e\sum_i \mathbf{r}_i = -e\mathbf{R}$. Because of the D_{3h} symmetry of the molecule, the diagonal elements of total dipole moment operator must vanish in the proper symmetry-adapted basis. However, the electric-dipole coupling can mix states with different chirality. The nonzero matrix elements are the ones that are invariant under the symmetry-transformations of the triangular magnet. Since the $|m_S, \pm 1\rangle$ states and the operators $\pm X + iY$ both transform as the irreducible representation E' of the group D_{3h}, it follows that the only nonzero components in the low-energy subspace of the triangular molecules are

$$\langle m_S, \pm 1| -eX|m'_S, \mp 1\rangle = i\langle m_S, \pm 1| -eY|m'_S, \mp 1\rangle \equiv d\delta_{m_S,m'_S}. \qquad (11.4)$$

Coupling to the $S = 3/2$ subspace is suppressed by the finite gap between the two subspaces. By its very nature, this symmetry analysis cannot yield any information on the magnitude of the effective electric dipole parameter d. This information will have to be extracted using other methods, such as ab initio modeling, Hubbard modeling, or experiments, something we will come back to later. We do note that a finite amount of asymmetry of the wave functions centered around each vertex of the triangle is required for the matrix elements in (11.4) to be nonzero. This asymmetry is caused by the small amount of delocalization of the electron states due to the exchange interaction with the states on the other vertices and creates the finite dipole moment of individual bonds. The dipole moment of the bonds, furthermore, must depend on the relative orientation of the two spins which are connected by that bond (i.e. whether they are parallel or anti-parallel) in order for the matrix elements in (11.4) to be nonzero.

Since the electric-dipole coupling connects states with different chirality, we can rewrite it in terms of the vector $\mathbf{C}_\parallel = (C_x, C_y, 0)$ as $H^{\text{eff}}_{\text{e-d}} = d\mathbf{E}'\cdot\mathbf{C}_\parallel$. The vector $\mathbf{E}'$ is given by $\mathbf{E}' = \mathcal{R}(7\pi/6 - 2\theta)\mathbf{E}$, where $\mathcal{R}(\phi)$ describes a rotation by an angle ϕ

around the z axis, and θ is the angle between $\mathbf{r}_1 - \mathbf{r}_2$ and $\mathbf{E}_\parallel = (E_x, E_y, 0)$. With the definition of the chirality operator as given in (11.4), we can rephrase the effective electric-dipole Hamiltonian in terms of exchange coupling between the individual spins

$$H^{\text{eff}}_{\text{e-d}} = \frac{4dE}{3} \sum_{i=1}^{3} \sin\left[\frac{2\pi}{3}(1-i) + \theta\right] \mathbf{S}_i \cdot \mathbf{S}_{i+1}, \tag{11.5}$$

where E is the magnitude of the in plane components of the electric field. Since the change in the exchange interaction $J_{i,i+1}$ is proportional to $|\mathbf{E}_\parallel \times (\mathbf{r}_{i+1} - \mathbf{r}_i)|$, only the component of the electric field that is perpendicular to the bond $\mathbf{r}_{i+1} - \mathbf{r}_i$ affects the exchange interaction $J_{i,i+1}$. This is consistent with the picture that the finite dipole moment of the bond between two vertices is caused by the deformation of the wave function due to exchange interaction. Otherwise, the strength of the coupling is completely determined by the parameter d. The fact that the change in $J_{i,i+1}$ is not uniform is crucial here, since therefore $[H_0, H^{\text{eff}}_{\text{e-d}}] \neq 0$ even in the absence of DM interaction, which allows the electric-dipole interaction to induce transitions between states with different chirality.

We have seen then that the electric-dipole coupling allows one to perform rotations of the chirality state about the x- and y axis, but not around the z axis (assuming a diagonal g-tensor). This is sufficient to perform arbitrary rotations in chirality space. However, so far the total spin does not couple to the electric field. This situation is remedied when we include spin-orbit interaction.

As with the electric-dipole coupling, one can deduct the form of the spin-orbit interaction from general symmetry considerations. Given the D_{3h} symmetry of the molecule, the most general form of the spin-orbit interaction is

$$H_{\text{SO}} = \lambda^{\parallel}_{\text{SO}} T_{A_2} S_z + \lambda^{\parallel}_{\text{SO}} (T_{E''_+} S_- + T_{E''_-} S_+). \tag{11.6}$$

Here, T_Γ denotes a tensor which acts on the orbital space and transforms according to the irreducible representation Γ. The nonzero elements in the low-energy subspace are then given by $\langle m_S, \pm 1 | H_{\text{SO}} | m'_S, \pm 1\rangle = m_S \lambda^{\perp}_{\text{SO}} \delta_{m_S, m'_S}$, which leads to the spin-orbit Hamiltonian $H_{\text{SO}} = \Delta_{\text{SO}} C_z S_z$, where $\Delta_{\text{SO}} = \lambda^{\parallel}_{\text{SO}}$. Alternatively, one can use the fact that the spin-orbit interaction can be described by the Dzyaloshinskii-Moriya term in (11.1). Because of the symmetry of the molecule, the only nonzero component of the DM vector $\mathbf{D}_{i,i+1}$ is the out-of-plane component, so that it takes the form $\mathbf{D}_{i,i+1} = (0, 0, D_z)$. This gives the same form for H_{SO} as the previous considerations, provided one identifies $\lambda^{\parallel}_{\text{SO}} = D_z$.

Combining the results from this section, it follows that the Hamiltonian describing a triangular magnet in the presence of a magnetic- and electric field can be written in terms of the chirality and total spin of the molecule as

$$H = \Delta_{\text{SO}} C_z S_z + \mu_B \mathbf{B} \cdot \bar{\bar{g}} \cdot \mathbf{S} + d\mathbf{E} \cdot \mathbf{C}_\parallel. \tag{11.7}$$

Hence, for a magnetic field in the z direction, the eigenstates are $| \pm 1/2, \pm 1\rangle$, and an electric field causes rotations of the chirality state, but does not couple states with opposite total spin. When $\mathbf{B}$ is not parallel to $\hat{\mathbf{z}}$, S_z is no longer a good quantum number, and hence an applied electric field can cause rotations in the total spin subspace through the electric-dipole and spin-orbit coupling. In this way it becomes possible to perform arbitrary rotations of the total spin state.

In Ref. [53], the authors were able to identify the parameters of the effective spin Hamiltonian with the parameters of the underlying Hubbard model. On the one hand, this has opened up the possibility to determine the parameters of the effective spin Hamiltonian by means of ab initio calculations [59, 60]. On the other hand, the description of the spin-electric effect in the language of the Hubbard model is useful because it gives an intuitive interpretation of the phenomena that we discussed so far. The Hubbard model description of a molecular magnet including spin-orbit interaction is given by

$$H_{\mathrm{H}} = \sum_{i,j} \sum_{\alpha,\beta} \left[c_{i\alpha}^{\dagger} \left(t\delta_{\alpha\beta} + \frac{i\mathbf{P}_{ij}}{2} \cdot \sigma_{\alpha\beta} \right) c_{j\beta} + \mathrm{H.c.} \right] + \sum_{j} U_j (n_{j\uparrow}, n_{j\downarrow}). \quad (11.8)$$

Here, $c_{i\alpha}^{\dagger}$ creates an electron with spin α whose wave function $|\phi_{i\sigma}\rangle$ is given by a Wannier function located around atom i. Furthermore, t describes spin-independent hopping. The vector $\mathbf{P}_{ij}$ describes spin-dependent hopping due to spin-orbit interaction and hence is proportional to the matrix element $\nabla V \times \mathbf{p}$ between Wannier states centered around atom i and j. The vector σ contains the Pauli matrices. Lastly, U describes the on-site repulsion. Typically, one considers a single-orbital model, and assumes that U is the largest energy scale. A perturbative expansion of (11.8) in $(|t|, |\mathbf{P}_{ij}|)/U$ allows one then to map the Hubbard model on a Heisenberg Hamiltonian with DM interaction [61, 62].

Equation (11.8) describes two scenarios. First, if the index i runs over the three magnetic atoms of the triangle only, it describes coupling between the magnetic atoms through direct exchange. Alternatively, (11.8) can describe the situation in which the coupling between two magnetic atoms is mediated by a non-magnetic bridge by adding a doubly-occupied non-magnetic atom on every line connecting two vertices. The former choice allows for a simpler description, whereas the latter choice is anticipated to be the more realistic one for molecular magnets. We will shortly discuss how either can be used to obtain more insight into the spin-electric effect.

The first thing one can show is that in the case of direct-exchange interaction the basis functions of the Hubbard model to first order in t and $\lambda_{\mathrm{SO}} \equiv \mathbf{P}_{ij} \cdot \mathbf{e}_z$ (due to symmetry $\mathbf{P}_{ij} = \lambda_{\mathrm{SO}} \mathbf{e}_z$) are

$$\left|\Phi_{A_2'}^{1\sigma}\right\rangle = \left|\psi_{A_2'}^{1\sigma}\right\rangle \quad (11.9)$$

$$\left|\Phi_{E_\pm'}^{1\sigma}\right\rangle = \left|\psi_{E_\pm'}^{1\sigma}\right\rangle + \frac{(e^{-2\pi i/3} - 1)(t \pm \sigma\lambda_{\mathrm{SO}})}{\sqrt{2}U} \left|\psi_{E_\pm'^1}^{2\sigma}\right\rangle + \frac{3e^{2\pi i/3}(t \pm \sigma\lambda_{\mathrm{SO}})}{\sqrt{2}U} \left|\psi_{E_\pm'^2}^{2\sigma}\right\rangle, \quad (11.10)$$

where $|\psi_{\Gamma}^{n\sigma}\rangle$ denotes the symmetry-adapted eigenstate of the Hubbard model with three electrons, total spin σ, and either single- ($n = 1$) or double ($n = 2$) occupancy that transforms according to the irreducible representation Γ. Specifically, the spin part of $|\psi_{E'_{\pm}}^{1\sigma}\rangle$ is given by the states $|\sigma, \pm 1\rangle$ in (11.2). It follows that in the limit of $t, \lambda_{\text{SO}} \ll U$ (the limit in which the spin model gives an accurate description) the eigenstates of the Hubbard model are indeed the chirality states. At finite t, λ_{SO}, the eigenstates contain small contributions from doubly-occupied states.

Within the direct-exchange model, the electric field couples to the state of the molecule via two different mechanisms. The first term that has to be added to the Hubbard Hamiltonian comes from the fact that the electric potential takes different values at the positions of the magnetic centers in a molecule, which affects the on-site energy of the electrons as

$$H_{\text{e-d}}^{0} = -e\sum_{\sigma} \frac{E_y a}{\sqrt{3}} c_{1\sigma}^{\dagger} c_{1\sigma} - \frac{a}{2}\left(\frac{E_y}{\sqrt{3}} + E_x\right) c_{2\sigma}^{\dagger} c_{2\sigma} + \frac{a}{2}\left(\frac{E_x}{\sqrt{3}} - E_y\right) c_{3\sigma}^{\dagger} c_{3\sigma}. \tag{11.11}$$

Here, a is the distance between two magnetic atoms. The second contribution is given by

$$H_{\text{e-d}}^{1} = \sum_{i,\sigma} t_{ii+1}^{\mathbf{E}} c_{i\sigma}^{\dagger} c_{i+1\sigma} + \text{H.c.}, \tag{11.12}$$

which describes the modification of the hopping strength due to the electric field. The electric field-dependent hopping is given by $t_{ii+1}^{\mathbf{E}} = -\langle \phi_{i\sigma}| e\mathbf{r}\cdot\mathbf{E}|\phi_{i+1\sigma}\rangle$, and is hence related to the matrix elements of the electric dipole moment which mix the different Wannier functions. As before, a symmetry analysis tells us that the only nonzero matrix elements within the total spin-1/2 subspace are those proportional to

$$\langle \phi_{E'_{+}}^{\sigma}| ex|\phi_{E'_{-}}^{\sigma}\rangle = -i\langle \phi_{E'_{+}}^{\sigma}| ey|\phi_{E'_{-}}^{\sigma}\rangle \equiv d_{EE}. \tag{11.13}$$

Here, $|\phi_{\Gamma}^{\sigma}\rangle$ describes the linear combination of Wannier states with total spin σ which transforms according to the irreducible representation Γ. One can then calculate the matrix elements of both the electric-dipole coupling as well as the spin-orbit Hamiltonian perturbatively in $(t, eaE, d_{EE}E)/U$. Furthermore, since the electrons are localized, the off-diagonal elements of the dipole moment, d_{EE}, satisfy $d_{EE} \ll ea$. To lowest order the results are

$$|\langle \Phi_{E'_{-}}^{1\sigma}| H_{\text{e-d}}^{0}|\Phi_{E'_{+}}^{1\sigma}\rangle| \propto \left|\frac{t^3}{U^3} eEa\right|, \tag{11.14}$$

$$|\langle \Phi_{E'_{-}}^{1\sigma}| H_{\text{e-d}}^{1}|\Phi_{E'_{+}}^{1\sigma}\rangle| \approx \left|\frac{4t}{U} E d_{EE}\right|, \tag{11.15}$$

$$|\langle \Phi_{E'_{-}}^{1\sigma}| H_{\text{SO}}|\Phi_{E'_{+}}^{1\sigma}\rangle| = \pm\frac{5\sqrt{3}\lambda_{\text{SO}} t}{2U} \text{sgn}(\sigma). \tag{11.16}$$

These first two matrix elements can be identified with the matrix elements in (11.7) that mix the states with different chirality, and hence determine the parameter d. The last matrix element determines D_z. Therefore, all parameters of the effective spin model in (11.7) can be determined from the underlying microscopic model. In Ref. [60], Nossa et al. utilized the presented analysis to determine the value of D_z and J in the molecular magnet Cu_3 using spin-density functional theory.

It is known that in molecular magnets the direct exchange mechanism is often suppressed due to the localized nature of the electrons that determine the magnetic properties (which are typically of a d-wave nature) combined with the fact that the magnetic atoms are typically separated by non-magnetic bridge atoms. In Cu_3, for instance, exchange interaction between two Cu atoms follows a superexchange path along a Cu-O-W-O-W-O-Cu bond, which makes the Cu atoms third nearest neighbors [54]. A more accurate description on a microscopic basis of the spin-electric effect in a triangular magnet is therefore given by a model which includes a doubly-occupied non-magnetic atom on every line connecting two vertices, so that the mechanism behind the exchange interaction is superexchange. This is further strengthened by the expectation that the orbitals of the magnetic atoms do not deform easily in an electric field, whereas the bridge orbitals are expected to change their shape more easily.

In Ref. [53], the authors analyzed the behavior of a single Cu-Cu bond, including the non-magnetic bridge atom that connects the two Cu atoms, under the application of an electric field. By performing a fourth-order Schrieffer-Wolf transformation [63] on the Hamiltonian (11.8) for such a bond (using $(|t|, |\mathbf{P}_{ij}|)/U$ as small parameter) one can map the Hubbard model on the spin model

$$H_{12} = J\mathbf{S}_1 \cdot \mathbf{S}_2 + \mathbf{D} \cdot (\mathbf{S}_1 \times \mathbf{S}_2) + \mathbf{S}_1 \cdot \boldsymbol{\Gamma} \cdot \mathbf{S}_2. \tag{11.17}$$

Here, $\boldsymbol{\Gamma}$ is a traceless- and symmetric matrix. Equation (11.17) describes the most general quadratic spin Hamiltonian possible. The parameters $J, \mathbf{D}, \boldsymbol{\Gamma}$ can be determined from the parameters of the Hubbard model. Assuming that the bond angle between the Cu atom and the bridge atom is finite, the largest possible symmetry of a single bond with bridge atom is C_{2v}. This determines which spin parameters can be nonzero. If the electric field breaks the C_{2v} symmetry, extra terms can be generated. However, from the C_{2v} symmetry it follows that the strongest spin-electric coupling will be in the plane spanned by the Cu atoms and the bridge atom, and perpendicular to the Cu-Cu bond. This is due to the fact that this is the only direction in which the bond can have a finite dipole moment in the absence of an electric field (due to the molecular field), which gives rise to linear electric-dipole coupling. Indeed, it is this coupling that causes the effective Hamiltonian in (11.5), with effective electric-dipole moment given by

$$d = \frac{4}{U^3}\big[\big(48t^3 - 20tp_z^2\big)\kappa_t + \big(-20t^2 p_z + 3p_z^3\big)\kappa_{pz}\big]. \tag{11.18}$$

Here, t is the hopping parameter, p_z is the z component of the spin-orbit hopping, and $\kappa_t = \delta t/E$ and $\kappa_{pz} = \delta p_z/E$ relate the changes in t and p_z to the electric field E.

Using ab initio methods, the authors in Ref. [59] calculated the effective electric-dipole moment d in Cu_3. They found the value $d = 3.38 \times 10^{-33}$ C m. This corresponds to $d \approx 10^{-4} ea$, where a is the length of the Cu-Cu bond, and leads to Rabi oscillation times $\tau \approx 1$ ns for electric field $E \approx 10^8$ Vm^{-1}.

So far, we have only discussed single-qubit rotations. However, for a complete set of quantum gates, we also need a two-qubit gate. In the next section, we will discuss different proposals that have been made on how to implement such a two-qubit gate.

11.4 Two-Qubit Gates

Suppose we chose to encode our qubit states in the spin degrees of freedom of a system. Two-qubit gates such as the CNOT- or the $\sqrt{\text{SWAP}}$-gate can then be implemented by turning on the Heisenberg exchange interaction between two spins for a certain time [64]. For spins in quantum dots, this is relatively simply done by applying appropriate voltage pulses to the gate that controls the tunneling between two quantum dots. In contrast, in molecular magnets the exchange interaction between two molecules is typically determined by the chemistry of the molecule, and one has to search for more sophisticated ways to implement two-qubit gates.

The first method to couple the state of two qubits that we will discuss is based on coupling of two triangular molecular magnets through a quantum mechanical electric field in a cavity or stripline [32]. Such electric fields offer long-range and switchable coherent interaction between two qubits. The electric field of a phonon with frequency ω in a cavity of volume V is given by $\mathbf{E}_0(b^\dagger_\omega + b_\omega)$, where $b^\dagger_\omega$ creates a photon with frequency ω and the amplitude of the field is $|\mathbf{E}_0| \propto \sqrt{\hbar\omega/V}$. The coupling of such a photon to the in plane component of the chirality $\mathbf{C}_\parallel$ of a triangular molecule is then given by $\delta H_\text{E} = d\mathbf{E}'_0 \cdot \mathbf{C}_\parallel(b^\dagger_\omega + b_\omega)$. In the rotating wave approximation, the Hamiltonian that describes the low-energy subspace of N triangular molecular magnets which interact with the photon field is given by $H_\text{s-ph} = \sum_j H^{(j)} + \hbar\omega b^\dagger_\omega b_\omega$, with

$$H^{(j)} = \Delta_\text{SO} C_z^{(j)} S_z^{(j)} + \mathbf{B} \cdot \bar{\bar{g}} \cdot \mathbf{S}^{(j)} + d|\mathbf{E}_0|\big[e^{i\phi_j} b^\dagger_\omega C_-^{(j)} + \text{H.c.}\big]. \tag{11.19}$$

Here, $\phi_j = 7\pi/6 + \theta_j$. Application of a magnetic field $\mathbf{B}$ with an in plane component allows one to couple both the chirality as well as the total spin degrees of freedom of spatially separated molecules. This coupling can be turned on and off by bringing the molecules in resonance with the photon mode, by applying an additional local electric field. One difficulty in using cavities is that the electric fields are weaker than those at an STM tip. A typical value is $|\mathbf{E}_0| \approx 10^3$ V m^{-1}, which leads to Rabi times $\tau \approx 0.01$–100 μs.

For the discussion of another proposed implementation of an electrically controlled two-qubit gate (in this case the $\sqrt{\text{SWAP}}$-gate), we turn our attention to the polyoxometalate $[PMo_{12}O_{40}(VO)_2]^{q-}$. This molecule consists of a central mixed-valence core based on the $[PMo_{12}O_{40}]$ Keggin unit, capped by two vanadyl groups containing one localized spin each [31]. In such a molecule, one can encode a two-qubit state in the spins of the vanadyl groups. The spins of the two vanadyl groups are weakly exchange coupled via indirect exchange interaction mediated by the core. The crucial property of the core is that one can tune the number of electrons it contains, since the exchange interaction between the vanadyl spins depends on the number of electrons on the core. Namely, if the core contains an odd number of electrons, the spin of the unpaired electron on the core couples to those of the vanadyl groups, and the effective interaction between the two qubits is relatively strong. In contrast, for an even number of spins on the core, the spins on the core pair up to yield a ground state with total spin 0. In this case, the exchange interaction between the pair of vanadyl spins is strongly reduced as compared to the situation with an odd number of electrons on the core. Since the redox flexibility of such polyoxometalates is typically rather high, the number of electrons n_C on the core can be tuned by electric means, by bringing the molecule near the tip of an STM. The system is then described by the Hamiltonian

$$\begin{aligned} H = & -J(n_C)\mathbf{S}_L \cdot \mathbf{S}_R - J_C(\mathbf{S}_L + \mathbf{S}_R) \cdot \mathbf{S}_C \\ & + (\epsilon_0 - eV)n_C + Un_C(n_C - 1)/2. \end{aligned} \tag{11.20}$$

Here, $\mathbf{S}_{L/R}$ are the spin operators of the two vanadyl groups, and $\mathbf{S}_C$ is the spin of the core. $J(n_C)$ denotes the exchange interaction between the two vanadyl spins. Given the previous discussion, $J(0) \approx 0$. The orbital energy of the electron on the core is given by ϵ_0, and V is the electric potential at the core. Lastly, U is the charging energy of the molecule, which defines the largest energy scale in the problem. We consider the subspace of only $n_C = 0$ or $n_C = 1$ electrons on the core.

The two-qubit $\sqrt{\text{SWAP}}$ is now implemented as follows: One starts out with an electric potential such that the stable configuration has $n_C = 0$ electrons on the core. That way, the two qubits are decoupled. By applying a voltage pulse V_g to the STM tip, one can switch to the state with $n_C = 1$ electrons. The Hamiltonian that describes the spin-state of the molecule is then given by [31]

$$H_1 = -\big[J(1) - J_C\big]\mathbf{S}_L \cdot \mathbf{S}_R - \frac{J_C}{2}\mathbf{S}^2. \tag{11.21}$$

Here, $\mathbf{S} = \mathbf{S}_L + \mathbf{S}_R + \mathbf{S}_C$ is the total spin of the molecule. The time-evolution of the system is determined by (11.21) for the duration τ_g of the pulse, afterwards the two vanadyl spins will be decoupled again. The first part of this Hamiltonian contains the wanted exchange coupling, and one can implement different two-qubit gates depending on the pulse length τ_g. For the $\sqrt{\text{SWAP}}$-gate, this time is given by the condition

$$\big[J(1) - J_C\big]\frac{\tau_g}{\hbar} = \frac{\pi}{2} + 2\pi n, \tag{11.22}$$

where n is an integer. The second term in (11.21) depends on the spin-state of the core, and is unwanted. However, we can get rid of it by choosing the pulse-length such that the unitary evolution associated with the second term is equal to the unit operator. This condition turns out to be satisfied for times

$$\tau_g = \frac{4\pi}{3} \frac{\hbar}{|J_C|} m, \tag{11.23}$$

where m is an integer. Together, these last two equations give a requirement on $J(1)$ and J_C, namely

$$\frac{J(1)}{|J_C|} = \text{sgn}(J_C) + \frac{3}{8} \frac{1-4n}{m}. \tag{11.24}$$

So far, we have assumed that switching between states with $n_C = 0$ and $n_C = 1$ can be perfectly controlled and is instantaneous. In reality, however, this transition is governed by quantum processes, and is a probabilistic process governed by the tunneling rate Γ between STM tip and molecule. Therefore, τ_g is inherently a stochastic quantity. To analyze these quantum effects, the authors in Ref. [31] numerically calculated the averaged fidelity $\mathcal{F} = \sqrt{\rho_{\text{real}} \rho_{\text{ideal}}}$ between the idealized $\sqrt{\text{SWAP}}$-gate with instantaneous switching and the real $\sqrt{\text{SWAP}}$-gate with the stochastic tunneling ($\rho_{\text{real}}/\rho_{\text{ideal}}$ denote the obvious density matrices at the end of the $\sqrt{\text{SWAP}}$-gate operation here). They found that the fidelity can be as high as $\mathcal{F} = 0.99$.

11.5 Decoherence in Molecular Magnets

Up to this point, we have assumed that the evolution of the quantum state of any qubit is unitary, and hence the information content of the qubit is infinitely long-lived. This assumption is only valid for a perfectly isolated system. In reality, however, any qubit will be coupled to its environment. Fluctuations in the environment can then lead to decoherence: The process whereby information about a quantum state is lost due to interaction with an environment. Decoherence of a single qubit typically takes place on two different time scales. The longitudinal decoherence time, or T_1-time, describes the average time it takes the environment to induce random transitions from $|0\rangle$ to $|1\rangle$, and vice versa. The transverse decoherence time, the T_2-time, describes the time it takes a systems to lose its information about the coherence between the $|0\rangle$ and $|1\rangle$ state. In other words, the T_2-time is the time it takes for a system initially in the pure quantum state described by the density matrix $\hat{\rho}_0 = |\psi_0\rangle\langle\psi_0|$, where $|\psi_0\rangle = \alpha|0\rangle + \beta|1\rangle$, to transform into the classical state $\hat{\rho}(t) = |\alpha|^2|0\rangle\langle 0| + |\beta|^2|1\rangle\langle 1|$. In this sense, decoherence is the cause of the transition from the quantum- into the classical regime. The T_1-time sets on upper limit on the time a system can be used as a classical bit, whereas a system can only be used as a qubit for times $T \ll T_1, T_2$. The T_1- and T_2-time of a system are not unrelated, and can indeed become of comparable magnitude in certain systems. For molecular magnets at low temperatures, however, typically $T_2 \ll T_1$.

The first measurement of the T_2-time of a system consisting of molecular magnets was performed by Ardavan et al. in 2007 (Ref. [65]). The measurements were performed on Cr_7M heterometallic wheels (M denotes Ni or Mn), and the authors found T_2-times of 3.8 μs for perdeuterated diluted Cr_7Ni solutions. The typical way to measure relaxation times is to use standard spin-echo techniques [66]. The T_2-time can be obtained from the decay with τ of a 2-pulse Hahn-echo measurement, consisting of the sequence: $\pi/2 - \tau - \pi - \tau -$ echo. In a similar manner, the T_1-time can be determined using the sequence $\pi - T - \pi/2 - \tau - \pi - \tau -$ echo. Here, T is varied, and τ is constant and short. One of the difficulties in measuring the T_2-times in magnetic clusters is the fact that, in a crystal, the different molecules are coupled by dipole-dipole interactions. This limits the T_2-time. The natural approach to avoid this problem is to consider molecules in solution. However, here the problem is that many magnetic clusters with high spin display strong axial anisotropy, with relatively large zero-field splitting. In a solution, these clusters will orient in a random matter. This problem is circumvented by using Cr_7Ni-clusters, which have a $S = 1/2$ ground state (and hence no zero-field splitting), and small anisotropy of the g-factor.

It was found that the main mechanism limiting the T_2 -time of the Cr_7Ni-clusters was coupling to protons. To increase the decoherence time, the authors therefore considered the perdeutered analogue compound. Indeed, according to expectations (^{2}D has a gyromagnetic ratio which is about 1/6 of that of ^{1}H), this increased the coherence time roughly by a factor of 6, leading to a T_2-time of 3.8 μs at 1.8 K.

Our remaining discussion of decoherence in molecular magnets follows that of Ref. [67]. In spin systems, the two most common sources of decoherence are fluctuations in the electric environment (which couple to the spin state via spin-orbit interaction) and fluctuations of the spin state of the N nuclear spins $\mathbf{I}_p$ in the host material of the qubit, which are coupled to the system spins $\mathbf{S}_i$ due to hyperfine interaction. We will mainly focus on the latter mechanism, since it typically limits the decoherence time [56, 65]. The hyperfine interaction between nuclear spins and system spins is due to dipole-dipole interaction as well as contact interaction

$$H_{\mathrm{HF}} = D_{\mathrm{HF}} \sum_i \sum_p \frac{\mathbf{S}_i \cdot \mathbf{I}_p - 3(\mathbf{S}_i \cdot \hat{\mathbf{r}}_{ip})(\mathbf{I}_p \cdot \hat{\mathbf{r}}_{ip})}{r_{ip}^3} + \sum_i a_i \mathbf{S}_i \cdot \mathbf{I}_{q(i)}. \tag{11.25}$$

Here, $D_{\mathrm{HF}} = (\mu_0/4\pi) g_I \mu_I g_S \mu_S$, and $\mathbf{r}_{ip} = \mathbf{r}_i - \mathbf{r}_p$. The contact interaction strength a_i is due to the finite overlap of the wave functions of the system spin and nuclear spins located at the same magnetic center. For small clusters, the latter term only leads to oscillations of the coherence, and hence we can neglect it [67]. To see how the hyperfine interaction leads to decoherence, let us consider a system in which the state of the qubit and that of the bath are initially uncorrelated. Furthermore, let the initial state of the qubit be given by $|\psi(0)\rangle = \frac{1}{\sqrt{2}}(|0\rangle + |1\rangle)$, and let the bath be prepared in the (mixed or pure) state described by the density operator $\hat{\rho}_n(0) = \sum_{\mathcal{I}} p_{\mathcal{I}} |\mathcal{I}\rangle\langle\mathcal{I}|$. Here, $|\mathcal{I}\rangle = |m_1^{\mathcal{I}}, \ldots, m_N^{\mathcal{I}}\rangle$ with $m_i^{\mathcal{I}}$ the projection of the nuclear spin operator $\mathbf{I}_i$ along the magnetic field. Two examples of possible states the bath may be prepared in are the spin-polarized (pure) state with polarization P, and the equal superposition (mixed) state. In the first case, $p_{\mathcal{I}} = \delta_{\mathcal{I},n}$, where $|n\rangle$ is the state such that $\sum_p I_p^z|n\rangle = \frac{P}{2}|n\rangle$. In the latter case, $p_{\mathcal{I}} = 1/2^N$. This is the initial state of the bath in the absence of an external magnetic field, ignoring interactions between the nuclear spins. Over time, interactions between the bath and the qubit will introduce correlations between the two subsystems, evolving the state $|\Psi_{\mathcal{I}}(0)\rangle = |\psi(0)\rangle \otimes |\mathcal{I}\rangle$ into the state $|\Psi_{\mathcal{I}}(t)\rangle = \frac{1}{\sqrt{2}}(|0,\mathcal{I}_0\rangle + |1,\mathcal{I}_1\rangle)$ (if we consider only loss of phase coherence). In general, the states $|\mathcal{I}_0\rangle$ and $|\mathcal{I}_1\rangle$ will not be the same. Therefore, the reduced density matrix of the qubit, given by $\hat{\rho}_S(t) = \mathrm{Tr}_n[\sum_{\mathcal{I}} p_{\mathcal{I}}|\Psi_{\mathcal{I}}(t)\rangle\langle\Psi_{\mathcal{I}}(t)|]$, may have a decreased degree of coherence (i.e. smaller off-diagonal elements), since the nuclear spins are correlated with the spins of magnetic centers that encode the qubit. The degree of coherence can be quantified by $r(t) = \sum_{\mathcal{I}} P_{\mathcal{I}} r_{\mathcal{I}}(t)$, where $r_{\mathcal{I}}(t) = \langle\mathcal{I}_1(t)|\mathcal{I}_0(t)\rangle$, and $\langle 0|\hat{\rho}_S(0)|1\rangle = r_{\mathcal{I}}/2$. It is known that the decoherence rate depends on the initial state of the nuclear spin bath. For example, it has been shown that techniques such as narrowing of the nuclear state can drastically increase the decoherence times in quantum dot systems [68].

Next, we want to show in what way (11.25) leads to decoherence in a spin-cluster qubit (such as is realized in the triangular magnet in Sect. 11.2) in more detail. We have shown before that in spin clusters the qubit state is typically not encoded in the $\mathbf{S}_i$'s themselves, but instead in quantities like the total spin $\mathbf{S}$ or the chirality $\mathbf{C}$. However, we can always denote the basis states of the qubit by $|0\rangle$ and $|1\rangle$. Quite generally then, by projecting the spin operators $\mathbf{S}_i$ on the space spanned by $|0\rangle$, $|1\rangle$, and performing a second order Schrieffer-Wolff transformation on the resulting Hamiltonian, one can transform (11.25) into the Hamiltonian $H = \sum_{k=0,1} |k\rangle\langle k| \otimes H_k$, with

$$H_k = \sum_{p=1}^{N} \omega_p^k I_p^{z'} + \sum_{p\neq q} \left(A_{pq}^k I_p^{z'} I_q^{z'} + B_{pq}^k I_p^+ I_q^-\right), \tag{11.26}$$

where $\hat{\mathbf{z}}' = \mathbf{B}/|B|$. In the derivation of (11.26), we ignored terms that do not conserve energy. $\omega_p^0 - \omega_p^1$ is linear in H_{HF}, and the quantities $A_{pq}^0 - A_{pq}^1$ and $B_{pq}^0 - B_{pq}^1$ are quadratic in H_{HF}. The fastest contribution to decoherence is due to inhomogeneous broadening due to the terms $\propto I_p^{z'}$ in (11.26). These terms describes the magnetic field due to the nuclear spins, which is called the Overhauser field. The Overhauser field depends on the specific realization of the nuclear spin state (for times $t \ll \tau_n$, where τ_n is the typical evolution time of the nuclear spin state, the magnetic field is static). Therefore, if the nuclear spins are in a mixture of states, the coherence of the state $|\psi(0)\rangle$ is washed out due interference of the states that undergo time-evolution under different effective magnetic fields. This can be seen from the decoherence factor $r(t)$, which for $t \ll \tau_n$ evolves as $r(t) \approx e^{i(E_0-E_1)t}\sum_{\mathcal{I}} P_{\mathcal{I}} e^{i\delta_{\mathcal{I}}t}$, where

$$\delta_{\mathcal{I}} \approx g_S\mu_S \sum_i \mathbf{B}_{\mathrm{HF}}^{\mathcal{I}}(\mathbf{r}_i)\cdot\left[\langle 0|\mathbf{S}_i|0\rangle - \langle 1|\mathbf{S}_i|1\rangle\right]. \tag{11.27}$$

The sum is over the spins in the spin cluster. Furthermore, $\mathbf{B}^{\mathcal{I}}_{\mathrm{HF}}(\mathbf{r}_i) = D_{\mathrm{HF}} \sum_p m^{\mathcal{I}}_p [\hat{\mathbf{z}}' - 3(\hat{\mathbf{z}}' \cdot \hat{\mathbf{r}}_{ip})\hat{\mathbf{r}}_{ip}]/r^3_{ip}$ is the Overhauser field. It has been shown, that decoherence of a qubit encoded in the total spin $\mathbf{S} = \sum_{i=1}^{3} \mathbf{S}_i$ of a triangular cluster due to the distribution of the Overhauser field for the equal superposition mixed state typically takes place on time scales of 100 ns. The second order terms in (11.26) give contributions to the decoherence times that are several orders of magnitude smaller.

We have seen that due to hyperfine interaction, both the qubit state as well as the nuclear spin state evolve in time. Furthermore, even in the absence of hyperfine interaction the nuclear spin state itself evolves in time, according to the Hamiltonian $H_n = \hat{\mathbf{B}} \cdot \sum_p \omega_p \mathbf{I}_p + D_n \sum_{p<q} [\mathbf{I}_p \cdot \mathbf{I}_q = 3(\mathbf{I}_p \cdot \hat{\mathbf{e}}_{pq})(\mathbf{I}_q \cdot \hat{\mathbf{e}}_{pq})]/r^3_{pq}$. This dynamics of the nuclear bath can lead to additional broadening of the Overhauser field, and has been shown to lead to decoherence on the μs-time scale for a qubit state encoded in the total spin.

An interesting possibility to increase the decoherence time of a qubit is a triangular spin cluster was put forward in Ref. [67]. The idea is to use the chirality of cluster as qubit, instead of the total spin. In that case, the states $|0\rangle$ and $|1\rangle$ of this section become $|0\rangle_{C_z} = |-1/2, 1\rangle$, $|0\rangle_{C_z} = |-1/2, -1\rangle$. The crucial property of these state that causes the increased decoherence time is that since

$$\langle 1|S_{z,i}|1\rangle = \langle 0|S_{z,i}|0\rangle = -1/6, \tag{11.28}$$

the Overhauser field from (11.27) does not couple to the qubit. Therefore, decoherence processes in (11.26) are second order only. This can lead to decoherence times approaching milliseconds.

11.6 Initialization and Read-out

Initialization of a qubit in its ground state is arguably the DiVincenzo criterion that is most routinely realized. Therefore, we will not spend a lot of time discussing it here. The way to prepare a qubit in its ground state is by cooling it down to temperatures that are much smaller than the gap between the ground state in which one wants to prepare the system and the first excited state. This gap, which could for instance be due to magnetic anisotropy, is typically of the order of a few Kelvin, and may be controlled by external means, such as placing the molecular magnet in a magnetic field. This limits the temperature at which experiments can be done to several mK to K.

The read-out of the spin state is a topic on itself, and we refer the reader to the literature for an overview of the different techniques that are used [69].

11.7 Grover's Algorithm Using Molecular Magnets

One special topic that we wish to discuss in this chapter is the implementation of Grover's algorithm using molecular magnets [27]. Grover's algorithm can be used to find an entry in an unsorted database with N entries. A typical situation in which this would be required is if we were given a phone number, and wanted to find the associated name in a phone book. Classically, we would have to start with the first entry, and work our way down the list. Finding the name in this manner requires on average $N/2$ queries. If we had encoded the information in the phone book in a quantum state, we would have been able to find the correct entry with high probability in $O(N^{1/2})$ queries using Grover's algorithm. A crucial requirement for this algorithm is the possibility to generate arbitrary superpositions of eigenstates (and in particular the superposition where all eigenstates have approximately the same weight).

In large-spin magnetic molecules, the eigenstates are labeled by the quantum number m_S, the z projection of the total spin $S \gg 1/2$. The Hamiltonian describing a single spin S with easy-axis along the z direction is given by

$$H = -AS_z^2 - BS_z^4 + V, \tag{11.29}$$

where $V = g\mu_B \mathbf{H} \cdot \mathbf{S}$. This gives rise to the typical double-well spectrum with non-equidistant level spacing. Such level spacing is crucial for the proposal in Ref. [27], as will become clear shortly. Suppose one starts out by preparing the system in the ground state $|\psi_0\rangle = |s\rangle$, and wishes to create an equal superposition of all the states $|m_0\rangle, |m_0 + 1\rangle, \ldots, |s - 1\rangle$, where $m_0 = 1, 2, \ldots, s - 1$. This corresponds to using $n - 1$ states for Grover's algorithm, where $n = s - m_0$. In principle, one can create superpositions by applying a weak transverse magnetic field $\mathbf{H}_\perp$ (whose effect can be described using perturbation theory) which drives multiphoton transitions via virtual states through its coupling to S^+, S^-. However, to create the equal superposition that is required for Grover's algorithm, the amplitudes of all k-photon processes (here $k = 1, 2, \ldots, s - m_0$) must be equal. Clearly, perturbation theory is not valid in this regime. Therefore, a more sophisticated scheme is required.

The scheme that is proposed in Ref. [27] to create an equal superposition uses a single coherent magnetic pulse of duration T with a discrete frequency spectrum $\{\omega_m\}$. It contains n high-frequency components and a single low-frequency component ω_0, chosen such that $\hbar\omega_0 \ll \epsilon_{m_0} - \epsilon_{m_0+1}$. Here, ϵ_m is the energy of the eigenstate $|m\rangle$. The frequencies of the n high-frequency components are given by $\hbar\omega_{s-1} = \epsilon_{s-1} - \epsilon_s - \hbar(n-1)\omega_0$ and $\omega_m = \epsilon_m - \epsilon_{m+1} + \hbar\omega_0$ for $m = m_0, \ldots, s-2$. For the molecular magnet Mn_{12}, the high-frequency components have frequencies between 20-120 GHz, and ω_0 is around 100 MHz. Because of the non-equidistant splitting of the energy levels, all frequencies are different. The low-frequency component is applied along the easy axis, the high frequency components are in plane, so that the coupling is given by

$$V_{\text{low}}(t) = g\mu_B H_0(t)\cos(\omega_0 t)S_z, \tag{11.30}$$

$$V_{\text{high}}(t) = \sum_{m=m_0}^{s-1} g\mu_B H_m(t)\big[\cos(\omega_m t + \Phi_m)S_x - \sin(\omega_m t + \Phi_m)S_y\big]$$
$$= \sum_{m=m_0}^{s-1} \frac{g\mu_B H_m(t)}{2}\big[e^{i(\omega_m t+\Phi_m)}S^+ + e^{-i(\omega_m t+\Phi_m)}S^-\big]. \qquad (11.31)$$

Hence, absorption (emission) of a high-frequency σ^--photon induces a transition with $\Delta m = -1$ (1); the low-frequency π-photons do not change m, instead they supply the energy required to fulfill the resonance condition for allowed transitions. The phases Φ_m can be chosen freely, we will come back to this point later. With this setup, the lowest order transition between the ground state $|s\rangle$ and all states $|m\rangle$ (for $m_0 \le m < s$) is n'th order in $V(t) = V_{\text{low}}(t) + V_{\text{high}}(t)$.

To see this, let us consider an explicit example where $s = 10$, $m_0 = 5$, and hence $n = 5$. The lowest order transition from $|s\rangle$ to $|s-1\rangle$ uses 4 π-photons of energy $\hbar\omega_0$ and 1 σ^--photon with energy $\hbar\omega_{s-1}$. The transition from $|s\rangle$ to $|s-2\rangle$ uses 3 π-photons of energy $\hbar\omega_0$, 1 σ^--photon with energy $\hbar\omega_{s-1}$, and 1 σ^--photon with energy $\hbar\omega_{s-2}$; and so on for the other transitions. ω_0 can be chosen such that lower order transitions are forbidden due to the requirement of energy conservation. The amplitude of higher order transitions is small in the perturbative regime.

Since all transition amplitudes are the same order in $V(t)$, they are all approximately equal. To make them exactly equal requires some fine-tuning. For rectangular pulses with $H_k(t) = H_k$ for $T/2 < t < T/2$, the n'th order contribution to the S-matrix for the transition between $|s\rangle$ and $|m\rangle$, denoted by $S^{(n)}_{m,s}$, is given by

$$S^{(n)}_{m,s} = \sum_F \Omega_m \frac{2\pi}{i}\left(\frac{g\mu_B}{2\hbar}\right)^n \frac{\Pi_{k=m}^{s-1} H_k e^{i\Phi_k} H_0^{m-m_0} p_{m,s}(F)}{(-1)^{q_F} q_F! r_s(F)! \omega_0^{n-1}}$$
$$\times\, \delta^{(T)}\left(\omega_{m,s} - \sum_{k=m}^{s-1}\omega_k - (m-m_0)\omega_0\right). \qquad (11.32)$$

The sum runs over all Feynman diagrams F. $\Omega_m = (m-m_0)!$, $q_F = m - m - r_s(F)$, $p_{m,s}(F) = \Pi_{k=m}^{s}\langle k|S_z|k\rangle^{r_k(F)}\Pi_{k=m}^{s-1}\langle k|S^-|k+1\rangle$, with $r_k(F) = 0, 1, 2, \ldots \le m - m_0$ the number of π-transitions in the transition belonging to the Feynman diagram F. $\delta^{(T)}(\omega) = 1/(2\pi)\int_{-T/2}^{T/2} \mathrm{d}t e^{i\omega T}$ is the delta-function of width T. It ensures energy conservation. For the example above, the requirement $|S^{(n)}_{m,s}| \approx |S^{(n)}_{-1,s}|$ for all $m \ge m_0$ (which corresponds to the equal superposition) is satisfied for parameters

$$H_8/H_0 = 0.04, \qquad H_7/H_0 = -0.25, \qquad H_6/H_0 = -0.61, \qquad H_5/H_0 = -1.12. \qquad (11.33)$$

H_9 can be chosen independently. For numerical estimates, we refer to the original paper, Ref. [27]. This concludes the discussion of generating the equal superposition required for Grover's algorithm.

With some adaptions, a single step in Grover's algorithm can be used to read-in and decode quantum information. This opens up the possibility to use molecular magnets as dense and efficient memory devices. The phases Φ_m in (11.30)–(11.31) play a crucial role here. We denote $\Phi_m = \sum_{k=s-1}^{m+1} \Phi_k + \phi_m$. As we have seen before, we can irradiate the system with a coherent magnetic pulse of duration T such that all $S_{m,s}^{(n)} = \pm\eta$. In other words, the state after the pulse is $|\psi\rangle = \sum_{m=m_0}^{s} a_m |m\rangle$, where the amplitudes $a_1 = 1$ and $a_m = \pm\eta$. By identifying the amplitude $\pm\eta$ with the logical-1, respectively logical-0, we see that this state encodes a n-bit state. Because of the Φ_m dependence of the S-matrix (see (11.32)), we can switch between the $\pm\eta$ amplitude by choosing $\phi_m = 0, \pi$. This allows us to encode a general state between 0 and $2^n - 1$ in the quantum state of the molecular magnet. The set $\{\phi_m\}$ that one uses depends on the number that has to be encoded. For instance, encoding $12_{10} = 1101_2$ requires $\phi_9 = \phi_8 = \phi_7 = 0$ and $\phi_6 = \phi_5 = \pi$. Here, the states with $m = 9, 8, 7, 6, 5$ represent respectively the binary digits $2^0, 2^1, 2^2, 2^3, 2^4$.

To decode the state of the molecule, one applies a pulse for which $S_{m_0,s}^{(n)} = S_{m_0+1,s}^{(n)} = \cdots = S_{s-1,s}^{(n)} = -\eta$. This pulse amplifies the bits which have amplitude $-\eta$, and suppresses those with amplitude η. The accumulated error in this procedure is approximately $n\eta^2$. Read-out of this decoded state can be done by measuring the occupation of the different levels by standard spectroscopy, for instance using pulsed ESR. Irradiation with a pulse which contains the frequency $\hbar\omega_{m-1,m} = \epsilon_{m-1} - \epsilon_m$ drives transitions that are given by $S_{m-1,m}^{(1)}$. If the state $|m\rangle$ is occupied (meaning that its amplitude was $-\eta$), we would observe stimulated absorption when irradiating with frequency $\omega_{6,7}$ and stimulated emission when irradiating with frequency $\omega_{7,8}$. Since the energy levels are non-equidistant, this uniquely identifies the level.

Acknowledgements The authors would like to acknowledge financial support from the Swiss NSF, the NCCR Nanoscience Basel, and the FP7-ICT project "ELFOS".

References

1. C.M. Papadimitriou, *Computational Complexity* (Addison-Wesley, Reading, 1994)
2. D.P. DiVincenzo, Fortschr. Phys. **48**, 771 (2000)
3. D. Deutsch, R. Jozsa, Proc. R. Soc. Lond. Ser. A, Math. Phys. Sci. **439**, 553 (1992)
4. R. Cleve, A. Ekert, C. Macchiavello, M. Mosca, Proc. R. Soc. Lond. Ser. A, Math. Phys. Sci. **454**, 339 (1998)
5. L.K. Grover, in *Proceedings of the 28th Annual ACM Symposium on the Theory of Computing*, (1996), p. 212
6. P. Shor, SIAM J. Comput. **26**, 1484 (1997)
7. D. Deutsch, Proc. R. Soc. Lond. A **400**, 97–117 (1985)
8. J. Church, Am. J. Math. **58**, 435 (1936)
9. A.M. Turing, Proc. Lond. Math. Soc. **442**, 230 (1936)
10. M.A. Nielsen, I.L. Chuang, *Quantum Computation and Quantum Information* (Cambridge University Press, New York, 2000)
11. W.K. Wootters, W.H. Zurek, Nature **299**, 802 (1982)
12. A. Barenco, C.H. Bennett, R. Cleve, D.P. DiVincenzo, N. Margolus, P. Shor, T. Sleator, J.A. Smolin, H. Weinfurter, Phys. Rev. A **52**, 3457 (1995)
13. D. Loss, D.P. DiVincenzo, Phys. Rev. A **57**, 120 (1998)
14. A. Imamoglu, D.D. Awschalom, G. Burkard, D.P. DiVincenzo, D. Loss, M. Sherwin, A. Small, Phys. Rev. Lett. **83**, 4204 (1999)

15. J.I. Cirac, P. Zoller, Phys. Rev. Lett. **74**, 4091 (1995)
16. Q.A. Turchette, C.J. Hood, W. Lange, H. Mabuchi, H.J. Kimble, Phys. Rev. Lett. **75**, 4710 (1995)
17. N.A. Gershenfeld, I.L. Chuang, Science **275**, 350 (1997)
18. A. Shnirman, G. Schön, Z. Hermon, Phys. Rev. Lett. **79**, 2371 (1997)
19. B.E. Kane, Nature (London) **393**, 133 (1998)
20. R. Vrijen, E. Yablonovitch, K.L. Wang, H.W. Jiang, A.A. Balandin, V. Roychowdhury, T. Mor, D.P. DiVincenzo, Phys. Rev. A **62**, 012306 (2000)
21. E. Knill, R. Laflamme, G.J. Milburn, Nature (London) **409**, 46 (2001)
22. F. Jelezko, J. Wrachtrup, Phys. Status Solidi A **203**, 3207 (2006)
23. R. Hanson, D.D. Awschalom, Nature (London) **453**, 1043 (2008)
24. P. Maletinsky, S. Hong, M.S. Grinolds, B. Hausmann, M.D. Lukin, R.L. Walsworth, M. Loncar, A. Yacoby, Nat. Nanotechnol. (2012). doi:10.1038/nnano.2012.50
25. D.V. Bulaev, B. Trauzettel, D. Loss, Phys. Rev. B **77**, 235301 (2008)
26. M. Trif, V.N. Golovach, D. Loss, Phys. Rev. B **77**, 045434 (2008)
27. M.N. Leuenberger, D. Loss, Nature (London) **410**, 789 (2001)
28. J. Tejada, E. Chudnovsky, E. del Barco, J. Hernandez, T. Spiller, Nanotechnology **12**, 181 (2001)
29. F. Meier, J. Levy, D. Loss, Phys. Rev. Lett. **90**, 047901 (2003)
30. F. Troiani, A. Ghirri, M. Affronte, S. Carretta, P. Santini, G. Amoretti, S. Piligkos, G. Timco, R.E.P. Winpenny, Phys. Rev. Lett. **94**, 207208 (2005)
31. J. Lehmann, A. Gaita-Arino, E. Coronado, D. Loss, Nat. Nanotechnol. **2**, 312 (2007)
32. M. Trif, F. Troiani, D. Stepanenko, D. Loss, Phys. Rev. Lett. **101**, 217201 (2008)
33. E.M. Chudnovsky, L. Gunther, Phys. Rev. Lett. **60**, 661 (1988)
34. D.D. Awschalom, J.F. Smyth, G. Grinstein, D.P. DiVincenzo, D. Loss, Phys. Rev. Lett. **68**, 3092 (1992)
35. R. Sessoli, D. Gatteschi, A. Caneschi, M.A. Novak, Nature (London) **365**, 141 (1993)
36. L. Thomas, F. Lionti, R. Ballou, D. Gatteschi, R. Sessoli, B. Barbara, Nature (London) **383**, 145 (1996)
37. J.R. Friedman, M.P. Sarachik, J. Tejada, R. Ziolo, Phys. Rev. Lett. **76**, 3830 (1996)
38. W. Wernsdorfer, E. Bonet Orozco, K. Hasselbach, A. Benoit, D. Mailly, O. Kubo, H. Nakano, B. Barbara, Phys. Rev. Lett. **79**, 4014 (1997)
39. J. Tejada, X.X. Zhang, E. del Barco, J.M. Hernández, E.M. Chudnovsky, Phys. Rev. Lett. **79**, 1754 (1997)
40. E. del Barco, A.D. Kent, E.M. Rumberger, D.N. Hendrickson, G. Christou, Phys. Rev. Lett. **91**, 047203 (2003)
41. A. Chiolero, D. Loss, Phys. Rev. Lett. **80**, 169 (1998)
42. F. Meier, D. Loss, Phys. Rev. Lett. **86**, 5373 (2001)
43. D. Gatteschi, A. Caneschi, L. Pardi, R. Sessoli, Science **265**, 1054 (1994)
44. C. Sangregorio, T. Ohm, C. Paulsen, R. Sessoli, D. Gatteschi, Phys. Rev. Lett. **78**, 4645 (1997)
45. D. Gatteschi, R. Sessoli, A. Cornia, Chem. Commun. **725** (2000). doi:10.1039/A908254I
46. D. Loss, D.P. DiVincenzo, G. Grinstein, Phys. Rev. Lett. **69**, 3232 (1992)
47. M.N. Leuenberger, D. Loss, Phys. Rev. B **61**, 1286 (2000)
48. M.N. Leuenberger, F. Meier, D. Loss, Monatsh. Chem. **134**, 217 (2003)
49. W. Wernsdorfer, R. Sessoli, Science **284**, 133 (1999)
50. M.N. Leuenberger, D. Loss, Phys. Rev. B **63**, 054414 (2001)
51. G. González, M.N. Leuenberger, Phys. Rev. Lett. **98**, 256804 (2007)
52. G. González, M.N. Leuenberger, E.R. Mucciolo, Phys. Rev. B **78**, 054445 (2008)
53. M. Trif, F. Troiani, D. Stepanenko, D. Loss, Phys. Rev. B **82**, 045429 (2010)
54. K.-Y. Choi, Y.H. Matsuda, H. Nojiri, U. Kortz, F. Hussain, A.C. Stowe, C. Ramsey, N.S. Dalal, Phys. Rev. Lett. **96**, 107202 (2006)
55. F.H.L. Koppens, C. Buizert, K.J. Tielrooij, I.T. Vink, K.C. Nowack, T. Meunier, L.P. Kouwenhoven, L.M.K. Vandersypen, Nature (London) **442**, 766 (2006)

56. S. Bertaina, S. Gambarelli, T. Mitra, B. Tsukerblat, A. Müller, B. Barbara, Nature (London) **453**, 203 (2008)
57. M. Borhani, V.N. Golovach, D. Loss, Phys. Rev. B **73**, 155311 (2006)
58. K.C. Nowack, F.H.L. Koppens, Yu.V. Nazarov, L.M.K. Vandersypen, Science **318**, 1430 (2007)
59. M.F. Islam, J.F. Nossa, C.M. Canali, M.R. Pederson, Phys. Rev. B **82**, 155446 (2010)
60. J.F. Nossa, M.F. Islam, C.M. Canali, M.R. Pederson, Phys. Rev. B **85**, 085427 (2012)
61. P.W. Anderson, Phys. Rev. **115**, 2 (1959)
62. T. Moriya, Phys. Rev. **120**, 91 (1960)
63. R. Winkler, *Spin-Orbit Coupling Effects in Two-dimensional Electron and Hole Systems* (Springer, Berlin, 2003)
64. G. Burkard, D. Loss, D.P. DiVicenzo, Phys. Rev. B **59**, 2070 (1999)
65. A. Ardavan, O. Rival, J.J.L. Morton, S.J. Blundell, A.M. Tyryshkin, G.A. Timco, R.E.P. Winpenny, Phys. Rev. Lett. **98**, 057201 (2007)
66. A. Schweiger, G. Jeschke, *Principles of Pulse Electron Paramagnetic Resonance* (Oxford University Press, New York, 2001)
67. F. Troiani, D. Stepanenko, D. Loss, Phys. Rev. B **86**, 161409 (2012)
68. W.A. Coish, D. Loss, Phys. Rev. B **70**, 195340 (2004)
69. D. Gatteschi, R. Sessoli, J. Villain, *Molecular Nanomagnets* (Oxford University Press, New York, 2006)

Chapter 12
Single-Molecule Spintronics

Enrique Burzurí and Herre S.J. van der Zant

Abstract During the last few years different techniques have become available to study transport through an individual magnetic molecule. In a spin transistor, the magnetic molecule links two electrodes that are used to apply a bias voltage; a third gate electrode controls the position of molecular levels such that resonant tunneling and different redox states become accessible. Sequential single-electron transport and current suppression (Coulomb blockade) are generally observed. In this chapter, we show that spectroscopic information obtained from these three-terminal measurements confirms the high-spin state and magnetic anisotropy of the robust Fe_4 single-molecule magnet incorporated in the junction. Moreover, we find that the electric gate field drastically modifies the magnetic properties of the oxidized or reduced molecule.

12.1 Introduction

In standard electronics we manipulate electrons and send them through different device components to, in the end, transport, read and write information. *Spintronics* [1, 2] aims at using not only the charge of the electrons but also controlling its spin states. The use of this additional degree of freedom as a relevant parameter in transport is expected to increase the speed and storage capacity of electronic devices. Most importantly, it opens the door to new functionalities like quantum computation arising from the quantum nature of the spin. Progress in this field seeks to use the spin state of individual magnetic molecules trapped between electrodes instead of, or in combination with, the spin of the flowing electrons. This emerging field is known as *molecular spintronics* [2, 3].

Among the different families of magnetic molecules, single-molecule magnets (SMM) discovered in the late 90s [4] are very promising candidates because of their

E. Burzurí (✉) · H.S.J. van der Zant
Kavli Institute of Nanoscience, Delft University of Technology, P.O. Box 5046, 2600 GA Delft, The Netherlands
e-mail: E.BurzuriLinares@tudelft.nl

H.S.J. van der Zant
e-mail: H.S.J.vanderZant@tudelft.nl

J. Bartolomé et al. (eds.), *Molecular Magnets*, NanoScience and Technology,
DOI 10.1007/978-3-642-40609-6_12, © Springer-Verlag Berlin Heidelberg 2014

magnetic anisotropy and usually large spin. Bulk properties have been extensively studied and now the challenge in the field lies in the detection and manipulation of their giant molecular spin at the individual level.

12.1.1 How to Detect Spin in Magnetic Molecules?

There are several well-established methods to detect the spin state of large ensembles of molecules. For instance, electron spin resonance (ESR) combines the use of electromagnetic waves (usually microwaves) and magnetic fields to induce and afterwards detect changes in the spin state of the molecules. Other examples include superconducting quantum interference devices (SQUIDS) that are made of superconducting rings to detect a tiny magnetic flux induced by magnetic particles lying around the coil. The sensitivity of these techniques is, however, limited to large numbers of molecules. By pushing the limits, groups have measured clusters of nanoparticles containing thousands of spins with highly sensitive micro-SQUIDS [5–7]. For single molecules, the route becomes complex as the sensitivity should be extraordinarily high. Some initiatives propose the use of SQUIDS based on carbon nanotubes [8].

An alternative approach is to use an electrical current to probe the magnetic properties of individual molecules. For applications, electrical control of the spin is faster and can be done locally in single molecules in contrast to other external stimuli like the magnetic field. Electrical control can be achieved in single-electron transistors where individual molecules are attached to conducting electrodes. From the perspective of a fundamental research, a three-terminal molecular *spin transistor* is a unique playground for exploring the interaction between the magnetic states of a molecule and the electrical current. Depending on the interaction strength between the current and the magnetic molecule, we can distinguish two different approaches: *indirect* (non-invasive) and *direct* (invasive) electrical probing described schematically in Fig. 12.1. With *indirect* probing, the current does not flow through the magnetic core of the molecule. Instead, the molecule is attached to a conducting intermediary channel like a carbon nanotube [9] or a graphene sheet [10] that is used as a probe to detect changes in the spin states of the attached magnetic molecule (see Fig. 12.1(b)). This method has very recently been used to read out the nuclear spin state of the double-decker SMM $TbPc_2$; the current through one of the Pb ligands probes the spin-state of the Tb atom [11].

When the current flows through the magnetic core of the molecule the spin-charge carrier interaction will be strong. It is an invasive method that in turn allows to charge the molecule and therefore explore its magnetic properties in different redox states (see Fig. 12.1(a)). In this chapter we discuss how spin information can be obtained in different charge states from *direct* probing of *individual magnetic molecules*. We stress that the field is still in rapid development and therefore we will include at the end a description of newly proposed methods to read and control the spin information. The chapter is organized as follows: we start with a section on

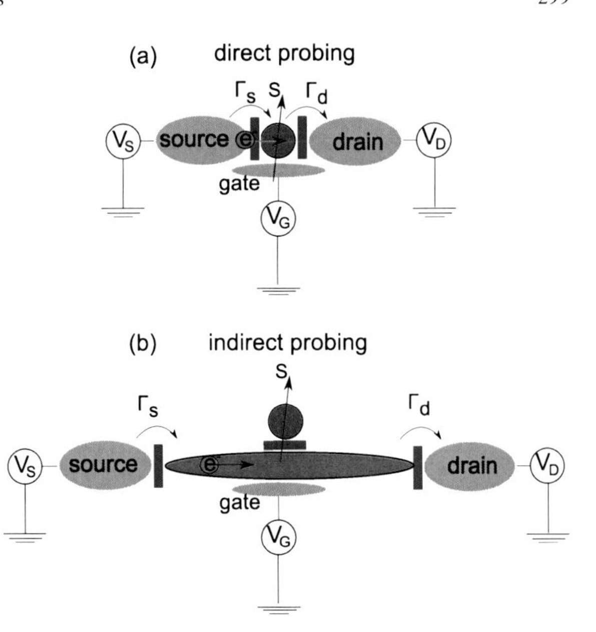

Fig. 12.1 Schematic picture of a molecular quantum dot (*red sphere*) in a three-terminal configuration. (**a**) With *direct* probing, the current flows through the magnetic core of the molecule. The molecular quantum dot is connected to source and drain electrodes via tunnel barriers. (**b**) With *indirect* probing the current flows through an adjacent conductor like a carbon nanotube or graphene sheet, which is coupled to the magnetic core by for example a tunnel barrier

the theory of charge transport followed by a section describing the fabrication of a spin transistor. We will then discuss measurements performed on a particular case, the Fe_4 single-molecule magnet. The chapter will finish with a description of future experiments.

12.2 Coulomb Blockade

At low temperature, a single molecule in a three-terminal configuration can be seen as a confined electronic system or quantum dot that is coupled by tunnel barriers to the source and drain electrodes as shown schematically in Fig. 12.1(a). The electrons can hop from the source to the molecule and from it to the drain by tunneling through the barriers. The tunneling rate depends on the coupling between the molecular wave function and that of the conduction electrons in the leads. It is given by Γ_s and Γ_d for the source and drain electrodes respectively. $\Gamma_{s,d}$ is typically expressed in meV and ranges from 1 to 100 meV in single-molecule transport experiments.

The metallic leads act as a reservoir of electrons available at the Fermi levels μ_s and μ_d of, respectively, the source and drain electrodes. In contrast, the quantum dot possesses a discrete electronic spectrum. Figure 12.2(a) shows schematically this electrochemical description of the magnetic molecule-electrodes system. The energy required to add or subtract an electron to the quantum dot, known as the addition energy E_{add}, is defined by the energy spacing $\Delta(N)$ between the Lowest Unoccupied Molecular Orbital (LUMO) and the Highest Occupied Molecular Orbital (HOMO) plus two times the charging energy $E_c = e^2/2C$ (see Fig. 12.2(a)). Depending on the relative strength of $\Gamma_{s,d}$, or simply Γ if we assume $\Gamma_s \approx \Gamma_d$ for sim-

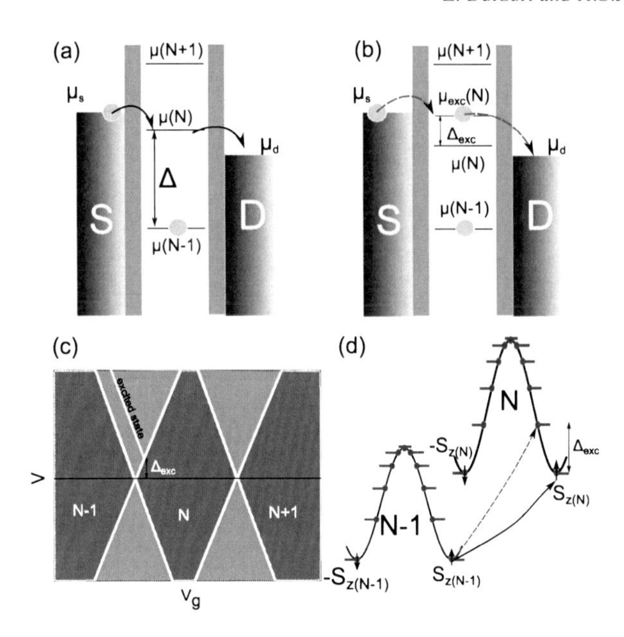

Fig. 12.2 First-order transport. (**a**) Electrochemical scheme of a first-order sequential tunneling process involving the ground state (*solid black arrow*) or (**b**) an excited spin state (*dashed blue arrow*). (**c**) Differential conductance map (or stability diagram) in which dI/dV is plotted in grey code as a function of V and V_g for a molecule in the weak coupling regime. *Dark areas* (Coulomb diamonds) are low-conductance areas where the charge is stabilized in the molecule. The different spin states participate in conduction and appear as *diagonal* (*white*) *lines* of high dI/dV. (**d**) Typical double-well potential for a SMM. The Coulomb edges correspond to a transition from $\pm S_z(N-1)$ to $\pm S_z(N)$ (*solid black arrow*). The SET excitation may correspond to a transition between the ground state $N-1$ to a spin excited state in the N charge state (*blue dashed arrow*)

plicity, we distinguish different transport regimes: weak coupling when $\Gamma \ll E_C$, $\Delta(N)$, k_BT and strong coupling when $\Gamma \gg E_C$, $\Delta(N)$, k_BT. The crossover between these two regimes is called the intermediate coupling regime and is of special interest as we will show below.

Figure 12.2(c) shows schematically a conductance map (or stability diagram) of a quantum dot in the weak coupling regime. A conductance map is built up from individual dI/dV traces plotted versus the bias voltage V as a function of the gate voltage V_g. Due to the discrete character of the molecular energy spectrum, electrons can only hop onto the molecule (at first order in Γ) when the Fermi levels of the electrodes are in resonance with a free energy level of the molecule. In the general case, these levels are out of resonance and transport is forbidden; the charge is stabilized within the magnetic molecule. This process is known as *Coulomb blockade* and shows up in conductance measurements as low conductance areas (dark areas in Fig. 12.2(c)). By applying a bias voltage the chemical potential difference between source and drain can be varied according to

$$eV = \mu_s - \mu_d. \tag{12.1}$$

Whenever a charge state ($\mu_{\text{mol}}(N)$) enters this bias window as sketched in Fig. 12.2(a), resonant tunneling occurs and transport through the molecule is allowed. Electrons now hop one by one on and off the molecule in an incoherent, *sequential electron tunneling (SET)* process: the number of electrons in the dot changes continuously between $N - 1$ to N. Coulomb blockade is lifted and a large increase in the conductance is observed as diagonal lines in the conductance map. These are the Coulomb edges shown in Fig. 12.2(c). The crossing of the Coulomb edges at zero bias is the charge degeneracy point where adjacent charge states have the same ground-state energy. An orbital can be brought into the bias window by increasing the bias window or by "pulling" the energy levels of the molecule with the gate voltage. For large enough gate fields, successive molecular orbitals can be brought into the bias window and thus different charge states are accessible. As long as the gate voltage remains fixed, these redox states are stable and they can therefore be characterized in great detail.

12.3 Spectroscopy of Magnetic Spin States

The magnetic molecular excitations of single-molecule magnets take part in the electronic transport and leave their fingerprint in the conductance of the molecule. A conductance map can therefore be used as a very sensitive spectroscopic tool. As we will show, in transport measurements one generally measures energy differences, and their interpretation therefore requires a description in terms of chemical potentials. One can furthermore infer information about any change in the charge (oxidation, reduction) or spin state that might occur upon charging the molecule. To study the magnetic nature of the excitations, we examine their evolution in the presence of an applied magnetic field. As a first approximation, the energy levels of an anisotropic single-molecule magnet can be described by the Hamiltonian

$$H = DS_z^2 + g\mu_B \vec{B} \cdot \vec{S}. \tag{12.2}$$

The first term is the anisotropy contribution where D is an axial magnetic anisotropy parameter and S_z is the z-component (parallel to the easy axis) of the spin. The second term is the Zeeman term which describes the interaction of the spin with the magnetic field B, where μ_B is the Bohr magneton and g is the Landé factor. Even in the absence of a magnetic field, the spin ground state multiplet of a SMM splits into $2S + 1$ levels distributed over an energy barrier (see Fig. 12.3(a)). The zero-field splitting (ZFS) is the energy difference (Δ_{exc}) between the two lowest-lying doublets (green arrow in Fig. 12.3(a)). Figure 12.3 shows the energy levels versus magnetic field for two different angles $\theta = 0$ (b) and $\theta = 90°$ (c) between the easy axis and the magnetic field orientation. The curves have been calculated by numerical diagonalization of the Hamiltonian (12.2) for a typical set of parameters.

Figure 12.3 shows that for an anisotropic molecule, the energy ($\Delta_{\text{exc}}(B, \theta)$) of the different spin states is a non-linear function of $\vec{B}$ and depends strongly on θ. The spectrum is different and the conductance map changes accordingly, thus, revealing information of the magnetic anisotropy and the spin excitations.

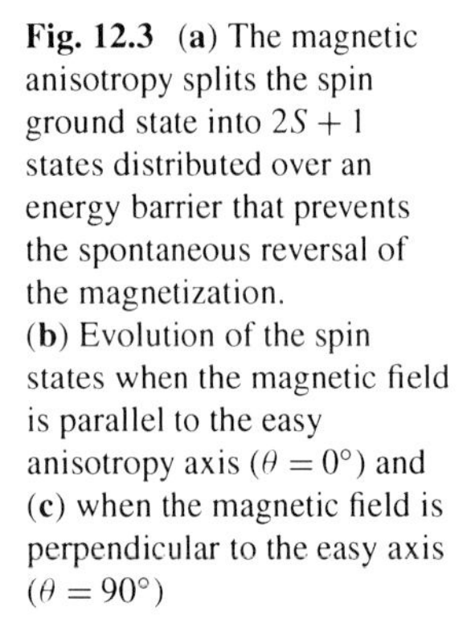

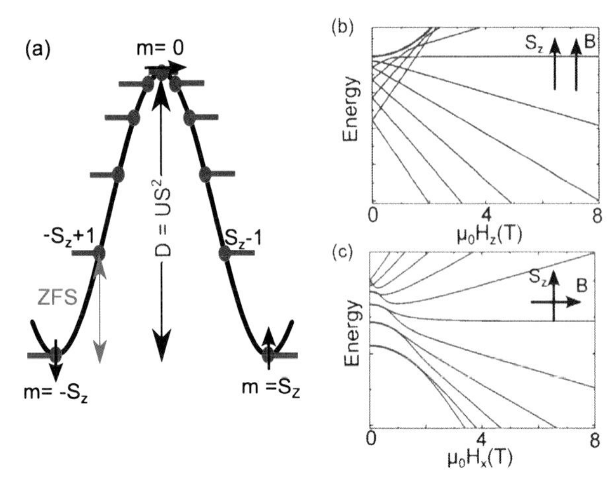

Fig. 12.3 (**a**) The magnetic anisotropy splits the spin ground state into $2S+1$ states distributed over an energy barrier that prevents the spontaneous reversal of the magnetization. (**b**) Evolution of the spin states when the magnetic field is parallel to the easy anisotropy axis ($\theta = 0°$) and (**c**) when the magnetic field is perpendicular to the easy axis ($\theta = 90°$)

12.3.1 Weak Coupling: SET Excitations

Figure 12.2(b) shows schematically the electrochemical picture of a quantum dot with one excited state within the N charge state (blue dashed arrow). The origin of this excited state can be, for instance, the excited spin states of a SMM in the ground state multiplet (see Fig. 12.2(d)). As soon as the excited state enters the bias window, the additional conduction channel induces an increase in the current and a corresponding peak in the differential conductance plot. In the stability diagram sketched in Fig. 12.2(c) these excitations appear as diagonal lines running parallel to the Coulomb edge ending in the charge state corresponding to the excitation (N in this case). The bias voltage at which the excited state and the Coulomb edges cross gives the energy of the excitation $\Delta_{exc}(B, \theta)$ that depends on B and θ.

Not all transitions between different states can be seen in a transport experiment. First, tunneling through an excited state is only possible when both excited and ground states lie in between the Fermi levels of the electrodes. In addition, since the spin of a single electron is 1/2, conservation of the spin imposes additional selection rules together with the conservation of the charge. The spin excitations can only be observed in SET when $\Delta S = \pm 1/2$ and $\Delta m = \pm 1/2$. In the double-well potential picture of a SMM depicted in Fig. 12.2(d), the excitation corresponds to a transition from the ground state in the $N-1$ charge state to the excited state in the N state (blue dashed arrow). It is also important to note that the first transition to be reached with the bias voltage is the ZFS which has the largest energy difference. The consequence is that once the bias voltage is increased such that the ZFS is accessible, all other transitions within the multiplet are triggered. In the dI/dV maps generally one single excitation line remains, namely the one corresponding to the ZFS [12]. Finally, at low bias, only excited states of the spin ground state multiplet participate in the electronic transport. By further increasing the bias voltage, higher energy spin excited multiplets may appear in the conductance map (i.e. a multiplet carrying total spin $S-1$).

Importantly, the value of Γ is usually a limiting factor for the detection of excitations in the SET regime as Γ is also a measure of the level broadening. Therefore, Γ has to be lower than the level spacing $\Delta_{\mathrm{exc}}(\theta, B)$ in order to observe the magnetic excitations as resolved lines in the conductance. Typically, the zero-field splitting (energy difference between the two lowest-lying states) for a SMM is in the range of the meV, so that Γ should be < 1 meV. However, Γ cannot be too low because the current is proportional to it. In practice this means that there is a rather small window to resolve the ZFS in the SET regime.

Finally, we emphasize that first-order SET processes are incoherent. This means that the electron tunnels onto the molecule and interacts with it so that stays there long enough to lose its phase. Chemically speaking, the molecule gets charged (oxidized or reduced) and for this reason SET processes always involve transitions between two adjacent charge states. The properties of the charged molecule are thus of great importance in describing transport. For single-molecule magnets, however, little is known about properties such as the spin value or the orientation of the easy axis of a charged molecule. As we will see, some of these magnetic properties can be inferred from transport characteristics in the SET regime (see e.g. Sect. 12.3.4). The situation must be contrasted to higher-order tunneling processes which take place in the coherent regime. They only involve one particular charge state, namely the one that is stabilized by Coulomb blockade. The next two subsections deal with this regime and in Sect. 12.3.4, we will come back to first-order SET transport.

12.3.2 Intermediate Coupling: Inelastic Spin-Flip Co-tunneling Process

In case the coupling is of the order of the level spacing and the thermal energy ($\Gamma \sim E_C, \Delta(N), k_B T$), high-order tunneling processes become relevant. These are of particular interest as they can lead to well defined features in the conductance maps. Depending on the final energy state of the molecule, two co-tunneling regimes can be distinguished: *elastic* and *inelastic* co-tunneling.

In an elastic co-tunneling process, an electron from the molecule hops into the drain leaving the molecule in a virtual forbidden state. Due to the Heisenberg uncertainty principle, this event is possible if the molecular state remains unoccupied for a period of time $\Delta t \leq \hbar/(E_c + \Delta)$. Another electron from the source hops into the molecule within Δt and occupies the same state leading to a net transport of current. This process can occur at an arbitrarily low bias and leads to a non-zero background conductance within the Coulomb blockade regime.

A second scenario is inelastic co-tunneling. It occurs when the molecule ends in an excited state at the end of the co-tunneling process, as sketched in Fig. 12.4(a). In this case, the electron hopping from the source occupies an excited state of the molecule within the same charge state. In contrast to elastic co-tunneling, it occurs when $V_b \simeq E_{\mathrm{exc}}$, and shows up in the conductance map as an horizontal line inside the Coulomb blockade regime that ends at the Coulomb edges (see Fig. 12.4(c)).

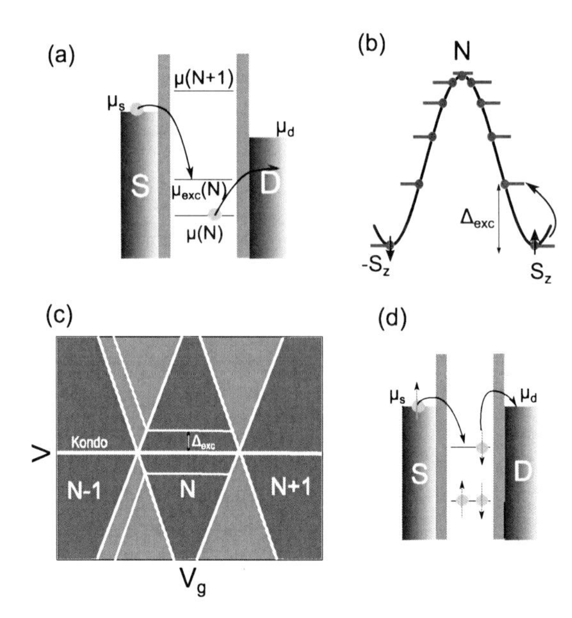

Fig. 12.4 High-order transport. (**a**) Electrochemical diagram of an inelastic co-tunneling process. (**b**) Typical double-well energy diagram of a SMM showing the excitation corresponding to an inelastic co-tunneling excitation. (**c**) Stability diagram showing high-order tunneling processess. Inelastic co-tunneling excitations appear as horizontal lines in the Coulomb blockade regime that end in the Coulomb edges of the SET regimes. Kondo appears as a zero-bias excitation. If the molecule is in a high spin state ($S \geq 1$), Kondo appears in succesive adjacent states. (**d**) Electrochemical diagram of a Kondo co-tunneling process

Since now two electrons participate in the transport, the corresponding spin selection rules are $\Delta S = 0, \pm 1$ and $\Delta m = 0, \pm 1$ for a second-order process. For a single-molecule magnet, this transition for example corresponds to spin-flip excitations inside the potential well of a single charge state (see Fig. 12.4(b)). As stated before, this energy depends on B and θ, so that the anisotropy of a particular charge state can be quantified by measuring its dependence on these parameters. Note again that the ZFS is the largest energy scale and would therefore be the dominant feature in the conductance maps.

12.3.3 Kondo Correlations

A different high-order co-tunneling process appears when the spin of the electron is taken into account. As a first approach, Kondo excitations appear when the molecule has orbitals with unpaired electrons. At zero-bias, the unpaired electron in the dot can hop into the drain by an elastic co-tunneling process and be replaced by an electron with the opposite spin orientation from the source as sketched in Fig. 12.4(d). This conduction process leads to a resonant conduction peak at zero-bias within the Coulomb diamond (see Fig. 12.4(c)). Such Kondo peaks have already been observed in single molecules [13–16]. In the case of a SMM, the scenario can be more complicated by the presence of several unpaired electrons in different ions of the molecule. In this case, Kondo resonances may appear in two or more consecutive charge states which is a fingerprint of a high-spin state ($S \geq 1$) in one of the charge states of the molecule. It can therefore be used as a strong indication of the presence of a SMM in the junction.

The conductance maximum of the Kondo resonance changes logarithmicaly with temperature following the empirical expression [17]:

$$G(T) = G_c + G_a\left[1 + 2^{1/s} - 1)(T/T_K)^2\right]^{-s}, \tag{12.3}$$

where G_a is the conductance at $T = 0$, G_c is a temperature-independent offset and s is a parameter that depends on the spin of the dot. For an electron with $S = 1/2$, $s = 0.22$ and it is lower for higher spin values in the SMM. The Kondo temperature T_K defines the onset of Kondo correlations. Note, that at low temperatures the width of the Kondo resonance is:

$$\text{FWHM} \simeq 4k_B T_K/e. \tag{12.4}$$

For typical vales of T_K of ~ 10 K (~ 1 meV), the Kondo peak width can be larger than the zero-field splitting and therefore may mask other co-tunneling excitations coming from spin excited states such as the ZFS. Spin excitations to higher multiplets may still be visible (e. g. the $S = 5$ to $S = 4$ transition in the Fe_4 SMM is about 4 meV). In high-magnetic fields the zero-bias peak splits due to the Zeeman effect in different components separated by $\Delta V_b = 2g\mu_B B$. This Zeeman splitting together with the logarithmic scale are hallmarks of the Kondo effect.

12.3.4 Ground State to Ground State: Gate Spectroscopy

A different route to obtain information on magnetic properties of SMMs involves ground state to ground state transitions rather than to excited spin states. The virtues of this method are, as we will see, that it is less sensitive to the strength of the coupling between the molecule and the electrodes and therefore it can be applied in a broader range of cases. Moreover, it is very sensitive to small changes of the magnetic anisotropy in different charge states. The method relies on measuring the position of the Coulomb peak in the conductance map. The Coulomb peak marks the ground-state S_N to ground-state S_{N+1} transition between two adjacent charge states N and $N + 1$ at zero-bias (see Fig. 12.2(d)). The position in gate voltage of this peak depends on the energy difference ΔE between those two states. Such energy difference can be varied by applying an external magnetic field, i.e. $\Delta E = \Delta E(B, \theta)$.

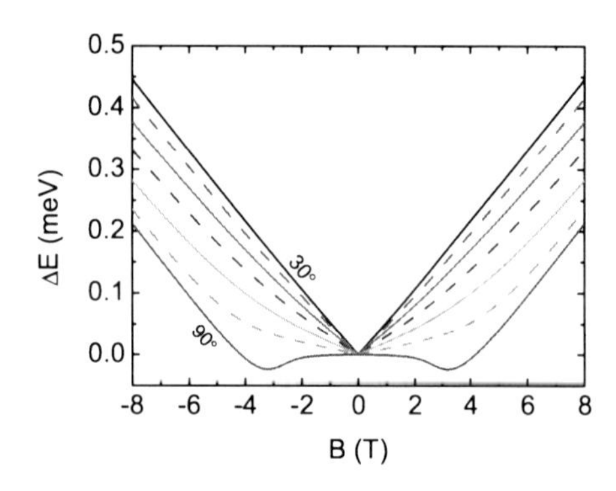

Fig. 12.5 ΔE calculated for seven different values of θ (from 30° to 90° every 10°) by numerical diagonalization of the Hamiltonian given in (12.2). The values of θ and D are the same for both charge states and are taken to be $D = -56\ \mu eV$

If the molecule is isotropic, the chemical potential depends linearly on the applied magnetic field (Zeeman effect):

$$\Delta E(B) = \big(E_{N+1}(B) - E_N(B)\big) - \big(E_{N+1}(0) - E_N(0)\big) = -g\mu_B B \Delta S \qquad (12.5)$$

where μ_B is the Bohr magneton and $\Delta S = S_{N+1} - S_N$. E_N and E_{N+1} are the energies of the S_N and S_{N+1} states. This change in the chemical potential corresponds to a change in the position of the zero-bias Coulomb peak:

$$\Delta V_g(B) = \Delta E(B)/\beta = -g\mu_B B \Delta S/\beta \qquad (12.6)$$

where β is the molecule-gate coupling.

In case of an anisotropic molecule, the energy spectrum is very sensitive to the relative orientation of the magnetic field with respect to the easy magnetization axis of the molecule. The energy ΔE can thus depend nonlinearly on B and this dependence can be used to study quantitatively the magnetic anisotropy of an individual molecule. Figure 12.5 shows ΔE calculated at different angles θ between the easy axis and magnetic field by numerical diagonalization of the Hamiltonian (12.2). For simplicity, we have taken $D_N = D_{N+1} = D$ and $\theta_N = \theta_{N+1} = \theta$.

At high magnetic fields, ΔE is linear with the magnetic field, meaning that the Zeeman contribution dominates over the magnetic anisotropy. The sign of the slope gives information about the change in the spin upon oxidation or reduction of the molecule. According to (12.6), the slope of the curve at high fields is positive when $\Delta S < 0$ and vice versa. Consequently, in the situation considered in Fig. 12.5 the spin ground state decreases upon reduction of the molecule. A negative slope would mean the opposite. In addition, the electron carries a spin $s = 1/2$ and then the difference in spin between two adjacent charge states is $|\Delta S| = 1/2$. Therefore, $S_{N+1} = S_N - 1/2$ for a positive slope. This information can be very valuable when no other reference of the charge states is known beforehand. A more complex scenario may appear in SMMs if $\Delta S \neq 1/2$ between ground states S_N and S_{N+1}. Selection rules forbid transport and the low-bias conductance should be suppressed (spin blockade, see for instance [18, 19]). Looking now at the low-field region in Fig. 12.5, we observe that ΔE is not linear with B and depends strongly on θ. In this case, the magnetic anisotropy dominates over the Zeeman effect. The shape of the

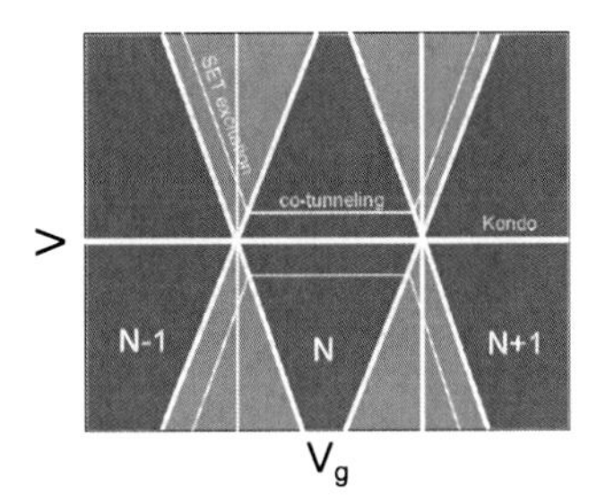

Fig. 12.6 Stability diagram of a high spin molecule schematically showing the Coulomb diamonds of three different charge states. SET excitations, co-tunneling lines and Kondo excitations provide information about the magnetic structure of individual molecules

curves, the evolution with the angle and the crossover field between the non-linear to linear behavior give quantitative information about the anisotropy parameters at the single-molecule level.

The previous curves are calculated for the simple case when the magnetic anisotropy does not change in magnitude or orientation when the molecules are charged. By making θ and D different in adjacent charge states, the curves undergo sizable differences that can be used to study subtle changes in the magnetic anisotropy of different charge states at the single molecule level.

12.3.5 Summary

To summarize this section, we have shown how *direct* probing of the magnetic molecule is a powerful and unique spectroscopic tool to study individual magnetic molecules in different charge states. Figure 12.6 compiles the main information we can obtain in a stability diagram. Co-tunneling and SET lines provide the energies of spin excited states. The field evolution of the degeneracy-point supplies information on the magnitude and orientation of the magnetic anisotropy. Finally, a Kondo resonance in adjacent charge states is a fingerprint of a high-spin state in the molecule.

12.4 Fabrication of a Spin Transistor

The fabrication of a three-terminal transistor involves various nanoscale techniques. The different components of the device are first defined by electron-beam lithography (EBL). Afterwards the selected metal for the electrodes (mostly gold) is deposited by evaporation. The fabrication of a nanometric gap (1–2 nm) between source and drain lies, however, beyond the resolution of EBL which is about 10 nm. For this reason, we use controlled electromigration [20, 21] with an active feedback to open a nanometric gap in the evaporated metal nanowire.

12.4.1 Electron-Beam Lithography

The four EBL steps required to fabricate the device are described in Fig. 12.7. The first step defines the alignment markers and the contact pads that will connect source

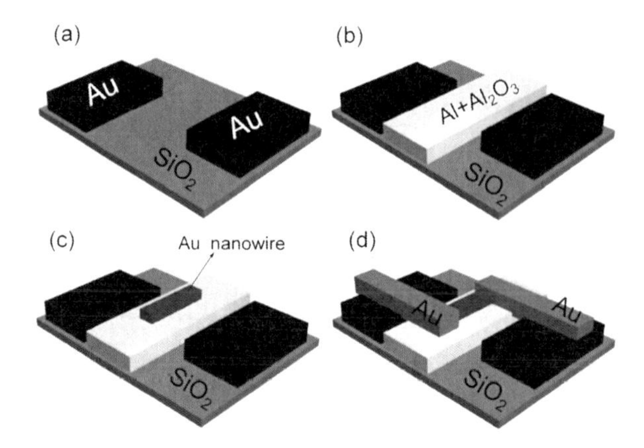

Fig. 12.7 The fabrication of the nanodevice consists of four electron-beam lithography steps. In the first step (**a**) the gold contact pads are defined. Second, the aluminum gate electrode is deposited. In a third step (**c**), on top of the oxidized gate, we pattern the gold nanobridge that will be electromigrated to form the source and drain electrodes. Finally (**d**), the strips that connect the gold nanowire with the contact pads are evaporated

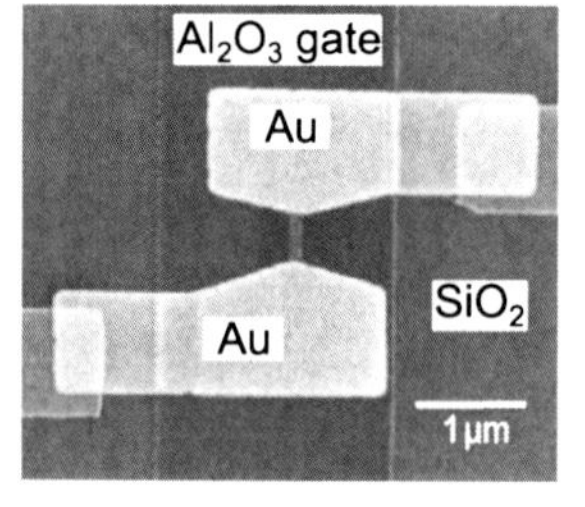

Fig. 12.8 Scanning electron microscopy image of the nano-junction showing the gold nanobridge before electromigration and the underlying aluminum gate electrode

and drain with the electronics. This step is followed by the evaporation of 3 nm of titanium and 50 nm of gold. Titanium is used as a sticking layer because of its good adhesion to silicon oxide.

The second lithography step defines the gate electrode that is common to all junctions. Afterwards, 75 nm of aluminum are evaporated and subsequently oxidized inside an O_2 chamber at 50 mTorr to obtain a 2–4 nm aluminum oxide layer. This fabrication step is critical since the oxide layer has to be thin enough to have a large molecule-gate coupling but thick enough to avoid any source/drain to gate current leakage. In the next EBL step, a thin (12 nm thick and 100 nm wide) gold nanowire is patterned on top of the oxidized gate electrode. This nanowire will be used to fabricate the source and drain electrodes afterwards. Finally, the last EBL step defines the strip (110 nm) that connects the gold nanowire with the contact pads defined in the first step.

A picture of a device is shown in Fig. 12.8. This scanning electron microscopy image shows the gold nanobridge before electromigration on top of the aluminum electrode used as the gate.

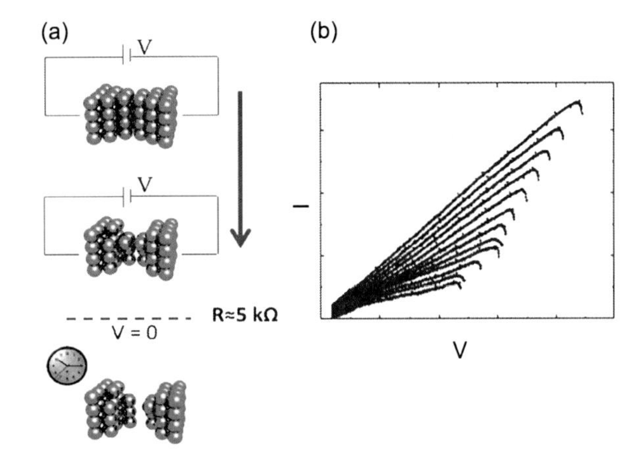

Fig. 12.9 (**a**) Scheme of the "self-breaking" electromigration technique. (**b**) I-V plot of a typical electromigration process

12.4.2 Electromigration

The fabrication of the nanogap consists of self-breaking electromigration of the gold nanowire. Electromigration is the movement of the ions in a metal caused by momentum transfer from the electrons onto the ions. In other words, the ions forming the metal are "pushed" away by the moving electrons. In practice, (see Fig. 12.9(a)), the voltage through the nanowire is increased over the electromigration threshold (around 100 meV in our gold wires) while continuously monitoring the wire resistance with fast-response electronics. Whenever a change in the resistance of typically 10 % of the original value is detected, the applied voltage is reduced to the starting value. The aim of this controlled electromigration is to avoid a too fast or violent breaking of the wire that in turn may create a too large gap not suitable to trap the molecules. The process is repeated until the resistance reaches around 5 kΩ. At this resistance, a nanoconstriction in the wire has been created. Due to the high mobility of gold atoms at room temperature, the resistance continues to increase and, eventually, the nanoconstriction "self-breaks" into a nanometric gap with a resistance of the order of a few MΩ. The time scale for self-breaking varies between a few minutes to a few hours [21]. Figure 12.9(b) shows typical voltage-current curves observed in the course of an electromigration process. The main advantage of self-breaking is that it avoids a sudden break of the gold nanowire which in turn can lead to the formation of gold nanoparticles within the gap. These unwanted gold particles can mask or even be mistaken with the target molecules.

Electromigration is carried out under solution of the magnetic molecules at room temperature. Once the target resistance (few MΩ) has been reached in as many junctions as possible on the chip, the molecular solution is pumped out and the whole system is cooled down to stop the self-breaking process and to allow detailed spectroscopic measurements to be performed.

12.4.3 Preliminary Characterization

The first electrical characterization is to distinguish whether the junction contains a single molecule or a few or even none. To do this, the bias voltage is fixed (typically at 50 meV) and the current through the junction is measured as a function of V_g between ± 3 V (gate leakage usually appears for higher voltages). Single-molecule magnets are expected to have charging energies of the order of eV whereas with the gate we can typically shift the energy levels by ± 0.4 V. For this reason, at most two redox states are usually accessible within this gate voltage window. The measured I–V_g characteristic would then show a single peak in current that marks the onset of SET between both charge states. In contrast, the occurrence of several peaks is probably the fingerprint of the presence of several molecules bridging the gap. Subsequently, a full stability diagram as a function of V_g and V is measured. Only those junctions showing a clear signature of a single molecule are used for further analysis (about 10 % of the junctions).

12.5 A Practical Example. The Fe_4 Single-Molecule Magnet

12.5.1 Why the Fe_4 Single-Molecule Magnet?

Mn_{12} is the archetypical and most studied single-molecule magnet since its synthesis in the early nineties [4]. For this reason, it was the first SMM studied in a three-terminal geometry and showed signatures of sequential electron transport and even magnetic properties in transport [22, 23]. However, subsequent experiments showed that the magnetic properties of the magnetic core may not be preserved under deposition of the molecule on gold [24]. Ideal SMMs for transport should have a magnetic core sufficiently protected by an outer shell to avoid any structural distortion that may alter its properties. In addition, the core has to be accessible for electrons. A large spin and magnetic anisotropy are paramount to observe other quantum phenomena such as quantum tunneling of the magnetization or quantum interference (Berry phase). Fe_4 SMM meets these characteristics and, importantly, X-ray experiments have shown that its magnetic properties, in particular the magnitude and orientation of the magnetic anisotropy, are preserved under deposition on gold surfaces [25, 26].

Fe_4 is made of four Fe^{3+} ions with spin $s = 5/2$ as illustrated in Fig. 12.10(a). The strong exchange antiferromagnetic interaction between the central and the peripheral ions gives rise to a total molecular spin $S = 5$. The magnetic anisotropy originated from the asymmetry of the molecule lifts the degeneracy of the spin ground state into five doublets and a singlet distributed over an energy barrier that hinders the spin reversal. The energy diagram is usually described by a double-well potential like the sketched in Fig. 12.10. The bulk value of the axial anisotropy parameter is $D = -56$ μeV resulting in a height of the barrier of $U = 1.4$ meV [27]. The zero-field splitting is 0.5 meV.

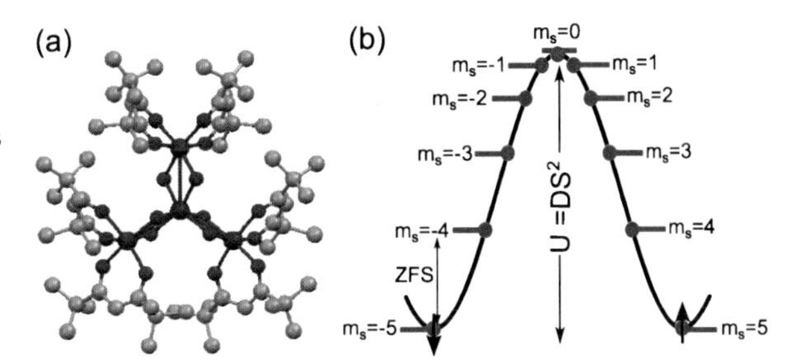

Fig. 12.10 (**a**) Magnetic core of the Fe_4 including some peripheral ligands. (**b**) Magnetic anisotropy splits the spin ground state into $2S + 1$ states distributed over an energy barrier that prevents the spontaneous reversal of the magnetization

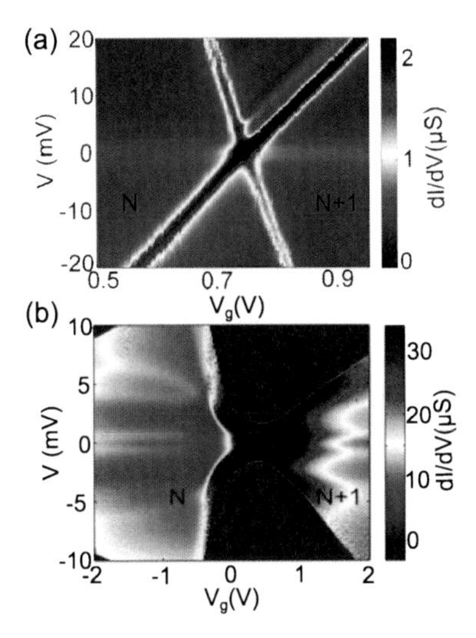

Fig. 12.11 dI/dV color maps versus V_g and V for two different junctions containing an individual Fe_4 SMM with different couplings Γ. (**a**) High-conductance lines mark the onset of SET that separates two charge states. Kondo correlations in adjacent charge states are fingerprint of a high spin state. (**b**) Charge-degeneracy point of a different sample. Γ is larger and co-tunneling excitations appear. The color scale is saturated as to highlight the co-tunneling lines. Measurements temperature is $T = 1.9$ K and $T = 1.6$ K respectively

As we have seen, different transport regimes provide different information on the magnetic properties of the molecule. The value of Γ cannot be controlled or changed during the experiments. We present therefore results obtained for different junctions in different transport regimes. Figure 12.11 shows differential conductance maps (dI/dV) of two different junctions containing an individual Fe_4 SMM with different couplings Γ to the electrodes. The upper plot shows lines of high dI/dV characteristic of single-electron transport through the molecule. The low-differential conductance regions (blue in the figure) on either side of the charge degeneracy point, correspond to charge states N and $N + 1$. At positive bias, a strong excitation is visible within the SET regime with energy 4.8 meV. This value is of the order of the calculated energy for the first excited spin multiplet ($S = 4$) of Fe_4 [26, 28]. High-order co-tunneling excitations are also visible around zero-

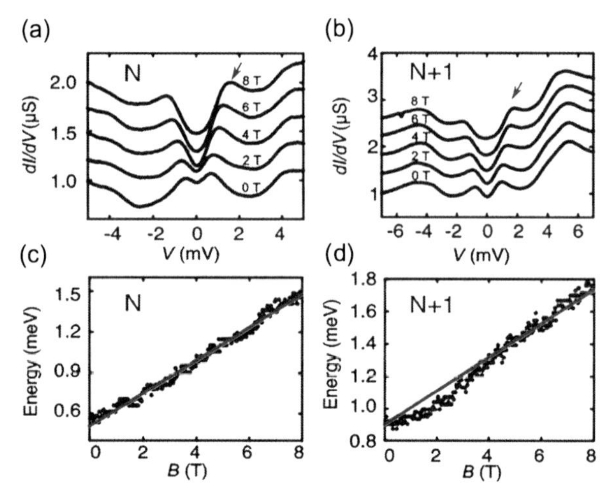

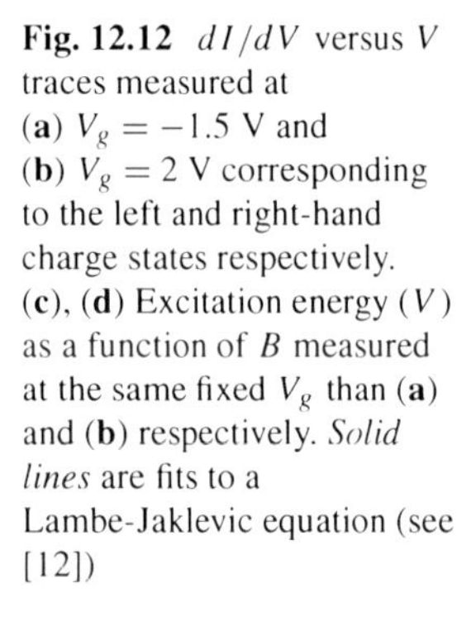
Fig. 12.12 dI/dV versus V traces measured at (**a**) $V_g = -1.5$ V and (**b**) $V_g = 2$ V corresponding to the left and right-hand charge states respectively. (**c**), (**d**) Excitation energy (V) as a function of B measured at the same fixed V_g than (**a**) and (**b**) respectively. *Solid lines* are fits to a Lambe-Jaklevic equation (see [12])

bias in both charge states, which is fingerprint of high-spin Kondo correlations. The value of Γ is relatively low and the Coulomb peak is well defined; we have performed gate-voltage spectroscopy in this sample (see Sect. 12.5.3). In the bottom dI/dV map of Fig. 12.11, the value of Γ is larger and inelastic co-tunneling lines are present in both charge states. We use this example to discuss spin excitations in the co-tunneling regime. Importantly, the results of both examples confirm the presence of an anisotropic magnetic molecule bridging source and drain electrodes [12, 29, 30].

12.5.2 Spin Excitations: Inelastic Spin Flip Spectroscopy

In Fig. 12.11(b), within the left-hand Coulomb diamond three clear inelastic co-tunneling excitations are observed at ±0.6 meV, ±4.6 meV and ±6.7 meV. Two other excitations appear in the right-hand charge state at ±0.9 meV and ±5 meV. The energy of the lowest excitation in both charge states is of the order of the zero-field splitting ($\sim$ 0.5 meV) of Fe_4. To gain a deeper insight on the magnetic nature of these excitations we study their evolution in the presence of a magnetic field. We measure dI/dV versus V at a fixed $V_g = -1.5$ V for different magnetic fields. The results are shown in Fig. 12.12(a). The energy of the excitation (marked with a red arrow) increases with B as shown in Fig. 12.12(c). The curve is symmetrical upon field reversal (see supplemental information in [12]). The observed behavior is characteristic of the existence of a finite ZFS, as described by the spin Hamiltonian (12.2). The linear dependence with B indicates a low value of θ. Figures 12.12(b) and (d) show similar data measured at $V_g = 2$ V corresponding to the excitations in the right-hand charge state. The low-field behavior shown in Fig. 12.12(d) is slightly non-linear with B which may be a fingerprint of magnetic anisotropy. Interestingly, the slightly different magnetic field dependence of the excitations in left and right

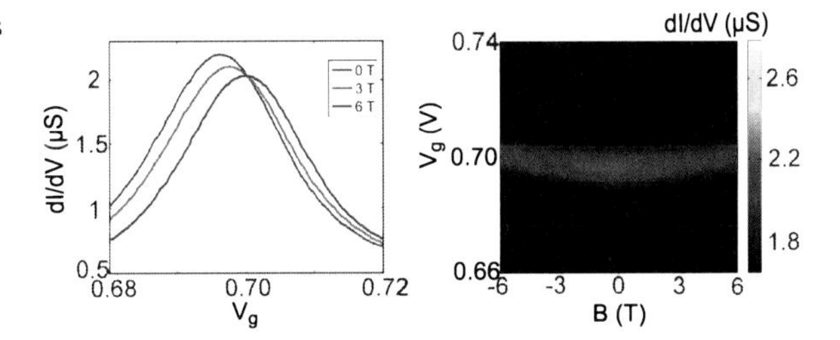

Fig. 12.13 (**a**) dI/dV versus V_g curves measured at zero-bias and $T = 1.9$ K for three different magnetic fields. (**b**) dI/dV color plot around the degeneracy-point as a function of B and V_g

charge states points to a misalignment of the easy axes when charging the molecule. In particular θ_N should then be larger than θ_{N+1}.

12.5.3 Gate-Voltage Spectroscopy

We now focus on the position of the Coulomb peak in gate voltage and its evolution under a magnetic field. We measure dI/dV curves as a function of V_g around the Coulomb peak shown in Fig. 12.11(a) [30]. A lock-in bias voltage modulation with amplitude 0.1 mV is used with no DC voltage applied between source and drain. The result are shown in Fig. 12.13(a) for different magnetic fields. These results show a shift of the peak towards higher gate voltages as B increases.

To understand in detail the evolution of the peak position with the applied magnetic field, we have carried out detailed measurements covering the field range $-6T \leq B \leq 6T$. The result is represented in Fig. 12.13(b) as a dI/dV color plot versus B and V_g. The position of the peak in V_g depends clearly non-linearly with B indicating the presence of magnetic anisotropy in this molecule [30]. For clarity, we show in Fig. 12.14(a) the position of the peak maximum as obtained from Gaussian fits of the curves shown in Fig. 12.13.

Further information about the anisotropy is obtained by changing in-situ the value of θ. This is done by rotating the whole chip with a piezo-driven rotator inside the cryostat. Figure 12.14 shows the position of the Coulomb peak in gate voltage at different angles of the rotation α.[1] The low-field behavior differs significantly. We observe a gradual change from an almost linear dependence in Fig. 12.14(a) to an almost flat field dependence at low fields (Fig. 12.14(d)). According to the model used in Fig. 12.5, the evolution of the curves suggest a rotation towards larger values of θ, approaching $\theta = 90°$

For a quantitative estimation of the magnetic anisotropy, we calculate ΔE by numerical diagonalization of Hamiltonian (12.2) following the model described in Sect. 12.3.4. The results are the solid lines in Fig. 12.14. For this calculation we assume that the neutral charge state is the left-hand Coulomb diamond and that the magnetic anisotropy is the same as in the bulk [26, 31]. Similar fits with slightly

---

[1] Note that α, the angle of rotation, is in general different from θ, the angle between the easy axis and the magnetic field.

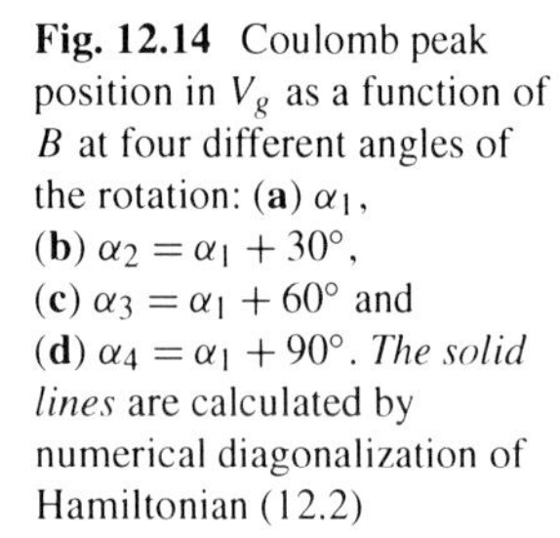

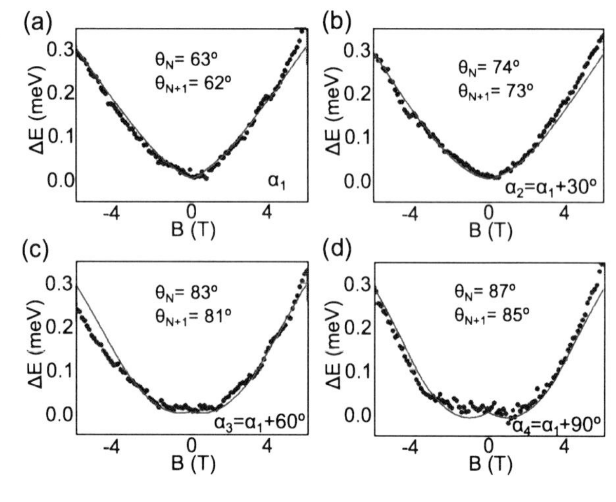

Fig. 12.14 Coulomb peak position in V_g as a function of B at four different angles of the rotation: (**a**) α_1, (**b**) $\alpha_2 = \alpha_1 + 30°$, (**c**) $\alpha_3 = \alpha_1 + 60°$ and (**d**) $\alpha_4 = \alpha_1 + 90°$. *The solid lines* are calculated by numerical diagonalization of Hamiltonian (12.2)

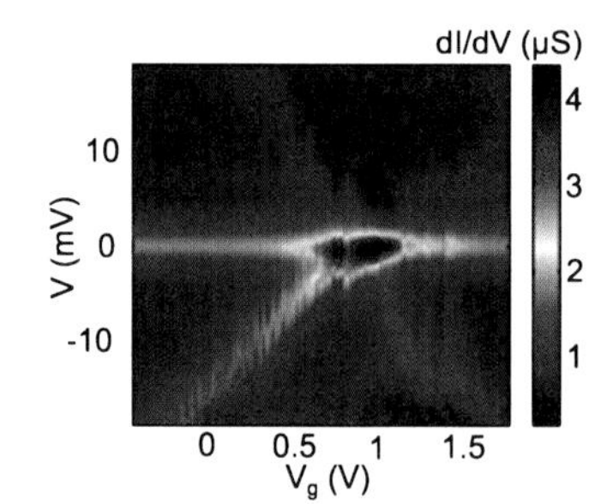

Fig. 12.15 dI/dV color map of a junction containing an Fe_4 SMM in the intermediate coupling regime. Zero-bias excitations are observed in adjacent charge states which is fingerprint of high-spin Kondo

different parameters can be obtained by assigning the neutral state to the right-hand state. From the positive slope at high fields, the spin in the reduced (right-hand) charge state is found to be $S_{N+1} = 9/2$. Free parameters are then D_{N+1}, θ_N and θ_{N+1}. The best fitting values are $D_{N+1} = 68$ μeV and the angles that appear in Fig. 12.14. It is therefore possible to conclude that the molecular ground state S decreases and the magnetic anisotropy is enhanced by reducing the molecule [30]. These results agree with the conclusions extracted from the co-tunneling lines [12, 29].

12.5.4 Kondo Excitations and High-Spin State

Figure 12.15 shows the stability diagram of a third molecular junction in the intermediate coupling regime. Strong zero-bias excitations appear in adjacent charge states, which is fingerprint of a high-spin ($S \geq 1$) Kondo effect. In order to reveal the Kondo origin of the excitation, we measure the dependence of the resonance with temperature and magnetic field [29].

Figure 12.16(a) shows a dI/dV trace versus V measured at a fixed $V_g = 0$ and for $B = 0$ (blue curve) and $B = 8$ T (red curve). Figure 12.16(c) shows the re-

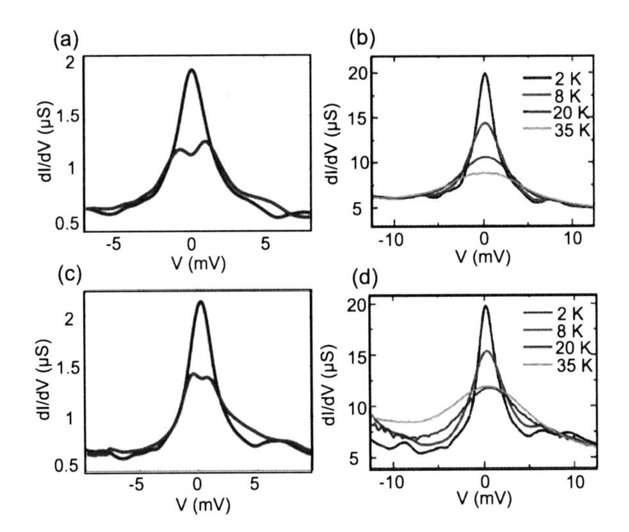

Fig. 12.16 dI/dV traces versus V. (**a**) dI/dV at $V_g = 0$ V measured at $B = 0$ T (*blue line*) and $B = 8$ T (*red line*). (**b**) dI/dV measured at $V_g = -0.6$ V and different temperatures. (**c**) dI/dV at $V_g = 1.7$ V measured at $B = 0$ T (*blue line*) and $B = 8$ T (*red line*). (**d**) dI/dV measured at $V_g = 1.4$ V and different temperatures

sults of measurements carried out at $V_g = 1.7$ V, corresponding to the right-hand charge state. The peak at zero field splits into two components at $B = 8$ T in both charge states. Using the Zeeman component of (12.2) we estimate the value of the Landé factor to be $g = 2$. Further proof of the Kondo resonance is obtained from the temperature dependence. Figure 12.16(b) shows the Kondo resonance measured at $V_g = -0.6$ V and at different temperatures. The peak width increases and its height decreases when increasing T. The same behaviour is observed in Fig. 12.16(d) for the right-hand charge state. The Kondo temperature is obtained (see [29] for more details) from the evolution of the total width half maximum (FWHM) with the temperature. The results are $T_K = 13$ K for the left-hand charge state and $T_K = 10$ K for the right-hand charge state (see [29]).

12.6 Future Directions

12.6.1 Quantum Tunneling of the Magnetization and Berry Phase

The sensitivity of the three-terminal transistor as a spectroscopic technique paves the way to study quantum phenomena at the individual molecule level. Some SMM present transverse anisotropy (perpendicular to the easy axis) that mixes the energy levels at both sides of the energy barrier. This overlap allows the spin to tunnel through the barrier instead of relaxing by "jumping" over it [32–34]. The probability of tunneling is usually small but it can be enhanced with the application of an external magnetic field perpendicular to the easy axis of the SMM [35]. If the magnetic field is applied along the "hardest" anisotropy axis within the hard anisotropy plane, the probability of tunneling oscillates and even quenches due to destructive interferences. This is known as Berry phase oscillation. Quantum tunneling and quantum

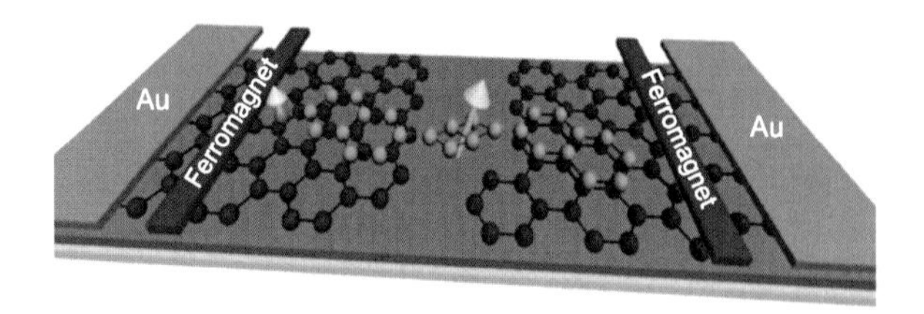

Fig. 12.17 Three-terminal transistor made of ferromagnetic electrodes on graphene. Gold is used to inject electrons into the graphene layer. A ferromagnetic metal is deposited on top of graphene to polarize the current. The magnetic molecule lies within the gap in the few-layers graphene gap

interferences may leave their fingerprint in the current flow through the magnetic molecule. A spin transistor may be used then to observe them at the single molecule level.

12.6.2 Ferromagnetic Electrodes

Theoretical and experimental approaches described in this chapter assume that non-polarized electrons are injected by the gold electrodes. Additional control over the molecular spin is predicted by using spin polarized electrons. Recent theoretical works [36] show that by making one of the linking electrodes ferromagnetic, the flowing electrons can flip the spin of the molecule to a final state where it is blocked leading to current suppression. That final state and the total current depend on the initial spin of the molecule and the magnetic anisotropy. This spin-charge conversion may be used to read out the spin information. Interestingly, it is predicted that by making ferromagnetic both electrodes the magnetic state of the molecule can be switched with the spin polarized current [37]. The switching process may be visible in the current by applying a voltage over time.

Usually, ferromagnetic metals oxidize in ambient conditions hindering electrical transport. For this reason, only a few examples of a spin polarized current addressing a molecule have been reported to date [38]. An alternative approach is to use an intermediary conductor between the ferromagnetic material and the gap with the molecule. Recently, it has been shown that graphene can be electromigrated in a similar way as gold to open a nanogap [39]. Graphene versatility allows to deposit a ferromagnetic metal on top that can act as a spin polarizer. Figure 12.17 shows schematically a possible configuration for these experiments.

12.6.3 Spin Crossover Molecules

Spin crossover (SC) molecules form a different family of magnetic molecules with interest for molecular spintronics. SC are known as molecular switches for their

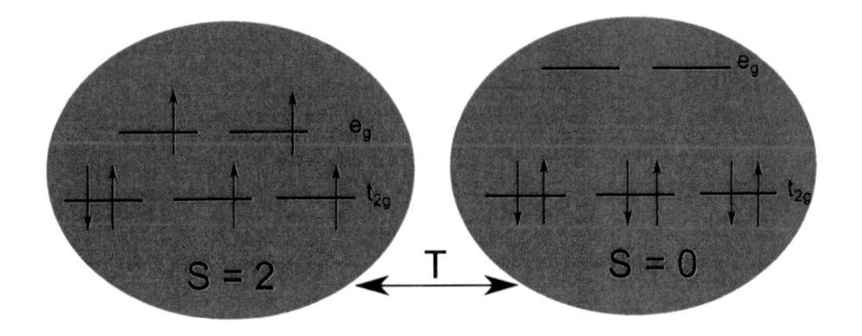

Fig. 12.18 Spin crossover mechanism in a $3-d$ metal. The molecule is in a *low-spin* state if Δ is larger than the spin pairing energy

ability to change their magnetic ground state under the action of external stimuli, such as temperature, light or pressure [40]. This phenomenon is widely observed in the first row transition metals like Fe.

Figure 12.18 shows the mechanism behind a spin crossover transition for a $3d$ metal with 6 electrons. In the presence of a crystal field the 5 levels of the d-orbital split in a doublet and a triplet separated by an energy Δ. When the splitting Δ is larger than the pairing energy, the electrons do not follow Hund's rule and fill the lower levels. The molecule is then in a *low-spin* state ($S = 0$). Under an external stimulus the symmetry around the $3d$ ion can change and Δ is reduced. The energy levels are quasi-degenerated and the orbital filling now follows Hund's rule. The molecule is in a *high-spin* state $S = 2$.

The crossover temperature of a temperature-driven SC transition can be tuned by tailoring the ligands around the metallic core. For instance some molecules show a SC transition close to room temperature [41]. Recent theoretical works [42] indicate that the conductance through the molecule may change because of a shifting of the molecular levels or a change in the coupling of the molecule to the electrodes. However this has not been established experimentally yet.

Acknowledgements This work was supported by FOM and the EU FP7 program under the Grant Agreement ELFOS. We thank A. Cornia, M. Wegewijs and K. Park for fruitful discussions.

References

1. I. Zutic, J. Fabian, S. Das Sarma, Rev. Mod. Phys. **76**, 323 (2004)
2. S. Sanvito, Chem. Soc. Rev. **40**, 3336 (2011)
3. L. Bogani, W. Wernsdorfer, Nat. Mater. **7**, 179 (2008)
4. A. Caneschi, D. Gatteschi, R. Sessoli, A.L. Barra, L.C. Brunel, M. Guillot, J. Am. Chem. Soc. **113**, 5873 (1991)
5. M. Jamet, W. Wernsodrfer et al., Phys. Rev. Lett. **86**, 4676 (2001)
6. L. Hao et al., Appl. Phys. Lett. **98**, 092504 (2011)
7. M.J. Martínez-Pérez et al., Appl. Phys. Lett. **99**, 032504 (2011)
8. J.-P. Cleuziou et al., Nat. Nanotechnol. **1**, 53 (2006)
9. M. Urdampilleta et al., Nat. Mater. **10**, 502 (2011)
10. A. Candini et al., Nano Lett. **11**, 2634 (2011)
11. R. Vincent et al., Nature **488**, 357 (2012)
12. A.S. Zyazin et al., Nano Lett. **10**, 3307 (2010)
13. J. Park et al., Nature **417**, 722 (2002)
14. D. Goldhaber-Gordon et al., Nature **391**, 156 (1998)

15. W.J. Liang et al., Nature **417**, 725 (2002)
16. L.H. Yu et al., Nano Lett. **4**, 79 (2004)
17. D. Goldhaber-Gordon et al., Phys. Rev. Lett. **81**, 5225 (1998)
18. D. Weinmann, W. Häusler, B. Kramer, Phys. Rev. Lett. **74**, 984 (1995)
19. C. Romeike et al., Phys. Rev. B **75**, 064404 (2007)
20. H. Park et al., Appl. Phys. Lett. **75**, 301 (1999)
21. K. O'Neill, E.A. Osorio, H.S.J. van der Zant, Appl. Phys. Lett. **90**, 133109 (2007)
22. H.B. Heersche et al., Phys. Rev. Lett. **96**, 206801 (2006)
23. M.H. Jo et al., Nano Lett. **6**, 2014 (2006)
24. M. Mannini et al., Chem. Eur. J. **14**, 7530 (2008)
25. M. Mannini et al., Phys. Rev. Lett. **8**, 194 (2009)
26. M. Mannini et al., Nature **468**, 417 (2010)
27. S. Accorsi et al., J. Am. Chem. Soc. **128**, 4742 (2006)
28. S. Carretta et al., Phys. Rev. B **70**, 214403 (2004)
29. A.S. Zyazin, H.S.J. van der Zant, M.R. Wegewijs, A. Cornia, Synth. Met. **161**, 591 (2011)
30. E. Burzuri, A. Zyazin, A. Cornia, H.S.J. van der Zant, Phys. Rev. Lett. **109**, 147203 (2012)
31. M. Mannini et al., Nat. Mater. **8**, 194 (2009)
32. J.M. Hernandez, X.X. Zhang, F. Luis, J. Bartolomé, J. Tejada, R. Ziolo, Europhys. Lett. **35**, 301 (1996)
33. J.R. Friedman et al., Phys. Rev. Lett. **76**, 3830 (1996)
34. L. Thomas et al., Nature **383**, 145 (1996)
35. E. Burzuri et al., Phys. Rev. Lett. **107**, 097203 (2011)
36. F. Elste, C. Timm, Phys. Rev. B **73**, 235305 (2006)
37. M. Misiorny, J. Barnás, Phys. Rev. B **75**, 134425 (2007)
38. A.N. Pasupathy, R.C. Bialczak et al., Science **306**, 86 (2004)
39. F. Prins, A. Barreiro et al., Nano Lett. **11**, 4607 (2011)
40. P. Gütlich, H.A. Goodwin, *Spin Crossover Transition Metal Compounds* (Springer, Berlin, 2004)
41. R. Gonzalez-Prieto et al., Dalton Trans. **40**, 7564 (2011)
42. D. Aravena, E. Ruiz, J. Am. Chem. Soc. **134**, 777 (2012)

Chapter 13
Molecular Quantum Spintronics Using Single-Molecule Magnets

Marc Ganzhorn and Wolfgang Wernsdorfer

Abstract The objective of molecular quantum spintronics is to combine the concepts of spintronics, molecular electronics and quantum computing in order to fabricate, characterize, and study molecular devices (for example molecular spin-transistors and molecular spin-valves) allowing the read-out and manipulation of the spin states of one or several molecules. The main first goal is to perform basic quantum operations. The visionary concept of molecular quantum spintronics is underpinned by worldwide research on molecular magnetism and supramolecular chemistry. Indeed, chemists have acquired a strong expertise in tuning, controlling and manipulating the properties of the molecules (spin, anisotropy, redox potential, light, electrical field...) allowing the creation of tuneable devices with new functionalities. This chapter summarizes the concepts and the first important results in this new research area, which open up prospects for new spintronic devices with quantum properties.

13.1 Introduction

Everyday life is full of useful magnets, solids, oxides, metals and alloys. On the contrary, molecules are most often considered as non-magnetic materials. However, recent discoveries show that molecules can bear large magnetic moments that can have a stable orientation like traditional magnets. They have therefore been called single-molecule magnets (SMMs) and they might be the ultimate limit for information storage. They do not only exhibit the classical macroscale property of a magnet, but also new quantum properties such as quantum tunnelling of magnetization and quantum phase interference, the properties of a microscale entity. Such quantum phenomena are advantageous for some challenging applications, e.g. molecular information storage or quantum computing. In this context, the objective of molecular quantum spintronics is to combine the concepts of three novel disciplines, spintronics, molecular electronics, and quantum computing. The resulting research field

M. Ganzhorn · W. Wernsdorfer (✉)
Institut Néel, CNRS & Université J. Fourier, BP 166, 25 rue des Martyrs, 38042 Grenoble Cedex 9, France
e-mail: wolfgang.wernsdorfer@grenoble.cnrs.fr

J. Bartolomé et al. (eds.), *Molecular Magnets*, NanoScience and Technology,
DOI 10.1007/978-3-642-40609-6_13,

aims at manipulating spins and charges in electronic devices containing one or more molecules [1]. The main advantage is that the weak spin-orbit and hyperfine interactions in organic molecules are likely to preserve spin-coherence over times and distances much longer than in conventional metals or semiconductors. In addition, specific functions (e.g. switchability with light, electric field etc.) could be directly integrated into the molecule [2].

This chapter summarizes the concepts and the first important results in this new research area of molecular quantum spintronics. It first discusses briefly the motivations for using molecular nanomagnets as magnetic center of spintronic devices. Several designs of supramolecular quantum spintronic devices are then presented, which are able to probe an individual molecular spin. The main part of the chapter focuses on a single-molecule magnet with a single magnetic center. It is a terbium ion Tb^{3+}, embedded between two phtalocyanines (Pc) ligand planes. The mononuclear complex is denoted as $TbPc_2$. We then describe various experimental approaches that have successfully probed the quantum mechanical nature of an isolated $TbPc_2$ single-molecule magnet. Landau-Zener tunneling, spin-lattice relaxation, single shot electrical read-out of a single nuclear spin, nuclear spin trajectories, level life times, and Rabi oscillations of a single nuclear spin are discussed in detail. The results contribute to the understanding of the electronic and magnetic properties of isolated molecular systems and they reveal intriguing new physics.

13.2 Molecular Nanomagnets for Molecular Spintronics

SMMs possess the right chemical characteristics to overcome several problems associated to molecular junctions. They are constituted by an inner magnetic core with a surrounding shell of organic ligands [3] that can be tailored to bind them on surfaces or into junctions [4–7] (Fig. 13.1). In order to strengthen magnetic interactions between the magnetic core ions, SMMs often have delocalized bonds, which can enhance their conducting properties. SMMs come in a variety of shapes and sizes and permit selective substitutions of the ligands in order to alter the coupling to the environment [3–5, 8]. It is also possible to exchange the magnetic ions, thus changing the magnetic properties without modifying the structure and the coupling to the environment [9, 10]. While grafting SMMs on surfaces has already led to important results, even more spectacular results will emerge from the rational design and tuning of single SMM-based junctions.

From a physics viewpoint, SMMs are the final point in the series of smaller and smaller units from bulk matter to atoms (Fig. 13.2). They combine the classic macroscale properties of a magnet with the quantum properties of a nanoscale entity. They have crucial advantages over magnetic nanoparticles in that they are perfectly monodisperse and can be studied in molecular crystals. They display an impressive array of quantum effects (that are observable up to higher and higher temperatures due to progress in molecular designs), ranging from quantum tunnelling of magnetization [11–14] to Berry phase interference [15, 16] and quantum coherence [17–19] with important consequences on the physics of spintronic devices. Although the magnetic properties of SMMs can be affected when they are

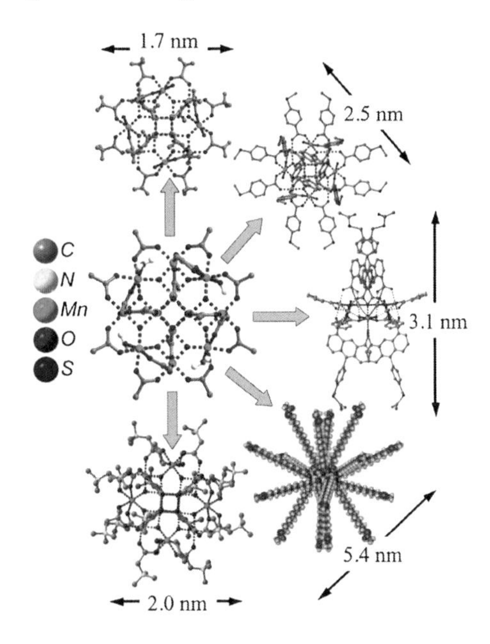

Fig. 13.1 Representative examples of the peripheral functionalization of the outer organic shell of the Mn_{12} SMM. Different functionalizations used to graft the SMM to surfaces are displayed [1]. Solvent molecules have been omitted. The atom color code is reported in the figure, as well as the diameter of the clusters

deposited on surfaces or between leads [8], these systems remain a step ahead of non-molecular nanoparticles, which show large size and anisotropy distributions, for a low structure versatility.

13.3 Introduction to Molecular Spintronics

Various designs for molecular spintronic devices using individual SMM's were proposed over the last decade [1, 2]. One can use for instance a scanning tunneling microscopy to probe an isolated SMM on a conducting surface [20–22]. Alternatively, one can built a three-terminal molecular spin-transistor where an individual SMM is bridging the gap between two non-magnetic leads [23, 24]. In such a configuration, the electric current is flowing directly through the molecule, leading to a strong coupling between the electrons and the magnetic core (see the Chap. 12). This *direct* coupling thus enables a readout of the molecule's magnetic properties with the electronic current, but also leads to a strong back-action on the molecule's magnetic core [1].

An less invasive approach consists in coupling the SMM to a second non-magnetic molecular conductor which is subsequently used as detector. For such an *indirect* coupling, the magnetic core of the molecule is only weakly coupled to the conductor but can still affect its transport properties, thus enabling an electronic readout with only minimal back-action.

Among the different possible detectors (nanowires, carbon nanotubes, quantum dots, ligands, mechanical resonators, nanoSQUIDs), the carbon nanotube stands out

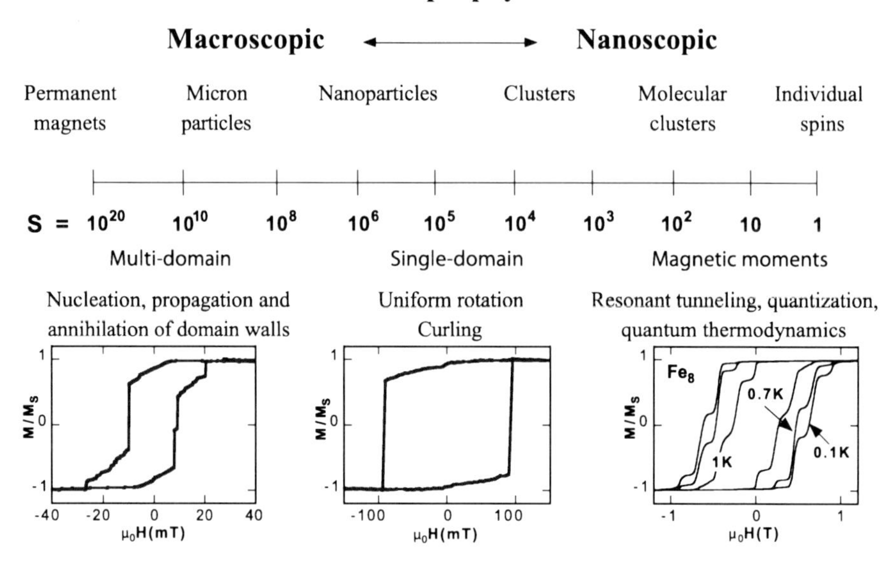

Fig. 13.2 Scale of size that goes from macroscopic down to nanoscopic sizes. The unit of this scale is the number of magnetic moments in a magnetic system (roughly corresponding to the number of magnetic atoms). At macroscopic sizes, a magnetic system is described by magnetic domains that are separated by domain walls. Magnetization reversal occurs via nucleation, propagation, and annihilation of domain walls (hysteresis loop on the *left*). When the system size is of the order of magnitude of the domain wall width or the exchange length, the formation of domain walls requires too much energy. Therefore, the magnetization remains in the so-called single-domain state, and the magnetization reverse by uniform rotation or nonuniform modes (*middle*). SMMs are the final point in the series of smaller and smaller units from bulk matter to atoms and magnetization reverses via quantum tunneling (*right*)

due to its unique structural, mechanical and electronic properties [1]. We will propose several designs of a *supramolecular quantum spintronic* device, where a carbon nanotube is used to probe an individual molecular spin *via* different coupling mechanism at cryogenic temperatures (flux coupling, electronic coupling, mechanical coupling).

13.3.1 Direct Coupling Scheme

The first scheme we consider is a magnetic molecule attached between two non-magnetic electrodes. One possibility is to use a scanning tunneling microscope tip as the first electrode and the conducting substrate as the second one. So far, only few atoms on surfaces have been probed in this way, revealing interesting Kondo effects [25] and single-atom magnetic anisotropies [26]. The next scientific step is to move from atoms to molecules in order to observe richer physics and to modify

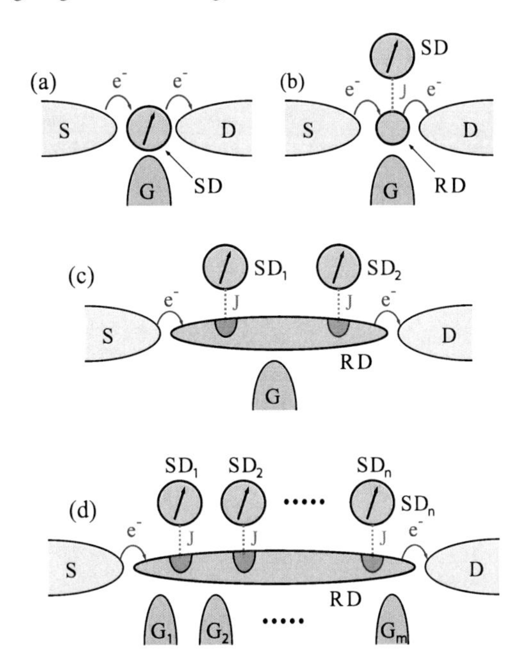

Fig. 13.3 Schematic representation of different device geometries for molecular spintronic devices. (**a**) Three-terminal spin dot device. In a direct coupling scheme, the current flows through the spin dot (SD). (**b**) Three-terminal double-dot device. In an indirect coupling scheme, the current flows through a second non-magnetic quantum dot, the readout dot (RD), which is coupled with the spin dot (SD) via exchange interaction (J). (**c**) Supramolecular spin valve device, with two SD coupled to the RD. The current in the RD is therefore sensitive to relative spin orientation in the two SD's. (**d**) Multi-terminal multi-dot device

the properties of the magnetic objects. Although isolated SMMs on gold have been obtained [4–7], the rather drastic experimental requirements, i.e. very low temperatures and high magnetic fields, represents a considerable technological challenge. Nevertheless, recent experiments performed by low temperature STM revealed a reversible chiral switching [21] and an electric current control of a local spin [20] in a bis(phthalocyaninato)terbium(III) SMM ($TbPc_2$ SMM) on a metal surface. Spin-polarized STM furthermore allows the real-space observation of spin-split orbitals in a $TbPc_2$ SMM [22].

Another possibility concerns break-junction devices [27], which integrate a gate electrode. Such a three-terminal transport device, called a molecular spin-transistor, consists of a single magnetic molecule, the "spin" dot, bridging the gap between two non-magnetic electrodes (Fig. 13.3(a)). The current passes through the magnetic molecule via the source and drain electrodes, and the electronic transport properties are tuned via a gate electrode.

The first experimental realization of this scheme has been achieved using Mn_{12} with thiol-containing ligands, which bind the SMM to the gold electrodes with strong and reliable covalent bonds [28]. An alternative route is to use short but weak-binding ligands [29]: in both cases the peripheral groups act as tunnel barriers and help conserving the magnetic properties of the SMM in the junction. As the electron transfer involves the charging of the molecule, we must consider, in addition to the neutral state, the magnetic properties of the negatively- and positively-charged species. Because crystals of the charged species can be obtained, SMMs permit direct comparison between spectroscopic transport measurements and more traditional characterization methods. In particular, magnetization measurements, electron paramagnetic resonance, and neutron spectroscopy can provide energy level spacings and anisotropy parameters. In the case of Mn_{12}, positively charged clusters possess a lower anisotropy barrier [30]. As revealed by the first current spectroscopy (Coulomb blockade) measurements, the presence of these states is fundamental to explain transport through the clusters [28, 29].

Studies in magnetic field showed a first evidence of the spin transistor properties [28, 29]. Degeneracy at zero field and nonlinear behavior of the excitations as a function of field are typical of tunneling via a magnetic molecule. However, follow up experiments indicate an alteration of the Mn_{12} properties during the deposition of the molecules on gold electrodes [31].

In contrast to Mn_{12} molecular magnets, the magnetic properties of Fe_4 SMM's are preserved upon deposition on gold electrodes due to the protection of the magnetic centers by an outer shell of organic ligands. Indeed, the magnetic anisotropy of the Fe_4 molecule has been observed *via* current spectroscopy in single molecule transistors with various electron transport regimes (SET, co-tunneling, Kondo) [32, 33].

13.3.2 Indirect Coupling Scheme

In the direct coupling scheme, the conduction electrons are tunneling directly through the magnetic center of the SMM. Although the strong coupling between the electrons and the magnetic center enables the detection of the molecular spin by the conduction electrons, a significant back-action acts on the molecule's magnetic core.

In order to reduce the back-action, one can couple the magnetic molecule to a second non-magnetic conductor serving as a readout dot (Fig. 13.3(b)). In this indirect coupling regime, the spin dot is only weakly coupled to the readout dot via an exchange interaction but can still affect its transport properties, thus enabling a readout of the molecular spin with only minimal back-action. Such a configuration was recently demonstrated in a $TbPc_2$ SMM spin transistor [23, 24], where the current tunneling through the organic ligand of the $TbPc_2$ (i.e readout dot) enables the detection of the nuclear spin on the magnetic Tb^{3+} ion core (i.e spin dot).

A molecular spin valve has a geometry similar to a spin transistor but contains at least two magnetic elements or spin dots. In analogy to a polarizer-analyzer setup, the current in the spin valve device depends on the relative spin orientation of the magnetic centers. In the simplest geometry, a molecular spin valve consists of a diamagnetic molecule integrated between two magnetic electrodes. A first experimental realization on a C_{60} molecule between two Ni electrodes reveals for instance considerable magnetoresistance effects [34]. Other geometries have also been proposed in the past, involving for example a magnetic molecule coupled to only one magnetic electrode or a molecular magnet with two magnetic centers sandwiched between non-magnetic electrodes [2, 35].

The most promising alternative however consists of coupling two molecular magnets to a readout dot connected between two non-magnetic electrodes (Fig. 13.3(c)).

The localized and highly anisotropic magnetic moment of the molecular magnets can influence the electric current in the readout dot via exchange interaction, which leads to a highly efficient spin polarization or very large magnetoresistance effects. Hence, a spin valve configuration enables the non-destructive readout of molecular spin states [1]. Among the different potential readout dots (nanowires, nanotubes, semiconductor quantum dots, mechanical resonators), carbon nanotube stands out due outstanding mechanical and electronic properties. In fact, the carbon nanotube is a one-dimensional molecular conductor with a cross section on the same order of magnitude than that of a molecular magnet, resulting in an almost ideal coupling between the readout and the spin dot. Indeed, in a supramolecular spin valve consisting of $TbPc_2$ SMM grafted to a carbon nanotube, the strongly localized magnetic moment of the SMM's leads to a magnetic field dependance of the electrical current in the carbon nanotube, resulting in a magnetoresistence ratio up to 300 % below 1 K [36].

Chemical engineering furthermore allows the synthesis of billions of perfectly identical molecular magnets. One can therefore functionalize a readout dot like a carbon nanotube with several molecular magnets and, using state-of-art nanofabrication techniques, integrate multiple gate electrodes in such a device (Fig. 13.3(d)). In such a multi-terminal device configuration, each molecular magnet can be addressed and manipulated independantly, for example by oscillating electric or magnetic fields. Furthermore, the gate electrodes can be used to modulate the exchange interaction between a spin dot and the readout dot. Using the readout dot as a mediator or quantum-information bus, one could therefore perform a highly efficient spin or quantum information transfer between two different spin dots. With the strong quantum coherence of certain molecular magnets, such a device configuration would enable quantum information processing and the implementation of certain quantum computation protocols [1].

13.3.3 Magnetic Torque Detector or Probing Via Mechanical Motion

Alternatively, one can also couple a SMM to a suspended carbon nanotube NEMS and probe the molecular nanomagnet with the carbon nanotube's mechanical motion (Fig. 13.4). Such a detection scheme is based on torque magnetometry.

We consider a SMM with a magnetic moment $\boldsymbol{\mu} = \frac{g\mu_B}{\hbar}\boldsymbol{S}$ to be rigidly grafted to the suspended carbon nanotube beam. Upon applying an external magnetic field $\boldsymbol{B}_{ext}$, the SMM magnetization will experience a torque given by

$$\boldsymbol{\Gamma}_{SMM} = \boldsymbol{\mu} \times \boldsymbol{B}_{ext} = \frac{g\mu_B}{\hbar}\boldsymbol{S} \times \boldsymbol{B}_{ext}, \qquad (13.1)$$

resulting in its rotation towards the magnetic field direction. In order to minimize its magnetic anisotropy energy the SMM starts to rotate, hence inducing a bending and mechanical strain in the suspended carbon nanotube beam. The additional tension in

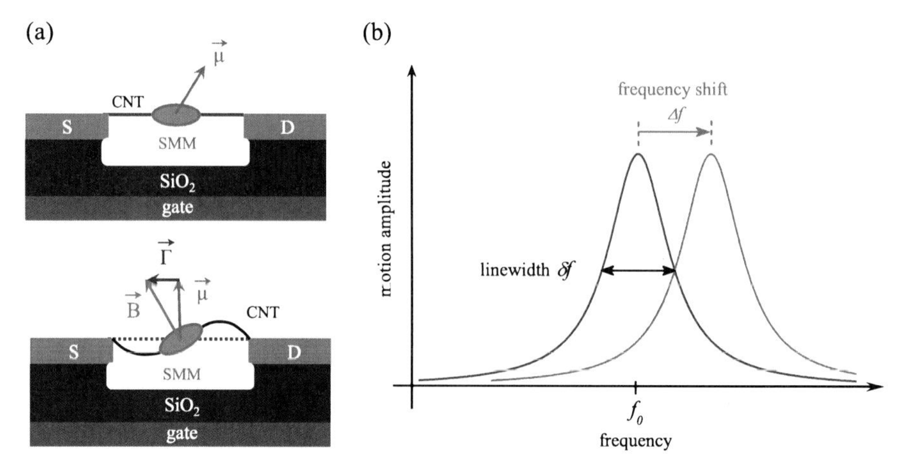

Fig. 13.4 Carbon nanotube NEMS as magnetometer for molecular magnets. (**a**) Schematic representation of a molecular magnet (*blue ellipse*) grafted to a carbon nanotube NEMS. The carbon nanotube is suspended between two non-magnetic electrodes over trench etched in a SiO_2/Si substrate. Under the influence of a magnetic field $\boldsymbol{B}$ (*green arrow*), the molecular magnet yielding a magnetic moment $\boldsymbol{\mu}$ (*blue arrow*) will experience a magnetic torque $\vec{\Gamma}$ (*red arrow*), given by equation (13.1). By changing the magnetic field $|\boldsymbol{B}|$, one can therefore induce a change in the molecule's magnetic torque $|\boldsymbol{\Gamma}|$. If the molecular magnet is rigidly grafted to the carbon nanotube NEMS, a change of the torque $\Delta|\boldsymbol{\Gamma}|$ will induce a bending and an additional tension in the carbon nanotube beam, resulting in a shift of its resonance frequency $f_0 \rightarrow f_0 + \Delta f$. (**b**) Amplitude of the nanotube motion as a function of frequence, before and after the magnetization reversal of the magnetic moment

the resonator will result in a shift the mechanical resonance frequency of the carbon nanotube, see Fig. 13.4(b).

Lassagne et al. studied the mechanical response of a carbon nanotube NEMS to the magnetization dynamics of a nanomagnet with a uniaxial magnetic anisotropy and a magnetic moment of $100\mu_B$ [37]. They determined the magnetic hysteresis of the nanomagnet (Stoner-Wolfahrt model, Fig. 13.5(a)) and the magnetic field dependance of the carbon nanotube NEMS resonance frequency (Euler-Bernoulli formalism, Fig. 13.5(b)) for different orientations of the magnetic field with respect to the nanomagnets easy axis. The model can be readily extended to a SMM, with a magnetic moment of a few μ_B.

The calculations reveal a discontinous jump in the nanotube's resonance frequency, induced by the magnetization reversal of the nanomagnet (highlighted by the black arrows in Fig. 13.5(b)). Furthermore, the field dependance of the resonance frequency reflects the hysteretic behaviour of the nanomagnet as well as its magnetic anistropy. For a magnetic field aligned close the nanomagnet's easy axis, a large hysteresis in the magnetization m and the frequency Δf is visible ($\theta_0 = \pi/50$, red loop in Fig. 13.5(a) and (b)). The hysteresis gradually disappears upon rotating the magnetic field into the hard plane of the molecule ($\theta_0 = \pi/2$, black loop in Fig. 13.5(a) and (b)). The maximum frequency shift reaches 90 kHz for a magnetic moment of $100\mu_B$.

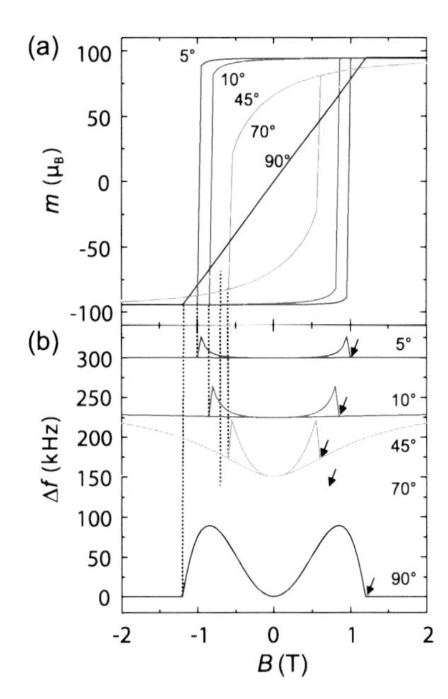

Fig. 13.5 Mechanical response of carbon nanotube NEMS to the magnetization reversal of a nanomagnet with uniaxial anisotropy. (**a**) Magnetic hysteresis loop of the nanomagnet for five different orientations of the magnetic field, with respect to its easy axis of magnetization $\theta_0 = 90°, 70°, 45°, 10°, 5°$. One can observe the uniaxial anisotropy upon rotating the magnetic field away from the easy axis. (**b**) Frequency shift Δf as a function of the magnetic field. *The curves* translate the hysteretic behavior and the magnetic anisotropy of the nanomagnet. Modified figure from [37]

The sensitivity for such a magnetic torque detector is limited by the frequency noise induced by thermomechanical fluctuations and yields

$$\delta f_{\text{th}} = \frac{1}{2\pi}\left(\frac{k_B T}{k x_0^2}\frac{2\pi f_0 f_{BW}}{Q}\right)^{1/2} \tag{13.2}$$

For a resonance frequency $f_0 = 50$ MHz and quality factor $Q = 10^5$, [37, 38] a spring constant $k \approx 10^{-4}$ N/m of the carbon nanotube NEMS, [37] one obtains a sensitivity of $\delta f_{\text{th}} = 150\ \text{Hz}/\sqrt{\text{Hz}}$ at 40 mK.

Furthermore, the carbon nanotube NEMS should provide a strong coupling with an individual molecular magnet in order to achieve the maximum sensitivity δf_{th}. Indeed, previous experiments on SMM grafted to a carbon nanotube NEMS revealed a strong coupling on the order of 1 MHz between an individual molecular spin and the nanotube's mechanical motion [39].

One should therefore be able to reach a sensitivity of $1\mu_B$ at cryogenic temperatures with such a carbon based torque magnetometer, thus providing a mechanical readout scheme for a single (molecular) spin.

13.3.4 NanoSQUID or Probing Via Magnetic Flux

Finally, one can also probe the magnetic flux emanating from a single molecule magnet grafted onto a SQUID magnetometer. The maximum magnetic flux generated by SMM's like lanthanide complexes $LnPc_2$ was estimated to be on the order

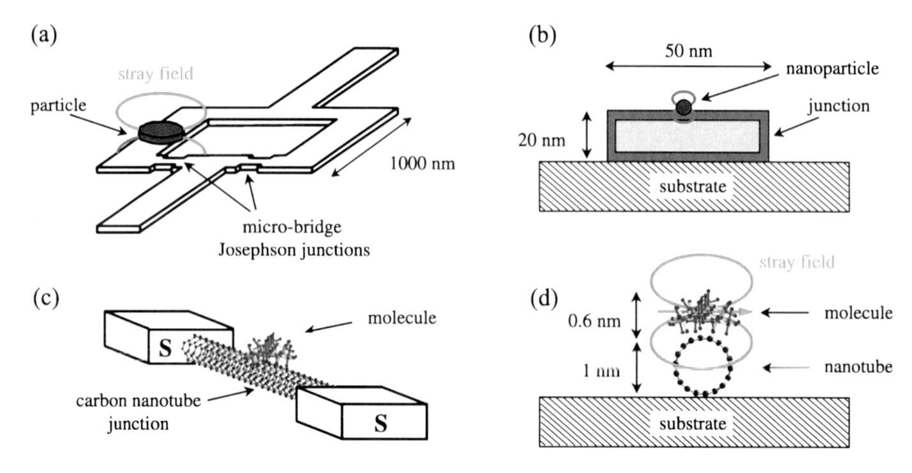

Fig. 13.6 Schemes of the flux coupling between a magnetic particle or a molecule and a SQUID. (**a**) Schematic drawing of a planar nano-bridge DC-SQUID on which a ferromagnetic particle is placed. The SQUID detects the flux through its loop produced by the sample magnetisation. Due to the close proximity between sample and SQUID, a very efficient and direct flux coupling is achieved. (**b**) Cross-section (50×20 nm^2) of a nano-bridge junction on which a 3-nm-sized particle is placed. The flux coupling is rather poor because of the large mismatch between the particle size and the junction cross-section. (**c**) Schematics of one of the two carbon nanotube junctions of a nano-SQUID. A nanometer-sized molecule sits on top of the carbon nanotube. (**d**) Cross-section (1 nm^2) of a single-walled carbon nanotube junction on which a 0.6-nm-sized molecule is placed. The flux coupling is optimized because the molecule size and the junction cross-section are comparable

of $\Phi_{\mathrm{SMM}} \approx 10^{-5}\Phi_0$, which is well within the flux sensitivity of a SQUID, given by the quantum limit $\Phi_q \approx 10^{-8}\Phi_0$ [40, 41].

However, only the magnetic flux penetrating the SQUID's cross section will contribute to the measured flux quantity in the SQUID loop. The cross section of a SQUID is usually orders of magnitude larger than the size of the molecule, which typically results in a very weak flux coupling in this case (Fig. 13.6(a)–(b)).

In order probe the flux of a single molecule it is therefore essential to reduce the cross section of the SQUID, ideally to the same size than the SMM. Indeed, the cross section of a carbon nanotube is on the order of 1 nm^2 which is comparable to the size of the molecule (Fig. 13.6(c)–(d)). One obtains an almost ideal flux coupling in a carbon nanotube SQUID with a flux sensitivity of $S_\Phi \approx 10^{-4}\Phi_0/\sqrt{\mathrm{Hz}}$ [40, 41].

Although the flux of an individual SMM is within the NanoSQUID's sensitivity at measurement frequency about a few hundreds of Hz, only preliminary results have been published so far [42].

13.4 Magnetism of the TbPc$_2$ Molecular Nanomagnet

In this section we describe a SMM with a single magnetic center, in this case a terbium ion Tb^{3+}, embedded between two phtalocyanines (Pc) ligand planes

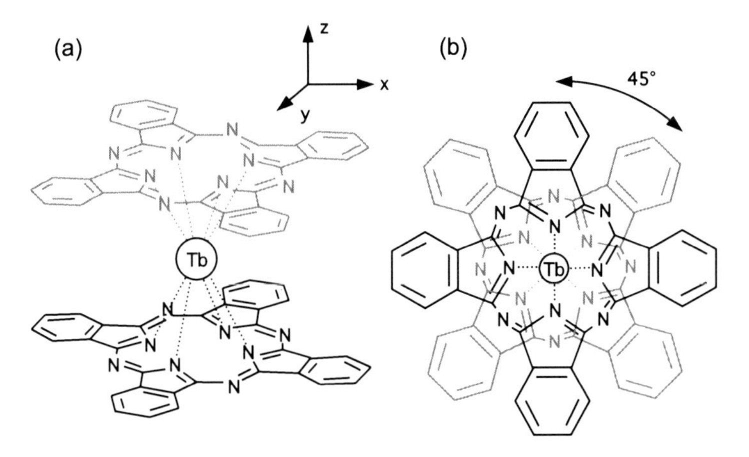

Fig. 13.7 Mononuclear Terbium complex $TbPc_2$. Figure from [10] (**a**) Side view and (**b**) top view of the molecule. The upper Pc ligang (*light grey*) is a mirror image of the lower Pc ligand (*black*), rotated by 45°

(Fig. 13.7). The mononuclear complex will be denoted as $TbPc_2$ in the following and was designed, synthesized and characterized by Svetlana Klyatskaya and Mario Ruben at the Karlsruhe Institute of Technology (KIT) in Germany [43].

13.4.1 Molecular Structure

The magnetism of the terbium(III) or other lanthanide(III) is mainly determined by the strongly anisotropic and partially filled $4f$ orbitals in these ions. A Tb^{3+} ion exhibits an electronic structure of $[Xe]4f^8$ which corresponds to a spin of $S = 3$ and an orbital momentum of $L = 3$. Due to the inherentely strong spin-orbit coupling in rare earth atoms, L and S are no good quantum numbers. In consequence, the magnetic moment is described by a total angular momentum $J = L - S \ldots L + S$, where $L + S$ is the ground state for more than half-shell filling (Hund's rule). The ground state of the Tb^{3+} ion thus yields $J = L + S = 6$ and is separated from the first excited state $J = 5$ by an energy of 2900 K due to the large spin-orbit coupling in the Tb^{3+} ion [44]. We can therefore restrict the discussion to the ground state multiple $J = 6$, yielding $2J + 1$ (degenerate) substates $|J, J_z\rangle$.

13.4.2 Spin Hamiltonian

The Tb^{3+} ion is embedded between two parallel Pc ligand planes, with a quantization axis z oriented perpendicular to the Pc ligand planes. The Tb^{3+} ion is coordinated by 8 nitrogen atoms and the upper Pc ligand (blue in Fig. 13.7) is a mirror

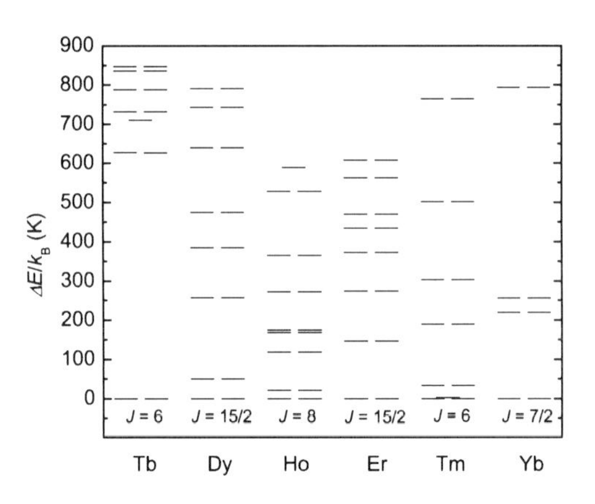

Fig. 13.8 Ligand field splitting of the ground state multiplet J in different mononuclear lanthanide complexes $LnPc_2$. Modified from [45]

reflection of the lower Pc ligand with respect to the (x-y) plane (black in Fig. 13.7), rotated by 45° around the z-axis. The Tb^{3+} is therefore exposed to a ligand electric field with a antiprismatic symmetry D_{4d} which is described by the following Hamiltonian [45]

$$\mathcal{H}_{\mathrm{lf}} = \alpha A_2^0 O_2^0 + \beta \left(A_4^0 O_4^0 + A_4^4 O_4^4 \right) + \gamma \left(A_6^0 O_6^0 + A_6^4 O_6^4 \right) \tag{13.3}$$

where $\alpha = -1/99$, $\beta = 2/16335$ and $\gamma = -1/891891$ are constant parameters related to the ion [46], O_q^k the equivalent Stevens operators and A_q^k the ligand field parameters.

The Stevens operators O_q^k are defined as linear combinations of J_z, J_-, J_+ and are listed in Refs. [44, 46]. The ligand field parameters A_q^k can be determined experimentally by NMR or magnetic susceptibility, yielding [47]

$$\begin{aligned} &A_2^0 = 595.7\ \mathrm{K} \\ &A_4^0 = -328.1\ \mathrm{K}, \qquad A_4^4 = 47.5\ \mathrm{K} \\ &A_6^0 = 14.4\ \mathrm{K}, \qquad A_6^4 = 0 \end{aligned} \tag{13.4}$$

The diagonalization of $\mathcal{H}_{\mathrm{cf}}$ in the $|J, m_J\rangle$ eigenbasis then reveals that the ligand field partially lifts the degeneracy of the $2J + 1$ substate in the ground state multiplet $J = 6$. Indeed, the degenerate ground state doublet $m_J = \pm 6$ of the Tb^{3+} ion is now separated from the first excited state doublet $m_J = \pm 5$ by approximately 600 K (see Ref. [45] and Fig. 13.8). The $TbPc_2$ single molecule magnet thus behaves as an Ising spin system at cryogenic temperatures.

Furthermore, each m_J-doublet splits in an external magnetic field due to the Zeeman effect as described by the following Hamiltonian

$$\mathcal{H}_{\mathrm{TbPc_2}} = \mathcal{H}_{\mathrm{lf}} + g \mu_B \mu_0 H_z J_z \tag{13.5}$$

where $g = 3/2$ is the g-factor of the terbium. The resulting Zeeman diagram is depicted in Fig. 13.9 [10].

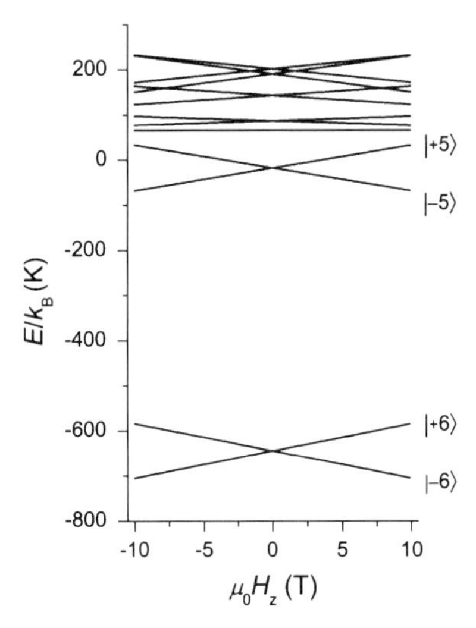

Fig. 13.9 Calculated Zeeman diagram for the ground state multiplet $J = 6$ in a $TbPc_2$ SMM extracted from [10]

In the following we will therefore restrict the discussion to the ground state doublet $m_J = \pm 6$.

The Tb^{3+} ion also carries a nuclear spin of $I = 3/2$ with a natural abundance of 100 %, yielding $(2I + 1)$ substates $m_I = 3/2, 1/2, -1/2, -3/2$. Due to a strong hyperfine interaction between the nuclear spin $I = 3/2$ and the electronic angular momentum $J = 6$, the Hamiltonian of $TbPc_2$ contains two additional terms corresponding to nuclear dipole and quadrupole interactions, respectively,

$$\mathcal{H}_{\mathrm{TbPc}_2} = \mathcal{H}_{\mathrm{lf}} + g\mu_{\mathrm{B}}\mu_0 H_z J_z + A_{\mathrm{dip}}\boldsymbol{I} \cdot \boldsymbol{J} + P_{\mathrm{quad}}\left(I_z^2 + \frac{1}{3}I(I+1)\right) \quad (13.6)$$

$$\boldsymbol{I} \cdot \boldsymbol{J} = J_z I_z + \frac{1}{2}(J_+ I_- + J_- I_+) \quad (13.7)$$

where $A_{\mathrm{dip}} = 24.5$ mK is the hyperfine constant and $P_{\mathrm{quad}} = 14.4$ mK the quadrupole constant [10].

The dipolar term $A_{\mathrm{dip}}\boldsymbol{I} \cdot \boldsymbol{J}$ thus splits the electronic states from the ground state doublet $m_J = \pm 6$ into the four nuclear spin states $m_I = 3/2, 1/2, -1/2, -3/2$. The quadrupole term $P_{\mathrm{quad}} I_z^2$ results in a non-equidistant spacing of the nuclear spin states. The *excited* nuclear spin states thus have energies of 120 mK, 270 mK and 450 mK with respect to the nuclear spin *ground* state (Fig. 13.10) [10].

Finally, it should be noted that the resulting intersections are indeed energy level crossings except the four intersections that conserve the nuclear spin (black circles Fig. 13.10). For the latter ones, an anti-crossing on the order of a few μK is observed [10]. These anti-crossings are due to the transverse anisotropy term O_4^4 in (13.3) which yields off diagonal terms in the ligand field Hamiltonian. In fact, the operator O_4^4 is a linear combination of J_+^4 and J_-^4, which mixes states with m_J and states with $m_J \pm 4n$. Hence, the $m_J = +6$ state mixes with the $m_J = -6$ state, generating an avoided level crossing at the four highlighted intersections close to zero magnetic field (Fig. 13.10).

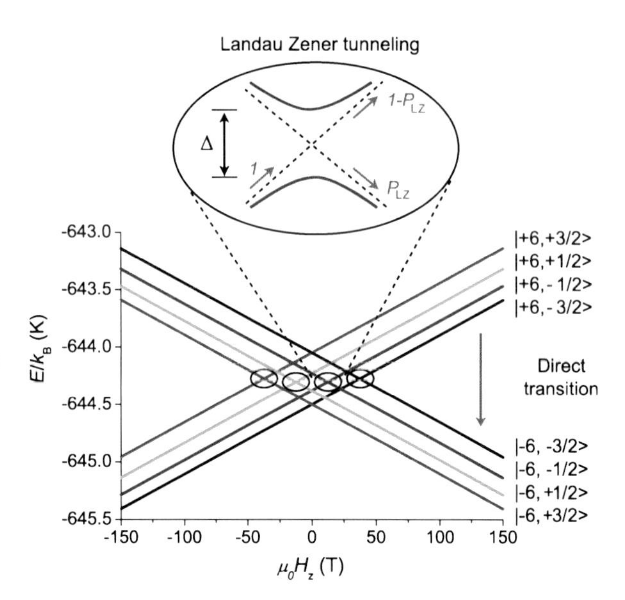

Fig. 13.10 Zeeman diagram for the ground state multiplet $J = 6$ in a $TbPc_2$ SMM. Avoided level crossings are highlighted by *the black circles*. The enlarged region shows an avoided level crossing with tunnel splitting Δ caused by the mixing of the $m_J = -6$ and $m_J = +6$ states. *The dotted lines* shows the two states without mixing and the blue arrows depict the QTM process. Phonon assisted relaxation or direct transitions of the $TbPc_2$ magnetization occurs at higher magnetic field

By sweeping the magnetic field applied to the $TbPc_2$ SMM one can then induce a magnetization reversal from $|m_J = -6, m_I\rangle$ to $|m_J = +6, m_I\rangle$ (and *vice versa*) either by QTM at one of the four avoided level crossings around zero field or by phonon-assisted or spin-lattice relaxation at large magnetic fields (Fig. 13.10).

Figure 13.11 shows a magnetic hysteresis loop measured by a microSQUID technique [48] on an assembly of $TbPc_2$ SMMs arranged in a matrix of non-magnetic YPc_2 SMM, with a $[TbPc_2]/[YPc_2]$ ratio of 2 %. Upon sweeping the magnetic field from -1 T up to 1 T, approximately 75 % of the SMM in the crystal reverse their magnetization by QTM around zero field, resulting in sharp steps in the crystal's magnetization. The remaining SMM reverse their magnetization by phonon-assited relaxation at larger magnetic fields.

13.4.3 Quantum Tunneling of Magnetization and Landau-Zener Model

At the four avoided level crossings highlighted by the black circles in Fig. 13.10, the nuclear spin is conserved and the small transverse anisotropy mixes the state $m_J = -6$ with the state $m_J = +6$, due to the J_+^4 and J_-^4 terms in (13.3). By sweeping the magnetic field through one of these avoided crossing one can therefore tunnel from the state $m_J = -6$ to the state $m_J = +6$ (and *vice versa*). Such a process is called quantum tunneling of magnetization (QTM).

According to Landau and Zener [49], the probability P_{QTM} for such a QTM process depends on the magnitude of the mixing, i.e. the tunnel splitting Δ, and the

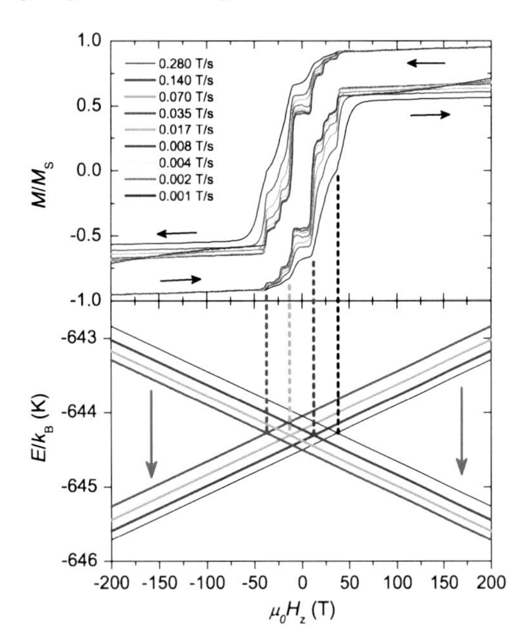

Fig. 13.11 Magnetic hysteresis loop of an assembly of $TbPc_2$ SMM measured with a microSQUID in a diluted single crystal (2 % $TbPc_2$ in an YPc_2 matrix) for different field sweep rates and temperature of 40 mK (*top panel*). Zeeman diagram of a $TbPc_2$ SMM (*bottom panel*). The characteristic steps in the magnetization curve around zero field correspond to QTM events in the single crystal (highlighted by the *dashed lines*), whereas the continuous change of the magnetization at high field is attributed to phonon-assisted relaxation (*blue arrows* in the Zeeman diagram)

sweep rate of the magnetic field ν, thus yields

$$P_{LZ} = 1 - \exp\left(-\alpha \frac{\Delta^2}{\nu}\right) \qquad (13.8)$$

For a given (nonzero) Δ, if the magnetic field is swept adiabatically slow in the vicinity of the avoided crossing, QTM from $m_J = -6$ to $m_J = +6$ (and inversely) will occur with a high probability. One observes the characteristic steps in the magnetization curves, as described in the previous section (dark blue cycle Fig. 13.11). As the magnetic field sweep rate increases the probability P_{QTM} for QTM becomes exponentially smaller. The steps associated with QTM become less pronounced and eventually vanish at large sweep rates (black cycle Fig. 13.11).

If $\Delta = 0$, the levels cross and no quantum tunneling of magnetization can occur irrespective of the sweep rate.

13.4.4 Spin-Lattice Relaxation

Alternatively, the magnetization reversal of the $TbPc_2$ SMM from $|m_J = -6, m_I\rangle$ to $|m_J = +6, m'_I\rangle$ can occur by phonon-assisted or spin-lattice relaxation (Fig. 13.10). Spin-lattice relaxation in an SMM is due to the modulation of the molecule's ligand electric field by phonons from a surrounding thermal bath, for example the lattice vibration in a diluted SMM single crystal [44]. Due to the strong spin-orbit coupling, the modulation of the ligand field will result in transitions between different spin

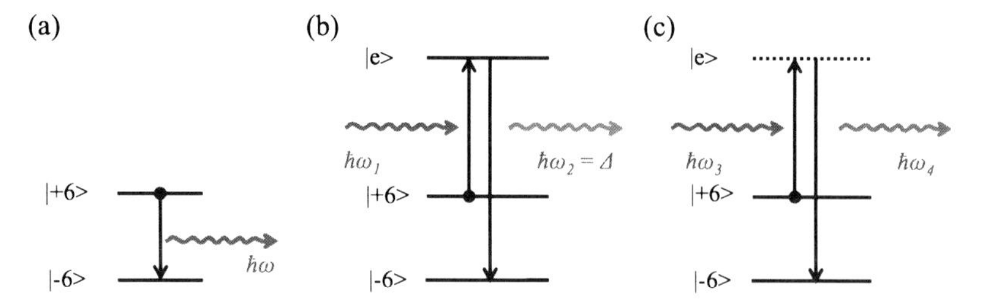

Fig. 13.12 Spin-lattice relaxation processes. (**a**) Direct relaxation into the ground state involving the emission of a phonon with an energy $\hbar\omega$. (**b**) Two-phonon Orbach process. The electron is excited into a vibrational (lattice) mode by absorption of a phonon $\hbar\omega_1$, before relaxing into the ground state under the emission of a phonon $\hbar\omega_2$ with $\hbar\omega = \hbar\omega_2 - \hbar\omega_1$. (**c**) Two-phonon Raman process. The relaxation occurs by the absorption and emission of virtual phonons. The hyperfine splitting was omitted for simplicity

states of the Tb^{3+} ion. The magnetization reversal can be assisted by one or multiple phonons [44] (see Fig. 13.12).

Direct Relaxation Process At very low temperatures, the magnetization of the $TbPc_2$ SMM is reversed in a direct relaxation process under the emission of a phonon into the surrounding bath (Fig. 13.12(a)). The relaxation rate $1/\tau$ for a such a direct relaxation can therefore be expressed in terms of the absorption and emission rates of the phonon, given by Fermi's golden rule

$$\begin{aligned} \omega_{\text{abs}} &= \frac{2\pi}{\hbar}|\langle -6|\mathcal{H}'|+6\rangle|^2 \\ \omega_{\text{em}} &= \frac{2\pi}{\hbar}|\langle -6|\mathcal{H}'|+6\rangle|^2 \exp\left(\frac{\hbar\omega}{k_B T}\right) \end{aligned} \tag{13.9}$$

where $\langle -6|\mathcal{H}'|+6\rangle$ is the matrix element of the pertubation Hamiltonian $\mathcal{H}'$ between the states $|m_J = -6\rangle$ and $|m_J = +6\rangle$, $\hbar\omega$ the phonon energy and T the temperature of the phonon bath. The relaxation rate then holds

$$\frac{1}{\tau} = \omega_{\text{em}} + \omega_{\text{abs}} = \frac{2\pi}{\hbar}|\langle -6|\mathcal{H}'|+6\rangle|^2 \left[1 + \exp\left(\frac{\hbar\omega}{k_B T}\right)\right] \tag{13.10}$$

The relaxation from $|m_J = -6\rangle$ to $|m_J = +6\rangle$ is induced by the modulation of the ligand field by phonons, and the modified ligand field Hamiltonian yields [44]

$$\mathcal{H}_{\text{lf}} = \mathcal{H}_{\text{lf}}^0 + \epsilon\mathcal{H}_{\text{lf}}^1 + \epsilon^2\mathcal{H}_{\text{lf}}^2 \tag{13.11}$$

where ϵ refers to the mechanical strain in the ligand.

In first order, the pertubation Hamiltonian gives $\mathcal{H}' = \epsilon\mathcal{H}_{\text{lf}}^1$ and (13.10) transforms to

$$\frac{1}{\tau} = \frac{2\pi}{\hbar}\epsilon^2|\langle -6|\mathcal{H}_{\text{lf}}^1|+6\rangle|^2 \left[1 + \exp\left(\frac{\hbar\omega}{k_B T}\right)\right] \tag{13.12}$$

The strain can be related to the phonon density $D(\omega)$ by $\epsilon^2 = D(\omega)/2\rho v^2$, where v is the phonon group velocity and ρ the mass density of the phonon bath [44]. Finally one obtains

$$\frac{1}{\tau_d} \sim D(\omega)|\mathcal{H}_{\mathrm{lf}}^1|^2\left[1+\exp\left(\frac{\hbar\omega}{k_{\mathrm{B}}T}\right)\right] \tag{13.13}$$

Therefore, the relaxation rate is essentially limited by the phonon density $D(\omega)$, the matrix element coupling the spin to phonon $|\mathcal{H}_{\mathrm{lf}}^1|$ as well as the phonon energy $\hbar\omega$. Also, the relaxation rate intrinsically depends on the dimensionality of the phonon bath [39].

Two Phonon Relaxation Processes Upon increasing the temperature higher excited phonon states become accessible and a so called two-phonon Orbach process contributes to the relaxation of the SMM's magnetization. As depicted in Fig. 13.12(b), the electron is excited from $|m_J = +6\rangle$ into a vibrational state $|e\rangle$ under absorption of a phonon with an energy $\hbar\omega_1$. In a second step, the electron then relaxes into the ground state $|m_J = -6\rangle$ under emission of a second phonon with an energy $\hbar\omega_2 = \hbar\omega_{\mathrm{ph}}$. The process can occur if $\omega_2 - \omega_1 = \omega$, where ω is the energy separation of the two spin states. The relaxation rate yields [44]

$$\frac{1}{\tau_o} \sim |\mathcal{H}_{\mathrm{lf}}^{1'}|^2\Delta^3\frac{1}{\exp(\Delta/k_{\mathrm{B}}T)-1} \tag{13.14}$$

where $|\mathcal{H}_{\mathrm{lf}}^{1'}|$ is the product of the matrix elements between the states $|m_J = -6\rangle$, $|m_J = +6\rangle$ and $|e\rangle$. The process is therefore thermally activated if $k_{\mathrm{B}}T \gtrsim \Delta$.

Finally, relaxation can occur *via* a Raman like process (Fig. 13.12(c)). The mechanism is similar to the Orbach process, however it involves a *virtual* excited state, resulting in a relaxation time of the form [44]

$$\frac{1}{\tau_r} \sim |\mathcal{H}_{\mathrm{lf}}^2|^2\left(\frac{k_{\mathrm{B}}T}{\hbar}\right)^7 \tag{13.15}$$

where $|\mathcal{H}_{\mathrm{lf}}^2|$ is the second order pertubation from (13.11).

13.5 Molecular Quantum Spintronics with a Single $TbPc_2$

Various molecular quantum spintronic devices have been recently proposed to probe the quantum mechanical nature of an isolated $TbPc_2$ single-molecule magnet (Sect. 13.3). The most promising approach consists in probing the spin dynamics of a $TbPc_2$ SMM with the electrical current tunneling through a quantum dot, which is weakly coupled to the molecule (Sect. 13.3).

As we will describe in the following, such an indirect coupling scheme enables in fact the electronic readout of both the electronic and the nuclear spin of the $TbPc_2$, ultimately revealing the lifetime and coherence time of a single nuclear spin.

Fig. 13.13 Carbon nanotube based supramolecular spin valve. Courtesy of M. Urdampilleta

13.5.1 Read-out of the Electronic Spin

The magnetization reversal of the electron spin in an isolated $TbPc_2$ SMM can occur via two different processes at cryogenic temperatures (Sect. 13.4). Around zero magnetic field, the avoided level crossings allow for quantum tunneling of magnetization $|J_z, I_z\rangle \rightarrow |-J_z, I_z\rangle$. At high magnetic fields the magnetization reverses through a direct relaxation process involving non-coherent tunneling events combined with the emission or absorption of phonons.

One can couple a SMM to a state-of-the-art carbon nanotube transistor and use the electric current in the nanotube to probe and manipulate the spin of the SMM. A carbon nanotube behaves as a quantum dot at very low temperatures, showing an impressive array of electronic properties ranging from Coulomb blockade [50] to Kondo effect [51]. In this regime, a carbon nanotube is sensitive to very small charge fluctuations in its environment which results in a modulation of the conductance in the carbon nanotube quantum dot. For instance, the nanotube's conductance can be altered by the magnetization reversal of a SMM grafted to the carbon nanotube's sidewall as we will describe in the following.

Indeed, Urdampilleta et al. [36] reported a supramolecular spin valve behaviour without magnetic leads in a carbon nanotube quantum dot functionnalized with $TbPc_2$ SMMs (Fig. 13.14(a)).

They showed that two SMMs, coupled to a carbon nanotube via a π-π interaction, act as spin-polarizer and analyzer for the conduction electrons in the carbon nanotube channel. Mediated by exchange interactions, the magnetic moment of each molecule induces a localized spin polarized dot in the carbon nanotube quantum dot, which can be controlled by a magnetic field (Fig. 13.14(b) and (c)).

At large negative magnetic fields, both molecular spins are oriented in parallel to each other and the quantum dot is in a high conductance state. Upon increasing the magnetic field (following the red trace in Fig. 13.14(a)), the molecular spin A is reversed by quantum tunneling of magnetization close to zero field, resulting in an antiparallel spin orientation and a current blockade in the quantum dot (Fig. 13.14(b)). By further increasing the field, the second spin B is reversed by phonon assisted relaxation, restoring a parallel spin orientation and the high conductance regime in the quantum dot (Fig. 13.14(c)). After reversing the sweep direction, one obtains the characteristic butterfly hystersis loop of a spin valve device with a magnetoresistance ratio $(G_P - G_{AP})/G_{AP}$ up to 300 % (Fig. 13.14(a)). A detailed description

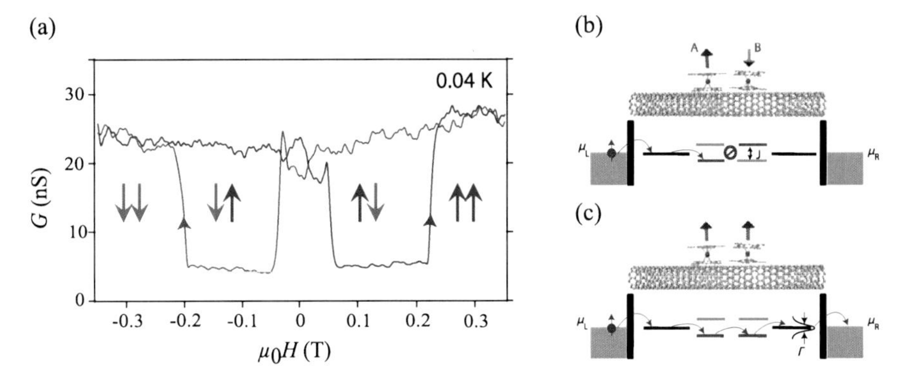

Fig. 13.14 Spin valve behaviour in a supramolecular spintronic device based on a carbon nanotube quantum dot functionnalized with $TbPc_2$ SMM's. Figures from [36, 52] (**a**) Butterfly hysteresis loop at $T = 40$ mK. (**b**) Antiparallel spin configuration: the spin state in dot A is reversed with respect to that of dot B. The energy mismatch between levels with identical spin results in a current blockade. (**c**) Parallel spin configuration for both molecules A and B. Energy levels with same spin are aligned allowing electron transport through the carbon nanotube

of the mechanism can be found in [36, 52]. Each current switching event, or switching field, can be attributed to the magnetization reversal of a SMM thus providing an electronic readout scheme for a molecular spin.

Uniaxial Magnetic Anisotropy The spin valve effect and the current switching exhibits the fingerprint like characteristics of the $TbPc_2$ SMM. Figure 13.15(a) shows the difference between upwards (red trace in Fig. 13.14(a)) and downwards sweeps of the magnetic field (blue trace in Fig. 13.14(a)) during a hysteresis loop as function of the applied magnetic field orientation. In the white region both molecules have the same polarization, while in the red region their polarizations are antiparallel to each other. The limit between both regions corresponds to the switching field associated with the phonon assisted relaxation of the second molecule B. The projection of the switching field along the H_x-direction, defined as easy axis, is constant, which is consistent with the Ising like uniaxial magnetic anisotropy of the $TbPc_2$.

Also, the inherently stochastic character of the SMM magnetization reversal can be revealed by repeating a hysteresis loop a certain number of times (Fig. 13.15(b)). Finally, the spin valve effect becomes less pronounced upon increasing temperature and disappears above a temperature of 1 K (Fig. 13.15(c)), which is also consistent with the thermally activated magnetization reversal of a SMM. For a more detailed description, the reader may refer to Urdampilleta et al. [36].

Landau-Zener Tunneling One can probe the Landau-Zener tunneling of the electron spin in an individual $TbPc_2$ SMM using a carbon nanotube quantum dot as readout system [52] (Fig. 13.13). As described above, the conductance through the carbon nanotube quantum dot exhibits a typical spin valve behaviour consistent with such a supramolecular spintronic device. Figure 13.16(a) shows measurements

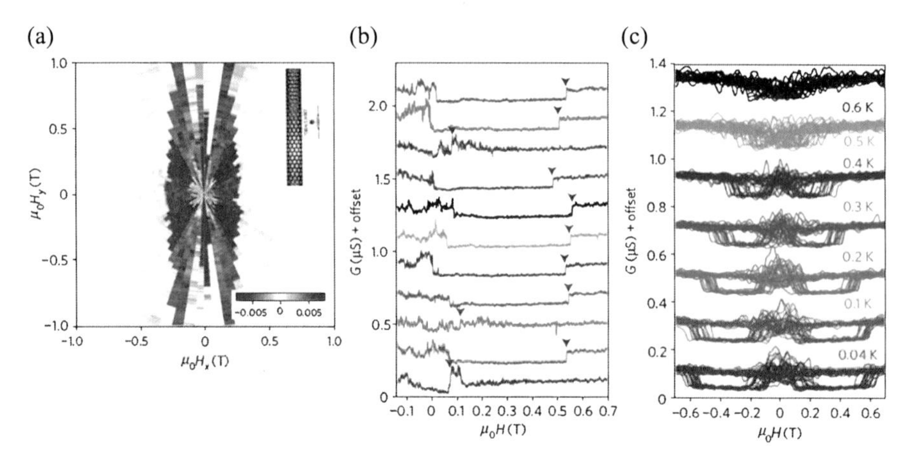

Fig. 13.15 Supramolecular spin valve and molecular fingerprint. Figures from [36] (**a**) Angular dependance of the spin valve behaviour. The difference in conductance between the up- and downwards magnetic field sweep in a hysteresis loop is plotted as a function the magnetic field angle. For spins oriented parallel, the difference is negligeable (*white color*), whereas in an antiparallel configuration the difference is non zero (*red color*). The red-white boundary corresponds to the switching field of the second molecule B, and is consistent with the uniaxial anisotropy of the $TbPc_2$ SMM. (**b**) Stochastic switching of molecule B. Three times out of 11, the molecule can switch its magnetization by quantum tunneling magnetization, while 8 times out of 11 the reversal occurs by phonon assisted magnetic relaxation as predicted by the Landau Zener model. (**c**) 20 hysteresis loops at different temperatures

recorded with a constant transverse field of 0.35 T and presenting only quantum tunneling of magnetization (QTM) features, the ones presenting a direct transition (DT) being rejected. The conductance suddenly decreases, between the magnetic field values of −50 mT and +50 mT (QTM of B), and then increases in all cases above approximately +50 mT (QTM of A). This measurement was repeated 3500 times with a 100 $\mathrm{mT\,s^{-1}}$ sweep rate. Whenever DT failed to occur, the longitudinal position of the QTM in molecule B was stored in the histogram plotted in Fig. 13.16(b): four peaks emerge with a FWHM of approximately 10 mT, and a mean peak-to-peak separation of 25 mT.

In order to explain these results, we compare the position of these peaks with the Zeeman diagram. Figure 13.16(b) shows a very good correspondence between the four peaks and the avoided-level crossings of the Zeeman diagram. This diagram is slightly different from the one presented in Fig. 13.10 since we took into account that the easy axis is not lying exactly in the plane ($H_\parallel$, $H_\perp$). As evident from the comparison between the histogram and the diagram, each of these peaks can be attributed to a particular nuclear spin state.

Similar results have recently been demonstrated at the single molecule level by Vincent et al. [23], in a molecular transistor configuration. It is important to note that in the present case the FWHM is larger than in the case of Vincent et al. leading to a lower fidelity in the single-shot read-out measurement [24]. One reason for this could be that the current tunneling through the carbon nanotube is interacting

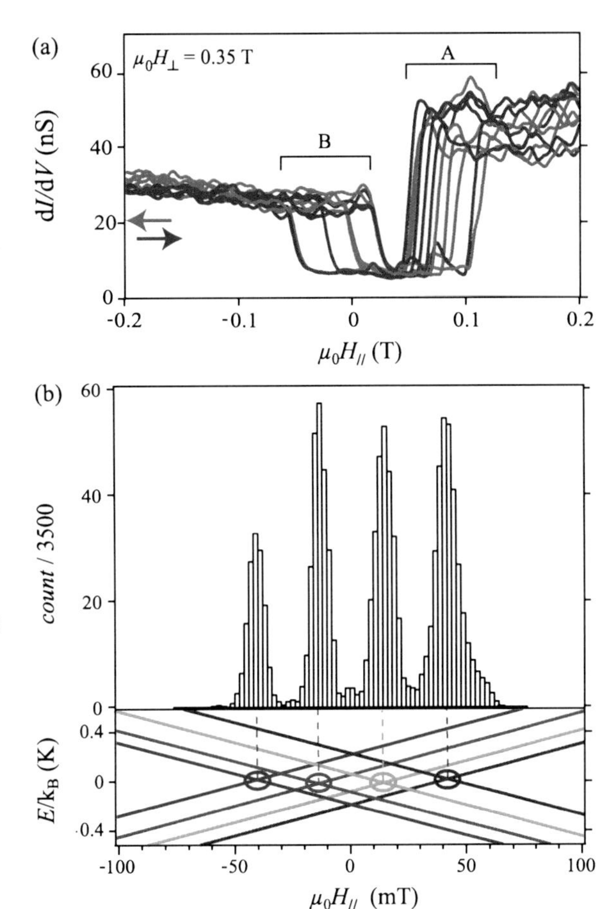

Fig. 13.16 Landau-Zener tunneling of a single $TbPc_2$ electronic spin. (**a**) Magneto-conductance curves recorded under a 0.35 T transverse field, at a 100 mT s^{-1} sweep rate. The relative QTM positions for molecules A and B are clearly split, the angle between their easy axes is then estimated to be about 15°. (**b**) Histogram of QTM jumps for molecule B, determined from 3500 consecutive traces recorded under the same conditions as those applied in (**a**). Each peak corresponds to an avoided level-crossing in the Zeeman diagram of the ground doublet $J_z = \pm 6$ split by the hyperfine coupling with the nuclear spin $I = 3/2$

more strongly with the $TbPc_2$ molecules. Nevertheless, one advantage of the present device is the very large variation of conductance induced by the spin reversal (200 % in the present case versus 1 % in the work of Vincent et al.[23]), which makes the measurement very easy since it does not require any specific filtering (physical or numerical).

In order to confirm that only one QTM position exists per nuclear spin state, we measured the tunneling probability as a function of the sweep rate and compared it with the Landau-Zener theory. 100 magneto-conductance measurements were recorded for a given sweep rate. The tunneling probability P_{QTM} of molecule B can be obtained by $P_{LZ} = 1 - P_{DT}$, where P_{DT} is the probability of a DT. P_{LZ} is plotted in Fig. 13.17 as a function of the reciprocal sweep rate. The experimental data were fit with (13.8), from which a tunnel splitting $\Delta = 1.7$ μK was extracted. The exponential behavior clearly demonstrates that only the four circled level crossings in Fig. 13.16(b) are avoided level-crossings. This is in agreement with the work of Vincent et al. [23] but not with the measurement done on a single crystal of $TbPc_2$. The latter case presents QTM at all intersections of the diagram

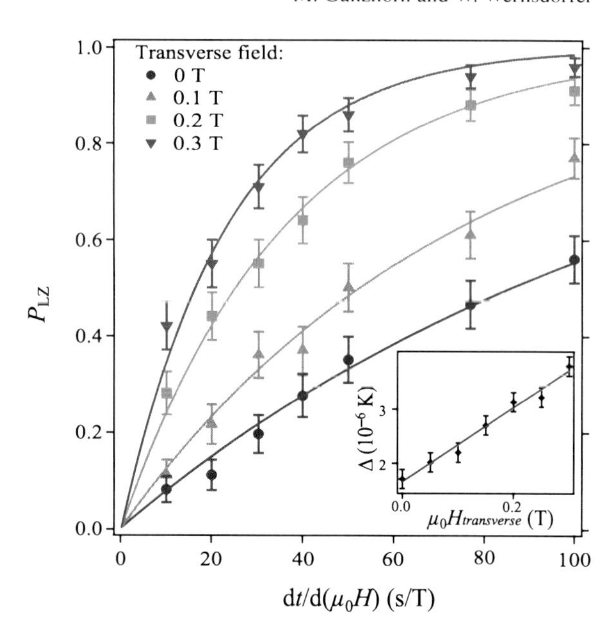

Fig. 13.17 Landau Zener tunnel probability as a function of the inverse of the sweep rate measured at transverse magnetic fields $\mu_0 H_\perp = 0, 0.1, 0.2, 0.3$ T. The experimental data are fitted by exponential curves, in accordance with equation (13.8). *The inset* indicates the variation of tunnel splitting as a function of the transverse field, *the red line* is a guide for the eye

depicted in Fig. 13.16(b), see the work of Ishikawa et al. [10]. In an assembly of molecules, coupled with weak dipole interactions, multi-spin tunnel effects[53, 54] might be responsible for this observation but further investigations are needed to better understand this issue.

An applied transverse field tunes the tunnel splittings via the $H_\perp(J_+ + J_-)$ term of the Hamiltonian. In order to study this effect on the different level crossings, we measured the tunneling probability for several constant transverse fields and field sweep rates. The symbols in Fig. 13.17 correspond to the experimental points and the continuous lines are least-square fits using (13.8). The data agree very well with the Landau-Zener behavior, which suggests that no other measurable avoided-level crossings are induced by the application of a transverse field. The tunnel splitting amplitudes were extracted from the fits, and then plotted as a function of the transverse field (inset of Fig. 13.17). This behavior cannot be explained by using the parameters of Ishikawa et al. [10] Firstly, the ligand field might be different from the mean bulk value because the grafted molecules are probably slightly distorted and their anisotropy modified [55]. Secondly, the ligand field Hamiltonian does not predict a linear increase as observed for our measurements, which were confirmed on other molecular devices. This observation needs further experimental and theoretical investigations. In particular, we believe that, in the case of single molecules, the angular moment conservation has to be taken into account. When the latter is not conserved, the Landau-Zener equation is not valid[39]. Nevertheless, our studies showed that the hyperfine interaction is a robust feature allowing us to read the nuclear spin state regardless the deposition and measurement techniques. Moreover, the possibility of tuning the tunnel splitting is very convenient for experiments of coherent nuclear spin manipulation since the read-out mechanism needs the right

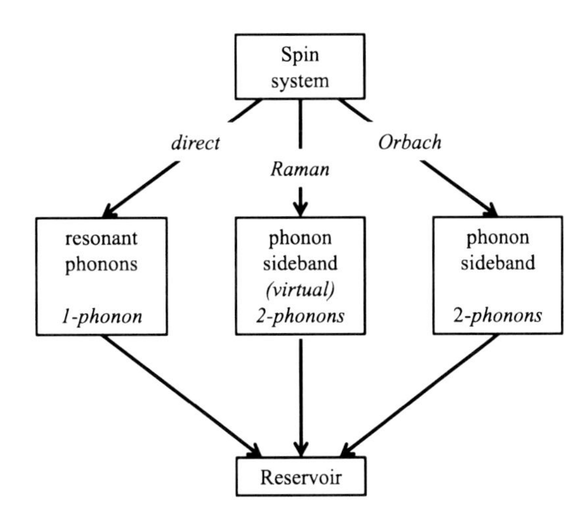

Fig. 13.18 Spin-phonon relaxation processes in $TbPc_2$* single molecule magnets: A *direct* one-phonon process dominates at lowest temperatures, whereas two-phonon *Orbach* and *Raman* processes take over as the temperature increases. The phonon energy is then transfered to a thermal reservoir. *The arrows* indicate the direction of the energy transfer. Figure inspired from [44]

value: not too small (no read-out) and not too large (DT possible, reducing the read-out fidelity).

Spin-Lattice Relaxation Spin-phonon relaxation processes in $TbPc_2$ SMM are due to a modulation of the molecule's crystal electric or ligand field by the vibrations of charged ions in a surrounding lattice, so called phonons. Due to strong spin-orbit interactions in the Tb^{3+} ion, the fluctuating ligand field can induce transitions between different spin states. Different transition mechanisms can contribute to the molecule's magnetization reversal, involving the emission or absorption of one or two phonons as described in Sect. 13.4 and in Fig. 13.18. At the lowest temperatures, spin-lattice relaxation is typically dominated by the one-phonon *direct* process. As temperature increases, higher order two-phonon *Orbach* and *Raman* processes, involving a real and virtual excited state (or phonon sideband) respectively, contribute to magnetic relaxation [44]. In the following, we discuss the time scale for spin-lattice relaxation in a $TbPc_2$ SMM coupled to a carbon nanotube [56].

The spin relaxation time associated to a *direct* transition is essentially given by the phonon energy and the phonon density at the spin resonance as well as the spin-phonon coupling (see (13.13)). In a $TbPc_2$ coupled to a non-suspended carbon nanotube, the direct relaxation is mainly enabled by 3D bulk phonons in the amorphous SiO_2 [56]. The coupling is mediated by the carbon nanotube, which is mechanically coupled to the $TbPc_2$ spin and is in thermal contact with the SiO_2 phonons [56]. The energy distribution of 3D bulk phonons in SiO_2 is continuous and the phonon density for a longitudinal acoustic phonon mode is given in the Debye approximation by

$$D(\omega) = \frac{\hbar\omega^3}{2\pi^2 v_l^3 [\exp(\frac{\hbar\omega}{k_B T}) - 1]} \tag{13.16}$$

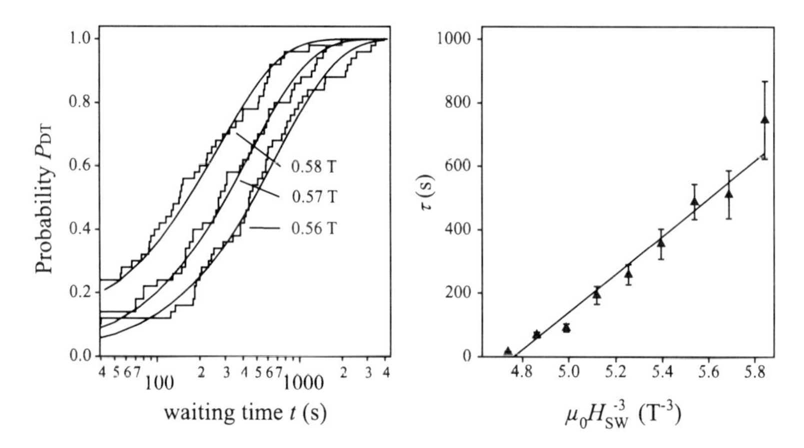

Fig. 13.19 Relaxation time τ of the electron spin in an individual $TbPc_2$ SMM [56]. *Left panel*: Cumulative histogram showing the probability of electron spin reversal P_{DT} as a function of the waiting time t (see text). The measurement is performed for 3 different switching fields H_{SW}. A fit with an exponential law $P = 1 - \exp(-t/\tau)$ gives the relaxation time τ. *Right panel*: Relaxation time τ vs. switching field H_{SW}. The magnetic field is oriented with the easy axis of the $TbPc_2$ SMM

where v_l is the group velocity of a longitudinal phonon. Combining (13.16) and (13.13) one finds

$$\frac{1}{\tau_d} \sim (\hbar\omega)^3 \coth\left(\frac{\hbar\omega}{2k_B T}\right) \sim H_{SW}^3 \coth\left(\frac{\beta H_{SW}}{2k_B T}\right) \tag{13.17}$$

with the phonon energy $\hbar\omega = g\mu_0\mu_B \Delta m_J H_{SW} = \beta H_{SW}$ and H_{SW} the magnetic switching field.

The relaxation time of the electronic spin can be estimated using the following measurement protocol described by Urdampilleta et al. [56]: starting at a magnetic field value of $H_{init} = -1$ T, the field is ramped with a sweep rate of 20 mT/s to a magnetic field $H_{SW} > 0$. The electron spin then switches after a certain waiting time t and the measurement is repeated 100 times. The waiting time t is then extracted from each measurement and compiled in an cumulative histogram showing the switching probability P_{DT} of the electron spin as function of the waiting time t. Finally the measurement is performed for different switching fields H_{SW}.

The relaxation time τ is obtained by fitting the cumulative histogram, i.e. the reversal probability with an exponential law $P = 1 - \exp(-t/\tau)$ for each switching field H_{SW}. As depicted in Fig. 13.19, the relaxation τ indeed follows the predicted H_{SW}^{-3} dependance.

As described in the previous section, higher order spin relaxation processes (Orbach, Raman) are activated upon increasing temperature to the system. Both processes effectively correspond to a thermal activation of the magnetization over an energy barrier. In agreement with (13.14) and (13.15), the relaxation time τ of the electron spin should therefore decrease upon increasing temperature.

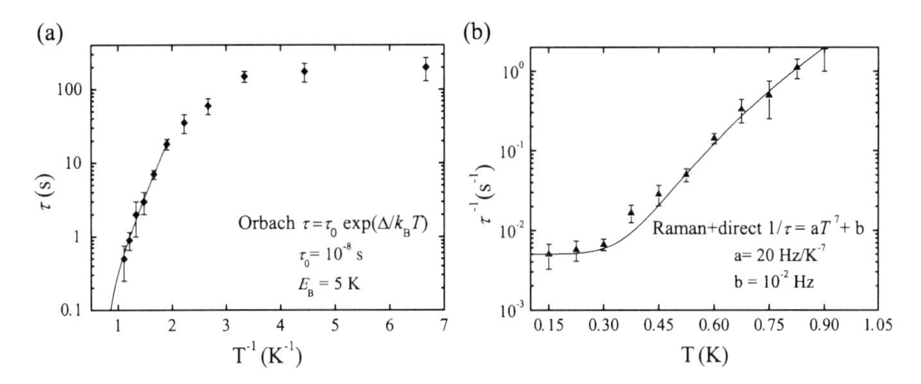

Fig. 13.20 Relaxation time τ of the electron spin in a $TbPc_2$ SMM as a function of the temperature. Figures reproduced from Ref. [56]. (**a**) Scaling law of an Orbach process. *Solid line* corresponds to a fit with equation (13.14). (**b**) Scaling law of a Raman and direct relaxation process. *Solid line* corresponds to a fit with $1/\tau = 1/\tau_r + 1/\tau_d = aT^7 + b$, where the first term corresponds to the Raman process and the second to the direct relaxation

Urdampilleta et al. [56] observed a constant relaxation time τ below 300 mK, consistent with a direct transition process. Above 300 mK however, a significant decrease of the relaxation time is found with increasing temperature (Fig. 13.20). The experimental data are indeed in qualitative agreement with the Orbach and Raman scaling laws above 300 mK (Fig. 13.20(a) and (b), respectively). However, a quantitative analysis of the fitting parameters reveals significant differences between the magnetization reversal in an individual SMM's described here and SMM's in molecular crystals [44].

Considering the Orbach scaling parameters (Fig. 13.20(a)), the time constant τ_0 can be related to the energy barrier E_B by the relation $\tau_0^{-1} = CE_B^3$, where C is the coupling the electron spin of the $TbPc_2$ and the SiO_2 phonons involved in the process [56]. The obtained value $C \approx 400$ kHz/K^3 is three orders of magnitude larger than that found for SMM's in a molecular crystal [56]. Such a strong coupling can indeed be mediated by the carbon nanotube, which yields a strong mechanical coupling to the $TbPc_2$ spin (see Ref. [39] and Sect. 13.5.3) and is in good thermal contact with the SiO_2 phonons. Also, the energy barrier E_B is two orders of magnitude smaller than the expected barrier in molecular crystals [56].

The comparison of the experimental data with the Raman process (Fig. 13.20(b)) furthermore reveals a ratio of the relaxation times τ_d and τ_r for an individual SMM of the form

$$\frac{\tau_r}{\tau_d} = \frac{b}{a} = \left(\frac{|\mathcal{H}_{\text{lf}}^1|}{|\mathcal{H}_{\text{lf}}^2|}\right)^2 10^5 \tag{13.18}$$

with the crystal field matrix elements $|\mathcal{H}_{\text{lf}}^2|$ and $|\mathcal{H}_{\text{lf}}^1|$. One therefore finds $|\mathcal{H}_{\text{lf}}^2| = 10^4 |\mathcal{H}_{\text{lf}}^1|$ for the relaxation of an individual SMM, which stands in contrast to the findings for SMM crystals yielding $|\mathcal{H}_{\text{lf}}^2| \sim |\mathcal{H}_{\text{lf}}^1|$ [56].

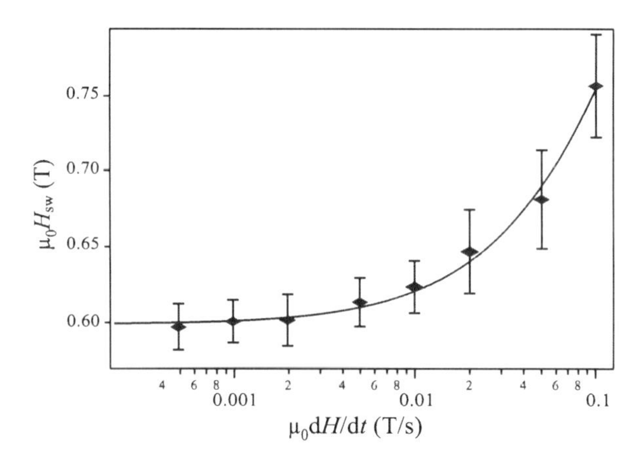

Fig. 13.21 Mean switching field H_{SW} of the electron spin in a $TbPb_2$ as a function of the magnetic field sweep rate dH/dt [56]. The error bars correspond to the standard deviation from the mean value and *the solid line* corresponds to a fit with (13.20)

Finally, the switching field associated with a direct transition of the $TbPc_2$ depends on the magnetic field sweep rate, in analogy to the magnetization reversal of ferromagnetic nanoparticles [57]. Assuming a magnetization reversal probability of $P = 1 - \exp(-t/\tau)$, the probability density function is maximum for $\frac{d^2P}{dt^2} = 0$, i.e. for

$$\frac{d^2P}{dt^2} = -\frac{d\tau}{dt}\frac{1}{\tau^2}(1-P) - \frac{1}{\tau}\frac{dP}{dt} = \frac{P-1}{\tau^2}\left(\frac{d\tau}{dt} + 1\right) = 0 \tag{13.19}$$

Equation (13.19) implies $\frac{d\tau}{dt} = -1$. Using (13.17), one gets $\frac{d\tau}{dt} \sim H_{SW}^{-4}\frac{dH}{dt}$ and therefore

$$H_{SW} \sim \left(\frac{dH}{dt}\right)^{1/4} \tag{13.20}$$

where dH/dt corresponds to the magnetic field sweep rate.

Urdampilleta et al. performed [56], magnetization reversal measurements for different field sweep rates. Figure 13.21 depicts the mean value of the switching fields from 100 consecutive measurements performed at 8 different field sweep rates, as well as the standard deviation. A fit with (13.20) shows indeed an excellent agreement between the experimental data and theoretical predictions.

13.5.2 Read-out of the Nuclear Spin

Single Shot Electrical Read-out of a Single Nuclear Spin So far, only signatures accounting for the electronic magnetic moment have been addressed, and there has been no quantitative comparison with the expected theoretical magnetic behaviour of an individual SMM. Recently, transport measurements taken through a single bis(phthalocyaninato)terbium(III) SMM were studied in a three-terminal geometry obtained by electromigration [23] (see Fig. 13.22). It has been reported

Fig. 13.22 Artist view of the molecular spin-transistor, consisting of a $TbPc_2$ molecular magnet, connected to source and drain gold electrodes, and a back-gate underneath. The Pc-ligands (*white*) are acting as a read-out quantum dot. The terbium ion (*pink*) posses an electronic spin with $J = 6$ (*orange*) and a nuclear spin with $I = 3/2$ (*green*). The uniaxial anisotropy axis of the Tb ion is perpendicular to the Pc-plane

that $TbPc_2$ SMMs conserve both their structural integrity and their magnetic properties even when sublimated at 820 K on a copper surface [23]. In addition, the redox state of the Tb^{3+} ion is very stable, suggesting that a current flow through the Tb^{3+} ion is highly unlikely (see Supplementary Information in [23]). The two Pc ligands have a conjugated π system, which can easily conduct electrons.

Transport measurements through a single $TbPc_2$ SMM were previously performed by scanning tunnelling spectroscopy (STS) experiments [20], where the electronic transport occurred through the Pc ligands and exhibited Coulomb blockade and Kondo effects depending on its charge state (spin states $S = 0$ or $1/2$). However, no signature of the magnetic moment carried by the Tb^{3+} ion was observed in this experiment. In order to detect the reversal of the magnetic moment, the $TbPc_2$ SMM was directly inserted into an electromigrated gold junction (see Fig. 13.1). The differential conductance, dI/dV, is shown in Fig. 13.23(a) as a function of the drain-source voltage, V_{ds}, and the gate voltage, V_g. It exhibits a single charge-degeneracy point with a weak spin $S = 1/2$ Kondo effect. A detailed study of the Kondo peak as a function of the applied magnetic field is presented in the Supplementary Information of [23]. A ferromagnetic exchange interaction of about 0.35 T was measured between the spin 1/2 of the quantum dot and the magnetic moment carried by the Tb^{3+} ion. Alternative coupling mechanisms such as dipolar, magneto-Coulomb, mechanical, or flux coupling were also considered [23], but the relatively high value of the exchange interaction strongly indicates the latter being the major contribution to the coupling mechanism. As a consequence, the read-out quantum dot is spatially located close to the Tb^{3+} ion. This is indirect proof that the electronic transport occurs through the aromatic Pc ligands, and that the most favourable geometric configuration is the one depicted in Fig. 13.22.

In summary, the Pc ligands form a molecular quantum dot and the anisotropic magnetic moment of the Tb^{3+} ion is coupled to the electron path only indirectly, mainly through a ferromagnetic exchange interaction. Moreover, the presence of a gate allows fine-tuning from the Coulomb blockade to the Kondo regimes of the

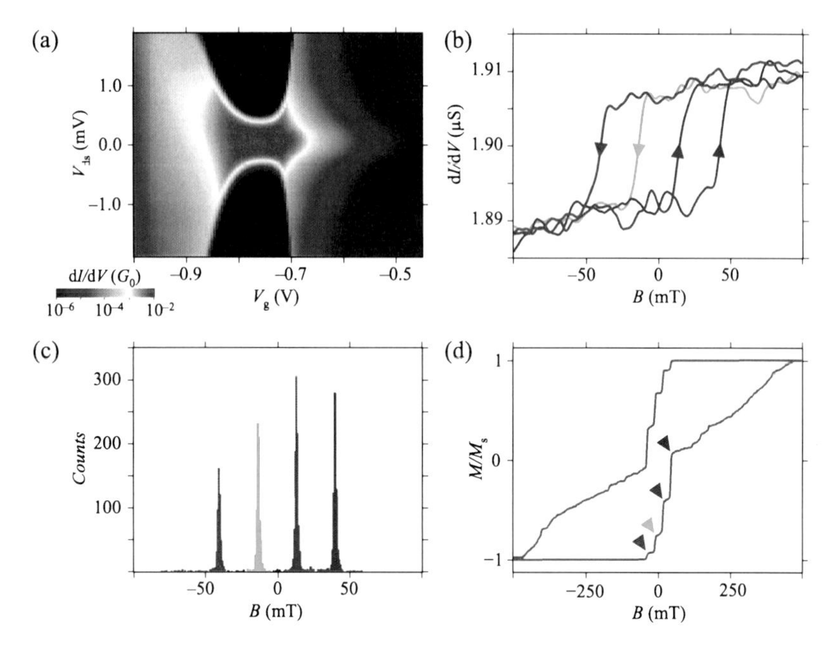

Fig. 13.23 Conductance characteristics and electronic read-out procedure [23]. (**a**) Stability diagram of the Pc read-out quantum dot exhibiting the differential conductance, dI/dV, in units of the quantum of conductance, G_0, as a function of gate voltage, V_g, and bias voltage, V_{sd}, at 0.1 K. (**b**) dI/dV measurements for a given working point ($V_g = -0.9$ V; $V_{ds} = 0$ V) as function of the magnetic field B. The arrows indicate the field-sweep direction. Abrupt jumps in the differential conductance, attributed to the switching of the Tb^{3+} magnetic moment, are visible for all traces of B, showing a clear hysteresis in the dI/dV characteristics. (**c**) Histogram of switching field obtained for 11000 field sweeps showing four preferential field values that are assigned to QTM events. (**d**) Normalized hysteresis loop of a single $TbPc_2$ SMM obtained by integration of 1000 field sweeps and performed for trace and retrace on a larger magnetic field range than in c. The four arrows on the trace curve show the four preferential field values associated to QTM (*red*, −40 mT; *green*, −14 mT; *blue*, 14 mT; *purple*, 40 mT)

molecular quantum dot, revealing that the magnetic properties of the Tb^{3+} ion are then independent of the charge state of the Pc quantum dot (see Supplementary Information in [23]). Finally, owing to the exchange coupling, we can use the Pc ligands as a read-out quantum dot to detect the reversal of the electronic magnetic moment carried by the Tb^{3+} ion spin dot.

To achieve the electronic read-out of the single spin carried by the spin dot, we chose experimental conditions close to the charge degeneracy point ($V_g = -0.9$ V and $V_{ds} = 0$ V in Fig. 13.23(a)). When sweeping the magnetic field from negative to positive values (upwards sweep), we observed a single abrupt jump of the differential conductance, which is reversed if the field is swept in the opposite direction (downwards sweep) as depicted in Fig. 13.23(b). These jumps and the corresponding magnetic field, the switching field, can be associated with the reversal of the Tb^{3+} magnetic moment, which slightly influences the chemical potential of the read-out quantum dot through the magnetic interactions.

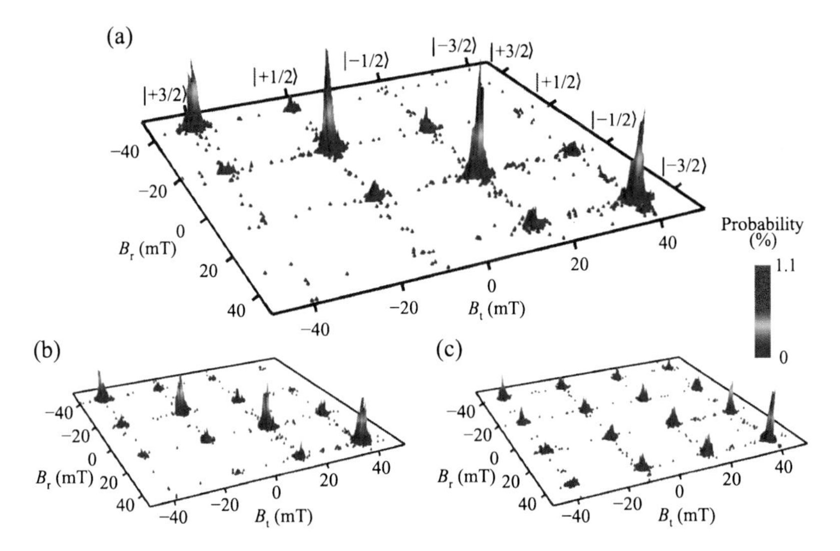

Fig. 13.24 Transition matrix of the QTM events as a function of the waiting time. The switching fields of the Tb^{3+} magnetic moment of subsequent field sweeps are plotted in two-dimensional histograms for three waiting times, tw: (**a**) $t_w = 0$ s; (**b**) $t_w = 20$ s; and (**c**) $t_w = 50$ s. The two axes correspond to the trace and retrace field sweeps, B_t and B_r, respectively. Two successive measurements with the same nuclear spin states are situated on the diagonal of the matrix, whereas the off-diagonal positions correspond to nuclear spin-state changes of $\Delta m_I = \pm 1, \pm 2$ and ± 3. The predominance of diagonal terms up to $t_w = 20$ s indicates the long level lifetime of the nuclear spin states. For $t_w = 50$ s, the diagonal terms vanish owing to nuclear spin-flip processes. Furthermore, the high amplitude of the bottom-right ($B_t = B_r = 40$ mT) matrix element accounts for the relaxation of the nuclear spin towards a thermal equilibrium

In order to obtain a magnetization reversal statistic, 11000 upwards and downwards sweeps were performed at a sweep rate of 50 mT/s. All switching fields were recorded and compiled into a histogram, revealing a magnetization reversal at four distinct values of magnetic field(Fig. 13.23(c)). These are in perfect quantitative agreement with theoretical predictions [10] of QTM of a Tb^{3+} magnetic moment at the four avoided energy-level crossings with the nuclear spin states $|-3/2\rangle$, $|-1/2\rangle$, $|1/2\rangle$, $|3/2\rangle$ (Fig. 13.10). Moreover, the histograms of the four switching fields do not overlap, revealing the high efficiency of this electronic read-out procedure. The field integration of the normalized switching histograms yields the magnetic hysteresis loop (Fig. 13.23(d)), which is in excellent accordance with micro-SQUID measurements of assemblies of $TbPc_2$ SMMs (see Ref. [23]). At higher magnetic fields, the reversal of the magnetization occurs stochastically, as predicted for a direct relaxation process involving a non-coherent tunnelling event combined with a phonon emission.

The lifetimes of the four nuclear spin states could be measured by studying the correlations between subsequent measurements as a function of the waiting time, t_w, between field sweeps. Figure 13.24 presents two-dimensional histograms (transition matrices) obtained from 22000 field sweeps. The two axes correspond to the

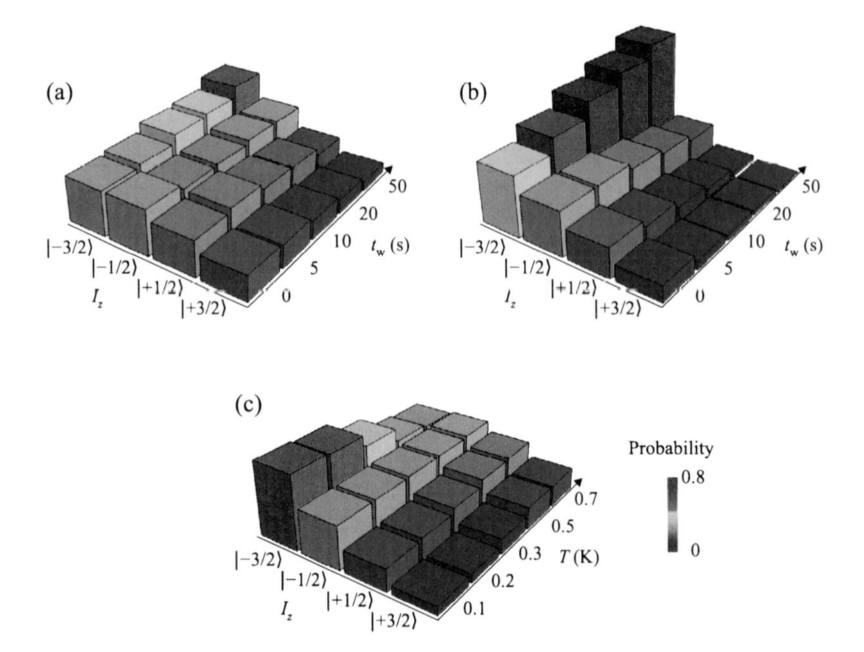

Fig. 13.25 Spin-flip dynamics and nuclear spin-state occupancy of the Tb^{3+} nuclear spin states. Evolution of the nuclear spin-state occupancy as a function of the waiting time, t_w, for two different working points: (**a**) gate voltage $V_g = -0.9$ V and (**b**) $V_g = -0.1$ V (both at bias voltage $V_{sd} = 0$ V). The measurements clearly show that the populations evolve towards different thermal equilibriums. (**c**) Spin dynamics for a fixed waiting time of 10 s, as a function of temperature, T. With increasing temperature, the population of the different spin states evolves towards equal occupancy

trace and retrace field sweeps, B_t and B_r. Two subsequent measurements with the same nuclear spin states are situated on the diagonal of the matrix, whereas off-diagonal positions correspond to nuclear spin state changes of $\Delta m_I = \pm 1, \pm 2, \pm 3$. For zero waiting time ($t_w = 0$ s, Fig. 13.24(a)), the diagonal positions are predominant, highlighting the robustness of the nuclear spin states and long level lifetimes for the individual Tb^{3+} nuclear spin states. The diagonal positions persist even for a waiting time of $t_w = 20$ s (Fig. 13.24(b)). However, for $t_w = 50$ s, the off-diagonal positions start to be populated, which suggests the occurrence of nuclear spin-flip processes during the waiting time (Fig. 13.24(c)). From this series of measurements, we conclude that the level lifetime (T_1) on the nuclear spin states is on the order of tens of seconds, confirming that the invasiveness of the measurement procedure is low (see Supplementary Information of Ref. [23]).

A more detailed insight into the spin-flip dynamics of an individual nuclear spin can be gained by measuring the population of nuclear spin states as a function of waiting time and temperature (Fig. 13.25). To this end, we determined the nuclear spin-state occupancy for two different working points corresponding to two different charge states of the read-out quantum dot ($V_g = -0.9$ V in Fig. 13.25(a) and $V_g = -0.1$ V in Fig. 13.25(b), both at $V_{sd} = 0$ V). It is clear that the state occupancy, relaxing towards a thermal equilibrium, depends heavily on the transport characteristics (the current flowing through the Pc quantum dot and/or the electrostatic environment modulated by the gate voltage). Indeed, the electron tunneling through the Pc-ligands read-out dot gives rise to small fluctuations of the local electric field, which could modify nuclear spin-flip processes through the quadrupole interaction [44]. A detailed study addressing this problem is in progress and will be described in the next section. We also determined the population of nuclear spin states as a function of the temperature (Fig. 13.25(c)). The strong temperature dependence of this occupancy demonstrates that a single nuclear spin can be thermalized down to at least 0.2 K, which is close to the electronic temperature of our dilution refrigerator (0.08 K).

In order to obtain a deeper understanding of the relaxation mechanism of an isolated nuclear spin, one can for instance follow the time-trajectory of the Tb^{3+} nuclear spin via the previously described electrical readout. The device, a $TbPc_2$ single-molecule magnet spin-transistor (Fig. 13.22), detects the four different nuclear spin states of the Tb^{3+} ion with high fidelities of 95 %, allowing us to measure individual relaxation times (T_1) of several tens of seconds. A good agreement with quantum Monte Carlo simulations suggests that the relaxation times are limited by the current tunneling through the transistor, which opens up the possibility to tune T_1 electrically by means of bias and gate voltages [24].

Nuclear Spin Trajectories One can follow the time-trajectory of an isolated nuclear spin in $TbPc_2$ SMM using the single shot electrical readout presented earlier. Whenever the Tb electronic spin is reversed, the chemical potential of the readout dot located on the Pc ligands of the SMM is shifted due to a ferromagnetic exchange interaction, resulting in sharp conductance jump [23].

The external magnetic field was aligned, using a home-made 3D vector magnet, with the easy-axis of the $TbPc_2$ and constantly ramped it up and down between ± 60 mT (Fig. 13.26(a)). By simultaneously monitoring the conductance of the read-out dot, one observes conductance jumps happening at four distinct magnetic fields, which can be accounted for QTM transitions of the Tb electronic spin. From the four unique positions of the jumps, one can determine the four nuclear spin states and thus reconstruct the nuclear spin trajectory (Fig. 13.26(b)). For statistical analysis this procedure was repeated 80000 times. The first 500 seconds of the nuclear spin trajectory are shown in Fig. 13.27(a). Note that due to the probabilistic nature of the tunnel mechanism, QTM transitions were observed in ≈ 51 % of all sweeps. By plotting all detected jumps in histograms (Fig. 13.27(b)), one obtains 4 non-overlapping Gaussian distributions. The fidelity of nuclear spin read-out is about 95 % and only limited by the noise-floor. The widths of the histograms were dominated by electronic noise and the finite response time of the lock-in amplifier used to measure the conductance jumps.

Level Life Times The time-average population of each state was obtained by integration of the Gaussian distributions (colored bars Fig. 13.27(b)). Since with every QTM transition the energy levels of the ground state and the excited states

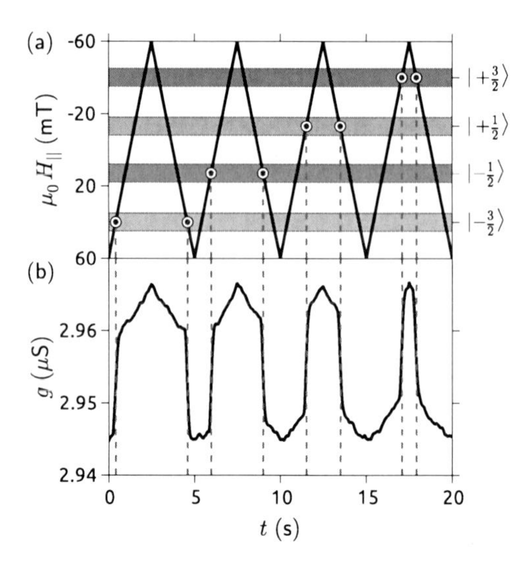

Fig. 13.26 Scheme of the measurement procedure. (**a**) The magnetic field $H_{||}$ is swept up and down as a function of time t over the four avoided level crossings and (**b**) the conductance g through the read-out dot is simultaneously measured. Whenever the electronic spin undergoes a QTM transition, a conductance jump is observed (indicated by *dashed lines*), revealing the nuclear spin state

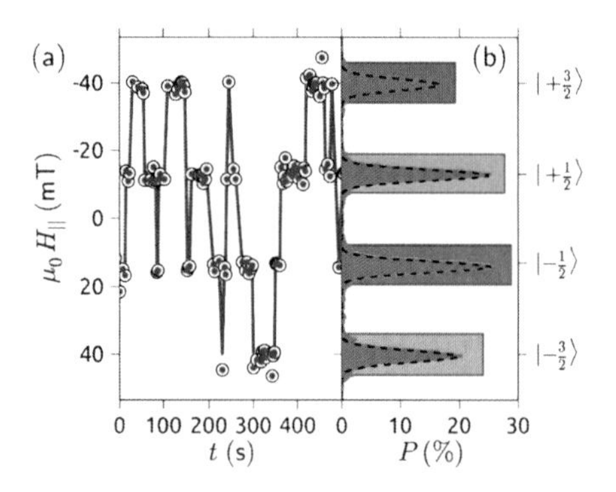

Fig. 13.27 Nuclear spin trajectory. (**a**) By continuously ramping the magnetic field up and down, the conductance jumps reveal the nuclear spin states (*grey dots*) as a function of time, yielding the nuclear spin trajectory (*red curve*). We found that the nuclear spin quantum number changes only by $\Delta m_I = \pm 1$ (see Supplementary Material of Ref. [24]). (**b**) Histograms (*grey*) of about 40000 conductance jumps, showing four non-overlapping Gaussian distributions (*dashed lines*) and yielding a 95 % fidelity of the nuclear spin state read-out. *The colored bars*, obtained by integrating over the Gaussian distribution, show the time-average population P of each nuclear spin state

were inverted, the population was found far from its thermal equilibrium. By plotting the expectation value of each nuclear spin state versus time, one obtains a perfect exponential decay, yielding the relaxation times T_1 of each nuclear spin state (Fig. 13.28(a)–(d)). The close to perfect exponential decay indicates an ideal Marcovian behaviour, i.e. there is no hidden memory effect and the time evolution of the system solely depends on its current state. Furthermore the obtained lifetimes were an order of magnitude larger than the measurement interval, which is a proof of the quantum non-demolition nature of the detection scheme.

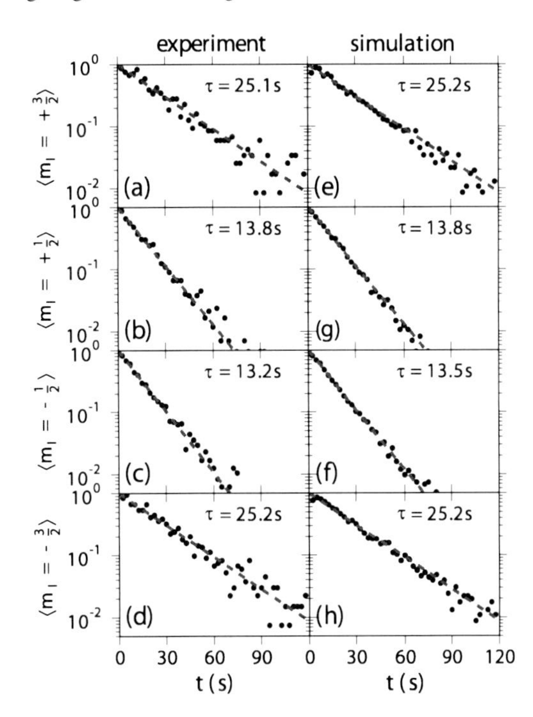

Fig. 13.28 Nuclear spin relaxation times. (**a**)–(**d**) Measured expectation value and (**e**)–(**f**) QMC simulations of each nuclear spin state versus time. The given T_1 values were obtained by fitting the data to an exponential function $\exp(-t/T_1)$ (*grey dashed lines*)

In order to get a deeper understanding of the nuclear spin trajectory obtained under non-equilibrium conditions, we performed quantum Monte Carlo (QMC) simulations using the QMC wave function approach [58–60]. The nuclear spin was modeled as a four-level system ($2I + 1$ states) which was coupled to a thermal bath of temperature $T = 150$ mK (cryostat temperature). The Hamiltonian of the isolated nuclear spin H_0 is mainly determined by its quadrupole moment, resulting in unequal nuclear level spacings of $\omega_{0,1} = 121$ mK, $\omega_{1,2} = 149$ mK, and $\omega_{2,3} = 178$ mK [10]. All environmental contributions were combined in an effective Hamiltonian with non-hermitian perturbation H_1:

$$H_1 = -\frac{i\hbar}{2}\sum_m C_m^{\dagger} C_m, \tag{13.21}$$

where m represents all possible transition paths. We further assumed that transitions between those three levels are only allowed if $\Delta m_I = \pm 1$, leading to one relaxation and one excitation path for each transition (i, j), modeled by $C_1^{i,j}$ and $C_2^{i,j}$ respectively:

$$C_1^{i,j} = \sqrt{\Gamma_{i,j}(1 + n(\omega_{i,j}, T)}\delta_{i,j+1}, \tag{13.22}$$

$$C_2^{i,j} = \sqrt{\Gamma_{i,j}(n(\omega_{i,j}, T)}\delta_{i+1,j}, \tag{13.23}$$

where $n(\omega_{i,j}, T) = [\exp(\hbar\omega_{i,j}/k_B T) - 1]^{-1}$ is the Bose-Einstein distribution function and $\Gamma_{i,j}$ is the transition rate between the i-th and the j-th nuclear spin state. Note that latter was the only adjustable parameter in the simulation. The transition probability dp for each level and time step dt is then calculated as:

$$dp = \langle \Psi(t)|\boldsymbol{C}_1^{\dagger}\boldsymbol{C}_1 + \boldsymbol{C}_2^{\dagger}\boldsymbol{C}_2|\Psi(t)\rangle dt. \tag{13.24}$$

The non-equilibrium dynamics is introduced by sweeping the magnetic field in intervals of $\Delta t = 2.5$ s back and forth. Every time we reached one of the four avoided level crossings we swapped the ground state and the excited states with the experimentally obtained QTM probability. Since $\Delta t < T_1$ we get a non-equilibrium distribution. To compute a nuclear spin trajectory of several days, we repeated this procedure 2^{24} times.

From the simulated data we extracted the relaxation times T_1 of each nuclear spin state (Fig. 13.28(e)–(h)) and obtained a perfect agreement with our experiment. The difference in lifetime between the $\pm 3/2$ and $\pm 1/2$ states comes from the fact that the nuclear spin in the $\pm 3/2$ states has only one escape path (excitation *or* relaxation), whereas if the nuclear spin in the $\pm 1/2$ has two escape paths (excitation *and* relaxation). Since the lifetime is roughly inversely proportional to the number of transition paths, the T_1's show a difference of approximately two. The exact ratio depends of course on temperature and the individual transition rates.

In order to reveal the dominant relaxation mechanism, we considered spin-lattice interactions and nuclear spin diffusion. The latter mechanism was found to be very weak in Tb crystals [61] and can hence be neglected for rather isolated and non-aligned SMMs. Concerning the spin-lattice relaxation mechanism, we examined closer the $\Gamma_{i,j}$ extracted from the simulation. Depending on its proportionality to the nuclear level spacing $\omega_{i,j}$ we can distinguish between three types of mechanisms.

(i) The Korringa process in which conduction electrons polarize the inner lying s-electrons. Since these couple with the nuclear spins via contact interaction, an energy exchange over this interaction chain is established, leading to $\Gamma_{i,j} \propto |\langle i|I_x|j\rangle|^2$ [62].

(ii) The Weger process, which suggests that the spin-lattice relaxation is dominated by the intra-ionic hyperfine interaction and the conduction electron exchange interaction [63]. It is a two-stage process, where the energy of the nucleus is transmitted to the conduction electrons via the creation and annihilation of a virtual spin wave. This process is similar to the Korringa process but results in $\Gamma_{i,j} \propto |\langle i|I_x|j\rangle|^2\omega_{i,j}^2$.

(iii) The magneto-elastic process, which leads to a deformation of the molecule due to a nuclear spin relaxation yields $\rightarrow \Gamma_{i,j} \propto |\langle i|I_x|j\rangle|^2\omega_{i,j}^4$ [64]. The term $|\langle i|I_x|j\rangle|^2$ arises from the fact, that only rotations of the spin perpendicular to the z-directions are responsible for longitudinal transitions [65].

A comparison between the $\Gamma_{i,j}$'s and the different mechanisms is shown in (Fig. 13.29). The almost perfect agreement with the Weger process suggests that the dominant relaxation process is caused by the conduction electrons. Since they are ferro-magnetically coupled to the Tb electronic spin which in turn is hyperfine

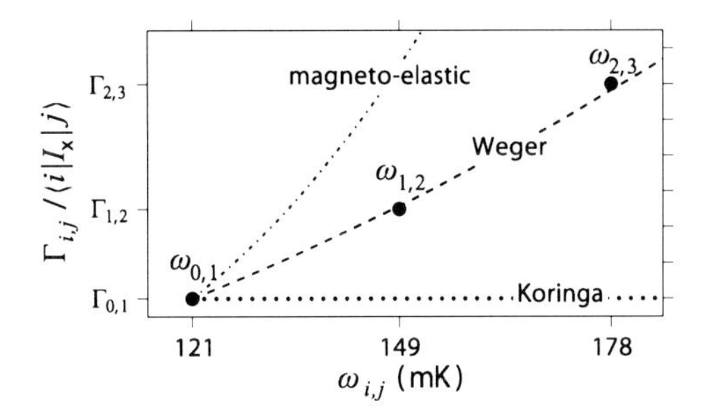

Fig. 13.29 The transition rates $\Gamma_{i,j}$ obtained from the QMC simulations exhibit a quadratic dependence on the nuclear spin level spacing $\omega_{i,j}$. This behaviour is expected from a Weger relaxation process in which the nuclear spin is coupled via virtual spin waves to conduction electrons

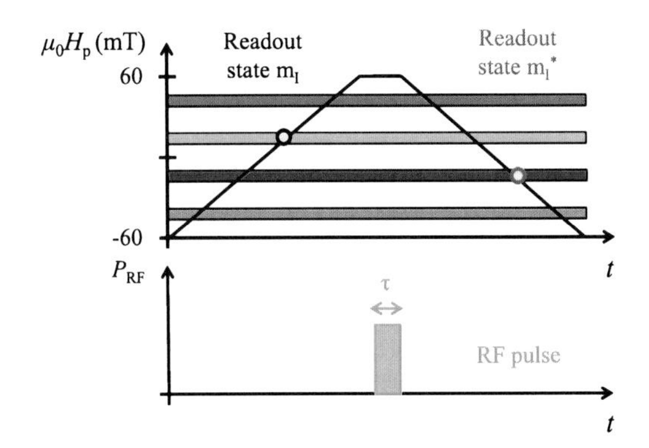

Fig. 13.30 Measurement protocol for the Rabi oscillations of the $TbPc_2$ nuclear spin. The magnetic field is swept upwards over the four avoided levels to read the nuclear spin state m_I before the following RF pulse (frequency 2.451 GHz and a pulse length τ). The magnetic field is then swept downwards to read the nuclear spin state m_I^* after the RF pulse. *The dots* in *the top panel* depict the QTM of the Tb, revealing the nuclear spin state

coupled to the nuclear spin, an energy and momentum exchange via virtual spin waves could be possible. This suggests that by controlling the amount of available conduction electrons per unit time the relaxation rate and thus T_1 can be changed. Hence, an electrically control by means of the bias and gate voltages is possible. We performed such experiments and were able to significantly reduce the T_1 of the nuclear spin [24].

Rabi Oscillations With nuclear spin lifetime T_1 of tens of seconds, one can now perform coherent manipulation i.e. Rabi oscillations of the $TbPc_2$'s nuclear spin.

For this purpose, an RF antenna was mounted in close proximity to the device to address the SMM with a microwave signal. The dc magnetic field applied to the SMM is then swept upwards over the four avoided level crossing to read out the nuclear spin state m_I prior to a microwave pulse. A microwave pulse with a frequency of 2.451 GHz and a pulse length τ is then applied to induce a coherent oscillation between the nuclear spin ground state and the first excited state. The magnetic field is then swept downwards to read the nuclear spin state m_I^* after the microwave pulse (Fig. 13.30). This measurement is then performed 400 times in

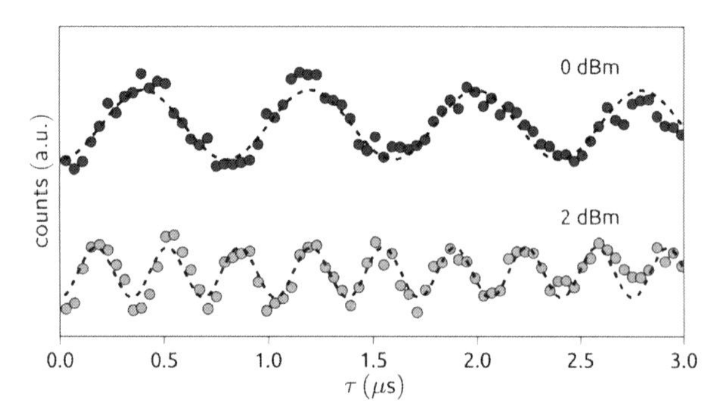

Fig. 13.31 Rabi oscillation of a single nuclear spin. The transition probability from a nuclear spin state m_I to a nuclear spin state $m_I^* = m_I \pm 1$, obtained from 400 repetitions of the measurement procedure in Fig. 13.30, is plotted as a function of the applied pulse length τ for a microwave power of 0 and 2 dBm

order to obtain the transition probability of the nuclear spin from the state m_I to the state $m_I^* = m_I \pm 1$.

Figure 13.31 represents the probability to observe a transition of the nuclear spin from m_I to $m_I^* = m_I \pm 1$ as a function of the microwave pulse length τ at two different microwave powers. The preliminary measurements show Rabi oscillations with frequencies on the order of a few MHz.

In the future, further experiments such as Ramsey or Hahn-echo measurement would allow to the determine the coherence times T_2^* and T_2. Based on the findings from Thiele et al. indicating an electrical tuning of nuclear spin relaxation [24], one could also explore the possibility of controlling the quantum coherence by electrical means using gate and bias voltages.

13.5.3 Coupling of a Single TbPc$_2$ SMM to a Carbon Nanotube's Mechanical Motion

The spin-lattice relaxation time for a direct phonon induced transition (see Sect. 13.4 above) is essentially limited by the phonon density of states at the spin resonance and the spin-phonon coupling [44]. In a macroscopic 3D system, like a SMM crystal, the phonon energy spectrum is continuous whereas in a low dimensional quantum system, like a suspended carbon nanotube (CNT), the energy spectrum is discrete and can be engineered to an extremely low density of states [66]. An individual SMM, coupled to a suspended carbon nanotube, should therefore exhibit extremely long relaxation times [66] and the reduced size and dimensionality of the system should result in a strongly enhanced spin-phonon coupling [39, 67, 68].

Carbon nanotubes (CNT) have become an essential building block for nanoelectromechanical systems (NEMS). Their low mass and high Young's modulus give rise to high oscillation frequencies for transverse [69, 70] and longitudinal modes [71–73], therefore enabling ground state cooling with state of the art cryogenics and a large zero point motion in the quantum regime [74]. Moreover, the strong coupling between nanomechanical motion and single-electron tunneling in high-Q CNT NEMS allows an electronic actuation and detection of its nanomechanical motion [38, 75–77]. As such, a CNT NEMS can be used for ultrasensitive mass sensing [78–80] or as magnetic torque detectors for single spin systems [37].

A single spin, strongly coupled with a CNT NEMS in the quantum regime, could serve as elementary qubit in quantum information processing. It has been recently suggested that a strong coupling between a quantum CNT NEMS and a single electron spin would enable basic qubit control and the implementation of entangled states [66]. In this framework, coupling a single-molecule magnet (SMM) to the quantized nanomechanical motion of a CNT NEMS is a very attractive alternative [1]. Moreover, it was recently predicted that the spin-phonon coupling between an SMM and a quantum CNT NEMS is strong enough to perform coherent spin manipulation and quantum entanglement of a spin and a resonator [67, 68].

In Ref. [39] the magnetization reversal of a $TbPc_2$ SMM grafted to a CNT resonator was studied (Figs. 13.4(a) and 13.32(a)). At cryogenic temperatures, the magnetization reversal can occur via either quantum tunneling of magnetization $|J_z, I_z\rangle \rightarrow |-J_z, I_z\rangle$ [10, 36], or, at high magnetic fields, through a direct relaxation process involving non-coherent tunneling events combined with the emission or absorption of phonons [36] (Fig. 13.10). Previous experiments on $TbPc_2$ single crystals [10] and on $TbPc_2$ coupled to a non-suspended CNT [36] showed that the spin relaxation was mainly enabled by bulk phonons in the environment of an individual SMM (crystal or substrate). An individual $TbPc_2$ SMM grafted on a suspended CNT is however physically decoupled from the bulk phonons in the substrate or the transistor leads. As a consequence the $TbPc_2$ SMM can only couple to one-dimensional phonons, associated with the nanomechanical motion of the CNT. It was recently demonstrated that high frequency and high-Q transverse [69] and longitudinal phonon [71–73] modes in a CNT at cryogenic temperatures ought to be quantized, thus yielding a discrete phonon energy spectrum.

Single-electron tunneling (SET) onto the CNT quantum dot shifts the equilibrium position of the CNT along the CNT's axis by an amount proportional to the electron-phonon coupling g [72]. For an intermediate electron-phonon coupling $g \sim 1$, the electron therefore effectively tunnels into an excited vibrational state (Fig. 13.32(b)). If the tunnel rate Γ_{out} is larger than the relaxation rate γ into the vibrational ground state, the electron tunnels out of the dot, resulting in equidistant excited states in the region of SET, running parallel to the edge of the Coulomb diamond (black arrows Fig. 13.32(c)) [71, 72]. For large electron-phonon coupling $g \gg 1$, one observes additionally a current suppression at low bias, a phenomenon known as Franck-Condon blockade [73]. It was also demonstrated, that one can pump excited vibrational states in a Coulomb-blockade regime by higher-order co-tunneling processes [71].

Indeed, one observes such quantized longitudinal phonon mode for the device studied in Ref. [39]. Bias spectroscopy reveals a spectrum of equidistant excited states in the region of SET (black arrows in Fig. 13.32(c)) with an average energy separation and phonon energy of $\hbar\omega_{\mathrm{ph}} = 140$ μeV $= 1.5$ K (Fig. 13.32(d)). According to the relation $\hbar\omega_{\mathrm{ph}} = 110$ μeV/L [μm], the obtained phonon energy corresponds to a carbon nanotube length $L_{\mathrm{CNT}} = 850$ nm, which is consistent with the *measured* carbon nanotube length $L_{\mathrm{CNT}}^{\mathrm{meas}} = 800$ nm (Fig. 13.32(a)). The quality factor is given by $Q = \omega/(2\pi\gamma)$, where γ is the relaxation rate into the ground state.

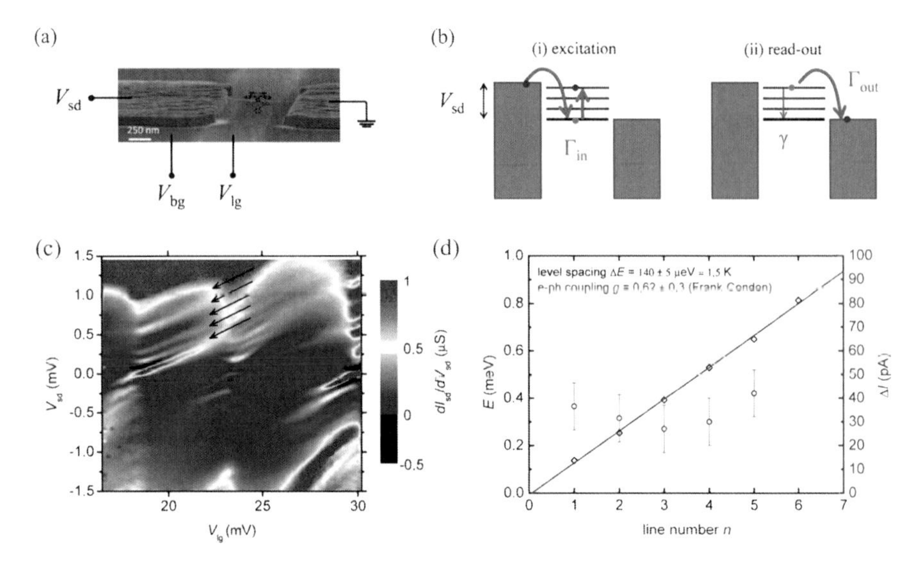

Fig. 13.32 Longitudinal stretching modes in CNT NEMS functionnalized with $TbPc_2$ SMMs. (**a**) False color SEM image of a CNT NEMS with local metallic (V_{lg}, *blue*) and Si++ (V_{bg}, *gray*) back gate. The SMM is shown as a chemical structure overlaid on the image. (**b**) Actuation and detection of longitudinal stretching modes (LSM) of a CNT NEMS: (i) Single Electron Tunneling onto the suspended CNT shifts the equilibrium position of the CNT along his axis and proportional to the electron-phonon coupling g, leaving the electron in an excited vibrational state (*red*). (ii) If $\Gamma_{out} > \gamma$, the electron tunnels out of the dot, resulting in equidistant excited states running parallel to the edge of the Coulomb diamond (indicated by *the black arrows* in (**c**)). (**c**) Stability diagram of the CNT showing the differential conductance as a function of gate and bias voltages at 20 mK and 1.4 T. *The black arrows* indicate the excited vibrational states attributed to a LSM, as described above. (**d**) Energy of excited vibrational states (*black squares*) vs. the excitation line number n at $V_g = 18$ mV. A linear fit suggests a LSM frequency of $\hbar\omega_{ph} = 140$ μeV, which is consistent with CNT length of 850 nm (see text). From the maximum current intensity ΔI (*blue dots*) of the excited vibrational states we can estimate the LSM electron-phonon coupling factor to $g \sim 0.6 \pm 0.3$. Due to measurement uncertainties in (**c**), the relative error on ΔI is estimated to be about 10 %

In order for the vibrational states to be visible in SET, we should have $\Gamma_{out} > \gamma$. The tunnel current at the edge of the Coulomb diamond I gives an approximation of the tunneling rate $\Gamma_{out} = I/e$. For tunnel current at the edge of the Coulomb diamond of $I \approx 50$ pA, we find a tunneling rate $\Gamma_{out} \approx 310$ MHz. We can determine a lower boundary for the quality factor $Q > \omega_{ph}/2\pi\Gamma_{out} \approx 110$, equivalent to an upper boundary for the linewidth of excited phonon states of $\delta E_{ph} \lesssim 15$ mK.

In a regime of Coulomb blockade and strong spin-phonon coupling, the magnetization reversal of an individual $TbPc_2$ SMM via direct transitions can induce the excitation of this longitudinal stretching mode phonon on the CNT quantum dot, which effectively behaves as a two-level system (Fig. 13.33(a)). The linewidth of the high-Q phonon was estimated to be on the order of $\delta E_{ph} \lesssim 15$ mK and is therefore smaller than the energy level spacing $\Delta E_I = 120, 150, 180$ mK between the nuclear spin states of the Tb^{3+} ion [10, 23]. As a consequence we should observe four different direct transitions matching the phonon energy $\hbar\omega_{ph} = 1.5$ K, i.e. four

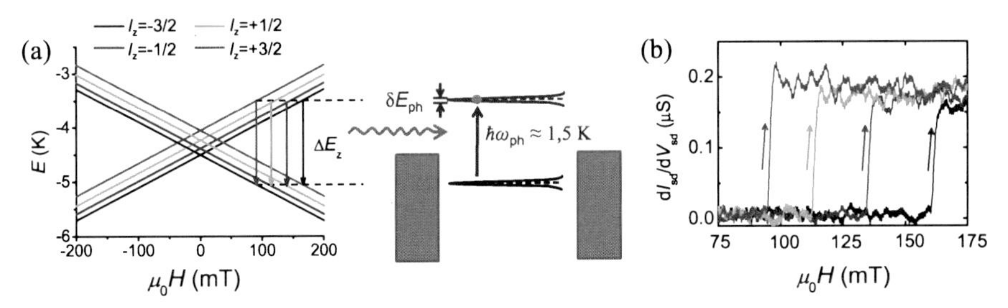

Fig. 13.33 Coupling mechanism between a single $TbPc_2$ spin to a quantized longitudinal resonator mode, described as a two-level system. The magnetization reversal of the $TbPc_2$ via *direct* transition results in the excitation of the electron into a vibrational LSM state in the carbon nanotube resonator. In contrast to bulk phonons, the energy spectrum of a LSM phonon in 1D carbon nanotubes is discretized and yields high quality factors around $Q \sim 100$. The corresponding phonon linewidth $\delta E_{\text{ph}} \approx 15$ mK is smaller than the energy separation between the Tb^{3+} nuclear spin states $\Delta E_{\text{I}} > 120$ mK. Hence, we observe four different transitions, corresponding to the four nuclear spin states as depicted by *the colored arrows*

switching fields at 89, 113, 137, 161 mT corresponding to the different nuclear spin states of the Tb^{3+} ion (Fig. 13.3(a)). By sweeping the magnetic field component parallel to the $TbPc_2$ easy axis, we can induce the magnetization reversal of the Tb^{3+} ion. As described in Sect. 13.5.1 the SMM's reversal causes an abrupt increase of the differential conductance through the CNT quantum dot, thus enabling an electronic readout and revealing the four switching fields of the Tb^{3+} ion (Fig. 13.33(b)).

In Ref. [39], magnetic field sweeps along the $TbPc_2$ easy axis were performed with a rate of 50 mT/s, from negative to positive field values at different transverse magnetic fields $\mu_0 H_{\|}$ while monitoring the differential conductance in the CNT. The magnetization reversal of the Tb^{3+} ion in a sweep translates as a jump in the CNT's differential conductance, as described above (see also Fig. 13.33(b)). The corresponding magnetic field, the switching field $\mu_0 H_{\text{SW}}$, is extracted and plotted as a function of the transverse magnetic field $\mu_0 H_{\perp}$ (Fig. 13.34(a)). The measurement reveals four switching fields between 80 and 160 mT, which are independant of $\mu_0 H_{\perp}$. By reversing the field sweep direction, the $TbPc_2$ magnetization is reversed symmetrically between -80 mT and -160 mT. In order to obtain a magnetization reversal statistic, we performed 200 back and forth sweeps at zero transverse magnetic field. The histogram of the extracted switching fields shows four dominant switching events at 88 mT, 112 mT, 137 mT and 160 mT with an average FWHM of 2 mT (Fig. 13.34(b)). The model described in Fig. 13.33 predicts that in case of a strong coupling between the $TbPc_2$ spin and the observed quantized longitudinal phonon mode in the CNT ($\hbar\omega_{\text{ph}} = 1.5$ K and linewidth $\delta E_{\text{ph}} \lesssim 15$ mK), the magnetization reversal of the $TbPc_2$ will indeed occur from each of the four nuclear spin states of the Tb^{3+} ion at magnetic fields of 89, 113, 137, 161 mT (Fig. 13.33). Also, the Zeeman energy corresponding to the FWHM of each switching event in Fig. 13.34(b) is approximately 30 mK, which is in close agreement with the phonon linewidth of $\delta E_{\text{ph}} \lesssim 15$ mK determined above. Moreover, all four switching events in the $TbPc_2$ have a transition energy of $\Delta E_{\text{z}} = 1.5$ K, which is in perfect agreement

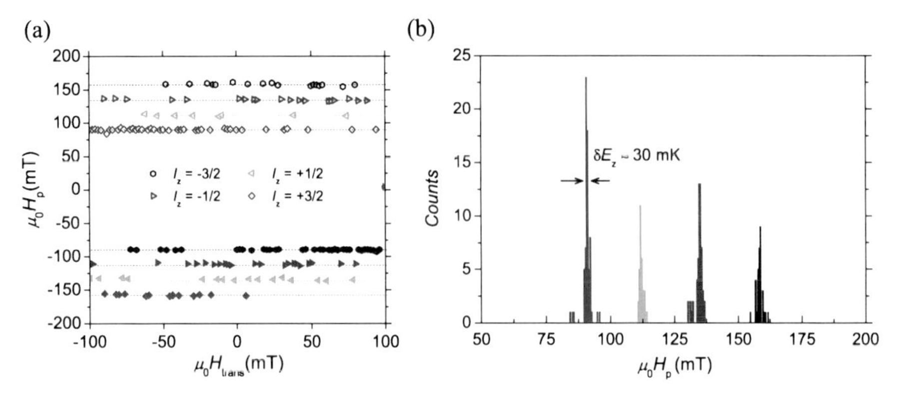

Fig. 13.34 Nuclear spin dependent magnetization reversal of a single $TbPc_2$ coupled to a CNT NEMS. (**a**) Switching field $\mu_0 H_{SW}$ vs. transverse magnetic field component $\mu_0 H_\perp$ for a sweep rate of 50 mT/s. We observe four distinct switching fields at positive values when sweeping the field from negative to positive values (*open symbol*) and at negative values when reversing the sweep direction (*full symbols*). The switching does not depend on the transverse magnetic field component. We therefore attribute these events to the magnetization reversal of the Tb^{3+} ion via a nuclear spin dependent direct transition enabled by a strong spin-phonon coupling. No quantum tunneling of magnetization was observed in our measurement. *The dotted lines* are to guide the eye. (**b**) Histogram of the switching fields obtained for 200 field sweeps with a zero transverse magnetic field component for one sweep direction and a sweep rate of 50 mT/s. We observe four dominant switching events at 88 mT, 112 mT, 137 mT and 160 mT corresponding to the switching fields for the $+3/2$, $+1/2$, $-1/2$ and $-3/2$ nuclear spin states, respectively. The FWHM of the switching field yields $\delta B = 2$ mT and the corresponding Zeeman energy of $\delta E_z \sim 30$ mK is consistent with phonon linewidth $\delta E_{ph} \lesssim 15$ mK

with the LSM phonon energy of $\hbar\omega_{ph} = 1.5$ K. The experimental findings are thus in excellent agreement with the provided model.

Upon coupling the magnetic moment of a $TbPc_2$ SMM to a quantized longitudinal phonon mode in the carbon nanotube NEMS, i.e. a two-level system, the quantization of the latter enables the detection of the four nuclear spin states of the Tb^{3+} in magnetization reversal measurements. Those findings suggests a strong coupling between the molecular spin and the quantized phonon mode, which was indeed estimated to 1.5 MHz [39]. The value is comparable to the coupling strength predicted for a carbon nanotube based spin qubit coupled to a carbon nanotube's nanomechanical motion [66] or to a superconducting coplanar waveguide [81]. In the following, we will show that other low dimensional, two-level systems such as quantum dots, may also enable the magnetic relaxation and the detection of the nuclear spin in an individual molecular magnet.

13.5.4 Coupling of a Single $TbPc_2$ SMM to a Quantum Dot

It has been demonstrated in Sect. 13.5.2 and Ref. [23], that the magnetic moment of $TbPc_2$ SMM couples to a spin $S = 1/2$ quantum dot located on the SMM's Pc ligands *via* a strong exchange interaction. The quantum dot effectively behaves as

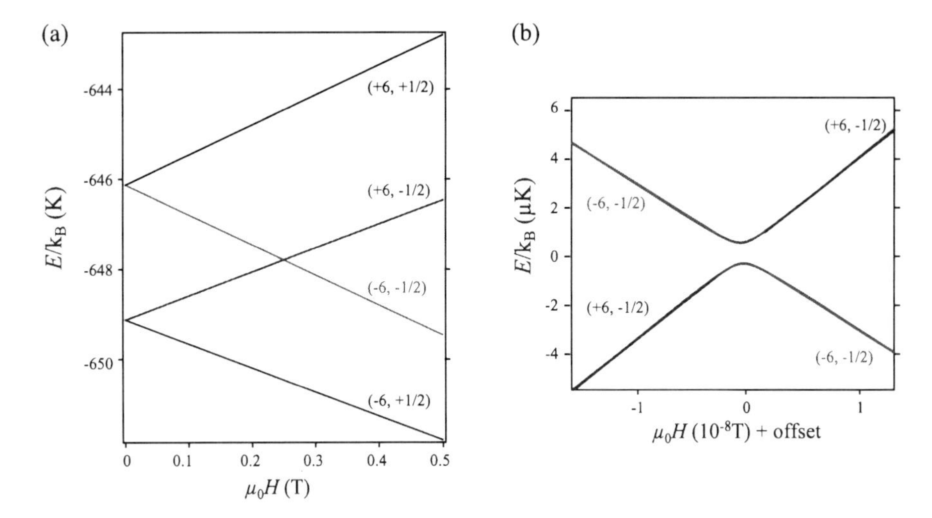

Fig. 13.35 Zeeman diagram of a system coupling the magnetic moment $J = 6$ of the $TbPc_2$ and the spin $S = 1/2$ of a quantum dot. (**a**) The coupling leads to two ground state doublets with an exchange interaction of 0.5 K. (**b**) A zoom on the crossing around 250 mT reveals a tunnel splitting of 1 μK, enabling a tunneling event from $(+6, -1/2)$ to $(-6, -1/2)$

a two-level system and could therefore enable the magnetic relaxation in the $TbPc_2$ SMM, in analogy to the magnetization reversal of the $TbPc_2$ SMM coupled to a quantized carbon nanotube phonon described in the previous section.

A system coupling the magnetic moment $J = 6$ of the $TbPc_2$ and the spin $S = 1/2$ of a quantum dot (electron) can be described by the following Hamiltonian (hyperfine interaction is omitted for simplicity)

$$\mathcal{H} = \mathcal{H}_{cf} + (g_j \mu_B J_z + g_s \mu_B S_z)\mu_0 H_{||} + (g_j \mu_B J_x + g_s \mu_B S_x)\mu_0 H_{\perp} + C\boldsymbol{J}\cdot\boldsymbol{S} \quad (13.25)$$

where $\mathcal{H}_{cf}$ describes the crystal electric field of the $TbPc_2$ and accounts for the magnetic anisotropy energy of the $TbPc_2$, g_j and g_s the gyromagnetic factors for the Tb^{3+} and the quantum dot respectively, $J_{z,x}$ and $S_{z,x}$ the spin operators of the Tb^{3+} and the quantum dot respectively, and C the magnitude of the exchange coupling between the $TbPc_2$'s magnetic moment and the spin 1/2.

Assuming an exchange interaction of $C = 0.5$ K and $g_s = 2$, the diagonalization of $\mathcal{H}$ leads to the Zeeman diagram depicted in Fig. 13.35. At a magnetic field of 250 mT the coupled system undergoes a tunnel event from $(+6, -1/2)$ to $(-6, -1/2)$, due to a tunnel splitting of a few μK. Moreover, the field position of this avoided level crossing should evolve linearly with an applied transverse magnetic field $\mu_0 H_{\perp}$ for each nuclear spin state according to (13.25).

Urdampilleta et al. performed magnetization reversal measurements on $TbPc_2$ SMM grafted to a non-suspended carbon nanotube [56] using an electronic readout described in Sects. 13.5.1 and 13.5.3. Magnetoresistance curves recorded along a field direction parallel to the SMM's easy axis show a spin-valve behaviour which

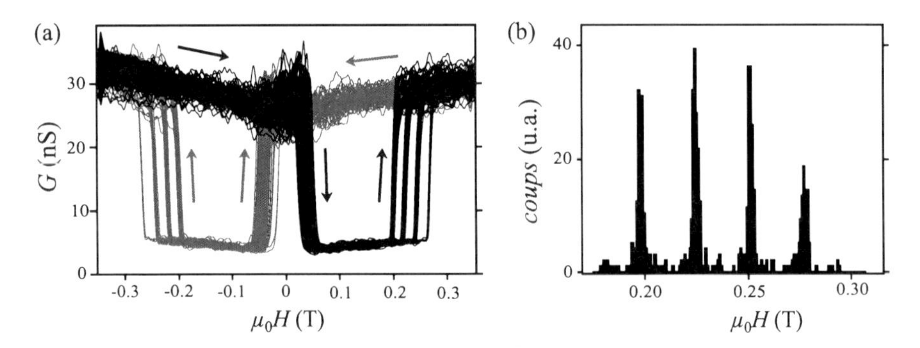

Fig. 13.36 Direct relaxation of a $TbPc_2$'s magnetic moment coupled to a spin 1/2 quantum dot. (**a**) 50 butterfly hysteresis loop in the conductance of a carbon nanotube functionnalized with $TbPc_2$ SMM at a sweep rate of 50 mT/s. (**b**) Histogram of the switching events at higher magnetic field extracted from the hysteresis loops in (**a**). The measurements reveal four distinct switching events, characteristic for the magnetization reversal from the four nuclear spin sates in the Tb^{3+} ion in the presence of a spin 1/2 system

is characteristic for such a supramolecular spintronic device (Fig. 13.36(a)). Magnetization reversal statistics furthermore reveal four distinct conductance switching events at high magnetic field, which can be associated with the direct relaxation of the $TbPc_2$'s magnetic moment from its four nuclear spin states (Fig. 13.36(a) and (b)).

Finally, the switching fields $\mu_0 H_{||,\mathrm{SW}}$ associated with the relaxation from each nuclear spin state exhibit a linear dependance on a transverse magnetic field component $\mu_0 H_\perp$ (Fig. 13.37(a)), in very good agreement with theoretical predictions (Fig. 13.37(b)).

In conclusion, a strong coupling on the order of 0.5 K was found between the $TbPc_2$'s magnetic moment and a $S = 1/2$ quantum dot, which consequently enables the magnetization reversal in the $TbPc_2$ SMM from its four nuclear spin states. A similar behavior was also observed in $TbPc_2$ spin transistor by Vincent et al., indicating that the $S = 1/2$ quantum dot is in fact located on the molecule, most likely on Pc ligands planes.

13.6 Conclusion

The achievements in the field of molecular magnetism and spintronics proved to be milestones on the ambitious path towards the implementation of a quantum computer. The unique physical and chemical properties of molecular magnets provide a large variety of quantum mechanical effects ranging from tunneling processes, interference phenomena to large quantum coherence, making them an ideal candidate for a so called spin qubit system. In order to explore these possibilities, new and very precise setups are currently built and new methods and strategies are developed. This chapter resumed the new research field of molecular quantum spintronics, which is

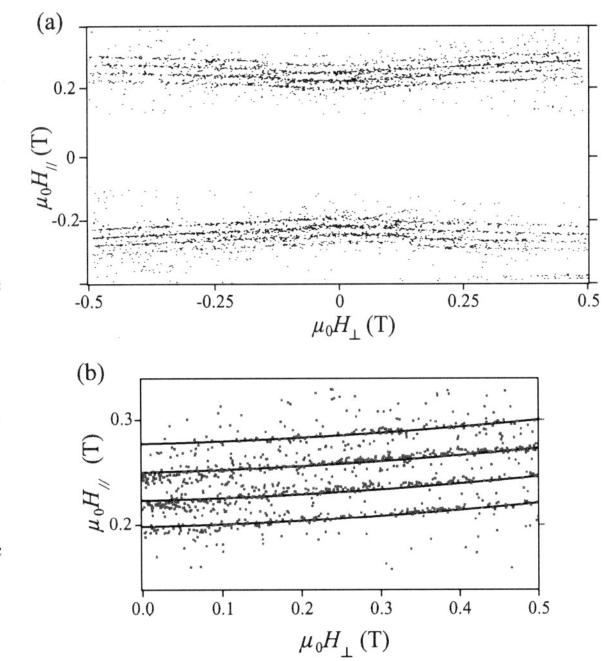

Fig. 13.37 Transverse magnetic field dependance of the $TbPc_2$ switching field. (**a**) Switching field $\mu_0 H_{||,\mathrm{sw}}$ vs. transverse magnetic field component $\mu_0 H_\perp$ for a sweep rate of 50 mT/s. We observe four distinct switching fields at positive values when sweeping the field from negative to positive values and at negative values when reversing the sweep direction. The switching field associated with each nuclear spin state depends linearly on the transverse magnetic field component. (**b**) Experiment vs. model. The fit with the model (*black curves*) are in very good agreement with the experimental data (*black dots*) extracted from (**a**)

a emerging field of nanoelectronics with a strong potential impact for the realization of new functions and devices helpful for information storage as well as quantum information. Such devices will lead to enormous progress in the understanding of the electronic and magnetic properties of isolated molecular systems and they will reveal intriguing new physics.

Acknowledgements The authors acknowledge financial support from the ANR-PNANO project MolNanoSpin No. ANR-08-NANO-002 and ERC Advanced Grant MolNanoSpin No. 226558. M. Ganzhorn acknowledges the financial support from the RTRA Nanosciences Foundation. Samples were fabricated in the NANOFAB facility of the Neel Institute. The authors are indebted to F. Balestro, E. Bonet, J.P. Cleuziou, E. Eyraud, V. Nguyen, M. Urdampilleta, S. Thiele, C. Thirion, R. Vincent.

References

1. L. Bogani, W. Wernsdorfer, Nat. Mater. **7**(3), 179 (2008)
2. S. Sanvito, A.R. Rocha, J. Comput. Theor. Nanosci. **3**, 624 (2006)
3. G. Christou, D. Gatteschi, D. Hendrickson, R. Sessoli, Mater. Res. Soc. Bull. **25**, 66 (2000)
4. A. Cornia, A.C. Fabretti, L. Zobbi, A. Caneschi, D. Gatteschi, M. Mannini, R. Sessoli, Struct. Bond. **122**, 133 (2006)
5. B. Fleury, L. Catala, V. Huc, C. David, W. Zhong, P. Jegou, L. Baraton, S.P.P. Albouy, T. Mallah, Chem. Commun. 2020 (2005). doi:10.1039/B419271K

6. A. Naitabdi, J. Bucher, P. Gerbier, P. Rabu, M. Drillon, Adv. Mater. **17**, 1612 (2005)
7. E. Coronado, A. Forment-Aliaga, A. Gaita-Arino, C. Giménez-Saiz, F. Romero, W. Wernsdorfer, Angew. Chem., Int. Ed. Engl. **43**, 6152 (2004)
8. L. Bogani, L. Cavigli, M. Gurioli, R. Novak, M. Mannini, A. Caneschi, F. Pineider, R. Sessoli Clemente-Léon, E. Coronado, A. Cornia, D. Gatteschi, Adv. Mater. **19**, 3906 (2007)
9. N. Ishikawa, M. Sugita, W. Wernsdorfer, J. Am. Chem. Soc. **127**, 3650 (2005)
10. N. Ishikawa, M. Sugita, W. Wernsdorfer, Angew. Chem., Int. Ed. Engl. **44**, 2 (2005)
11. J.R. Friedman, M.P. Sarachik, J. Tejada, R. Ziolo, Phys. Rev. Lett. **76**, 3830 (1996)
12. L. Thomas, F. Lionti, R. Ballou, D. Gatteschi, R. Sessoli, B. Barbara, Nature (London) **383**, 145 (1996)
13. C. Sangregorio, T. Ohm, C. Paulsen, R. Sessoli, D. Gatteschi, Phys. Rev. Lett. **78**, 4645 (1997)
14. W. Wernsdorfer, M. Murugesu, G. Christou, Phys. Rev. Lett. **96**, 057208 (2006)
15. W. Wernsdorfer, R. Sessoli, Science **284**, 133 (1999)
16. W. Wernsdorfer, N.E. Chakov, G. Christou, Phys. Rev. Lett. **95**, 037203 (2005)
17. A. Ardavan, O. Rival, J.J. Morton, S.J. Blundell, A.M. Tyryshkin, G.A. Timco, R.E.P. Winpenny, Phys. Rev. Lett. **98**, 057201 (2007)
18. S. Carretta, P. Santini, G. Amoretti, T. Guidi, J.R. Copley, Y. Qiu, G.A. Timco, R.E.P. Winpenny, Phys. Rev. Lett. **98**, 167401 (2007)
19. S. Bertaina, S. Gambarelli, T. Mitra, B. Tsukerblat, A. Muller, B. Barbara, Nature **453**, 203 (2008)
20. T. Komeda, H. Isshiki, J. Liu, Y. Zhang, N. Lorente, K. Katoh, B.K. Breedlove, M. Yamashita, Nat. Commun. **2**, 217 (2011)
21. Y.S. Fu, J. Schwöbel, S.W. Hla, A. Dilullo, G. Hoffmann, S. Klyatskaya, M. Ruben, R. Wiesendanger, Nano Lett. **12**(8), 3931 (2012)
22. J. Schwoebel, Y.S. Fu, J. Brede, A. Dilullo, G. Hoffmann, S. Klyatskaya, M. Ruben, R. Wiesendanger, Nat. Commun. **3**, 953 (2012)
23. R. Vincent, S. Klyatskaya, M. Ruben, W. Wernsdorfer, F. Balestro, Nature **488**, 357 (2012)
24. S. Thiele, R. Vincent, M. Holzmann, S. Klyatskaya, M. Ruben, F. Balestro, W. Wernsdorfer, Phys. Rev. Lett. **111**, 037203 (2013)
25. P. Wahl, P. Simon, L. Diekhoner, V. Stepanyuk, P. Bruno, M. Schneider, K. Kern, Phys. Rev. Lett. **98**, 056601 (2007)
26. C.F. Hirjibehedin, C.Y. Lin, A.F. Otte, M. Ternes, C.P. Lutz, B.A. Jones, A.J. Heinrich, Science **317**, 1199 (2007)
27. H. Park, A.K.L. Kim, A.P. Alivisatos, J. Park, P.L. McEuen, Appl. Phys. Lett. **75**, 301 (1999)
28. H.B. Heersche, Z. de Groot, J.A. Folk, H.S.J. van der Zant, C. Romeike, M.R. Wegewijs, L. Zobbi, D. Barreca, E. Tondello, A. Cornia, Phys. Rev. Lett. **96**, 206801 (2006)
29. M.H. Jo, J. Grose, K. Baheti, M. Deshmukh, J. Sokol, E. Rumberger, D. Hendrickson, J. Long, H. Park, D. Ralph, Nano Lett. **6**, 2014 (2006)
30. N.E. Chakov, M. Soler, W. Wernsdorfer, K.A. Asbboud, G. Christou, Inorg. Chem. **44**, 5304 (2005)
31. M. Mannini, P. Sainctavit, R. Sessoli, C. Cartier dit Moulin, F. Pineider, M.A. Arrio, A. Cornia, D. Gatteschi, Chemistry **14**(25), 7530 (2008)
32. E. Burzurí, A.S. Zyazin, A. Cornia, H.S.J. van der Zant, Phys. Rev. Lett. **109**, 147203 (2012)
33. A.S. Zyazin, J.W.G. van den Berg, E.A. Osorio, H.S.J. van der Zant, N.P. Konstantinidis, M. Leijnse, M.R. Wegewijs, F. May, W. Hofstetter, C. Danieli, A. Cornia, Nano Lett. **10**(9), 3307 (2010)
34. A. Pasupathy, R. Bialczak, J. Martinek, J. Grose, L. Donev, P. McEuen, D. Ralph, Science **306**, 86 (2004)
35. N. Roch, S. Florens, V. Bouchiat, W. Wernsdorfer, F. Balestro, Nature **453**, 633 (2008)
36. M. Urdampilleta, S. Klyatskaya, J.P. Cleuziou, M. Ruben, W. Wernsdorfer, Nat. Mater. **10**, 502 (2011)
37. B. Lassagne, D. Ugnati, M. Respaud, Phys. Rev. Lett. **107**, 130801 (2011)
38. M. Ganzhorn, W. Wernsdorfer, Phys. Rev. Lett. **108**, 175502 (2012)
39. M. Ganzhorn, S. Klyatskaya, M. Ruben, W. Wernsdorfer, Nat. Nanotechnol. **8**, 165 (2013)

40. J.P. Cleuziou, W. Wernsdorfer, V. Bouchiat, T. Ondarcuhu, M. Monthioux, Nat. Nanotechnol. **1**, 53 (2006)
41. R. Maurand, T. Meng, E. Bonet, S. Florens, L. Marty, W. Wernsdorfer, Phys. Rev. X **2**, 011009 (2012)
42. R. Maurand, Carbon nanotube squid: Josephson junction quantum dots,junction, Kondo effect and detection of single molecule magnets. Ph.D. Thesis, Joseph Fourier University, Grenoble (2011)
43. S. Kyatskaya, J.R.G. Mascarós, L. Bogani, F. Hennrich, M. Kappes, W. Wernsdorfer, M. Ruben, J. Am. Chem. Soc. **131**(42), 15143 (2009)
44. A. Abragam, B. Bleaney, *Electron Paramagnetic Resonance of Transition Ions* (Clarendon Press, Oxford, 1970)
45. N. Ishikawa, M. Sugita, T. Okubo, N. Tanaka, T. Iino, Y. Kaizu, Inorg. Chem. **42**(7), 2440 (2003)
46. K.W.H. Stevens, Proc. Phys. Soc. A **65**(3), 209 (1952)
47. H. Konami, M. Hatano, A. Tajiri, Chem. Phys. Lett. **160**(2), 163 (1989)
48. W. Wernsdorfer, Supercond. Sci. Technol. **22**(6), 064013 (2009)
49. C. Zener, Proc. R. Soc. Lond. Ser. A, Math. Phys. Sci. **137**, 696 (1932)
50. J.C. Charlier, X. Blase, S. Roche, Rev. Mod. Phys. **79**(2), 677 (2007)
51. J. Nygard, D.H. Cobden, P.E. Lindelof, Nature **408**, 342 (2000)
52. M. Urdampilleta, S. Klyatskaya, M. Ruben, W. Wernsdorfer, Phys. Rev. B **87**, 195412 (2013)
53. W. Wernsdorfer, S. Bhaduri, R. Tiron, D.N. Hendrickson, G. Christou, Phys. Rev. Lett. **89**, 197201 (2002)
54. W. Wernsdorfer, D. Mailly, G.A. Timco, R.E.P. Winpenny, Phys. Rev. B **72**, 060409(R) (2005)
55. L. Sorace, C. Benelli, D. Gatteschi, Chem. Soc. Rev. **40**, 3092 (2011)
56. M. Urdampilleta, Molecular spintronics: from spin-valves to single spin detection. Ph.D. Thesis, Joseph Fourier University, Grenoble (2012)
57. W. Wernsdorfer, E.B. Orozco, K. Hasselbach, A.B.B. Barbara, N. Demoncy, A. Loiseau, D. Boivin, H. Pascard, D. Mailly, Phys. Rev. Lett. **78**, 1791 (1997)
58. J. Dalibard, Y. Castin, K. Mølmer, Phys. Rev. Lett. **68**, 580 (1992)
59. K. Mølmer, Y. Castin, J. Dalibard, J. Opt. Soc. Am. B **10**(3), 524 (1993)
60. K. Mølmer, Y. Castin, Quantum Semiclassical Opt. **8**(1), 49 (1996)
61. N. Sano, J. Itoh, J. Phys. Soc. Jpn. **32**(1), 95 (1972)
62. J. Korringa, Physica **16**(7–8), 601 (1950)
63. M. Weger, Phys. Rev. **128**, 1505 (1962)
64. N. Sano, S. Kobayashi, J. Itoh, Prog. Theor. Phys. Suppl. **46**, 84 (1970)
65. M. McCausland, I. Mackenzie, Adv. Phys. **28**(3), 305 (1979)
66. A. Pályi, P.R. Struck, M. Rudner, K. Flensberg, G. Burkard, Phys. Rev. Lett. **108**, 206811 (2012)
67. A.A. Kovalev, L.X. Hayden, G.E.W. Bauer, Y. Tserkovnyak, Phys. Rev. Lett. **106**, 147203 (2011)
68. D.A. Garanin, E.M. Chudnovsky, Phys. Rev. X **1**, 011005 (2011)
69. E.A. Laird, F. Pei, W. Tang, G.A. Steele, L.P. Kouwenhoven, Nano Lett. **12**(1), 193 (2012)
70. H.B. Peng, C.W. Chang, S. Aloni, T.D. Yuzvinsky, A. Zettl, Phys. Rev. Lett. **97**, 087203 (2006)
71. A.K. Hüttel, B. Witkamp, M. Leijnse, M.R. Wegewijs, H.S.J. van der Zant, Phys. Rev. Lett. **102**, 225501 (2009)
72. S. Sapmaz, P. Jarillo-Herrero, Y.M. Blanter, C. Dekker, H.S.J. van der Zant, Phys. Rev. Lett. **96**, 026801 (2006)
73. R. Leturcq, C. Stampfer, K. Inderbitzin, L. Durrer, C. Hierold, E. Mariani, M.G. Schultz, F. von Oppen, K. Ensslin, Nat. Phys. **5**, 327 (2009)
74. M. Poot, H.S. van der Zant, Phys. Rep. **511**(5), 273 (2012)
75. B. Lassagne, Y. Tarakanov, J. Kinaret, D. Garcia-Sanchez, A. Bachtold, Science **325**(5944), 1107 (2009)

76. G.A. Steele, A.K. Hüttel, B. Witkamp, M. Poot, H.B. Meerwaldt, L.P. Kouwenhoven, H.S.J. van der Zant, Science **325**(5944), 1103 (2009)
77. A.K. Hüttel et al., Nano Lett. **9**(7), 2547 (2009)
78. B. Lassagne, D. Gárcia-Sánchez, A. Aguasca, A. Bachtold, Nano Lett. **8**(11), 3735 (2008)
79. H.Y. Chiu, P. Hung, H.W.C. Postma, M. Bockrath, Nano Lett. **8**(12), 4342 (2008)
80. K. Jensen, K. Kim, A. Zettl, Nat. Nanotechnol. **3**(9), 533 (2008)
81. A. Cottet, T. Kontos, Phys. Rev. Lett. **105**, 160502 (2010)

Chapter 14
Molecule-Based Magnetic Coolers: Measurement, Design and Application

Marco Evangelisti

Abstract The recent progress in molecule-based magnetic materials exhibiting a large magnetocaloric effect at liquid-helium temperatures is reviewed. The experimental methods for the characterization of this phenomenon are described. Theory and examples are presented with the aim of identifying those parameters to be addressed for improving the design of new refrigerants belonging to this class of materials. Advanced applications and future perspectives are also discussed.

14.1 Introduction

Magnetic refrigeration exploits the magnetocaloric effect (MCE), which can be described as either an isothermal magnetic entropy change (ΔS_m) or an adiabatic temperature change (ΔT_{ad}) following a change of the applied magnetic field (ΔH). The roots of this technology date back to 1881, when Warburg experimentally observed that an iron sample heated a few milliKelvin when moved into a magnetic field and cooled down when removed out of it [1]. In 1918, Weiss and Piccard explained the magnetocaloric effect [2]. In the late 1920s, Debye and Giauque independently proposed adiabatic demagnetization as a suitable method for attaining sub-Kelvin temperatures [3, 4]. In 1933, Giauque and MacDougall applied this method to reach 0.25 K by making use of 61 grams of $Gd_2(SO_4)_3 \cdot 8H_2O$, starting from 1.5 K and applying $\mu_0 \Delta H = 0.8$ T [5]. Since then, magnetic refrigeration is a standard technique in cryogenics, which has shown to be useful to cool down from a few Kelvin [6, 7]. Applications include, among others: superconducting magnets, helium liquifiers, medical instrumentation, in addition to many scientific researches. So-called adiabatic demagnetization refrigerators (ADR) are used as ultra-low-temperature platforms in space borne missions, where the absence of gravity prevents cooling by methods based on ^{3}He-^{4}He dilutions. Magnetic refrigeration at liquid-helium temperatures provides a valid alternative to the use of helium itself, specially for the

M. Evangelisti (✉)
Instituto de Ciencia de Materiales de Aragón and Departamento de Física de la Materia Condensada, CSIC–Universidad de Zaragoza, C/Pedro Cerbuna 12, 50009 Zaragoza, Spain
e-mail: evange@unizar.es

J. Bartolomé et al. (eds.), *Molecular Magnets*, NanoScience and Technology,
DOI 10.1007/978-3-642-40609-6_14, © Springer-Verlag Berlin Heidelberg 2014

Table 14.1 Selection of molecule-based magnetic refrigerants and corresponding references. A rough chronological order is followed from *top* to *bottom*

Compound	Ref.	Compound	Ref.
$\{Mn_{12}\}$	[9]	$\{Fe_8\}$	[9]
$\{Cr_7Cd\}$	[10]	$\{Fe_{14}\}$	[11–13]
$\{NiCr\}_n$	[14, 15]	$\{Cr_3Gd_2\}_n$	[14, 15]
$\{Mn_6^{3+}Mn_4^{2+}\}$	[16]	$\{Mn_6^{3+}Mn_4^{2+}\}$	[17]
$\{Mn_6^{3+}Mn_8^{2+}\}$	[17]	$\{Mn_4^{3+}Gd_4\}$	[18, 19]
$\{Mn_8^{4+}Mn_{24}^{2+}\}$	[20]	$[Gd_2(fum)_3(H_2O)_4] \cdot 3H_2O$	[21]
$\{Mn_6^{3+}Mn_4^{2+}\}$	[22]	$\{Mn_{11}^{3+}Mn_6^{2+}\}$	[22]
$\{Mn_{12}^{3+}Mn_7^{2+}\}$	[22]	$\{Gd_7\}$	[23]
$\{Mn_{19}\}$	[24]	$\{Na_2Mn_{15}\}$	[24]
$\{Ni_6Gd_6P_6\}$	[25]	$\{Ni_{12}Gd_{36}\}$	[26]
$\{Co_6Gd_8\}$	[26]	$[Gd_2(OAC)_6(H_2O)_4] \cdot 4H_2O$	[27, 28]
$\{Cu_5Gd_4\}$	[29]	$\{Ni_3Gd\}$	[30]
$\{Co_8Gd_8\}$	[31]	$\{Co_4Gd_6\}$	[31]
$[Gd_2(OAC)_3(H_2O)_{0.5}]_n$	[32]	$[Gd_4(OAC)_4(acac)_8(H_2O)_4]$	[32]
$[Gd_2(OAC)_3(MeOH)]_n$	[32]	$[Gd_2(OAC)_2(Ph_2acac)_6(MeOH)_2]$	[32]
$\{Co_4Gd_6\}$	[33]	$\{Co_8Gd_4\}$	[33]
$\{Co_4Gd_2\}$	[33]	$\{Co_8Gd_2\}$	[33]
$\{Co_8Gd_8\}$	[33]	$\{Co_8Gd_4\}$	[33]
$\{Fe_5Gd_8\}$	[34]	$\{Cu_{36}Gd_{24}\}$	[35]
$\{Gd_{10}\}$-POM	[36]	$\{Gd_{30}\}$-POM	[36]
$\{Zn_8Gd_4\}$	[37]	$\{Ni_8Gd_4\}$	[37]
$\{Cu_8Gd_4\}$	[37]	$\{Gd_5Zn(BPDC)_3\}_n$	[38]
$\{Cr_2Gd_3\}$	[39]	$\{Cr_2Gd_2\}$	[40]
$\{Co_{10}Gd_{42}\}$	[41]	$\{Ni_{10}Gd_{42}\}$	[41]
$[Gd_2(N\text{-}BDC)_3(dmf)_4]$	[42]	$[Mn(H_2O)_6][MnGd(oda)_3]_2 \cdot 6H_2O$	[43]
$\{Cu_6^{2+}Gd_6^{3+}\}$	[44]	$Co_4(OH)_2(C_{10}H_{16}O_4)_3$	[45]
$\{NiNb^{4+}\}_n$	[46]	$\{Mn^{2+}Nb^{4+}\}_n$	[46]
$[Gd(HCOO)(C_8H_4O_4)]$	[47]	$\{Gd_{12}Mo_4\}$	[48]
$\{Fe_{17}\}$	[49]	$\{Gd_{24}\}$	[50]
$\{Ni_{12}Gd_5\}$	[51]	$\{Co_6Gd_4\}$	[52]
$\{Gd(OOCH)_3\}_n$	[53]	$\{MnGd\}$	[54]

rarer ^{3}He isotope, whose cost has been increasing dramatically during the past few years.

All magnetic materials show the MCE, although the intensity of the effect depends on the properties of each material. Since the initial proposition that magnetic molecular clusters are promising systems for refrigeration at low temperatures [8], very large values of ΔS_m and ΔT_{ad} have been repeatedly reported for several molecule-based magnetic materials. Table 14.1 lists a wide up-to-date selection of such refrigerants, with their corresponding references. An extensive investigation is currently under way, with a view to finding or synthesizing new molecule-based materials capable of record performances in terms of the MCE.

In order to efficiently exploit the MCE for a realistic application, this effect should be maximized within the working temperature range of interest. In order to do so at cryogenic temperatures, the molecule-based magnetic coolers must possess a combination of a large spin ground state with negligible anisotropy, weak ferromagnetic exchange between the constituent magnetic ions, in addition to a relatively large metal:non-metal mass ratio, i.e., a large magnetic density. This chapter describes the underlying physics of magnetic refrigeration with molecule-based coolers. Section 14.2 defines the MCE and provides its theoretical framework. Section 14.3 addresses which experimental techniques should be applied, and how to correctly do it, in order to characterize the MCE of this class of materials. Several case examples are provided in Sect. 14.4, with the aim of highlighting the characteristics which are known to influence the performance of these coolers. Section 14.5 introduces the reader into the field of on-chip microrefrigeration—an advanced application which starts from the challenging magnetic characterization of molecule-based magnetic coolers deposited over a substrate. Concluding remarks are presented in Sect. 14.6.

14.2 Theoretical Framework

In order to explain the origin of the magnetocaloric effect, we use thermodynamic relations which relate the magnetic variables (magnetization M and magnetic field H) to entropy S_E and temperature T. Let us recall [55] that the definition of the entropy of a system having Ω accessible states is $S_E = k_B \ln(\Omega)$. Since a magnetic moment of spin s has $2s + 1$ magnetic spin states, the entropy content per mole of substance associated with the magnetic degrees of freedom at $T = \infty$ is

$$S_m = R \ln(2s + 1), \tag{14.1}$$

where $R = N_A k_B$ is the gas constant. The spin s should be considered as an effective spin describing the multiplicity of relevant magnetic states.

When a material is magnetized by the application of a magnetic field, the magnetic entropy is changed as the field changes the magnetic order of the material. The MCE and the associated principle of adiabatic demagnetization is readily understood by looking at Fig. 14.1. The system, assumed to be a paramagnetic material, is initially in state $A(T_i, H_i)$, at temperature T_i and field H_i. Under adiabatic conditions, i.e., when the total entropy of the system remains constant, the magnetic entropy change must be compensated for by an equal but opposite change of the entropy associated with the lattice, resulting in a change in temperature of the material. That is, the adiabatic field change $H_i \to H_f$ brings the system to state $B(T_f, H_f)$ with the temperature change $\Delta T_{ad} = T_f - T_i$ (horizontal arrow in Fig. 14.1). On the other hand, if the magnetic field is isothermally changed to H_f, the system goes to state $C(T_i, H_f)$, resulting in the magnetic entropy change ΔS_m (vertical arrow). Both ΔS_m and ΔT_{ad} are the characteristic parameters of the MCE. It is easy to see that if the magnetic change ΔH reduces the entropy ($\Delta S_m < 0$), then ΔT_{ad} is positive, whereas if ΔH is such that $\Delta S_m > 0$, then $\Delta T_{ad} < 0$.

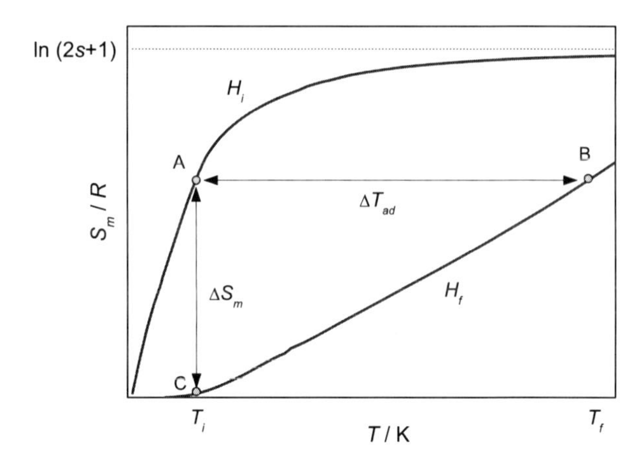

Fig. 14.1 Temperature-dependence of the molar magnetic entropy of a (super)paramagnet with spin s per formula unit, for magnetic field H_i and $H_f > H_i$. AB process: adiabatic magnetization (A → B) or demagnetization (B → A), providing ΔT_{ad}. AC process: isothermal magnetization (A → C) or demagnetization (C → A), providing ΔS_m

14.3 Experimental Evaluation of the MCE

A widely-accepted approach used to evaluate the MCE consists in obtaining ΔS_m exclusively from magnetization measurements as function of temperature and applied magnetic field, by adopting the procedure described in Sect. 14.3.1. Although some care should be taken for collecting and then analyzing the experimental data correctly, this approach has the clear advantages of being simple and relatively fast. No other experimental tool is needed but a conventional magnetometer. A far more complete characterization of the MCE is accomplished by means of heat capacity measurements collected for varying temperature and applied magnetic field, which permit to compute both ΔS_m and ΔT_{ad}. For the practical cases, these two *indirect* approaches rely on the numerical evaluation of integrals that, by their nature, can produce heavy errors, as made evident by Pecharsky and Gschneidner [56]. To overcome any inherent shortfall, a third and more reliable option is the *direct* measurement of the physical effect. Although several experimental methods succeeded in measuring directly the MCE, a higher degree of sophistication is required and therefore this option is restricted within a few specialized laboratories.

14.3.1 Indirect Methods

In order to establish the relationship between H, M and T to the MCE terms, ΔT_{ad} and ΔS_m, we consider the Maxwell equation for the magnetic entropy

$$\left(\frac{\partial S_m(T,H)}{\partial H}\right)_T = \left(\frac{\partial M(T,H)}{\partial T}\right)_H. \tag{14.2}$$

Integrating (14.2) for an isothermal process, we obtain

$$\Delta S_m(T,\Delta H) = \int_{H_i}^{H_f} \left(\frac{\partial M(T,H)}{\partial T}\right)_H \mathrm{d}H. \tag{14.3}$$

This equation indicates that ΔS_m is proportional to both the derivative of magnetization with respect to temperature at constant field and to the field variation. The accuracy of ΔS_m calculated from magnetization experiments using (14.3) depends on the accuracy of the measurements of the magnetic moment, T and H. It is also affected by the fact that the exact differentials are replaced by the measured variations (ΔM, ΔT and ΔH). Furthermore, it is worth mentioning that the Maxwell equation does not hold for first order phase transitions, since $\partial M/\partial T \to \infty$.

By replacing the specific heat at constant field

$$C = T\left(\frac{\partial S_m}{\partial T}\right)_H,$$

in the expression of the infinitesimal change of $S_m(T, H)$, we have

$$\mathrm{d}S_m = \left(\frac{\partial S_m}{\partial T}\right)_H \mathrm{d}T + \left(\frac{\partial S_m}{\partial H}\right)_T \mathrm{d}H = \frac{C}{T}\mathrm{d}T + \left(\frac{\partial S_m}{\partial H}\right)_T \mathrm{d}H. \tag{14.4}$$

For an adiabatic process, $\mathrm{d}S_m = 0$. Thus, we obtain

$$\mathrm{d}T_{ad} = -\frac{T}{C}\left(\frac{\partial S_m}{\partial H}\right)_T \mathrm{d}H, \tag{14.5}$$

where T_{ad} is the adiabatic temperature. Therefore, taking into account (14.2), the adiabatic temperature change is expressed by

$$\Delta T_{ad}(T, \Delta H) = \int_{H_i}^{H_f} \left(\frac{T}{C(T, H)}\right)_H \left(\frac{\partial M(T, H)}{\partial T}\right)_H \mathrm{d}H. \tag{14.6}$$

From the experimental specific heat, the temperature dependence of the magnetic entropy $S_m(T)$ is obtained by integration, i.e., using

$$S_m(T) = \int_0^T \frac{C_m(T)}{T}\mathrm{d}T, \tag{14.7}$$

where $C_m(T)$ is the magnetic specific heat as obtained by subtracting the lattice contribution from the total measured C. Hence, if $S_m(T)$ is known for H_i and H_f, both $\Delta T_{ad}(T, \Delta H)$ and $\Delta S_m(T, \Delta H)$ can be obtained.

The accuracy in the evaluation of MCE using specific heat data depends critically on the accuracy of the C measurements and data processing, e.g., the use of ΔT instead of $\mathrm{d}T$. Indeed, small errors in C can produce important differences in ΔS_m and ΔT_{ad} at high temperature due to the integration process. Moreover, C data measured by the heat pulse technique are less accurate near phase transitions due to the long relaxation times required for thermal equilibrium after each pulse. An additional source of uncertainty may eventually come from the fact that, in order to carry out the integration of (14.7), one has to extrapolate the experimental data to $T = 0$ and to $T = \infty$. The former extrapolation might become critical depending on the lowest experimentally-accessible temperature in comparison to the magnitude of the

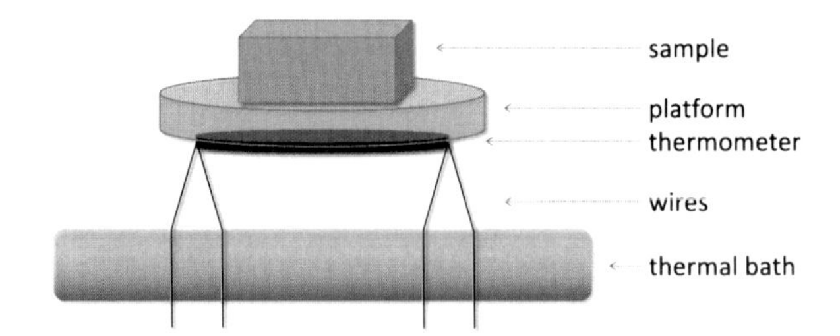

Fig. 14.2 Scheme of the thermal sensor used for the direct determination of the temperature change on magnetization and demagnetization

relevant energies involved in the magnetic ordering mechanism due to, e.g., interactions that are typically weak in molecule-based magnetic materials. For a magnetic system that undergoes a phase transition within the accessible T window, one can sometimes attempt the extrapolation to $T = 0$ by making use of spin-wave models [20]. Alternatively, the 'missing' entropy that characterizes the not-accessible lowest temperatures can be estimated from the expected full entropy content (14.1), after subtracting the result obtained by integrating (14.7) between the experimental lowest T and the high-T extrapolation ($\propto T^{-2}$). However, one has to consider that the drawback of such a method is an increasingly large uncertainty, which might jeopardize and even invalidate the analysis based on the specific heat data. For a molecule-based refrigerant containing Gd^{3+} spin centers, the zero-field magnetic specific heat shows up for temperatures lower than 2–3 K. It is then not sufficient to use a commercial calorimeter typically limited to ≈ 2 K as the lowest achievable T by pumping ^{4}He. An unfortunate example of poor analysis of specific heat data can be spotted in the recent literature [41]. As a rule of thumb, if the molecule-based refrigerant contains Gd^{3+} spin centers, then sub-Kelvin temperatures are needed for characterizing the MCE by specific heat experiments.

14.3.2 Direct Measurements

A far more elegant and reliable method for determining the MCE is by means of *direct* measurements. Clearly, the advantage resides in avoiding those drawbacks inherent to indirect methods, such as the poor accuracy associated to the data processing and the $T \to 0$ extrapolation of the specific heat (see Sect. 14.3.1). However, any experimental set-up designed for direct MCE measurements has to deal with unavoidable heat dissipations, i.e., the lack of ideal adiabatic conditions. Most employed methods are based on a rapid change of the applied magnetic field during the, correspondingly short, time interval of a single measurement [57]. These measurements could be considered adiabatic experiments, at least to a first approximation. In the procedure described below, we go beyond this time interval by providing a full description of the physical process involved, which becomes relevant at a scale longer than the time needed for fully changing the applied field [27, 53].

In the experimental set-up, the sample-holder is a sapphire plate to which a resistance thermometer is attached (Fig. 14.2). The wires provide electrical connection, mechanical support and thermal contact to a controlled thermal bath at constant temperature T_0. Starting with the sample at zero field $H = 0$ and T_0, the measuring procedure comprises the following four steps: (a) gradual application of a magnetic field, up to a maximum H_0; (b) relaxation until the sample reaches again the thermal equilibrium with the bath; (c) gradual demagnetization down to $H = 0$; (d) relaxation at zero field until the sample reaches the equilibrium at T_0. During the whole procedure, the as-measured temperature T and applied magnetic field H are recorded continuously.

In order to cope with the unavoidable lack of ideal adiabaticity, one has to relate the as-measured T to the adiabatic temperature T_{ad}, i.e., the temperature if the sample would have been kept thermally isolated during the process. For this purpose, the experimental entropy gains (losses) of the sample which originate from heat dissipation from (to) the thermal bath should be evaluated. The thermal conductance κ of the wires is previously measured as a function of the temperature using a free-oxygen copper block as the sample. The specific heat at constant field, C, of the sample is also previously measured using another calorimeter.

The entropy change of the sample in an infinitesimal time interval is

$$\mathrm{d}S = \frac{k(T_0 - T)}{T}\mathrm{d}t. \tag{14.8}$$

Taking into account (14.4), we then have

$$\begin{aligned} \frac{k(T_0 - T)}{T}\mathrm{d}t &= \frac{C}{T}\mathrm{d}T + \left(\frac{\partial S}{\partial H}\right)_T \mathrm{d}H, \\ \frac{\mathrm{d}T}{\mathrm{d}t} &= \frac{k(T_0 - T)}{C} - \frac{T}{C}\left(\frac{\partial S}{\partial H}\right)_T \frac{\mathrm{d}H}{\mathrm{d}t}. \end{aligned} \tag{14.9}$$

By replacing $\mathrm{d}T_{ad}$ from (14.5), we obtain

$$\frac{\mathrm{d}T}{\mathrm{d}t} = \frac{k(T_0 - T)}{C} + \frac{\mathrm{d}T_{ad}}{\mathrm{d}t},$$

which finally results in

$$T_{ad}(t) = T(t) - \int_{t_0}^{t} \frac{k(T_0 - T)}{C}\mathrm{d}t. \tag{14.10}$$

Therefore, knowing κ and C, the adiabatic temperature can be precisely determined for the whole magnetization-demagnetization process. From (14.10), we note that the deviation of $T(t)$ from $T_{ad}(t)$, as in the ideal adiabatic process, increases with t. We then also note that $T(t) \simeq T_{ad}(t)$ when $t/t_0 \simeq 1$. Thus, if the measurement is based on a fast change of the applied magnetic field and it is limited to the short time scale corresponding to the interval needed for fully changing the applied field, then the as-measured T already provides a good determination of the adiabatic temperature. In this treatment, the entropy contribution due to the heat transferred

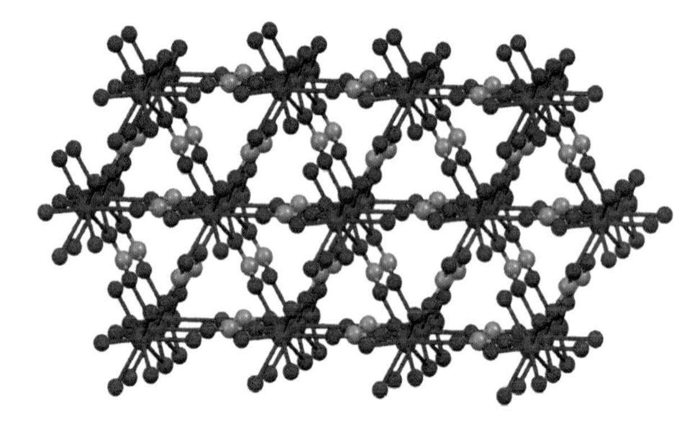

Fig. 14.3 Structure of the $\{Gd(OOCH)_3\}_n$ three-dimensional metal-organic framework material. Gd = purple, O = red, C = gray. H atoms are omitted for clarity

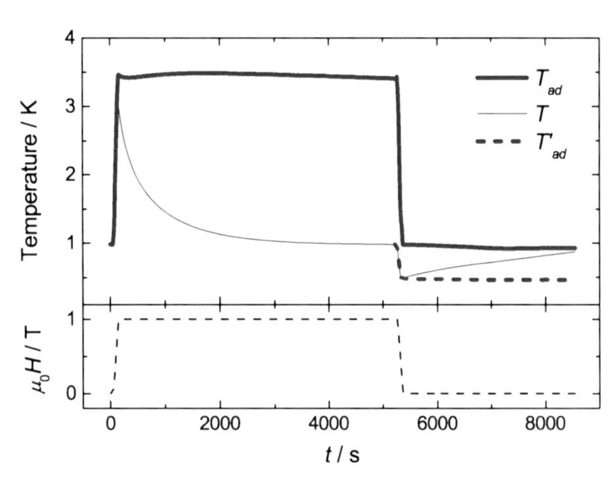

Fig. 14.4 *Top*: Direct measurement of the experimental temperature T and deduced adiabatic temperatures T_{ad} and T'_{ad} for $\{Gd(OOCH)_3\}_n$ on a magnetization and demagnetization cycle. *Bottom*: Time evolution of the corresponding applied magnetic field

from the sample holder to the refrigerant material, i.e., $\Delta S_{sh} = \int_{T_0}^{T} C_{sh}/T \mathrm{d}T$, is disregarded. This is acceptable since the specific heat of the sample holder C_{sh} typically is orders of magnitude lower, and therefore negligible, with respect to that of the sample at these liquid-helium temperatures.

Case Example: $\{Gd(OOCH)_3\}_n$ 3D Metal-Organic Framework Hereafter, we describe the direct measurements of the temperature changes, induced by ΔH, that were reported in Ref. [53] for gadolinium formate, whose chemical formula is $\{Gd(OOCH)_3\}_n$, which belongs to the class of metal-organic framework (MOF) materials (Fig. 14.3).

For $\{Gd(OOCH)_3\}_n$, Fig. 14.4 shows the time evolution of the field H, experimental temperature T and deduced adiabatic temperature T_{ad} for a representative magnetization-demagnetization full cycle, starting at $T_0 = 0.98$ K and reaching $\mu_0 H_0 = 1$ T. In sequential order, we can observe the following stages. The experimental temperature T increases while the field increases up to 1 T. Here T_{ad} increases more than T because the thermal losses to the bath are compensated to compute T_{ad}. The experimental temperature T decays back to $T_0 = 0.98$ K, but $T_{ad} = 3.5$ K is constant, since it corresponds to an hypothetical adiabatic process at constant H. In the demagnetization process, starting from $t_0 = 5270$ s in Fig. 14.4, T decreases below T_0 due to the magnetocaloric effect (Sect. 14.2). By 'resetting' T_0 to 0.98 K, we can here define a new, though equivalent, adiabatic temperature T'_{ad} (dashed line in the top panel). We observe that T_{ad} tends to recover the initial value T_0 corresponding to $t_0 = 0$, while T'_{ad} cools down to the new temperature of 0.47 K because of the MCE. Then, the experimental temperature T gradually relaxes back to the equilibrium value, while T_{ad} and T'_{ad} are constant. Specifically, T_{ad} is equal to the starting $T_0 = 0.98$ K for $t_0 = 0$, since the real entropy gain is exactly compensated for by the calculation. The fact that the final temperature T tends to agree with T_{ad} after demagnetization, indicates that entropy gains and losses have been correctly estimated throughout the whole process. Remarkably, the final adiabatic temperatures of 3.5 K and 0.47 K obtained after sweeping the 1 T field up and down, respectively, corroborate the results independently inferred from indirect methods [53].

Recollecting the discussion on the adiabaticity and its lack thereof, we finally note that the use of (14.10) is not essential if the MCE is sufficiently large, as for $\{Gd(OOCH)_3\}_n$, and the measuring time does not exceed the time needed for fully changing H. For instance for the above-mentioned demagnetization process, the as-measured T is 0.48 K at the precise time in which the field reaches zero value (i.e., $t = 5370$ s in Fig. 14.4). From (14.10), we obtain the corresponding $T_{ad} = 0.47$ K, which is equivalent to a 2 % correction, only. This means that the measured cooling for $T = 0.48$ K and $\mu_0 \Delta H_0 = (1 - 0)$ T is given by $\Delta T = (0.98 - 0.48)$ K $= 0.50$ K, which is corrected to $\Delta T_{ad} = 0.51$ K after applying (14.10). Therefore, we conclude that this type of experiments can provide a direct estimate of the parameters which characterize the MCE.

14.4 Designing the Ideal Refrigerant

This section addresses the parameters which are known to influence the performance of a molecule-based material as a cryogenic refrigerant. We anticipate that the design of the ideal refrigerant requires the optimization of the following items:

- magnetic anisotropy,
- type and strength of the magnetic interactions,
- relative amount of non-magnetic ligand elements.

One further criterium to be considered is the type of spins involved since the magnetic entropy is determined by the spin according to (14.1). In this respect, gadolinium is the preferred constituent element because its $^8S_{7/2}$ ground state provides the largest entropy per single ion. Furthermore, it has no orbital angular momentum contribution to the ground state. This implies that its full magnetic entropy $R\ln(8)$, corresponding to a spin value $s = 7/2$, is readily available at liquid-helium temperatures. For the same reason, Mn^{2+} and Fe^{2+} ions are also often used for the synthesis of molecule-based refrigerants because of their next largest $5/2$ spin value.

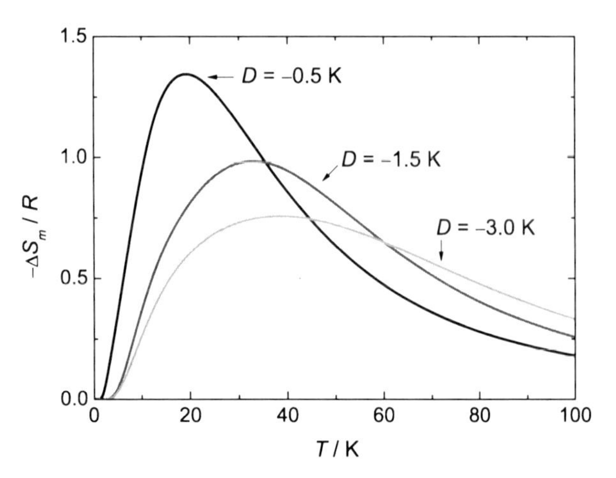

Fig. 14.5 Magnetic entropy changes ΔS_m, normalized to the gas constant R, calculated for spin $s = 10$ and varying anisotropy $D = -0.5$ K, -1.5 K and -3.0 K, following an applied field change of $\mu_0 \Delta H = (7 - 1)$ T

14.4.1 Magnetic Anisotropy

In addition to a large spin value, another condition that favors a large MCE is a relatively small magnetic anisotropy. The crystal-field effects arising from the metal oxidation states and surrounding organic ligands, concurrently with anisotropic magnetic interactions, set in a preferential direction for the spins. The larger is this anisotropy, the less sensitive to H is the polarization of the spins, or (equivalently) higher fields are needed, therefore yielding a lower MCE. This concept is further explained by the following example which is based on observing the evolution of the Schottky specific heat C_{Sch} as a function of temperature, field and anisotropy. Let us mention that the Schottky anomaly for a finite set of energy levels E_i and corresponding degeneracies g_i is defined by the expression

$$C_{\mathrm{Sch}} = \left(\frac{1}{k_B T} \right)^2 \frac{\sum_{i,j} g_i g_j (E_i^2 - E_i E_j) \exp[-(E_i + E_j)/k_B T]}{\sum_{i,j} g_i g_j \exp[-(E_i + E_j)/k_B T]}. \tag{14.11}$$

We consider a *hypothetical* fixed value $s = 10$ for the spin, while we vary the axial anisotropy as such to be $D = -0.5$ K, -1.5 K and -3.0 K. First, for each D we calculate the Schottky heat capacities C_{Sch} from (14.11) for two different values of the applied field, e.g., $\mu_0 H_i = 1$ T and $\mu_0 H_f = 7$ T. Then, we obtain the corresponding magnetic entropies $S_m(T, H)$ by making use of (14.7). As depicted in Fig. 14.1, we finally deduce the magnetic entropy changes $\Delta S_m(T, \Delta H) = [S_m(T, H_f) - S_m(T, H_i)]$ for the applied field change $\mu_0 \Delta H = (7 - 1)$ T. Figure 14.5 shows that the resulting T-dependence of $-\Delta S_m$ shifts to higher temperatures and, overall, decreases to lower values by increasing the value of D. Therefore, we can conclude that, if we target the highest MCE, we should design the molecule-based material as such to present the lowest anisotropy, which would permit the easy polarization of the spins in order to yield a large magnetic entropy change. This also demonstrates that, in order to be successful, the applicability of the (isotropic) molecule-based materials has to be at very low temperatures.

14.4.2 Magnetic Interactions

A common strategy to optimize the MCE is by playing with the magnetic interactions since these set the way in which the magnetic entropy is released as a function of temperature. Let us present the physics involved in the way in which magnetic ordering can lead to a partial concentration of the total magnetic entropy change into a limited range of temperature. For the sake of simplicity, we assume a magnetically-isolated molecule with a total spin $S_{tot} = ns$ for a finite number n of spins s, which are part of the same molecule. If it is paramagnetic, with n non-interacting spins s, the magnetic entropy per mole is

$$S_m = nR\ln(2s+1), \tag{14.12}$$

from (14.1). However, at low temperatures where the n spins s couple into $S_{tot} = ns$, the entropy to consider is $S'_m = R\ln(2S_{tot}+1) = R\ln(2ns+1)$, which is clearly different. Obviously the total magnetic entropy gain that can be reached between zero and *infinite* temperature remains equal to S_m, which is the maximum entropy gain. What does change is the way in which the magnetic entropy is released as a function of temperature. Indeed, the temperature dependence of the magnetic entropy shows a smooth gradual increase from zero at $T = 0$ to the maximum $R\ln(2s+1)$ in the paramagnetic case, while it changes into a more steep dependence in the temperature range where the interactions become important. This can be used advantageously to produce a large ΔS_m by means of a limited change in T and/or H, that is, a much larger change than can be produced in the absence of such interactions.

We note that the aforementioned argument is conceptually analogous for a bulk solid-state material. In the case of a magnetic phase transition at a critical temperature T_C, one could in principle play the same game, i.e., enhancing the entropy change in proximity of T_C by small changes in field or temperature. For most high-temperature solid-state refrigerant materials, the MCE is indeed driven by the mechanism of magnetic ordering [57], and so is also for molecule-based materials, namely Prussian blue analogues [14, 15]. In the case of liquid-helium temperatures, thermal fluctuations are typically stronger than magnetic fluctuations arising from intermolecular interactions, especially when the material contains Gd^{3+} spin centres. Therefore for such systems, one would expect the magnetic dimensionality to play no dominant role in the MCE, unless experiments are carried out deep in the sub-Kelvin regime [36]. We finally note that a drawback inherent to any magnetic phase transition is that the MCE steeply falls to near zero values below T_C, limiting the lowest temperature which can be attained in a process of adiabatic demagnetization. Therefore, particular attention should be devoted to 'control' the magnetic interactions depending on the target working temperature of the magnetic refrigerant.

14.4.2.1 Sign of Exchange Interaction

The MCE is heavily influenced by the type of magnetic interactions involved. This is particularly true in the case of antiferromagnetic interactions that tend to contribute

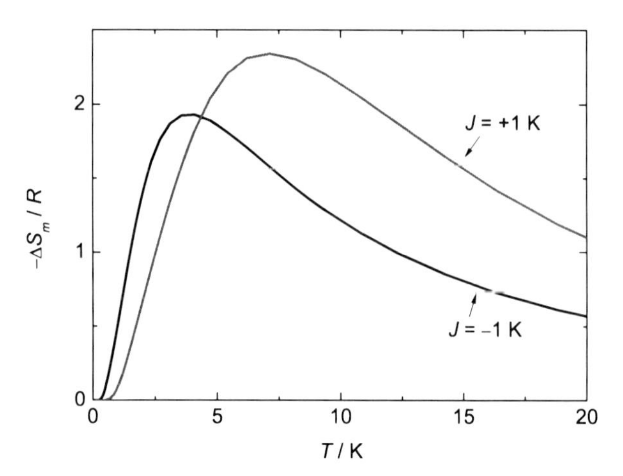

Fig. 14.6 For a dimer of spins $s_{1z} = s_{2z} = 7/2$, calculated magnetic entropy changes ΔS_m, normalized to the gas constant R, obtained for exchange constant $J = +1$ K (ferromagnetic interaction) and $J = -1$ K (antiferromagnetic interaction) following an applied field change $\mu_0 \Delta H = (7-1)$ T

negatively to the physical effect. To shed some light, let us present the model of a dimer of spins $\mathbf{s}_1$ and $\mathbf{s}_2$ that are magnetically coupled to each other by an exchange constant J. As a simplification, we restrict the spins to point along a z direction and we assume $s_{1z} = s_{2z} = 7/2$. Therefore, the Ising Hamiltonian accounting for the magnetic exchange and a Zeeman interaction is given by

$$\mathcal{H} = -J s_{1z} s_{2z} - g\mu_B (s_{1z} + s_{2z}) H, \tag{14.13}$$

where g is the Landé g-factor and μ_B is the Bohr magneton. Through numerical matrix diagonalization, one can compute the energy levels and eigenvectors, and hence the specific heat, for varying J and H. We consider ferromagnetic and antiferromagnetic exchange either by setting $J = +1$ K or $J = -1$ K, respectively. For each case, the calculation is repeated twice for applied field values $\mu_0 H = 1$ T and $\mu_0 H = 7$ T, respectively. Then by making use of (14.7), we obtain the magnetic entropy, which straightforwardly leads to the entropy change ΔS_m, depicted in Fig. 14.6 for $\mu_0 \Delta H = (7-1)$ T and both ferro- and antiferromagnetic interaction. It is easy to discern that the sign of J is highly relevant in the determination of the MCE, this being larger and shifted to higher temperatures for the case of ferromagnetic coupling. By further increasing ΔH, both ferro- and antiferromagnetic $-\Delta S_m(T, \Delta H)$ curves will gradually increase to ultimately reach the limit of the full entropy content, which corresponds to the entropy of two magnetically-independent spins, i.e. $2 \times R \ln(8) \simeq 4.16\ R$. We note that a larger ΔH is needed in the case of the antiferromagnetic interaction for reaching such a limit. Extrapolating the result of this simulation, we can conclude that ferromagnetism is to be preferred to antiferromagnetism since the former promotes a higher magnetocaloric effect.

14.4.2.2 Screening by Diluting: Ultra-Low Temperatures

Attaining temperatures in the range of milliKelvin dates back to the very beginning of the research field on cooling by adiabatic demagnetization. For the aforemen-

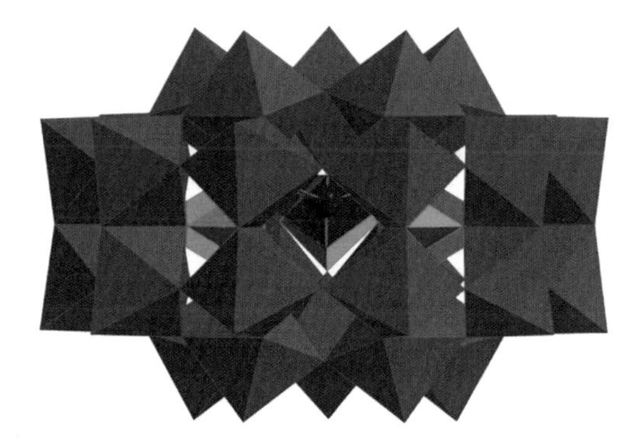

Fig. 14.7 Molecular structure of the GdW_{30} polyoxometalate salt

tioned reasons, this goal can only be achieved by avoiding any source of magnetic interactions. Diluted paramagnetic salts, like cerium magnesium nitrate (CMN) and chromic potassium alum (CPA), can achieve mK temperatures favored by the weak strength of the interactions between the paramagnetic ions [6]. However, these commercially-employed magnetic refrigerants are also characterized by a relatively strong magnetic anisotropy and low refrigeration power, which results from the small effective-spin values and spin-to-volume ratios. From Sect. 14.4.1 we have learnt that the lower the anisotropy, the less pronounced are the crystal field effects which, splitting the energy levels, result in MCE maxima at lower temperatures. This leads once again to consider gadolinium as a potentially interesting element for mK cooling. Gadolinium sulfate [5, 58] and gadolinium gallium garnet (GGG) [59, 60] are well-known low-temperature magnetic refrigerants, although they are limited by their magnetic ordering temperatures, and so is the gadolinium acetate tetrahydrate with $T_C \simeq 0.2$ K [27]. Likewise, one may expect a relatively large ordering temperature in the case of extended Gd^{3+}-based systems, such as one-dimensional chains [32]. Mixed Gd^{3+}-Mn^{3+}, Gd^{3+}-Co^{2+}, and Gd^{3+}-Ni^{2+} molecular nanomagnets have been considered as magnetic coolers (see Table 14.1) but they are not suitable for ultra-low temperatures due to the anisotropy induced by the Mn^{3+}, Co^{2+}, and Ni^{2+} ions, respectively.

Case Example: Mononuclear GdW_{10} and GdW_{30} POM Salts A recent research has focused on the magnetocaloric properties of two novel molecular nanomagnets based on polyoxometalate (POM) salts with general formula $Na_9[Gd(W_5O_{18})_2]\cdot 35H_2O$ (hereafter shortened as GdW_{10}) and $K_{12}(GdP_5W_{30}O_{110}) \cdot 54H_2O$ (hereafter shortened as GdW_{30}—see Fig. 14.7), respectively [36]. Both compounds are characterized by having a single Gd^{3+} ion per molecular unit, providing therefore a relatively large spin ground state and small magnetic anisotropy. Importantly, each magnetic ion is encapsulated by a closed POM framework, which acts as a capping ligand. The resulting intermolecular distances are exceptionally large, reaching 10 Å for GdW_{10} and 20 Å for GdW_{30}. By chemically engineering the molecules in such a way, one can achieve an effective screening of all magnetic interactions and therefore a suitable refrigerant for ultra-low temperatures. This is supported by experiments since magnetic ordering is reported to occur only at

36 mK in the case of GdW_{10}, while the more diluted GdW_{30} is the best realization of a paramagnetic single-atom gadolinium compound because it remains paramagnetic down to the accessed lowest $T \simeq 10$ mK. The inherent downside of such an approach is related to the heavy structural POM framework of each molecular unit that, being non-magnetic and anticipating the discussion presented in the following section, ultimately lowers the efficiency of these refrigerants. The search for other mononuclear molecular isotropic nanomagnets having lighter capping ligands, yet effective in screening all magnetic interactions, should motivate further studies.

14.4.3 Magnetic Density and Choice of Units

As the name tells, the magnetocaloric effect is 'magnetic'. For any refrigerant material, this obvious remark implies nothing but cooling driven by the magnetic elements solely, while the remaining majority of constituting elements participate passively in the physical process. The first step towards the application is the self cooling of the refrigerant material itself: the magnetic elements have to cool the non-magnetic ones, indeed. Therefore, in order to successfully design an ideal refrigerant material one should maximize the magnetic:non-magnetic ratio—for instance, by making use of light ligands interconnecting the magnetic centers.

So far, we have expressed the magnetic entropy change ΔS_m in terms of the *molar* gas constant $R \simeq 8.314\ \mathrm{J\,mol^{-1}\,K^{-1}}$ since this has facilitated us in focusing on parameters, such as anisotropy and interactions, that determine the MCE. However, the most common choice of units for ΔS_m is $\mathrm{J\,kg^{-1}\,K^{-1}}$. By including the *mass*, these units carry information on the relative amount of magnetic elements. Furthermore from a practical standpoint, an engineer dealing with the development of an adiabatic demagnetization refrigerator would prefer to know the amount of refrigerant material which can be employed per unit of *volume*. The third option, which is then better suited for assessing the applicability of a refrigerant material, consists in expressing the volumetric $\rho\Delta S_m$, where ρ is the mass density of the material, in terms of $\mathrm{mJ\,cm^{-3}\,K^{-1}}$ units. On this point, one could correctly argue that the MCE of molecule-based refrigerant materials is disfavored by their typically low ρ—though it is not always the case, as exemplified below.

The experimentally-observed maximum value of the entropy change has experienced a terrific escalation in the recent literature. Numerous publications break records and report comparison tables or graphs containing the ΔS_m of several compounds. However, we point out that the impression that the reader could get from such comparisons may be mislead by the choice of units employed for ΔS_m. To better illustrate how arbitrary and yet how important are the units of measurement, let us consider the following examples.

Case Example: $\{Mn_8^{4+}Mn_{24}^{2+}\}$ Molecular Nanomagnet Let us start by considering the high-nuclearity cluster $\{Mn_8^{4+}Mn_{24}^{2+}\}$, whose magnetically-relevant molecular structure [61] consists of eight planar "centered triangles" composed

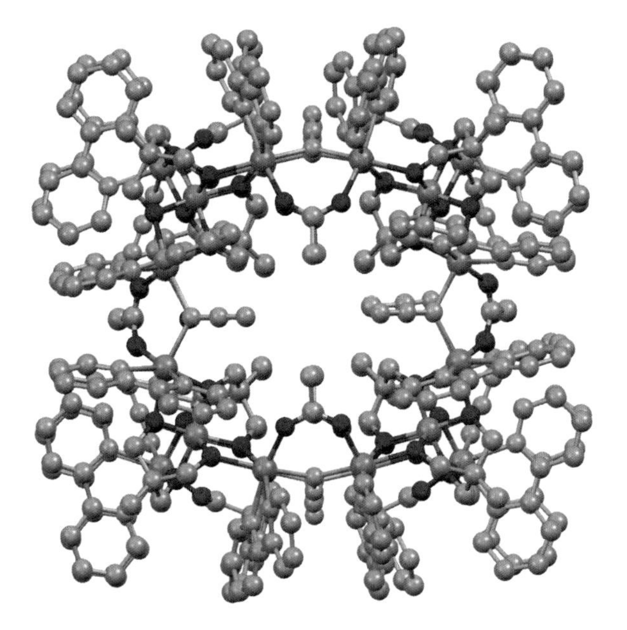

Fig. 14.8 Molecular structure of $\{Mn_8^{4+}Mn_{24}^{2+}\}$. Mn^{4+} and Mn^{2+} = violet, N = light blue, O = red, C = gray. H atoms are omitted for clarity

of a central Mn^{4+} spin center, with $s = 3/2$, antiferromagnetically coupled to three peripheral Mn^{2+} spin centers, each having $s = 5/2$ (Fig. 14.8). Within the molecule, eight triangular clusters are weakly coupled together in the form of a truncated cube by azide and carboxylate ligands. Each $\{Mn_8^{4+}Mn_{24}^{2+}\}$ core is also surrounded by one and a half non-coordinated $[Mn(bpy)_3]^{2+}$. The full formula of the complex reads $\{Mn(bpy)_3\}_{1.5}[Mn_{32}(thme)_{16}(bpy)_{24}(N_3)_{12}(OAc)_{12}](ClO_4)_{11}$ [61]. The magnetocaloric investigations of $\{Mn_8^{4+}Mn_{24}^{2+}\}$ reported a maximum value $-\Delta S_m = 23.2\ R$ at $T \simeq 1.6$ K for $\mu_0\Delta H = (7-0)$ T, which reduces to $-\Delta S_m = 10.0\ R$ at $T = 0.5$ K for $\mu_0\Delta H = (1-0)$ T field change [20]. For widespread applications, the interest is chiefly restricted to applied fields which can be produced with permanent magnets, viz., in the range $1-2$ T. The important remark here is the extremely large values for the entropy change in units of R. Obviously, this is the result of the high spin-nuclearity which favors a correspondingly large magnetic entropy according to (14.12). Taking into account the molecular mass $m = 11\,232.47\ \mathrm{g\,mol^{-1}}$ of $\{Mn_8^{4+}Mn_{24}^{2+}\}$, the 'new' though equivalent values of the entropy change read $-\Delta S_m = 18.2\ \mathrm{J\,kg^{-1}\,K^{-1}}$ and $7.5\ \mathrm{J\,kg^{-1}\,K^{-1}}$ for $\mu_0\Delta H = (7-0)$ T and $(1-0)$ T, respectively. Finally, to complete our analysis, we consider its mass density $\rho = 1.37\ \mathrm{g\,cm^{-3}}$ which provides $-\rho\Delta S_m \simeq 25.0\ \mathrm{mJ\,cm^{-3}\,K^{-1}}$ and $10.3\ \mathrm{mJ\,cm^{-3}\,K^{-1}}$ for $\mu_0\Delta H = (7-0)$ T and $(1-0)$ T, respectively.

Case Example: $\{Gd_2\}$ Molecular Nanomagnet Next, we focus on the gadolinium acetate tetrahydrate, $[Gd_2(OAc)_6(H_2O)_4]\cdot 4H_2O$, hereafter shortened as $\{Gd_2\}$ (see Fig. 14.9), which is a second example of a molecular cluster, though the nuclearity strongly decreases to just a mere Gd^{3+}-Gd^{3+} ferromagnetic dimer [27]. Because of the low nuclearity, $-\Delta S_m$ does not exceed $\approx 4.0\ R$ at $T \simeq 1.8$ K

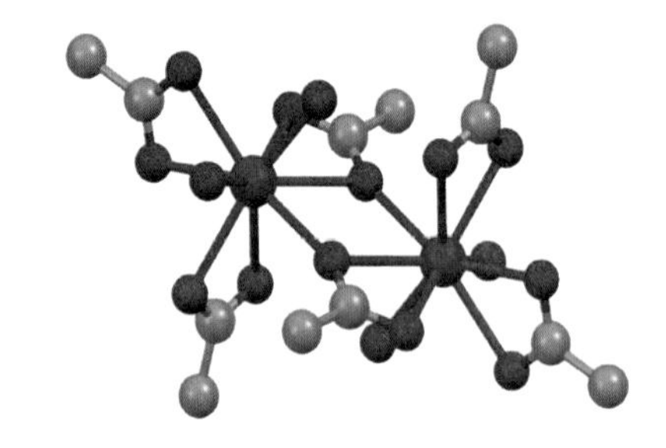

Fig. 14.9 Molecular structure of $\{Gd_2\}$. Gd = purple, O = red, C = gray. H atoms are omitted for clarity

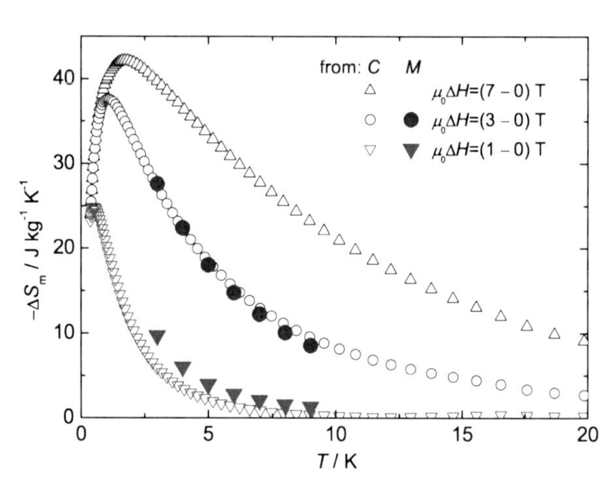

Fig. 14.10 Temperature-dependence of the magnetic entropy change ΔS_m for $\{Gd_2\}$, as obtained from magnetization and specific heat data [27] for the indicated applied-field changes ΔH

for $\mu_0\Delta H = (7-0)$ T, which is nearly six times smaller than in $\{Mn_8^{4+}Mn_{24}^{2+}\}$. However, this scenario changes drastically after considering the $\{Gd_2\}$ molecular mass $m = 812.89$ g mol^{-1}, since the latter yields $-\Delta S_m = 40.6$ J kg^{-1} K^{-1} and 27.0 J kg^{-1} K^{-1} for $\mu_0\Delta H = (7-0)$ T and $(1-0)$ T, respectively (Fig. 14.10). For the sake of information, $\rho = 2.04$ g cm^{-3} for $\{Gd_2\}$, which results in $-\rho\Delta S_m \simeq$ 82.8 mJ cm^{-3} K^{-1} and 55.1 mJ cm^{-3} K^{-1} for $\mu_0\Delta H = (7-0)$ T and $(1-0)$ T, respectively, i.e., definitely much higher than in $\{Mn_8^{4+}Mn_{24}^{2+}\}$. Really are these last values so exceptionally large?

Case Example: GGG Prototype Material Gadolinium gallium garnet (GGG) is *the* reference magnetic refrigerant material for the liquid-helium temperature region [59, 60]. Indeed, its functionality is commercially exploited in spite of a relatively modest maximum $-\Delta S_m = 20.5$ J kg^{-1} K^{-1} for $\mu_0\Delta H = (2-0)$ T. This apparent contradiction is resolved by measuring the entropy change in terms of equivalent volumetric units, which take into consideration the GGG mass density $\rho = 7.08$ g cm^{-3}. By so-doing, GGG achieves a record value $-\rho\Delta S_m \simeq$ 145 mJ cm^{-3} K^{-1} for the same applied field change of 2 T.

Case Example: $\{Gd(OOCH)_3\}_n$ 3D Metal-Organic Framework In Sect. 14.3.2, we have introduced the molecule-based $\{Gd(OOCH)_3\}_n$ metal-organic framework

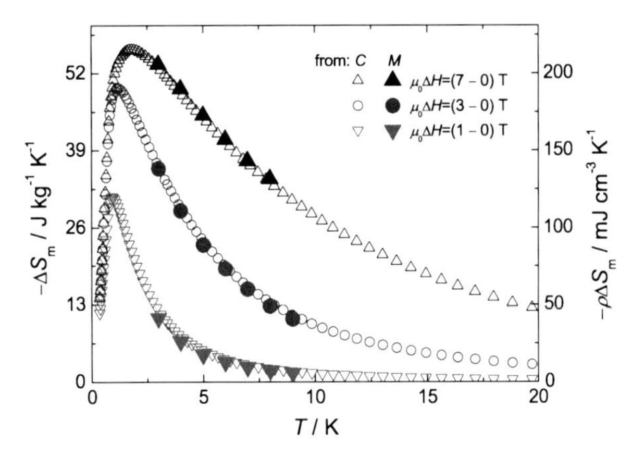

Fig. 14.11 Temperature-dependence of the magnetic entropy change ΔS_m for $\{Gd(OOCH)_3\}_n$, as obtained from magnetization and specific heat data [53] for the indicated applied-field changes ΔH. Vertical axes report units in $J\,kg^{-1}\,K^{-1}$ (*left*) and volumetric $mJ\,cm^{-3}\,K^{-1}$ (*right*)

material. The MCE of $\{Gd(OOCH)_3\}_n$ was recently determined down to sub-Kelvin temperatures by direct and indirect experimental methods [53]. This three-dimensional MOF is characterized by a relatively compact crystal lattice of weakly interacting Gd^{3+} spin centers interconnected via light formate ligands, overall providing a remarkably large magnetic:non-magnetic elemental weight ratio.

In units of R, the magnetic entropy change is reported to reach the value $-\Delta S_m \approx 2\ R$ at $T \simeq 1.9$ K for $\mu_0\Delta H = (7-0)$ T. Because of just one Gd^{3+} spin center per formula unit, the maximum experimental value is indeed consistent with the full magnetic entropy, which corresponds to $R\ln(2s+1) = 2.08\ R$, according to (14.1) for $s = 7/2$. This very modest $-\Delta S_m$ turns out spectacularly large after taking into account the molecular mass $m = 292.30\ g\,mol^{-1}$ and mass density $\rho = 3.86\ g\,cm^{-3}$ of $\{Gd(OOCH)_3\}_n$. As can be seen in Fig. 14.11, the MCE of $\{Gd(OOCH)_3\}_n$ is characterized by maxima $-\rho\Delta S_m \simeq 120\ mJ\,cm^{-3}\,K^{-1}$ and $189\ mJ\,cm^{-3}\,K^{-1}$ for $\mu_0\Delta H = (1-0)$ T and $(3-0)$ T, respectively. These values compare favorably with the ones obtained from GGG and are decidedly superior than in any other molecule-based refrigerant material.

Among the Aforementioned Examples, Which One Has the Largest MCE? It should be clear by now that there exist multiple and apparently contradictory answers. If we restrict ourselves to ΔS_m as expressed in R units, then there is no doubt that we should prefer $\{Mn_8^{4+}Mn_{24}^{2+}\}$. However adopting the $J\,kg^{-1}\,K^{-1}$ units, $\{Gd_2\}$ and $\{Gd(OOCH)_3\}_n$ perform largely better. Finally, GGG and again $\{Gd(OOCH)_3\}_n$ are far more appealing in the case of volumetric $mJ\,cm^{-3}\,K^{-1}$ units. As anticipated, the latter choice of units provides more information since it includes the mass density of the material. In this regard, we note that $\{Gd(OOCH)_3\}_n$ has a very large ρ among molecule-based magnetic materials, though yet smaller than that of GGG. As a matter of fact, the mass density of these two materials is effectively counterbalanced by the magnetic:non-magnetic weight ratio $nA_r/m = 0.54$ and 0.47 for $\{Gd(OOCH)_3\}_n$ and GGG, respectively, where $A_r = 157.25\ g\,mol^{-1}$ is the gadolinium relative atomic mass and n is number of Gd^{3+} ions per formula unit, which amounts to 1 in $\{Gd(OOCH)_3\}_n$ and to 3 in GGG. For comparison, the nA_r/m ratio further reduces to 0.39 in the case of $\{Gd_2\}$ for which $n = 2$.

That $\{Gd(OOCH)_3\}_n$ has a larger MCE than the other molecule-based refrigerant materials is also corroborated by the behavior of the adiabatic temperature change, which is strictly related to ΔS_m as we have learnt in Sect. 14.2. For $\mu_0 \Delta H = (7 - 0)$ T, we indeed observe a maximum $\Delta T_{ad} = 22.4$ K, 12.7 K and 6.7 K for $\{Gd(OOCH)_3\}_n$, $\{Gd_2\}$ and $\{Mn_8^{4+}Mn_{24}^{2+}\}$, respectively [20, 27, 53].

14.5 Towards Applications: On-Chip Refrigeration

Sub-Kelvin microrefrigeration is an emerging trend in cryogenic physics and technology since it allows for the reduction of large quantities of refrigerants [62, 63]. It also has the potential to open up new markets by making available cheap (^{3}He-free) cooling. On-chip devices are expected to find applications as cooling platforms for all those instruments where local refrigeration down to very-low temperatures is needed. These can include, although is not limited to, high-resolution X-ray and gamma-ray detectors for, e.g., astronomy, materials science, and security instrumentation.

In parallel, research on surface-deposited molecular aggregates has been evolving with the aim of assembling and integrating molecules into on-chip functional devices [64]. In this regard, the exploitation of the cooling properties of molecule-based materials is seen as a promising future technology. By developing a suitable silicon-based host device which is adiabatically isolated and has a negligible specific heat in the working temperature range, one could expect to cool from liquid-helium temperature down to milliKelvin, after having provided a field change of a few tesla. This could represent by far the best performance for on-chip cooling. Microrefrigerators based on solid-state electronic schemes, currently studied and developed for low-temperature applications, provide a cooling of the order of $\Delta T \approx 0.1$ K at the very best [63]—a value notably smaller than that promised by the magnetic molecules. Obviously, for this approach to become a reality, a relatively strong binding of the molecules to the surface and the preservation of their functionalities once deposited are sine-qua-non conditions.

The magnetic investigations on molecule-based coolers have so far been carried out for bulk materials. The target of extending these studies to include molecules deposited onto surfaces is challenging, both for the low temperatures required and, specially, for the relatively small amount of deposited material which results in a weak strength of the magnetic signal.

Case Example: Surface-Deposited $\{Gd_2\}$ Molecular Nanomagnet From the very recent literature [28], we report the first study by magnetic force microscopy (MFM) of molecular coolers deposited on a Si substrate, as an intermediate step towards the interfacing of these molecules with a future Si-based thermal sensor designed to function as a microrefrigerator. This work specifically refers to the $\{Gd_2\}$ molecular nanomagnet (Fig. 14.9) that we already met in Sect. 14.4.3.

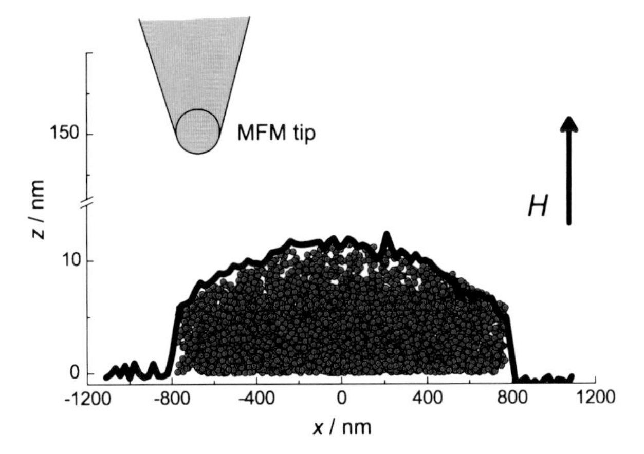

Fig. 14.12 Scheme representing a sampling of $\{Gd_2\}$ molecules (*dots*) positioned within a droplet delimited by an experimental profile and deposited on the Si substrate (xy plane). The sensing magnetic tip is at constant height $h \approx 150$ nm. The applied field H is oriented along z

The substrate consists of a Si wafer that is p-doped with boron to improve its conductivity and to permit its grounding, particularly important for preventing the accumulation of electric charges during MFM measurements. Previous to surface magnetic measurements, a rational organization of the $\{Gd_2\}$ molecules on the Si substrate is necessary to ensure a proper contrast between magnetic and non-magnetic areas as needed to estimate the magnetic stray field generated by the deposits. For this purpose, dip-pen nanolithography (DPN) is a suitable tip-assisted technique since it has already been shown to precisely place drops of a controlled size according to predefined patterns with sub-micrometer precision [65]. As a last step before the deposition, a clean writing surface is provided by ultrasound in acetonitrile, ethanol and deionized water. This last step also ensures the presence of a thin layer of native oxide, which in turn enables the adsorption of molecular species through hydrogen bonding with hydroxyl groups naturally present at the surface of oxides, even without specific pre-treatment. With its four terminal coordinated water molecules and acetate groups in various coordination modes, the neutral $\{Gd_2\}$ molecule may form a range of hydrogen bonds, either as donor or acceptor, with surface hydroxyls or adsorbed water, as it indeed does in its crystalline form with lattice water molecules [27]. A further advantage of this material resides in a relatively robust, yet light, structural framework surrounding each Gd^{3+} ion. $\{Gd_2\}$ is thus a good candidate to preserve its structure after an efficient grafting to hydrophilic surfaces without pre-functionalization.

Figure 14.12 shows the scheme of the measurements reported in Ref. [28], consisting of a MFM tip positioned at a constant height ≈ 150 nm from the Si substrate and a $\{Gd_2\}$ droplet, whose height is ≈ 10 nm, while the lengths of the two oval axes are ≈ 1.7 μm and ≈ 1.4 μm, respectively. The applied field is oriented perpendicular to the plane. Figure 14.13 shows the MFM images collected in the frequency shift mode (Δf) at $T = 5.0$ K. The frequency shift measures the gradient of the force acting on the MFM tip and it is here directly proportional to the stray field generated by the drop [28]. Each MFM image is accompanied by the corresponding profile along a line bisecting the droplet. For $H = 0$, no magnetic stray field is expected from the $\{Gd_2\}$ droplet. Therefore, in order to minimize van der Waals contributions, the

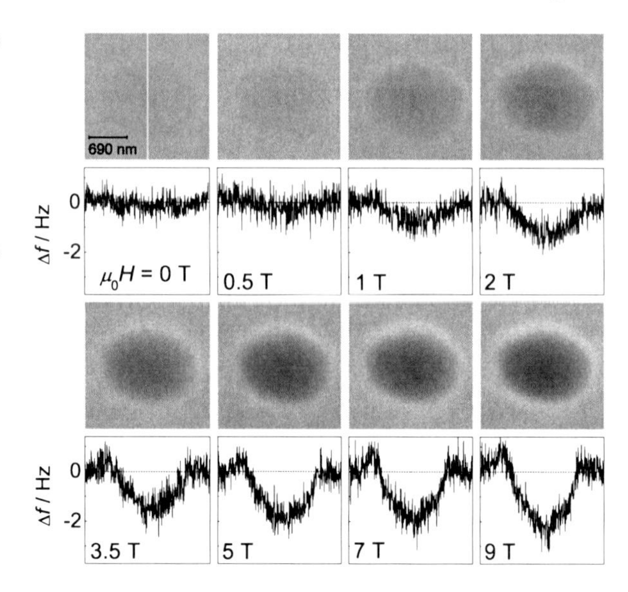

Fig. 14.13 Magnetic images (frequency shift, Δf) of an individual $\{Gd_2\}$ droplet taken at $T = 5.0$ K and different magnetic fields, as labeled. The images are represented in the same contrast scale, namely from -3.4 Hz to 1.5 Hz. Magnetic profiles are presented below each corresponding image, with the background zero-field level being represented by *a horizontal dotted line*

tip-to-sample distance is set as such to barely see any topography for zero-applied field (see the first panel in Fig. 14.13). The area external to the drop is the non-magnetic contribution of the substrate which constitutes the reference background (dotted lines in the profiles). The tip magnetization is constantly at saturation for all in-field MFM images, since the applied magnetic field largely exceeds the coercive field of the tip ($\approx 5 \times 10^{-2}$ T).

The evolution of magnetic contrast between the $\{Gd_2\}$ droplet and the non-magnetic substrate is well visible in Fig. 14.13, as a function of the applied magnetic field. Specifically, the inner area of the drop becomes darker, while the border brighter and thicker, by increasing the field. In order to explain the observed behavior, let us first consider the magnetic field generated by the $\{Gd_2\}$ droplet as represented by lines of induction or flux lines. One can easily understand that the stray-field flux lines gradually change their direction, till reaching the inversion, on approaching the border of the drop. Accordingly, the magnetic interaction between tip and sample changes from attractive to repulsive depending on the orientation of these flux lines, therefore shifting the resonance from lower (darker) to higher (brighter) frequencies, respectively. The profile lines provide further evidence for the dependence of the magnetic contrast on the applied field.

Importantly, a quantitative analysis of the magnetic contrast reveals that the (T, H)-dependence of the Δf measurements can be directly associated to the magnetization of bulk $\{Gd_2\}$, enabling us to conclude that the as-deposited molecules hold intact their magnetic characteristics and, consequently, the cooling functionality as well [28]. Transferring a known, excellent cryogenic magnetocaloric material, such as the $\{Gd_2\}$ molecular nanomagnet, from bulk crystal to a silicon substrate without deterioration of its properties paves the way towards the realization of a molecule-based microrefrigerating device for very low temperatures.

14.6 Concluding Remarks

Over the past couple of years there has been an upsurge in the number of molecule-based materials proposed as enhanced magnetic coolers for cryogenic temperatures. The research has been recently opened to extended three-dimensional structural frameworks, which will allow taking advantage of both the chemical variety and intrinsic robustness of MOF materials. However, in spite of the many efforts devoted so far to this end, there are still challenges to overcome before molecule-based magnetic coolers find widespread applications. For instance, sizeable intermolecular magnetic correlations and intrinsically low thermal conductivities are two issues that limit their applicability, especially at very low temperatures. New solutions are proposed and explored by combining chemical synthesis with materials science and advanced instrumentation techniques. There is a promising research future on grafting molecule-based magnetic coolers to substrates with a high thermal conductivity. One can envision that in a not-too-distant future, devices of reduced sizes will exploit the cooling functionality of these molecules.

Acknowledgements This work would not have been possible without the contribution of, in alphabetical order: E.K. Brechin, A. Camón, D. Collison, E. Coronado, L.J. de Jongh, G. Lorusso, F. Luis, E.J.L. McInnes, E. Palacios, O. Roubeau, D. Ruiz-Molina and R. Sessoli. Financial support by the Spanish MINECO through grant MAT2012-38318-C03-01 is acknowledged.

References

1. E. Warburg, Ann. Phys. **13**, 141 (1881)
2. P. Weiss, A. Piccard, Compt. Rend. Ac. Sc. **166**, 352 (1918)
3. P. Debye, Ann. Phys. **81**, 1154 (1926)
4. W.F. Giauque, J. Am. Chem. Soc. **49**, 1864 (1927)
5. W.F. Giauque, D.P. MacDougall, Phys. Rev. **43**, 768 (1933)
6. R.P. Hudson, *Principles and Application of Magnetic Cooling* (North-Holland, Amsterdam, 1972)
7. F. Pobell, *Matter and Methods at Low Temperatures*, 3rd edn. (Springer, Berlin, 2007)
8. Yu.I. Spichkin, A.K. Zvezdin, S.P. Gubin, A.S. Mischenko, A.M. Tishin, J. Phys. D, Appl. Phys. **34**, 1162 (2001)
9. F. Torres, J.M. Hernández, X. Bohigas, J. Tejada, Appl. Phys. Lett. **77**, 3248 (2000)
10. M. Affronte, A. Ghirri, S. Carretta, G. Amoretti, S. Piligkos, G.A. Timco, R.E.P. Winpenny, Appl. Phys. Lett. **84**, 3468 (2004)
11. M. Evangelisti, A. Candini, A. Ghirri, M. Affronte, E.K. Brechin, E.J.L. McInnes, Appl. Phys. Lett. **87**, 072504 (2005)
12. M. Evangelisti, A. Candini, A. Ghirri, M. Affronte, S. Piligkos, E.K. Brechin, E.J.L. McInnes, Polyhedron **24**, 2573 (2005)
13. R. Shaw, R.H. Laye, L.F. Jones, D.M. Low, C. Talbot-Eeckelaers, Q. Wei, C.J. Milios, S. Teat, M. Helliwell, J. Raftery, M. Evangelisti, M. Affronte, D. Collison, E.K. Brechin, E.J. McInnes, Inorg. Chem. **46**, 4968 (2007)
14. E. Manuel, M. Evangelisti, M. Affronte, M. Okubo, C. Train, M. Verdaguer, Phys. Rev. B **73**, 172406 (2006)
15. M. Evangelisti, E. Manuel, M. Affronte, M. Okubo, C. Train, M. Verdaguer, J. Magn. Magn. Mater. **316**, e569 (2007)

16. M. Manoli, R.D.L. Johnstone, S. Parsons, M. Murrie, M. Affronte, M. Evangelisti, E.K. Brechin, Angew. Chem., Int. Ed. Engl. **46**, 4456 (2007)
17. M. Manoli, A. Collins, S. Parsons, A. Candini, M. Evangelisti, E.K. Brechin, J. Am. Chem. Soc. **130**, 11129 (2008)
18. G. Karotsis, M. Evangelisti, S.J. Dalgarno, E.K. Brechin, Angew. Chem., Int. Ed. Engl. **48**, 9928 (2009)
19. G. Karotsis, S. Kennedy, S.J. Teat, C.M. Beavers, D.A. Fowler, J.J. Morales, M. Evangelisti, S.J. Dalgarno, E.K. Brechin, J. Am. Chem. Soc. **132**, 12983 (2010)
20. M. Evangelisti, A. Candini, M. Affronte, E. Pasca, L.J. de Jongh, R.T.W. Scott, E.K. Brechin, Phys. Rev. B **79**, 104414 (2009)
21. L. Sedláková, J. Hanko, A. Orendčov, M. Orendáč, C.L. Zhou, W.H. Zhu, B.W. Wang, Z.M. Wang, S. Gao, J. Alloys Compd. **487**, 425 (2009)
22. S. Nayak, M. Evangelisti, A.K. Powell, J. Reedijk, Eur. J. Chem. **16**, 12865 (2010)
23. J.W. Sharples, Y.Z. Zheng, F. Tuna, E.J. McInnes, D. Collison, Chem. Commun. **47**, 7650 (2011)
24. J.-L. Liu, J.-D. Leng, Z. Lin, M.-L. Tong, Asian J. Chem. **6**, 1007 (2011)
25. Y.Z. Zheng, M. Evangelisti, R.E.P. Winpenny, Angew. Chem., Int. Ed. Engl. **50**, 3692 (2011)
26. J.B. Peng, Q.C. Zhang, X.J. Kong, Y.P. Ren, L.S. Long, R.B. Huang, L.S. Zheng, Z.P. Zheng, Angew. Chem., Int. Ed. Engl. **50**, 10649 (2011)
27. M. Evangelisti, O. Roubeau, E. Palacios, A. Camón, T.N. Hooper, E.K. Brechin, J.J. Alonso, Angew. Chem., Int. Ed. Engl. **50**, 6606 (2011)
28. G. Lorusso, M. Jenkins, P. González-Monje, A. Arauzo, J. Sesé, D. Ruiz-Molina, O. Roubeau, M. Evangelisti, Adv. Mater. **25**, 2984 (2013)
29. S.K. Langley, N.F. Chilton, B. Moubaraki, T. Hooper, E.K. Brechin, M. Evangelisti, K.S. Murray, Chem. Sci. **2**, 1166 (2011)
30. A. Hosoi, Y. Yukawa, S. Igarashi, S.J. Teat, O. Roubeau, M. Evangelisti, E. Cremades, E. Ruiz, L.A. Barrios, G. Aromí, Eur. J. Chem. **17**, 8264 (2011)
31. Y.Z. Zheng, M. Evangelisti, R.E.P. Winpenny, Chem. Sci. **2**, 99 (2011)
32. F.S. Guo, J.D. Leng, J.L. Liu, Z.S. Meng, M.L. Tong, Inorg. Chem. **51**, 405 (2012)
33. Y.Z. Zheng, M. Evangelisti, F. Tuna, R.E.P. Winpenny, J. Am. Chem. Soc. **134**, 1057 (2012)
34. E. Cremades, S. Gomez-Coca, D. Aravena, S. Alvarez, E. Ruiz, J. Am. Chem. Soc. **134**, 10532 (2012)
35. J.D. Leng, J.L. Liu, M.L. Tong, Chem. Commun. **48**, 5286 (2012)
36. M.-J. Martínez-Pérez, O. Montero, M. Evangelisti, F. Luis, J. Sesé, S. Cardona-Serra, E. Coronado, Adv. Mater. **24**, 4301 (2012)
37. T.N. Hooper, J. Schnack, S. Piligkos, M. Evangelisti, E.K. Brechin, Angew. Chem., Int. Ed. Engl. **51**, 4633 (2012)
38. P.-F. Shi, Y.-Z. Zheng, X.-Q. Zhao, G. Xiong, B. Zhao, F.-F. Wan, P. Cheng, Eur. J. Chem. **18**, 15086 (2012)
39. T. Birk, K.S. Pedersen, C.Aa. Thuesen, T. Weyhermüller, M. Schau-Magnussen, S. Piligkos, H. Weihe, S. Mossin, M. Evangelisti, J. Bendix, Inorg. Chem. **51**, 5435 (2012)
40. C.Aa. Thuesen, K.S. Pedersen, M. Schau-Magnussen, M. Evangelisti, J. Vibenholt, S. Piligkos, H. Weihea, J. Bendix, Dalton Trans. **41**, 11284 (2012)
41. J.-B. Peng, Q.-C. Zhang, X.-J. Kong, Y.-Z. Zheng, Y.-P. Ren, L.-S. Long, R.-B. Huang, L.-S. Zheng, Z. Zheng, J. Am. Chem. Soc. **134**, 3314 (2012)
42. G. Lorusso, M.A. Palacios, G.S. Nichol, E.K. Brechin, O. Roubeau, M. Evangelisti, Chem. Commun. **48**, 7592 (2012)
43. F.-S. Guo, Y.-C. Chen, J.-L. Liu, J.-D. Leng, Z.-S. Meng, P. Vrábel, M. Orendáč, M.-L. Tong, Chem. Commun. **48**, 12219 (2012)
44. A.S. Dinca, A. Ghirri, A.M. Madalan, M. Affronte, M. Andruh, Inorg. Chem. **51**, 3935 (2012)
45. R. Sibille, T. Mazet, B. Malaman, T. Gaudisson, M. François, Inorg. Chem. **51**, 2885 (2012)
46. M. Fitta, M. Balanda, M. Mihalik, R. Pelka, D. Pinkowicz, B. Sieklucka, M. Zentkova, J. Phys. Condens. Matter **24**, 506002 (2012)
47. R. Sibille, T. Mazet, B. Malaman, M. François, Eur. J. Chem. **18**, 12970 (2012)

48. Y. Zheng, Q.-C. Zhang, L.-S. Long, R.-B. Huang, A. Müller, J. Schnack, L.-S. Zhenga, Z. Zhenga, Chem. Commun. **49**, 36 (2013)
49. I.A. Gass, E.K. Brechin, M. Evangelisti, Polyhedron **52**, 1177 (2013)
50. L.-X. Chang, G. Xiong, L. Wang, P. Cheng, B. Zhao, Chem. Commun. **49**, 1055 (2013)
51. Z.-Y. Li, J. Zhu, X.-Q. Wang, J. Ni, J.-J. Zhang, S.-Q. Liu, C.-Y. Duan, Dalton Trans. **42**, 5711 (2013)
52. E.M. Pineda, F. Tuna, R.G. Pritchard, A.C. Regan, R.E.P. Winpenny, E.J.L. McInnes, Chem. Commun. **49**, 3522 (2013)
53. G. Lorusso, J.W. Sharples, E. Palacios, O. Roubeau, E.K. Brechin, R. Sessoli, A. Rossin, F. Tuna, E.J.L. McInnes, D. Collison, M. Evangelisti, Adv. Mater. **25**, 4653 (2013)
54. E. Colacio, J. Ruiz, G. Lorusso, E.K. Brechin, M. Evangelisti, Chem. Commun. **49**, 3845 (2013)
55. C. Kittel, *Introduction to Solid-State Physics*, 8th edn. (Wiley, New York, 2005)
56. V.K. Pecharsky, K.A. Geschneidner Jr., J. Appl. Phys. **86**, 565 (1999)
57. V.K. Pecharsky, K.A. Gschneidner Jr., J. Magn. Magn. Mater. **200**, 44 (1999)
58. R.F. Wielinga, J. Lubbers, W.J. Huiskamp, Physica **37**, 375 (1967)
59. B. Daudin, R. Lagnier, B. Salce, J. Magn. Magn. Mater. **27**, 315 (1982)
60. T. Numazawa, K. Kamiya, T. Okano, K. Matsumoto, Physica B **329–333**, 1656 (2003)
61. R.T.W. Scott, S. Parsons, M. Murugesu, W. Wernsdorfer, G. Christou, E.K. Brechin, Angew. Chem., Int. Ed. Engl. **44**, 6540 (2005)
62. F. Giazotto, T.T. Heikkilä, A. Luukanen, A.M. Savin, J.P. Pekola, Rev. Mod. Phys. **78**, 217 (2006)
63. J.T. Muhonen, M. Meschke, J.P. Pekola, Rep. Prog. Phys. **75**, 046501 (2012)
64. N. Domingo, E. Bellido, D. Ruiz-Molina, Chem. Soc. Rev. **41**, 258 (2012)
65. A. Martínez-Otero, P. González-Monje, D. Maspoch, J. Hernando, D. Ruiz-Molina, Chem. Commun. **47**, 6864 (2011)

Index

J. Bartolomé et al. (eds.), *Molecular Magnets*, NanoScience and Technology,
DOI 10.1007/978-3-642-40609-6,

N

Printed by Publishers' Graphics LLC